Herbicides: Agricultural and Environmental Aspects

Herbicides: Agricultural and Environmental Aspects

Edited by Ben Davies

SYRAWOOD
PUBLISHING HOUSE
New York

Published by Syrawood Publishing House,
750 Third Avenue, 9th Floor,
New York, NY 10017, USA
www.syrawoodpublishinghouse.com

Herbicides: Agricultural and Environmental Aspects
Edited by Ben Davies

International Standard Book Number: 978-1-68286-389-3 (Hardback)

Cataloging-in-publication Data

Herbicides : agricultural and environmental aspects / edited by Ben Davies.
p. cm.
Includes bibliographical references and index.
ISBN 978-1-68286-389-3
1. Herbicides. 2. Herbicides--Environmental aspects. 3. Weeds--Control. 4. Agricultural ecology. I. Davies, Ben.
SB951.4 .H47 2017
632.954--dc23

Printed in China.

TABLE OF CONTENTS

PREFACE

This book outlines the processes and applications of herbicides in detail. The topics included in it are of utmost significance and are bound to provide incredible insights to readers. Herbicides are an important part of crop protection. They are referred to as the chemical materials used to kill weeds and other unwanted plants. Herbicides or weed killers are of two types, namely, selective herbicides, which are used on specific weeds and non-selective herbicides, which are used to control all unwanted plants. This text is compiled in such a manner, that it will provide in-depth knowledge about the science and usage of herbicides, and the effects they have on our environment. Readers would gain knowledge that would broaden their perspective about this field. For all those students, experts and researchers who are interested in the field of herbicides, the case studies included in this book will serve as an excellent guide to develop a comprehensive understanding.

Various studies have approached the subject by analyzing it with a single perspective, but the present book provides diverse methodologies and techniques to address this field. This book contains theories and applications needed for understanding the subject from different perspectives. The aim is to keep the readers informed about the progress in the field; therefore, the contributions were carefully examined to compile novel researches by specialists from across the globe.

Indeed, the job of the editor is the most crucial and challenging in compiling all chapters into a single book. In the end, I would extend my sincere thanks to the chapter authors for their profound work. I am also thankful for the support provided by my family and colleagues during the compilation of this book.

Editor

1

Habitat Availability Is a More Plausible Explanation than Insecticide Acute Toxicity for U.S. Grassland Bird Species Declines

Jason M. Hill[1]*, J. Franklin Egan[2], Glenn E. Stauffer[1], Duane R. Diefenbach[3]

1 Pennsylvania Cooperative Fish and Wildlife Research Unit, Pennsylvania State University, University Park, Pennsylvania, United States of America, **2** USDA-ARS Pasture Systems and Watershed Management Research Unit, University Park, Pennsylvania, United States of America, **3** U.S. Geological Survey, Pennsylvania Cooperative Fish and Wildlife Research Unit, Pennsylvania State University, University Park, Pennsylvania, United States of America

Abstract

Grassland bird species have experienced substantial declines in North America. These declines have been largely attributed to habitat loss and degradation, especially from agricultural practices and intensification (the habitat-availability hypothesis). A recent analysis of North American Breeding Bird Survey (BBS) "grassland breeding" bird trends reported the surprising conclusion that insecticide acute toxicity was a better correlate of grassland bird declines in North America from 1980–2003 (the insecticide-acute-toxicity hypothesis) than was habitat loss through agricultural intensification. In this paper we reached the opposite conclusion. We used an alternative statistical approach with additional habitat covariates to analyze the same grassland bird trends over the same time frame. Grassland bird trends were positively associated with increases in area of Conservation Reserve Program (CRP) lands and cropland used as pasture, whereas the effect of insecticide acute toxicity on bird trends was uncertain. Our models suggested that acute insecticide risk potentially has a detrimental effect on grassland bird trends, but models representing the habitat-availability hypothesis were 1.3–21.0 times better supported than models representing the insecticide-acute-toxicity hypothesis. Based on point estimates of effect sizes, CRP area and agricultural intensification had approximately 3.6 and 1.6 times more effect on grassland bird trends than lethal insecticide risk, respectively. Our findings suggest that preserving remaining grasslands is crucial to conserving grassland bird populations. The amount of grassland that has been lost in North America since 1980 is well documented, continuing, and staggering whereas insecticide use greatly declined prior to the 1990s. Grassland birds will likely benefit from the de-intensification of agricultural practices and the interspersion of pastures, Conservation Reserve Program lands, rangelands and other grassland habitats into existing agricultural landscapes.

Editor: R. Mark Brigham, University of Regina, Canada

Funding: The authors have no support or funding to report.

Competing Interests: The authors have declared that no competing interests exist.

* E-mail: jmh656@psu.edu

Introduction

The U. S. Geological Survey (USGS) North American Breeding Bird Survey (BBS) has documented that grassland birds declined faster than any other habitat guild of birds in North America over recent decades, including our study period of 1980–2003 [1,2]. These declines have been largely attributed to habitat loss and degradation, particularly from an increasing scope and intensity of agricultural practices (the habitat-availability hypothesis) [3–5]. Total grassland losses from agriculture and other land uses have been dramatic – e.g., of >60 million ha of tallgrass prairie historically present in North America less than 14% remains in many areas [6]. While substantial acreage of grasslands were lost prior to the 1980s, native grassland in the tallgrass prairie region and elsewhere is still being lost to agriculture [6,7], succession [8], and other land use changes [9]. Indeed, between 1982 and 1997 (the approximate timeframe of our study) the United States lost approximately 97,000 km^2 of grasslands–primarily due to agricultural practices [6]. Intensive agricultural practices can also degrade habitat and adversely affect grassland bird populations via the overgrazing of grasslands, harvesting of forage crops earlier and more frequently, and through the conversion of heterogeneous rural landscapes into intensively-managed crop monocultures [10].

The influence of insecticides on grassland bird populations has received less attention than habitat loss, but researchers have long expressed concerns about excessive and widespread insecticide use [11,12]. Pesticide use, especially insecticides, can poison and kill birds and reduce reproductive success (e.g., through eggshell thinning) and food supplies [10,13,14]. Recently, Mineau and Whiteside [15] evaluated possible correlates, including several measures of insecticide exposure, of declines in BBS trends of North American grassland birds. They reported that exposure to insecticides was more strongly correlated with grassland bird declines (the insecticide-acute-toxicity hypothesis) in the conterminous U.S. from 1980–2003 than several other indices of agricultural practices and habitat availability, such as herbicide use and changes in permanent pasture area. The study was highlighted in the *New York Times* [16] and other popular media, which demonstrates public concern for the declines of grassland birds and the impacts of insecticides on the environment. Mineau and Whiteside [15] concluded that toxic insecticides "offer a more plausible explanation for overall declines than does the oft-cited 'habitat loss through agricultural intensification.'"

The conclusions of Mineau and Whiteside [15] have important implications for understanding grassland bird population dynamics, but these results are surprising for several reasons. While grassland birds have experienced substantial declines [1,2] insecticide use in the U.S. has been substantially reduced over the past thirty years [17]. Alternative chemistries that replaced the most toxic organophosphates were available starting in the mid-1970s [18], and most of the reduction in insecticide use occurred prior to the 1990s [17]. Acute poisoning events have also declined over the last 20 years [19]. Meanwhile, substantial grassland loss to agricultural use has continued to occur over that same timeframe [7]. The resulting smaller and fragmented grasslands are more likely to harbor smaller and less diverse grassland bird communities because many grassland birds are area- and edge-sensitive species [4,5,20] that respond strongly to the configuration and amount of grasslands in agricultural landscapes [21,22].

By suggesting that insecticide exposure is a more important correlate of grassland bird declines than changes in landscape structure and habitat availability, Mineau and Whiteside's [15] analysis could have substantial implications for agricultural policy and practice. Therefore, we reassessed whether habitat availability or insecticide acute toxicity offers a more plausible explanation of grassland bird declines in North America. Our analysis differed in two important ways from that of Mineau and Whiteside [15]. Firstly, Mineau and Whiteside [15] reduced individual grassland species' trends in each state to a dichotomous variable (1 = declining; 0 = increasing) and examined only two measures of habitat availability for grassland birds in agricultural landscapes. In contrast, we used the overall trend reported for the "grassland breeding" bird guild for each state [2], which accounts for the uncertainty among individual species' trends within a state [23]. Secondly, we identified and evaluated several additional covariates representing habitat availability and reexamined the correlation between U.S. grassland bird declines (1980–2003), habitat availability, and insecticide use and acute toxicity.

Methods

North American Grassland Bird Data

The USGS North American Breeding Bird Survey is the only long-term continental monitoring program of North American breeding bird species. Each year volunteers across North America count breeding birds at 50 locations along 34.9-km roadside routes. We used the BBS "grassland breeding" species trends produced for each of 45 states between 1980 and 2003 (http://www.mbr-pwrc.usgs.gov/bbs/trend/guild03.html) - the same time period examined by Mineau and Whiteside [15]. These composite statewide trends incorporate any grassland bird species detected on ≥14 survey routes within a state, and account for uncertainty among individual species' trends by adjusting them toward the mean trend of that state [23]. Additional details of BBS statistical procedure can be found in Link and Sauer [24,25]. Some states with few grassland birds detected during BBS surveys (Alaska, Delaware, Hawai'i, Iowa, and Maryland) did not have state-wide trend estimates and hence were not included in the analysis.

Agricultural Data and Covariates

We used publically available data primarily from the United States Department of Agriculture [USDA] Economic Research Service [26–28], the 1992 National Land Cover Data (NLCD [29]), the Natural Resources Conservation Service (NRCS [30]), and Mineau and Whiteside [15] to construct nine covariates describing agricultural intensity and landscape composition that we hypothesized would be relevant to grassland bird trends in the United States. Six of these covariates were used by Mineau and Whiteside [15], including the change in permanent pasture from 1978–2002, change in cropland for pasture from 1978–2002, agricultural intensity in 1992, agricultural herbicide use in 1992, agricultural insecticide use in 1992, and lethal insecticide risk to grassland birds in 1992. Change in permanent pasture or cropland for pasture, for each state, was calculated as the number of hectares in 2002 minus the number of hectares in 1978 divided by the number of hectares in 1978 [26,28]. Permanent pasture was defined as land not used for arable crops and composed of primarily introduced forage plants managed for intensive grazing, as defined by the U.S. Department of Agriculture [USDA] Economic Research Service [26,28]. In contrast, cropland for pasture (referred to as "cropped pasture" by Mineau and Whiteside [15]) is defined as land that was potentially suitable for arable crop production without improvements (e.g., adding tile drains or grading) but was used for short- or long-term livestock grazing rather than crop production in 1978 or 2002, respectively.

We indexed agricultural intensity as the percentage of all agricultural lands within a state allocated to active cropping (i.e., "cropland used for crops") in 1992; data were obtained from the USDA Economic Research Service [27]. For pesticide use, we used the herbicide and insecticide indices presented in Mineau and Whiteside [15], which were calculated with data obtained from the National Pesticide Use Database maintained by the National Center for Food and Agricultural Policy (http://www.ncfap.org/database/state.php). This database provides estimates of areas of crops treated with individual pesticides. The insecticide and herbicide use indices were created with summed areas of crops within a state treated with each insecticide or herbicide, respectively, divided by the total areas of farmland within that state in 1992 [15]. In some states the herbicide index greatly exceeded 1.0, because many crops were treated with multiple herbicides. Unlike Mineau and Whiteside [15], we did not restrain herbicide use values at 1.0. Such an approach would have both underestimated herbicide use estimates and assumed equal herbicide exposure across 14 states. We used the lethal insecticide risk covariate presented in Mineau and Whiteside [15]. Briefly, lethal insecticide risk was taken to be the percentage of insecticide-treated farmland within a state over which logistic exposure models predicted some avian mortality for birds [18,31]. State-specific estimates of pesticide use were only available in 1992 and 1997. Therefore, following Mineau and Whiteside [15], we used the lethal insecticide risk, agricultural intensity, herbicide use, and insecticide use data from only 1992 (the mid-point of the avian trend estimates). Thus our models, and those of Mineau and Whiteside [15], made the implicit assumption that these covariate values in 1992 were predictably related to other covariate values, such that states with high lethal insecticide risk in 1992 would have high insecticide risk in other years from 1980–2003.

Other herbaceous land cover types besides the two habitat types (permanent pasture and cropland for pasture) considered by Mineau and Whiteside [15] can benefit grassland birds [32,33]. We defined three such covariates: statewide grassland coverage in 1992, Conservation Reserve Program (CRP) area in 1992, and change in rangeland from 1982–2002. We calculated the percentage of grassland coverage within a state from the "grassland" land cover classification from the 1992 NLCD [29], which excluded all grass or herbaceous cover within a state subjected to intensive management (e.g., residential lawns, golf courses, and crop fields). The NLCD categorization of grasslands includes grasslands (e.g., reclaimed surface mine grasslands and other non-grazed grasslands) not included in our other covariate categories. While percent change values for state-wide grassland variables could be more informative than a static value for 1992 grassland coverage, there are unfortunately no other national land cover estimates that cover the timeframe of our study.

The CRP covariate was expressed as the area (km^2) of CRP lands within a state obtained from the NASS [34]. Change in rangeland was calculated as the rangeland area in 2002 minus the rangeland area in 1982 and expressed as a proportion of the 1982 area for each state using data from the NRCS [30]. Rangelands are extensively managed lands where the natural vegetation are mixtures of native grasses, forbs, and shrubs suitable for grazing as well as introduced forage species that have become naturalized [35].

Statistical Analysis

We used an information theoretic approach to compare linear regression models with Akaike's Information Criterion adjusted for sample size (AIC_c). We separated our analyses into two suites of regression models to reduce model complexity and to enable a more direct comparison of our results with Mineau and Whiteside [15]. In the first suite of models we compared models created with the six covariates used by Mineau and Whiteside [15]. In the second suite of models we compared models created with lethal insecticide risk and the three additional covariates not considered by Mineau and Whiteside [15]. Our linear regression models took the general form of

$$Y_i = \alpha + \beta * \text{covariate}_i + \varepsilon_i$$

where Y_i is the overall grassland bird trend (1980–2003) for state i, α is the global intercept for all states, β is the slope of the relationship for each covariate, and ε_i is the residual where $\varepsilon_i \sim N(0,\sigma^2)$.

Plots of the grassland bird trends and each covariate did not suggest any non-linear relationships, but at the suggestion of a reviewer we evaluated models with quadratic covariate terms. These models with quadratic terms yielded extraordinary large confidence intervals for coefficients (e.g., lethal insecticide risk) and hence we did not further consider such models. We examined residual plots from univariate linear regression models to determine if transformations of covariates were necessary. Based on these plots and the subsequent elimination of residual patterns, we square-root transformed the covariates representing agricultural intensity, lethal insecticide risk, insecticide use, herbicide use, and natural log (x+1) transformed the CRP area. We scaled and centered (i.e., standardized) all covariates to allow direct comparison of effect sizes among coefficients. Therefore, positive coefficients indicated a positive association with grassland bird trends and larger (absolute value) regression coefficients indicate a larger effect on grassland bird trends than coefficients closer to zero.

To reduce multicollinearity impacts [36], we excluded models from consideration (from all subsets of models) where combinations of predictors produced variance inflation factor (VIF) scores >5.0. In the first suite of models this criterion resulted in only models with ≤4 covariates and excluded models that contained both lethal insecticide risk and insecticide use. Lethal insecticide risk and insecticide use were correlated ($r = 0.95$, $P = <0.001$). At the suggestion of a reviewer we did evaluate models that contained both insecticide risk and insecticide use as predictors. Resulting coefficient estimates for these two covariates were highly uncertain with model-averaged 95% confidence intervals that were >4 times the width of the confidence intervals of any other covariate. Thus, correcting for multicollinearity allowed us to much more precisely estimate the effects of lethal insecticide risk. Consequently, we did not further consider models that included both insecticide use and insecticide risk. VIF scores did not exceed 5.0 for any of the models in the second model suite.

In the first model suite we considered 45 models (of all 56 possible models) that did not contain combinations of lethal insecticide risk and insecticide use; we considered all 16 possible models in the second set. When comparing models with AIC is it not uncommon for researchers to mistakenly assign importance to non-competitive models [37]. This often occurs when comparing a model with K+1 parameters to a better (i.e., lower AIC) nested model with only K parameters. The K+1 model may fall within 2 AIC units (ΔAIC≤2) of a simpler model, but such models "should not be interpreted as having any ecological effect" [37] unless they also substantially lower the model deviance [37,38]. Therefore, after running both suites of models we then removed any nested model with one additional parameter (K+1) with an AIC_c score greater than the simpler model with K parameters and a log-likelihood value <0.5 units smaller than the simpler model. We model-averaged coefficient estimates from the confidence model set (<4.0 ΔAIC_c units from the top model [38]) and produced unconditional standard errors and 95% confidence intervals with the MuMIn package [39] in program R [40]. We model-averaged coefficient estimates using only the coefficient estimates from models that included that parameter of interest in the confidence sets.

Results

First Model Suite: Reevaluation of Mineau and Whiteside

Our model selection procedure from the first model suite resulted in 10 competitive models, three of which were included in the final confidence set (Table 1). However, only change in cropland for pasture had an estimated effect for which the confidence interval did not include zero (Figure 1). Change in cropland for pasture occurred in two of the three models in the confidence set (Table 1) and was positively associated with increasing bird trends (Figure 1). Only considering the best-performing models representing both hypotheses, habitat availability had 21.0 times more support (ratio of AIC_c weights = 0.42/0.02) than lethal insecticide risk as the most parsimonious explanation of the data (Table 1).

Second Model Suite: Lethal Insecticide Risk & Additional Habitat Covariates

For the second model suite, our model selection procedure resulted in nine competitive models, six of which were included in the confidence set. All four possible covariates were included in the confidence set (Table 2), but again, confidence intervals for the estimated effect of most covariates included zero. CRP area had the largest effect size, was positively associated with grassland bird trends, and was the only covariate that had an estimated effect for which the confidence interval did not include zero (Figure 1). We had originally expressed the CRP covariate as the change in CRP area over time within a state, but we changed it to its current form at the request of a reviewer–our results and conclusions did not change. Although included in the confidence set, change in rangeland, percent grass, and lethal insecticide risk all had uncertain effects on grassland bird trends (Table 2, Figure 1). The most parsimonious model included only CRP area (Table 2), and this model had >1.3 times more support (ratio of AIC_c weights = 0.28/0.22) than the highest ranking model containing

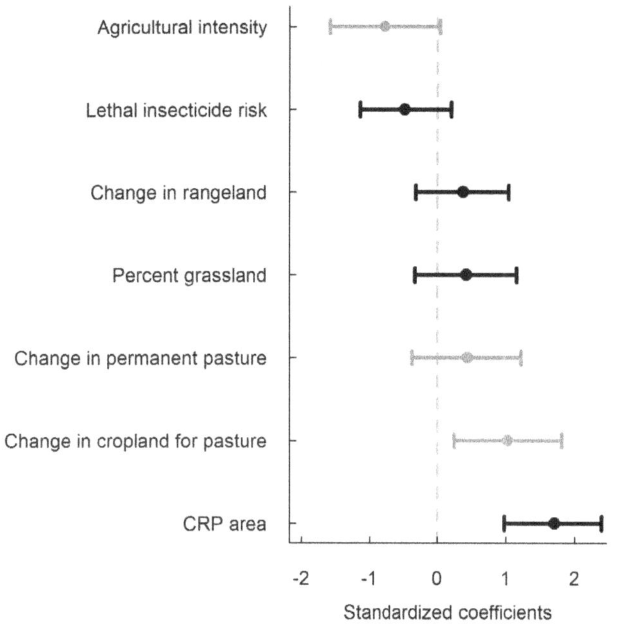

Figure 1. Model-averaged coefficients of covariates used to predict grassland breeding bird trends (1980–2003). Model-averaged coefficients (95% confidence intervals) occurring in the confidence sets of model suites one (gray bars) and two (black bars) that describe agricultural practices and habitat availability used to predict grassland breeding bird trends from 1980 to 2003 in the U.S.

lethal insecticide risk (Table 2). Comparing only univariate models, the CRP model had >2300 times more support than the lethal insecticide risk model. CRP had approximately 3.06 times the effect (ratio of standardized regression coefficients = 1.71/0.47) on grassland bird populations than did lethal insecticide risk.

Discussion

Our results indicate that population trends of grassland birds in the U.S. are primarily associated with habitat availability, rather than insecticide use or insecticide acute toxicity. Thus, in direct contrast to Mineau and Whiteside [15], our results do not support the insecticide-acute-toxicity hypothesis because we found stronger connections with declines in grassland bird abundance using habitat-based covariates. Pesticides may negatively influence grassland birds, but our results strongly support habitat associations over insecticide-acute-toxicity as the more plausible explanation for observed BBS trends.

We suggest three reasons why we arrived at different conclusions than Mineau and Whiteside [15]. First, Mineau and Whiteside's [15] use of a dichotomous response variable based on the point estimate of a species' trend means that species with steeply negative or precisely estimated trends are treated identically to species with near-zero or poorly estimated trends. In contrast, we used a single state trend for grassland bird species which explicitly accounted for the similarity and uncertainty among all grassland bird trends within that state [23]. Second, Mineau and Whiteside [15] analyzed individual grassland species' trends (1 = decreasing, 0 = increasing) without accounting for the lack of independence among trends from within a state (e.g., by treating state identity as a random effect). Trends of grassland birds from within a state are unlikely to be independent of one

Table 1. Linear models from the first model suite with standardized coefficients[a] that described the relationship between U.S. state grassland bird trends (1980–2003) and agriculture-related covariates (see Methods).

Change in cropland for pasture	Change in permanent pasture	Agricultural intensity	Herbicide use	Insecticide use	Lethal insecticide risk	k^b	log(*L*)	AIC_c	ΔAIC_c	w_i^c
1.02						2	−106.87	220.32	0.00[d]	0.42
1.06	0.43					3	−106.27	221.54	1.21[d]	0.23
		−0.77				2	−108.25	223.08	2.76[d]	0.11
	0.43	−0.82				3	−107.68	224.37	4.05	0.06
		−1.36	0.71			3	−107.74	224.49	4.16	0.05
				−0.49		2	−109.32	225.22	4.90	0.04
			0.41			2	−109.53	225.64	5.32	0.03
		−1.53	1.27	−0.63		4	−107.06	225.66	5.34	0.03
	0.33					2	−109.69	225.97	5.64	0.03
					−0.27	2	−109.79	226.16	5.83	0.02

[a]Model-averaged coefficients with unconditional standard errors: change in cropland for pasture ($\beta = 1.03$, SE = 0.40), change in permanent pasture ($\beta = 0.43$, SE = 0.40), and agricultural intensity ($\beta = -0.77$, SE = 0.41).
[b]No. parameters.
[c]Akaike model weights.
[d]Models appearing in the confidence set.

Table 2. Linear models with standardized coefficients[a] that appeared in the second model suite that described the relationship between U.S. state grassland bird trends (1980–2003), lethal insecticide risk, and three additional habitat covariates not considered by Mineau and Whiteside (2013).

CRP area	Percent grassland	Change in rangeland	Lethal insecticide risk	K^b	log(L)	AIC$_c$	AIC$_c$	w_i^c
1.71				2	−99.71	206.01	0.00[d]	0.28
1.76			−0.47	3	−98.73	206.46	0.45[d]	0.22
1.52	0.42			4	−99.08	207.15	1.14[d]	0.16
1.78		0.36		3	−99.15	207.30	1.28[d]	0.15
1.84		0.39	−0.49	4	−98.05	207.64	1.63[d]	0.12
1.59	0.42	0.36		4	−98.49	208.53	2.51[d]	0.08
	1.09			2	−106.34	219.28	13.26	<0.01
			−0.27	2	−109.79	226.16	20.15	<0.01
		0.02		2	−110.00	226.58	20.57	<0.01

[a]Model-averaged coefficients with unconditional standard errors: CRP area ($\beta = 1.71$, SE = 0.37), percent grassland ($\beta = 0.42$, SE = 0.38), change in rangeland ($\beta = 0.37$, SE = 0.35), and lethal insecticide risk ($\beta = -0.47$, SE = 0.34).
[b]No. parameters.
[c]Akaike model weights.
[d]Models appearing in the confidence set.

another due to similar land management, agricultural policies, pesticide exposure, weather conditions, and habitat availability that affect all species within that state. Failure to account for this lack of independence would not necessarily favor insecticide acute toxicity in the modeling process, but the violation of basic linear model assumptions can result in spurious results [41]. Third, we observed evidence of the effects of multicollinearity in Mineau and Whiteside's [15] analysis. For example, when insecticide use and lethal insecticide risk occurred within the same model (16 such occurrences in Mineau and Whiteside [15]) the coefficient for insecticide use was always negative, whereas it was always positive when it occurred in models without lethal insecticide risk (16 occurrences) (JMH unpublished data). Multicollinearity can complicate model selection, result in large variances, and cause coefficient estimates to reverse sign when another co-linear variable is included in the model [36,42–44]. We reduced multicollinearity effects by limiting model selection to combinations of linear predictors where VIF scores were <5.0.

Consistent with the prediction of the habitat-availability hypothesis we found that grassland bird trends were more strongly associated with (i.e., larger effect sizes of standardized coefficients) CRP area, agricultural intensity, and trends in cropland for pasture than they were for lethal insecticide risk (Tables 1 and 2). These results are consistent with the continuing decline of grassland birds over the study period [2] and the dramatic conversion of rangelands, pasturelands, and native grasslands to intensive croplands and developed land over the last 35 years [7,30]. For example, cropland for pasture area in the U.S. declined by 20% (6.3 million ha) from 1978 to 2002 (approximately the time frame of this study) and declined by a further 41% (10 million ha) from 2002 to 2007 alone [30,45]. Meanwhile overall pesticide use in the U.S. has likely remained relatively stable and insecticide use has likely declined since the 1980s [17]. Additionally, several researchers have compared the influence of habitat availability relative to the prevalence of organic (i.e. pesticide free) cropland. These studies have consistently found that the extent and configuration of grassland habitats has stronger effects on the richness and abundance of most grassland bird species than the amount of organic acreage [46–48]. As grassland area declines in agricultural landscapes the potential for negative impacts to grassland bird populations increases because many grassland bird species are sensitive to fragmentation and area and edge effects [4,5,20,49,50].

Our models suggest that grassland birds respond negatively to agricultural intensification [32,51] (i.e., increases in row-crop agriculture), but less-intensive agricultural practices may benefit some grassland bird species [52]. For example, light grazing pressure, and rangelands in general, have been positively linked to abundances of many species of grassland birds such as grasshopper sparrows (*Ammodramus savannarum*), Henslow's sparrows (*A. henslowii*), and upland sandpipers (*Bartramia longicauda*) [9,53–55]. Some grassland birds breed in hayfields and pasturelands [32,56], and adult bobolink (*Dolichonyx oryzivorus*) and savannah sparrow (*Passerculus sandwichensis*) apparent survival is negatively related to the intensity of agricultural practices in these hayfields [57]. The value of hayfields to breeding grassland birds, however, may be dependent upon the timing of mowing activities [58,59]. The point estimate for the effect of permanent pasture on grassland bird trends was positive, but the effect was uncertain and smaller than the effect of change in cropland for pasture area (Figure 1). This result may be related to the relatively small amount of permanent pasture changes observed across states from 1978 to 2002 (mean = −647 ha) compared to cropland for pasture changes (mean = −120,880 ha) over the same time period [26,28].

We found that CRP area was the strongest predictor (i.e., largest standardized coefficient; Figure 1) of grassland bird trends in the U.S. Other researchers have also documented positive effects of CRP area on some species of grassland birds [33,60,61], but see Pabian et al. [62]. We found nearly identical results when we expressed the CRP variable as the change in percent of acreage over time within a state (JMH unpub.). The Conservation Reserve Program is a type of agricultural de-intensification where farmers are paid to remove environmentally sensitive lands from agricultural production. However, as farmland values have risen sharply over the last decade [63], CRP enrolled hectares have dropped by 29% from 2006 to 2013 [64,65]. Our results suggest that CRP program enrollment may be an important component of grassland bird conservation strategy in the U.S. However, while CRP and other restored or non-native grasslands may benefit some species, these areas may not provide optimal habitat for some grassland bird species (e.g., Sprague's pipt [*Anthus spragueii*], and short-eared owl [*Asio flammeus*]) that are relatively intolerant to anthropogenic disturbance ([66,67]). The best conservation strategy for such species may be to protect remaining native grasslands from conversation to agricultural lands [5,68,69].

Our results are consistent with the scarcity of data linking pesticides to vertebrate population changes [19], but depend on several important assumptions. Following Mineau and Whiteside [15] we developed several covariates based on data from 1992 (the midpoint of our study period), because data for these covariates was not available circa 1980; thus our models assume that these 1992 covariate values were predictably related to other years in our analysis. Insecticides have direct and indirect negative effects on many groups of birds [12,14,70,71], but the phase-out of especially toxic active pesticide ingredients (e.g., organophosphates and cholinesterase-inhibiting carbamates) combined with the increased use of Bt crops, and reduced application rates of insecticides [17,18] may have lowered the risk of lethal avian exposures to North American grassland bird species over the past 20 years. If overall insecticide risk to grassland birds declined non-linearly between 1992 and 2003 then both we and Mineau and Whiteside [15] may have imprecisely estimated the risk of insecticides to grassland birds. Our models suggested that insecticide acute toxicity was negatively related to grassland bird trends, but insecticide acute toxicity was a relatively poor predictor of grassland bird trends on its own (Table 1). Our analyses could be improved if finer scale (temporal and spatial resolution) land use, pesticide application, and pesticide exposure data were available for the conterminous United States from 1980–2003.

Our results and those of others [15,32,33] that are based on broad-scale correlations of BBS and agricultural data clearly cannot prove or disprove the insecticide-acute-toxicity or habitat-availability hypotheses. However, the strong majority of the grassland bird research [5,20,61,72–76] suggests that grassland bird populations in the U.S. are most strongly related to grassland configuration and amount (the habitat-availability-hypothesis). Further grassland bird population declines are foreseeable given the continued loss of U.S. grasslands [7,30,45]. Building multi-functional landscapes that provide an abundant food supply while also conserving biodiversity and supplying ecosystem services, has become a central challenge for agriculture in the 21st century [77,78]. Farmers, conservation organizations, and governments have taken many approaches to build multifunctional working landscapes [79–82]. Our results, and those of others [5,61,72] suggest that an approach most effective for grassland bird conservation would include a focus on habitat creation and preservation, encouraging CRP acreages, pastures, and low-intensity grazing land within intensive agriculture landscapes, and strongly disincentivizing the conversion of remaining native grasslands to agricultural lands.

Acknowledgments

We thank three anonymous reviewers, Margaret Brittingham, and Pierre Mineau for editorial comments, and we thank Pierre Mineau and Mélanie Whiteside for sharing their data. Any use of trade, firm, or product names is for descriptive purposes only and does not imply endorsement by the U.S. Government.

Author Contributions

Conceived and designed the experiments: JMH JFE GES. Analyzed the data: JMH. Wrote the paper: JMH JFE GES DRD.

References

1. North American Bird Conservation Initiative, U.S. Committee (2011) The State of the Birds 2011 Report on Public Lands and Waters: United States of America. US Fish Wildl Publ: 48.
2. Sauer JR, Hines JE, Fallon JE, Pardieck KL, Ziolkowski DJ Jr, et al. (2012) The North American Breeding Bird Survey, Results and Analysis 1966–2011. Laurel, MD: USGS Patuxent Wildlife Research Center.
3. Herkert JR, Sample DW, Warner RE (1996) Management of Midwestern grassland landscapes for the conservation of migratory birds. St. Paul, MN: U.S. Dept. of Agriculture, Forest Service, North Central Forest Experiment Station.
4. Vickery PD, Tubaro PL, Cardoso da Silva JM, Peterjohn B, Herkert JR, et al. (1999) Conservation of grassland birds in the western hemisphere. Stud Avian Biol: 2–26.
5. Askins RA, Chávez-Ramírez F, Dale BC, Haas CA, Herkert JR, et al. (2007) Conservation of Grassland Birds in North America: Understanding Ecological Processes in Different Regions: "Report of the AOU Committee on Conservation." Ornithol Monogr: iii–46. doi:10.2307/40166905.
6. Samson F, Knopf FL, Ostlie W (2004) Great Plains Ecosystems: Past, Present, and Future. USGS Staff – Publ Res. Available: http://digitalcommons.unl.edu/usgsstaffpub/45.
7. Wright CK, Wimberly MC (2013) Recent land use change in the Western Corn Belt threatens grasslands and wetlands. Proc Natl Acad Sci 110: 4134–4139. doi:10.1073/pnas.1215404110.
8. Briggs JM, Knapp AK, Blair JM, Heisler JL, Hoch GA, et al. (2005) An Ecosystem in Transition: Causes and Consequences of the Conversion of Mesic Grassland to Shrubland. BioScience 55: 243. doi:10.1641/0006-3568(2005)055[0243:AEITCA]2.0.CO;2.
9. Powell AFLA (2006) Effects of Prescribed Burns and Bison (Bos bison) Grazing on Breeding Bird Abundances in Tallgrass Prairie. The Auk 123: 183–197. doi:10.2307/4090640.
10. Newton I (2004) The recent declines of farmland bird populations in Britain: an appraisal of causal factors and conservation actions. Ibis 146: 579–600. doi:10.1111/j.1474-919X.2004.00375.x.
11. Carson R (2002) Silent spring. Boston: Houghton Mifflin. 400 p.
12. Risebrough RW (1986) Pesticides and Bird Populations. In: Johnston RF, editor. Current Ornithology. Springer US. 397–427. Available: http://link.springer.com/chapter/10.1007/978-1-4615-6784-4_9. Accessed 30 September 2013.
13. Mineau P, Fletcher MR, Glaser LC, Thomas NJ, Brassard C, et al. (1999) Poisoning of raptors with organophosphorus and carbamate pesticides with emphasis on Canada, U.S. and U.K. United States Geological Survey. Available: http://pubs.er.usgs.gov/publication/70021842. Accessed 30 June 2013.
14. Bright JA, Morris AJ, Winspear R (2008) A Review of indirect effects of pesticides on birds and mitigating land-management practices. Research. Royal Society for the Protection of Birds. Available: http://www.rspb.org.uk/Images/bright_morris_winspear_tcm9-192457.pdf.
15. Mineau P, Whiteside M (2013) Pesticide acute toxicity is a better correlate of U.S. grassland bird declines than agricultural intensification. PLoS ONE 8: e57457. doi:10.1371/journal.pone.0057457.
16. Annonymous (2013) Toxic Threats to Grassland Birds. N Y Times. Available: http://www.nytimes.com/2013/03/12/opinion/toxic-threats-to-grassland-birds.html. Accessed 30 September 2013.
17. Osteen CD, Fernandez-Cornejo J (2013) Economic and policy issues of U.S. agricultural pesticide use trends. Pest Manag Sci 69: 1001–1025. doi:10.1002/ps.3529.
18. Mineau P, Whiteside M (2006) Lethal risk to birds from insecticide use in the United States–a spatial and temporal analysis. Environ Toxicol Chem 25: 1214–1222. doi:10.1897/05-035R.1.

19. Köhler H-R, Triebskorn R (2013) Wildlife Ecotoxicology of Pesticides: Can We Track Effects to the Population Level and Beyond? Science 341: 759–765. doi:10.1126/science.1237591.
20. Helzer CJ, Jelinski DE (1999) The relative importance of patch area and perimeter-area ratio to grassland breeding birds. Ecol Appl 9: 1448–1458. doi:10.1890/1051-0761(1999)009[1448:TRIOPA]2.0.CO;2.
21. Virkkala R, Luoto M, Rainio K (2004) Effects of landscape composition on farmland and red-listed birds in boreal agricultural-forest mosaics. Ecography 27: 273–284. doi:10.1111/j.0906-7590.2004.03810.x.
22. Wretenberg J, Pärt T, Berg Å (2010) Changes in local species richness of farmland birds in relation to land-use changes and landscape structure. Biol Conserv 143: 375–381. doi:10.1016/j.biocon.2009.11.001.
23. Link WA, Sauer JR (1996) Extremes in Ecology: Avoiding the Misleading Effects of Sampling Variation in Summary Analyses. Ecology 77: 1633–1640. doi:10.2307/2265557.
24. Link WA, Sauer JR (1997) Estimation of Population Trajectories from Count Data. Biometrics 53: 488–497. doi:10.2307/2533952.
25. Link WA, Sauer JR (1998) Estimating Population Change from Count Data: Application to the North American Breeding Bird Survey. Ecol Appl 8: 258–268. doi:10.1890/1051-0761(1998)008[0258:EPCFCD]2.0.CO;2.
26. Frey HT (1982) Major Uses of Land in the United States: 1978. Natural Resources and Environment Division, Economic Research Service, U.S. Department of Agriculture.
27. Daugherty AB (1995) Major Uses of Land in the United States: 1992. Agricultural Economic Report. Natural Resources and Environment Division, Economic Research Service, U.S. Department of Agriculture. Available: http://naldc.nal.usda.gov/download/CAT10825179/PDF.
28. Lubowski RN, Vesterby M, Bucholtz S, Baez A, Roberts MJ (2006) Major Uses of Land in the United States: 2002. Natural Resources and Environment Division, Economic Research Service, U.S. Department of Agriculture.
29. Fry J, Xian G, Jin S, Dewitz J, Homer C, et al. (2011) Completion of the 2006 National Land Cover Database for the Conterminous United States. Photogramm Eng Remote Sens 77: 858–864.
30. U.S. Department of Agriculture (2009) Summary Report: 2007 National Resources Inventory. Iowa State University, Ames, IA: Natural Resources Conservation Service and the Center for Survey Statistics and Methodology.
31. Mineau P (2002) Estimating the probability of bird mortality from pesticide sprays on the basis of the field study record. Environ Toxicol Chem 21: 1497–1506.
32. Murphy MT (2003) Avian population trends within the evolving agricultural landscape of Eastern and Central United States. The Auk 120: 20–34. doi:10.2307/4090137.
33. Riffell S, Scognamillo D, Burger L, Wes J, Bucholtz S (2010) Broad-scale relations between Conservation Reserve Program and grassland birds: do cover type, configuration and contract age matter? Open Ornithol J 3: 112–123. doi:10.2174/1874453201003010112.
34. U.S. Department of Agriculture (2004) 2002 Census of Agriculture. United States Summary and State Data. National Agricultural Statistics Service.
35. Nickerson C, Ebel R, Borchers A, Carriazo F (2011) Major Land Uses in the United States, 2007. USDA Economic Research Service.
36. Graham MH (2003) Confronting multicollinearity in ecological multiple regression. Ecology 84: 2809–2815. doi:10.1890/02-3114.
37. Arnold TW (2010) Uninformative parameters and Model selection using Akaike's information criterion. J Wildl Manag 74: 1175–1178. doi:10.2307/40801110.
38. Burnham KP, Anderson DR (2002) Model selection and multimodel inference: a practical information-theoretic approach. New York: Springer. 488 p.
39. Barton K. (2013) MuMIn: Multi-model inference. Available: http://cran.r-project.org/web/packages/MuMIn/MuMIn.pdf.
40. R Core Team (2013) R: A language and environment for statistical computing. Vienna, Austria: R Foundation for Statistical Computing.
41. Lazic SE (2010) The problem of pseudoreplication in neuroscientific studies: is it affecting your analysis? BMC Neurosci 11: 5. doi:10.1186/1471-2202-11-5.
42. Mansfield ER, Helms BP (1982) Detecting multicollinearity. Am Stat 36: 158–160. doi:10.2307/2683167.
43. Smith AC, Koper N, Francis CM, Fahrig L (2009) Confronting collinearity: comparing methods for disentangling the effects of habitat loss and fragmentation. Landsc Ecol 24: 1271–1285. doi:10.1007/s10980-009-9383-3.
44. Mela C, Kopalle P (2002) The impact of collinearity on regression analysis: the asymmetric effect of negative and positive correlations. Appl Econ 34: 667–677.
45. U.S. Department of Commerce (1981) 1978 Census of Agriculture. United States Summary and State Data. Bureau of the Census.
46. Piha M, Tiainen J, Holopainen J, Vepsäläinen V (2007) Effects of land-use and landscape characteristics on avian diversity and abundance in a boreal agricultural landscape with organic and conventional farms. Biol Conserv 140: 50–61. doi:10.1016/j.biocon.2007.07.021.
47. Smith HG, Dänhardt J, Lindström Å, Rundlöf M (2010) Consequences of organic farming and landscape heterogeneity for species richness and abundance of farmland birds. Oecologia 162: 1071–1079. doi:10.1007/s00442-010-1588-2.
48. Hiron M, Berg Å, Eggers S, Josefsson J, Pärt T (2013) Bird diversity relates to agri-environment schemes at local and landscape level in intensive farmland. Agric Ecosyst Environ 176: 9–16. doi:10.1016/j.agee.2013.05.013.
49. Renfrew RB, Ribic CA, Nack JL, Bollinger EK (2005) Edge Avoidance by Nesting Grassland Birds: a futile strategy in a fragmented landscape. The Auk 122: 618–636. doi:10.1642/0004-8038(2005)122[0618:EABNGB]2.0.CO;2.
50. Davis SK, Brigham RM, Shaffer TL, James PC, Stouffer PC (2006) Mixed-grass prairie passerines exhibit weak and variable responses to patch size. The Auk 123: 807–821. doi:10.1642/0004-8038(2006)123[807:MPPEWA]2.0.CO;2.
51. Geiger F, de Snoo GR, Berendse F, Guerrero I, Morales MB, et al. (2010) Landscape composition influences farm management effects on farmland birds in winter: A pan-European approach. Agric Ecosyst Environ 139: 571–577. doi:10.1016/j.agee.2010.09.018.
52. Best LB, Freemark KE, Dinsmore JJ, Camp M (1995) A review and synthesis of habitat use by breeding birds in agricultural landscapes of Iowa. Am Midl Nat 134: 1–29. doi:10.2307/2426479.
53. Smith CR (1997) Use of public grazing lands by Henslow's Sparrows, Grasshopper Sparrows, and associated grassland birds in New York State. In: Vickery PD, Dunwiddie PW, editors. Grasslands of Northeastern North America: Ecology and Conservation of Native and Agricultural Landscapes. Lincoln, MA: Massachusetts Audubon Society. 171–186.
54. Walk JW, Warner RE (2000) Grassland management for the conservation of songbirds in the Midwestern USA. Biol Conserv 94: 165–172. doi:10.1016/S0006-3207(99)00182-2.
55. Brennan LA, Kulesky WP (2005) Invited Paper: North American grassland birds: an unfolding conservation crisis? J Wildl Manag 69: 1–13. doi:10.2193/0022-541X(2005)069<0001:NAGBAU>2.0.CO;2.
56. Perlut NG, Strong AM, Donovan TM, Buckley NJ (2006) Grassland songbirds in a dynamic management landscape: behavioral responses and management strategies. Ecol Appl 16: 2235–2247. doi:10.1890/1051-0761(2006)016[2235:GSIADM]2.0.CO;2.
57. Perlut NG, Strong AM, Donovan TM, Buckley NJ (2008) Grassland Songbird Survival and Recruitment in Agricultural Landscapes: Implications for Source–Sink Demography. Ecology 89: 1941–1952. doi:10.1890/07-0900.1.
58. Perlut NG, Strong AM, Donovan TM, Buckley NJ (2008) Regional population viability of grassland songbirds: Effects of agricultural management. Biol Conserv 141: 3139–3151. doi:10.1016/j.biocon.2008.09.011.
59. Dale BC, Martin PA, Taylor PS (1997) Effects of Hay Management on Grassland Songbirds in Saskatchewan. Wildl Soc Bull 25: 616–626.
60. Johnson DH, Igl LD (1995) Contributions of the Conservation Reserve Program to populations of breeding birds in North Dakota. Wilson Bull 107: 709–718.
61. Herkert JR (2009) Response of Bird Populations to Farmland Set-Aside Programs. Conserv Biol 23: 1036–1040. doi:10.1111/j.1523-1739.2009.01234.x.
62. Pabian SE, Wilson AM, Brittingham MC (2013) Mixed responses of farmland birds to the Conservation Reserve Enhancement Program in Pennsylvania. J Wildl Manag 77: 616–625. doi:10.1002/jwmg.514.
63. Nickerson C, Morehart M, Kuethe T, Beckman J, Ifft J, et al. (2012) Trends in U.S. Farmland Vluaes and Ownership. USDA, Economic Research Service. Available: http://www.ers.usda.gov/media/377487/eib92_2_.pdf. Accessed 1 January 2014.
64. USDA Farm Service Agency (2007) Conservation Reserve Program summary and enrollment statistics FY 2006. Available: http://www.fsa.usda.gov/Internet/FSA_File/06rpt.pdf.
65. USDA Farm Service Agency (2013) Conservation Reserve Program November 2013 Monthly Report. Available: http://www.fsa.usda.gov/FSA/webapp?area=home&subject=copr&topic=rns-css.
66. Dechant JA, Sondreal ML, Johnson DH, Igl LD, Goldade CM, et al. (2003) Effects of managmement practices on grassland birds: short-eared owl. Northern Prairie Wildlife Research Center, Jamestown, ND. Available: http://www.npwrc.usgs.gov/resource/literatr/grasbird/seow/seow.htm. Accessed 15 March 2014.
67. Dechant JA, Sondreal ML, Johnson DH, Igl LD, Goldade CM, et al. (2003) Effects of management practices on grassland birds: Sprague's pipit. Northern Prairie Wildlife Research Center, Jamestown, ND. Available: http://www.npwrc.usgs.gov/resource/literatr/grasbird/sppi/sppi.htm. Accessed 15 March 2014.
68. Johnson DH (1996) Management of northern prairies and wetlands for the conservation of neotropical migratory birds. In: Thompson FRI, editor. Management of Midwestern Landscapes for the Conservation of Neotropical Migratory Birds. North Central Forest Experiment Station, St. Paul, MN: Northern Prairie Wildlife Research Center, Jamestown, ND. 53–67. Available: http://www.npwrc.usgs.gov/resource/habitat/neobird/index.htm.
69. Koford RR, Best LB (1996) Management of agricultural landscapes for the conservation of neotropical migratory birds. In: Thompson FRI, editor. Management of Midwestern Landscapes for the Conservation of Neotropical Migratory Birds. Nort: Northern Prairie Wildlife Research Center, Jamestown, ND. 68–88. Available: http://www.nrs.fs.fed.us/pubs/gtr/other/gtr-nc187/Page%2068%20RR%20Koford,%20NC-GTR-187.pdf.
70. Mineau P, Downes CM, Kirk DA, Bayne E, Csizy M (2005) Patterns of bird species abundance in relation to granular insecticide use in the Canadian prairies. Ecoscience 12: 267–278.
71. McEwen LC, Knittle CE, Richmond ML (1972) Wildlife Effects from Grasshopper Insecticides Sprayed on Short-Grass Range. J Range Manag 25: 188. doi:10.2307/3897053.

72. Veech JA (2006) A comparison of landscapes occupied by increasing and decreasing populations of grassland birds. Conserv Biol 20: 1422–1432. doi:10.1111/j.1523-1739.2006.00487.x.
73. Vickery PD, Hunter ML, Melvin SM (1994) Effects of Habitat Area on the Distribution of Grassland Birds in Maine. Conserv Biol 8: 1087–1097. doi:10.1046/j.1523-1739.1994.08041087.x.
74. Coppedge BR, Engle DM, Masters RE, Gregory MS (2001) Avian Response To Landscape Change In Fragmented Southern Great Plains Grasslands. Ecol Appl 11: 47–59. doi:10.1890/1051-0761(2001)011[0047:ARTLCI]2.0.CO;2.
75. Davis SK, Brittingham M (2004) Area Sensitivity In Grassland Passerines: Effects Of Patch Size, Patch Shape, And Vegetation Structure On Bird Abundance And Occurrence In Southern Saskatchewan. The Auk 121: 1130–1145. doi:10.1642/0004-8038(2004)121[1130:ASIGPE]2.0.CO;2.
76. Perlut NG (2014) Grassland birds and dairy farms in the northeastern United States. Wildl Soc Bull: n/a–n/a. doi:10.1002/wsb.415.
77. Boody G, Vondracek B, Andow DA, Krinke M, Westra J, et al. (2005) Multifunctional agriculture in the United States. BioScience 55: 27–38. doi:10.1641/0006-3568(2005)055[0027:MAITUS]2.0.CO;2.
78. Foley JA, DeFries R, Asner GP, Barford C, Bonan G, et al. (2005) Global Consequences of Land Use. Science 309: 570–574. doi:10.1126/science.1111772.
79. Jordan N, Warner KD (2010) Enhancing the Multifunctionality of US Agriculture. BioScience 60: 60–66. doi:10.1525/bio.2010.60.1.10.
80. Scherr SJ, McNeely JA (2003) Making space for wildlife in agricultural landscapes. In: EcoAgriculture: strategies to feed the world and save wild biodiversity. In: Harvest F, editor. Ecoagriculture: Strategies to Feed the World and Save Wild Biodiversity. Washington, D.C.: Island Press. 115–148.
81. Atwell RC, Schulte LA, Westphal LM (2011) Tweak, adapt, or transform: Policy scenarios in response to emerging bioenergy markets in the U.S. corn belt. Ecol Soc 16: 15.
82. Robertson GP, Gross KL, Hamilton SK, Landis DA, Schmidt TM, et al. (2014) Farming for Ecosystem Services: An Ecological Approach to Production Agriculture. BioScience: biu037. doi:10.1093/biosci/biu037.

2

A Novel W1999S Mutation and Non-Target Site Resistance Impact on Acetyl-CoA Carboxylase Inhibiting Herbicides to Varying Degrees in a UK *Lolium multiflorum* Population

Shiv Shankhar Kaundun*, Geraldine C. Bailly, Richard P. Dale, Sarah-Jane Hutchings, Eddie McIndoe

Syngenta, Jealott's Hill International Research Centre, Bracknell, United Kingdom

Abstract

Background: Acetyl-CoA carboxylase (ACCase) inhibiting herbicides are important products for the post-emergence control of grass weed species in small grain cereal crops. However, the appearance of resistance to ACCase herbicides over time has resulted in limited options for effective weed control of key species such as *Lolium* spp. In this study, we have used an integrated biological and molecular biology approach to investigate the mechanism of resistance to ACCase herbicides in a *Lolium multiflorum* Lam. from the UK (UK21).

Methodology/Principal Findings: The study revealed a novel tryptophan to serine mutation at ACCase codon position 1999 impacting on ACCase inhibiting herbicides to varying degrees. The W1999S mutation confers dominant resistance to pinoxaden and partially recessive resistance to cycloxydim and sethoxydim. On the other hand, plants containing the W1999S mutation were sensitive to clethodim and tepraloxydim. Additionally population UK21 is characterised by other resistance mechanisms, very likely non non-target site based, affecting several aryloxyphenoxyproprionate (FOP) herbicides but not the practical field rate of pinoxaden. The positive identification of wild type tryptophan and mutant serine alleles at ACCase position 1999 could be readily achieved with an original DNA based derived cleaved amplified polymorphic sequence (dCAPS) assay that uses the same PCR product but two different enzymes for positively identifying the wild type tryptophan and mutant serine alleles identified here.

Conclusion/Significance: This paper highlights intrinsic differences between ACCase inhibiting herbicides that could be exploited for controlling ryegrass populations such as UK21 characterised by compound-specific target site and non-target site resistance.

Editor: Pär K. Ingvarsson, University of Umeå, Sweden

Funding: The authors have no funding or support to report.

Competing Interests: All authors are employed by Syngenta, Jealott's Hill International Research Centre, the funder of this study. There are no patents, products in development or marketed products to declare. This does not alter the authors' adherence to all the PLOS ONE policies on sharing data and materials, as detailed online in the guide for authors.

* E-mail: deepak.kaundun@syngenta.com

Introduction

Acetyl-CoA carboxylase (ACCase) is an ubiquitous enzyme that catalyses the carboxylation of acetyl-CoA into malonyl-CoA [1]. Plants have two different forms of ACCase located in the cytoplasm and chloroplast [2]. The two different isoforms share around 60% similarity at the amino acid level and perform different functions [3]. Plastid ACCase is a key enzyme in *de novo* fatty acid synthesis whilst cytosolic ACCase provides malonyl-CoA for the elongation of long chain fatty acid and synthesis of secondary plant metabolites such as flavonoids and suberins [4]. The cytosolic ACCase of all plants are homomeric with the four subdomains biotin carboxyl carrier protein (BCCP), biotin carboxylase (BC) and carboxyl transferase α and β, all located on a single polypeptide [5]. In contrast, the structure of plastidic ACCase varies depending on plant groups. Chloroplastic ACCase is homomeric in the Poaceae, and heteromeric in most other plants with the four subunits encoded by four different genes co-ordinately expressed to form a functional enzyme [6].

This difference between grasses and broadleaved weeds has been exploited by the agro-industry with the development of three different classes of ACCase inhibiting herbicides including the aryloxyphenoxypropionates (FOPs), cyclohexanediones (DIMs) and phenylpyraxolin [7,8]. These herbicides are generally active on the chloroplast ACCase of most grass weeds but show very little to no efficacy on the cytoplasmic isoform [9]. Over the last 30 years, 18 ACCase herbicides have been marketed providing excellent control of grass weeds.

Given their cost-effectiveness and convenience for managing grass weeds post-emergence, ACCase herbicides were quickly adopted and often used year after year in monocotyledonous and the subsequent rotational dicotyledonous crops. The repeated use of ACCase herbicides has resulted in resistance evolution in a growing number of grass weeds. To date, 42 different weeds species have been reported to resist at least one ACCase herbicide [10]. These include rigid ryegrass (*Lolium rigidum*), the first species to have evolved resistance to an ACCase herbicide following six years of diclofop-methyl use in a wheat field in Australia [11].

Two types of mechanisms account for resistance to ACCase herbicides, namely, an insensitive target and/or non-target site based resistance [12]. Target site resistance results from changes in amino acid residues that are critical for the optimal binding of the herbicide. Mutations at codon positions 1781, 1999, 2027, 2041, 2078, 2088 and 2096 (*Alopecurus myosuroides* equivalent) are documented to confer resistance to ACCase herbicides [3,13]. Genotype to phenotype correlation experiments show that specific amino acid changes at the seven codon positions can have very diverse effects between and within ACCase chemical classes [14,15,16,17]. The levels of resistance depend on the herbicides, use rates, weed species and number of resistant alleles in individual plants [18]. Non-target site based resistance to ACCase herbicide appears to be widespread but less well understood than target site resistance [19,20]. It consists mainly of enhanced degradation of ACCase herbicides into non-toxic metabolites [21]. Metabolic resistance is unpredictable and can impact on herbicides with similar or even different herbicide modes of action [22]. Genetic studies have shown that metabolic resistance is complex with several loci acting additively to confer resistance to ACCase herbicides [23,24,25].

While resistance to ACCase herbicides has evolved in all major cropping systems worldwide, a significant proportion of grass weed populations are still sensitive to this class of herbicides [10]. In order to design weed management strategies that will delay the onset and sometimes overcome resistance, it is important to better understand the mechanisms by which weeds evolve resistance to ACCase herbicides. Here we have employed a holistic biological and molecular approach to investigate the mechanisms of resistance to a wide range of ACCase inhibiting herbicides in a *Lolium multiflorum* population (UK21) from the UK. We have subsequently developed a robust DNA marker to quickly and cost-effectively identify the novel target site resistance mutation detected in this study.

Materials and Methods

Materials

Seeds from a suspected pinoxaden resistant *Lolium multiflorum* population UK21 were collected from a wheat field in Chislet, Canterbury, UK in 2006. A standard sensitive population (STD1) was sourced from Herbiseed (Twyford, UK) for comparison in all biological and molecular experiments.

"No specific permissions were required for the location where the ryegrass seeds were collected. This study did not involve any endangered or protected species".

Methods

Pinoxaden resistance confirmation test. The suspected resistant and sensitive populations were sown in 5 inch pots in soil composed of peat and compost in 1:1 ratio. Around 50 seeds were sown per population per pot. The pots were watered daily, fertilized as required and placed in controlled glasshouse conditions set at 24°C/16 hr day, 18°C night, 65% relative humidity, and a photon flux density of approximately 250 µmol quanta m^{-2} s^{-1}. When the plants had reached the two leaf stage, they were sprayed with a single recommended rate of pinoxaden at 45 g ai ha^{-1} in a spray cabinet mounted with a single mobile Teejet flat fan nozzle (11002VS) calibrated to deliver 200 L ha^{-1} at 200 kPa. Three replicate pots were used per population and the percentage of herbicide damage compared with untreated pots was assessed visually three weeks after treatment.

Selection of pinoxaden sensitive and resistant plants. Around 100 seeds from each of UK21 and STD1 were sown in two rectangular trays and grown as described above. At the one leaf stage 48 seedlings from each of the two populations were individually transplanted in 3 inch pots and maintained under the same glasshouse conditions until they reached the two leaf stage. A 1 cm leaf fragment was collected from each plant for ACCase gene sequencing analysis prior to spraying with pinoxaden at the recommended field rate of 45 g ai ha^{-1}. Three weeks after treatment the plants were assessed for survival.

ACCase gene sequencing from selected sensitive and pinoxaden resistant plants. To determine whether a mutated target was responsible for pinoxaden resistance, the ACCase carboxyltranferase domain was sequenced for 8 UK21 plants that were killed and 8 UK21 plants that survived the pinoxaden treatment. Eight other plants from the standard sensitive population STD1 were analysed in a similar way for comparison. The RT-PCR procedure for generating and sequencing the ACCase gene fragment encompassing the carboxyl transferase domain was the same as described in Kaundun (2010) [26].

Development of a dCAPS assay for large scale genotyping of UK21 plants. A derived cleaved polymorphic sequence assay (dCAPS) [27] was developed for the positive identification of the wild type tryptophan and mutant serine alleles in *Lolium* spp. plants. The primers and corresponding restriction enzymes were chosen using the dCAPS freeware [28]. The assay consisted of a single PCR fragment and two different restriction enzymes, namely *Xcm*I for positively detecting the wild type 1999 tryptophan allele and *Mnl*I for identifying the mutant serine allele in two separate reactions. Changes in any one of the three bases of the wild 1999 codon would disrupt the *Xcm*I restriction site and result in an undigested band on the agarose gel. On the other hand, restriction with *Mnl*I would indicate presence of a mutant serine allele or nucleotides TC on the first and second bases of the 1999 codon. The recognition of the wild type tryptophan and mutant serine alleles was rendered possible by the introduction of three forced mutations (underlined bases) in the forward dCAPS primer (5′ cccatgagcggtccgttc**t**tcgtccagggcaagcc 3′). The forced mutations at positions N-13 (guanine to cytosine) and N-11 (thymine to adenine) with respect to the second variable base of the 1999 codon create the *Xcm*I recognition site CCA(N)9TGG for plants with the wild type tryptophan allele. The forced nucleotide change at position N-3 with respect to the causative resistance SNP introduces a recognition site for *Mnl*I (CCTC) for plants carrying the mutant 1999 serine allele. The nucleotide sequence for the reverse primer was: 5′ aatccttcaaaaaggtctctttgcccaccagagaagc**t**tctc-cagttagcaaggatgaaca**a**agg3′ for both dCAPS reactions. Noteworthy is three other forced mutations in all (in bold) were included on the forward (N-18) and reverse (N+90; N+66) primers with respect to the mutated second base of the 1999 codon to eliminate three non-specific restriction sites for *Mnl*1. If left unchanged, shorter and ambiguous digests which would be difficult to assess on a 2% agarose gel would be generated. The 8 plants each of sensitive UK21, resistant UK21 and STD1 groups previously sequenced for the entire ACCase CT domain were used to develop the *Xcm*I and *Mnl*I based assays. The three step dCAPS assays were carried out as follows: for each plant a 1 cm leaf segment was ground on a Spex Certiprep (Metuchen, NJ, USA) 2000 model Genogrinder; DNA from the ground material was subsequently extracted on a Beckman Coulter Biomek FX robot (High Wycombe, Buckinghamshire, UK) using a Wizard Magnetic 96 DNA Plant System kit (Promega, Madison, WI, USA). PCRs were carried out using

PuReTaq Ready-To-Go PCR beads (Amersham Biosciences, Bucks, UK) in a total volume of 25 µL containing 0.8 µM of each primer and from 10–50 ng genomic DNA. PCRs were performed on an Eppendorf Master Cycle Gradient Thermocycler Model 96 programmed for an initial denaturation step of 94°C of 2 min followed by 40 cycles of 30 s at 94°C, 30 s at 64°C and 1 min at 72°C. A final extension step for 10 min at 72°C was also included. For the positive identification of the wild type TGG codon characteristic of the tryptophan residue, 8 µl of the PCR product was then digested with 1 µL (5 units) of *Xcm*I at 37°C for 2 hours in a 40 µL reaction. Ten µL of the PCR product treated with *Xcm*I was then migrated on a 1xTBE 2% agarose gel containing 0.5 µg mL−1 ethidium bromide. The same procedure and conditions were used for the positive identification of the mutant serine 1999 allele except that *Xcm*I was replaced by the enzyme *Mnl*I.

Correlation between plant genotypes and phenotypes at single recommended rates of eight ACCase herbicides. To investigate the influence of the tryptophan to serine mutation at ACCase codon position 1999 and potential non-target site resistance mechanisms on eight other ACCase herbicides, 384 UK21 individual plants were grown to the two leaf stage as described above and genotyped using the *Xcm*I based dCAPS assay. Forty eight random UK21 plants each were subsequently sprayed with the recommended field rates of diclofop-methyl (1000 g ai/ha), clodinafop-propargyl (60 g ai/ha), fluazifop-P-butyl (200 g ai/ha), haloxyfop-P-methyl (100 g ai/ha), sethoxydim (200 g ai/ha), cycloxydim (200 g ai/ha), tepraloxydim (100 g ai/ha) and clethodim (120 g ai/ha). For comparison, 48 plants each from the standard sensitive population were also sprayed with the same eight herbicides. Survival to the herbicides was assessed 21 days after treatment. Plant survival data from the co-segregation experiment were arranged as a 2×2 contingency table for each genotype comparison and analysed using Fisher's Exact test (Table 1). Formal analyses were only carried out for haloxyfop-P-methyl, sethoxydim, cycloxydim and pinoxaden since all other compounds produced either 0% or 100% survival across all three WW1999 (homozygous wild type with two copies of the tryptophan allele), WS1999 (heterozygous mutant type with one copy each of the tryptophan and serine alleles) and SS1999 (homozygous mutant type with two copies of the serine allele) genotypes. A p-value of less than 0.05 indicates a statistically significant result at the 5% probability level and provides evidence that the true levels of survivorship in the two treatments in question are genuinely different.

Additionally given the complete resistance observed with the three FOP herbicides, 8 plants that have survived clodinafop-propargyl and that have shown a wild type allele at codon position 1999 with the dCAPS assay was sequenced with the RT-PCR procedure described above to determine whether any FOP specific target site mutations were present in population UK21.

Whole plant dose-response experiments using pre-determined 1999 ACCase genotypes. In order to determine the level of resistance conferred by target site and suspected non-target site resistance identified in UK21, whole plant dose-response experiments were conducted using known plant genotypes at ACCase codon position 1999. The genotypes were characterised with the dCAPS and sequencing assays described previously. Four mother plants for each of the WW1999, WS1999 and SS1999 genotypes were selected from population UK21. Four mother plants from the standard sensitive population STD1 (WW1999) were included for comparison. This amounted to four plant groups with four biological replicates each. Individual plants were subsequently tiller propagated over a period of one year under the same glasshouse conditions as described previously. Tillering was promoted by regularly fertilizing and trimming the plants to avoid flowering. One genetically identical tiller from each of the 16 mother plants were sprayed with several rates of diclofop-methyl (Hoelon 3EC, 347 gai.L^{-1}, Bayer Crop Science), sethoxydim (Grasidim, 213 gai.L^{-1}, BASF Corporation), haloxyfop-P-methyl (Gallant 535, 108 gai.L^{-1}, Dow AgroSciences), tepraloxydim (Aramo 50, 50 gai.L^{-1}, BASF Corporation) and pinoxaden (Axial, 100 gai.L^{-1}, Syngenta) (Table 2). Four tillers per mother plants were used as untreated control. The dose-responses were repeated two times in time (technical replicates). The pots were re-arranged in a randomized complete block design after herbicide application (blocking per mother plant). Twenty-one days after treatment (DAT) the tillers were cut above ground level, put into paper envelopes and dried in an oven at 70°C for at least three days. The dry mass for each individual was recorded to the nearest mg.

Since the mother plants represent the independent biological replicates in the experiment, the dry weight measurements were averaged across technical replicates (time) separately for each compound, rate and mother plant and divided by the average dry weight of the untreated tillers from the corresponding mother plant. GR_{50} estimates were derived from the resulting percent of untreated values separately for each mother plant by a non-linear least-squares regression model described by the equation:

$$P = \frac{100}{1 + e^{-\beta(x - \mu_{ij})}} + L$$

where x denotes log(Rate); μ_{ij} denotes the $\log GR_{50}$ for mother plant i of genotype j; β denotes the common slope and L denotes the common lower asymptote fitted to all regression lines. The regression line for each mother plant has 3 parameters (a slope, $\log GR_{50}$ and lower asymptote) resulting in a model with 18 parameters in total: 1 slope, 1 lower asymptote and 16 $\log GR_{50}$ values.

The estimated resistance factor (Rf) between a pair of genotypes is calculated as the ratio of their average GR_{50} values across mother plants. Confidence intervals around the estimated resistance factors were calculated using the error term derived from an analysis of variance carried out on the $\log GR_{50}$ values. All statistical analyses were carried out using SAS software, version 9.2.

Results

Initial whole plant pot test for confirming resistance to pinoxaden

Pinoxaden treatment at the recommended field rate provided 35±2.9% control (visual biomass reduction) of the suspected resistant population UK21. The individuals that have survived varied from being very stunted to being as healthy as untreated plants. As expected all the plants from the standard sensitive population STD1 were killed. Though pinoxaden is a cereal selective and metabolisable herbicide, to date most documented cases of product failures were due to a target site mutation in *Lolium* spp. [16,17,19,29]. Therefore, a mutated and insensitive target rather that non-target site resistance was suspected to being the most likely mechanism of resistance to pinoxaden in population UK21.

Table 1. Correlation between plant phenotypes and genotypes at ACCase codon position 1999.

			No. of survivors/Total		
Compound	**Rate (g/ha)**	**Comparison**	**Group1**	**Group2**	**P-value (2-sided)**
Pinoxaden	45	WW v WS	0/12	22/22	<**0.001**
		WW v SS	0/12	14/14	<**0.001**
		WS v SS	22/22	14/14	1
Haloxyfop-P-methyl	100	WW v WS	4/9	25/31	***0.047***
		WW v SS	4/9	8/8	***0.029***
		WS v SS	25/31	8/8	0.313
Sethoxydim	200	WW v WS	0/10	6/26	0.1567
		WW v SS	0/10	7/12	0.**0053**
		WS v SS	6/26	7/12	0.0636
Cycloxydim	200	WW v WS	0/13	4/21	0.144
		WW v SS	0/13	6/14	***0.016***
		WS v SS	4/21	6/14	0.251

A significant difference at the 5% probability level is shown in bold.

Segregation of the UK21 population into pinoxaden sensitive and resistant individuals and ACCase gene analysis

Out of 48 individual plants tested, 36 survived the pinoxaden treatment. Polymerase chain reaction from eight random plants each from the sensitive and resistant UK21 subpopulations and the standard sensitive population STD1 generated a 2272 gene fragment in all cases. Analysis of the nucleotide sequences from the 24 plants showed 95% homology with published data, thus confirming the identity of the ACCase carboxyl transferase domain amplified here. Sequence comparison identified 15 nucleotides changes in all among those 24 plants. Nine of the nucleotide substitutions were synonymous and thus did not amount to differences in amino acid sequences between and within the two ryegrass populations. Five single nucleotide polymorphisms were non-synonymous at amino acid positions L1701, E1874 E1946, R1995 and T2054 but these were evenly distributed among the three different plant groups. In contrast, a tryptophan to serine mutation at amino acid position 1999 (W1999S) was only present in the eight UK21 resistant plants but absent in the sensitive UK21 as well as the standard sensitive STD1 samples (Figure 1). Out of the eight resistant plants, five were heterozygous WS1999 and three were homozygous for the serine 1999 mutation.

dCAPS assay for the rapid and cost effective identification of the W1999S mutation

A dCAPS assay was developed for large scale genotyping of single UK21 plants in view of confirming the association between the 1999 mutation and pinoxaden resistance. Digestion with two different enzymes allowed the positive identification of the wild type tryptophan or mutant serine alleles. Typical dCAPS profiles as resolved on 2% agarose gel electrophoresis are provided in figures 2a and 2b. PCR generated a 164 bp fragment for all plants. When digested with the enzyme *Xcm*1, wild type homozygous plants showed a single restricted band of 134 bp (and 30 bp fragment not visible of 2% agarose gel) while plants that are homozygous for the serine 1999 allele displayed the undigested

Table 2. Summary of herbicide rates used in the whole-plant dose-response experiments.

Genotypes	Pinoxaden	Diclofop-methyl	Haloxyfop-P-methyl	Sethoxydim	Tepraloxydim
STD1-WW1999	1; 2; 4; 7; 13; 23; **45** and 95	10; 25; 56; 93; 155; 258; 431; 719; **1,000** and 2,233	0.1; 0.25; 0.63; 2.5; 12; 25; 64; **100** and 400	5; 13; 25; 37; 60; 95; 152; **200**; 314 and 400	0.1; 0.24; 0.59; 1; 5; 8; 20; 50; **100** and 300
UK21-WW1999	1; 2; 4; 7; 13; 23; **45** and 95	**1,000**; 1,351; 1,825; 2,233; 3,332; 4,502 and 6,082	5; 8; 12; 25; 39; 64; **100**; 151; 228 and 400	5; 13; 25; 37; 60; 95; 152; **200**; 314 and 400	0.1; 0.24; 0.59; 1; 5; 8; 20; 50; **100** and 300
UK21-WS1999	23; **45**; 95; 145; 279; 472; 800 and 1,000	**1,000**; 1,351; 1,825; 2,233; 3,332; 4,502; 6,082; 8,217; 11,102 and 15,000	25; 39; 64; **100**; 228; 400; 608; 1,035; 1,762 and 3,000	13; 25; 37; 60; 95; 152; **200**; 314; 604 and 1,000	5; 8; 12; 20; 31; 50; 77; **100**; 190 and 300
UK21-SS1999	**45**; 95; 145; 208; 346, 577, 1,000 and 1,600	**1,000**; 1,351; 1,825; 2,233; 3,332; 4,502; 6,082; 8,217; 11,102 and 15,000	25; 39; 64; **100**; 228; 400; 608; 1,035; 1,762 and 3,000	13; 25; 37; 60; 95; 152; **200**; 314; 604 and 1,000	5; 8; 12; 20; 31; 50; 77; **100**; 190 and 300

The rates are indicated in g ai.ha^{-1}. The rates in bold represent the recommended field rate for ryegrass control. All herbicides were formulated in dH_2O using their recommended adjuvant: 0.5% v/v Actipron for sethoxydim and diclofop-methyl; 0.375% v/v Output for haloxyfop-P-methyl and tepraloxydim and 0.5% v/v Adigor for pinoxaden.

```
Pinoxaden sensitive
UK21.3   CCTCGTGCTGGGCAAGTCTGGTTTCCAGATTCTGCTACC
UK21.5   .......................................
UK21.8   .......................................
UK21.14  .......................................
UK21.16  .......................................
UK21.24  .......................................
UK21.41  .......................................
UK21.44  .......................................

Pinoxaden resistant
UK21.2   ...................C...................
UK21.4   ...................C...................
UK21.6   ...................C...................
UK21.7   ...................S...................
UK21.13  ...................S...................
UK21.15  ...................S...................
UK21.28  ...................S...................
UK21.35  ...................S...................
```

Figure 1. ACCase gene alignment around the critical nucleotide position 5996 (second base of the 1999 codon) among eight plants sensitive and resistant to pinoxaden. All sensitive plants contained two copies of guanine (gg) whilst resistant plant had one (cg = s) or two copies of cytosine (cc) at this position.

PCR fragment of 164 bp. As expected, heterozygous plants showed a copy each of the 164 and 134 bp fragments. The inverse restriction profiles were obtained with the enzyme *MnlI* with plants containing the wild type tryptophan allele being unrestricted whilst individuals with the mutant serine allele showed a shorter digested band of 120 bp. It is noteworthy that the difference in size of restricted PCR fragments between *XcmI* and *MnlI* treatments is due to the fact that the second enzyme generally cuts 6/7 bp downstream of the CCTC recognition site while the first enzyme cleaves 4/5 base pairs upstream of the TGG 1999 codon. The dCAPS and sequencing results for the 16 UK21 and 8 STD1 plants were totally correlated demonstrating the accuracy of the dCAPS assay. Genotyping of the 32 remaining UK21 plants (that were phenotyped with pinoxaden at a single recommended rate) with the *XcmI* based dCAPS assay confirmed the complete association between the presence of the mutated serine 1999 ACCase allele and resistance to pinoxaden (Table 1).

Cross resistance patterns to a range of ACCase herbicides

Analysis of 384 UK21 individuals in all with the *XcmI* dCAPS marker revealed genotypic frequencies of 0.199, 0.598 and 0.203 respectively for wild type WW1999, heterozygous mutant WS1999 and homozygous mutant SS1999 plants in population UK21. These were randomly separated into eight lots of 48 plants and tested with eight different herbicides alongside the standard sensitive population. As expected, plants from the latter population were controlled with the eight ACCase herbicides. All UK21 plants survived the clodinafop-propargyl, diclofop-methyl and fluazifop-P-butyl treatments irrespective of their genotypes at ACCase codon position 1999. Sequencing of eight plants that have survived clodinafop-propargyl and containing a wild type allele at codon position 1999 did not reveal any other specific mutations compared to the sensitive population suggesting that additional resistance to the FOP herbicides in UK21 is very probably non-target site based. Conversely all the wild type and mutant UK21 plants were killed at the recommended field rates of tepraloxydim and clethodim. Intermediate responses were observed with haloxyfop-P-methyl, sethoxydim and cycloxydim (Table 1). Haloxyfop-P-methyl behaved in a very similar manner to pinoxaden, given that all the homozygous mutant SS1999 plants and a very large proportion of heterozygous WS1999 plants survived this FOP treatment. Additionally, around half of the wild type UK21-WW1999 plants also survived the haloxyfop-P-methyl treatment. Formal statistical analysis of the data revealed a significant difference between the heterozygous mutant UK21-WS1999 and the wild type UK21-WW1999 subpopulations implying that the W1999S mutation confers dominant resistance to haloxyfop-P-methyl in UK21. Sethoxydim and cycloxydim killed all the wild type UK21-WW1999 plants. In contrast to pinoxaden, only around half of the homozygous mutant UK21-SS1999 plants and an even lower proportion of heterozygous UK21-WS1999 plants survived sethoxydim (23%) and cycloxydim (19%) treatments. No convincing evidence of a difference in survivorship between heterozygous UK21-WS1999 and homozygous wild type UK21-WW1999 plants was identified with the Fisher exact test for these two DIM herbicides. On the other hand, there was evidence of a difference between homozygous wild type UK21-WW1999 and mutant UK21-SS1999 individuals suggesting that resistance could be classified as partially recessive to sethoxydim and cycloxydim.

Level of resistance conferred by the W1999S target site and non-target site resistance

Whole plant dose response tests on characterised UK21 genotypes were carried out to determine the precise levels of resistance conferred by target site and non-target site resistance to two FOPs, two DIMs and pinoxaden in UK21 (Tables 3 and 4). Dose response curves were obtained for all plant groups and herbicides except for diclofop-methyl on all three different UK21 genotypes (Figure 3). At the highest rate tested (15 kg ai/ha) all the plants survived the diclofop-methyl treatment. On average, the heterozygous and homozygous mutant plants (<20% biomass reduction) were less controlled than the homozygous wild type plants (42% biomass reduction) implying that the target site resistance at position 1999 did contribute to higher levels of resistance to diclofop-methyl in UK21. Compared to diclofop-methyl, lower levels of resistance were observed for haloxyfop-p-methyl with respect to the W1999S target site resistance and non-target site resistance contained in UK21 (Figure 4). One or two copies of the mutant serine allele amounted to a resistance factor (Rf) of around 3 in both cases. However this was sufficient to cause survivorship to the FOP herbicide at the recommended field rate. A higher level of resistance was identified between the wild type UK21-WW1999 and STD1-WW1999 plants. Resistance to pinoxaden was confirmed with the W1999S mutation resulting in resistance factors of 11 and 33 for one and two mutant serine alleles respectively (Figure 5). The shift between the wild type UK21-WW1999 and STD1-WW1999 plants was small (Rf of 3) with average GR50 values of 2.1 g ai/ha and 8.3 g ai/ha only, thus confirming the non-significance of non-target site resistance at the recommended field rate of pinoxaden. The W1999S mutation had a lesser effect on sethoxydim compared to pinoxaden reflecting the partially recessive and dominant nature of this mutation on the DIM and DEN herbicide respectively (Figure 6). And finally, the dose response assay ascertained the minimal impact of the W1999S mutation on tepraloxydim with almost overlapping curves for UK21-WW1999 and UK21-WS1999 (Figure 7). GR50 values for any one wild and mutant UK21 genotypes were well below the recommended field rate of the herbicide (Table 3).

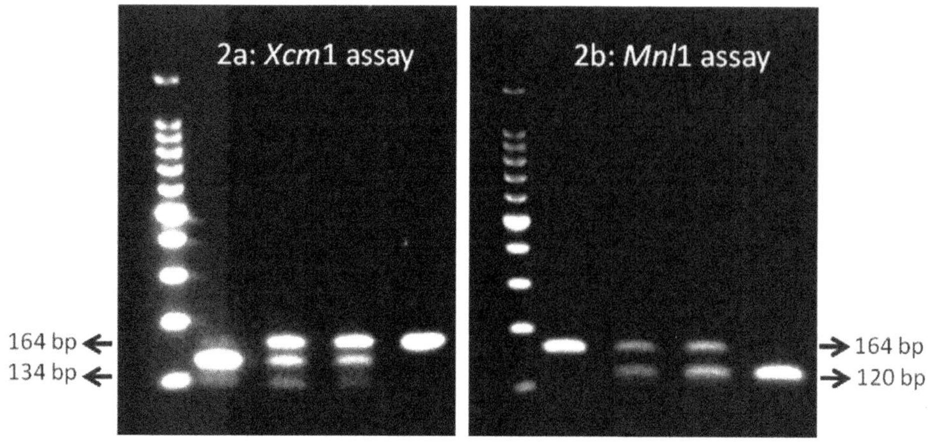

Figure 2. dCAPS procedures for the detection of the wild type tryptophan and mutant serine amino acid residues at ACCase codon position 1999: (a) ***XcmI*** **restricted (134 bp) and unrestricted (164 bp) fragments correspond to sensitive W1999 and resistant S1999 ACCase alleles respectively.** (b) *MnlI* restricted fragment is indicative of the mutant serine allele (120 bp) while the undigested band (164 bp) corresponds to the wild type tryptophan allele. Heterozygous plants display one each of the restricted and unrestricted PCR fragment in both assays.

Discussion

Integrated biological and molecular approach for revealing multiple herbicide resistance mechanisms

A growing number of *Lolium* spp. populations worldwide are evolving resistance to the few remaining post-emergence herbicides used to control this species. Additionally the development and registration of new herbicides has become increasingly problematic due to the difficulty of finding new modes of action and stricter regulatory hurdles. Thus, understanding the basis of resistance is imperative for a better stewardship of very effective herbicides such as inhibitors of acetyl-CoA carboxylase. This study reveals a novel target site resistance mutation and other mechanisms, likely non-target site based to several ACCase herbicides in population UK21. Additional underlying non-target site resistance has been inferred in many other *Lolium* spp. [19] *Alopecurus* spp.[30], *Avena* spp. [31] populations that have evolved resistance to ACCase inhibiting herbicides. An exception to this rule is found in a UK ryegrass population characterised by the C2088R ACCase resistance mutation only [32]. The presence of multiple resistance mechanisms often acting additively is expected in a highly polymorphic, outbreeding species such as *Lolium multiflorum*, especially when pressured with the same herbicide mode of action year after year [22]. However, in most previously published studies the contributions of target site and non-target site resistance were not properly evaluated due to heterogeneous

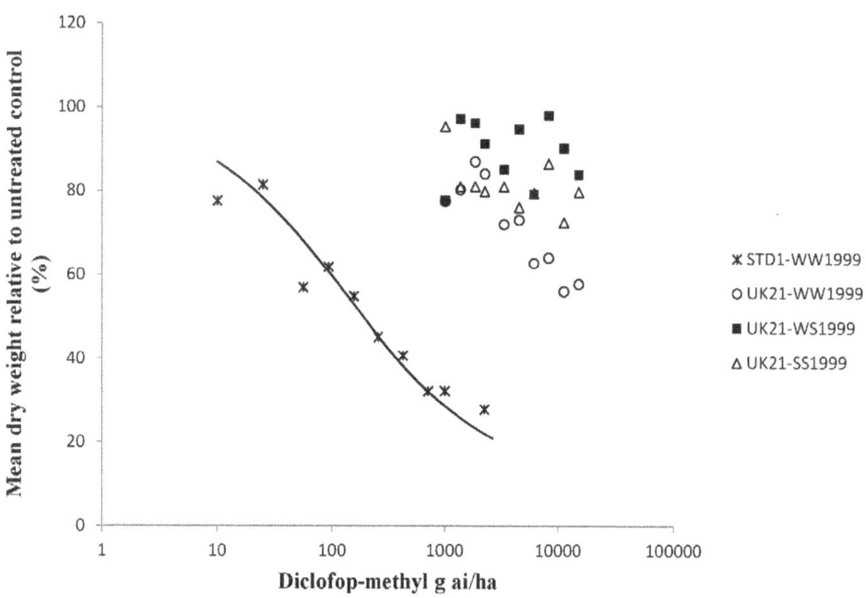

Figure 3. Diclofop-methyl dose response tests on four plant groups: homozygous wild type WW1999, heterozygous WS1999 and homozygous mutant SS1999 from the mixed resistant population UK21 and standard sensitive STD1 plants (WW1999) for comparison. Observed values represent dry weight relative to untreated (%) averaged across mother plants. The curves are constructed based on the average GR50 across biological and technical replicates.

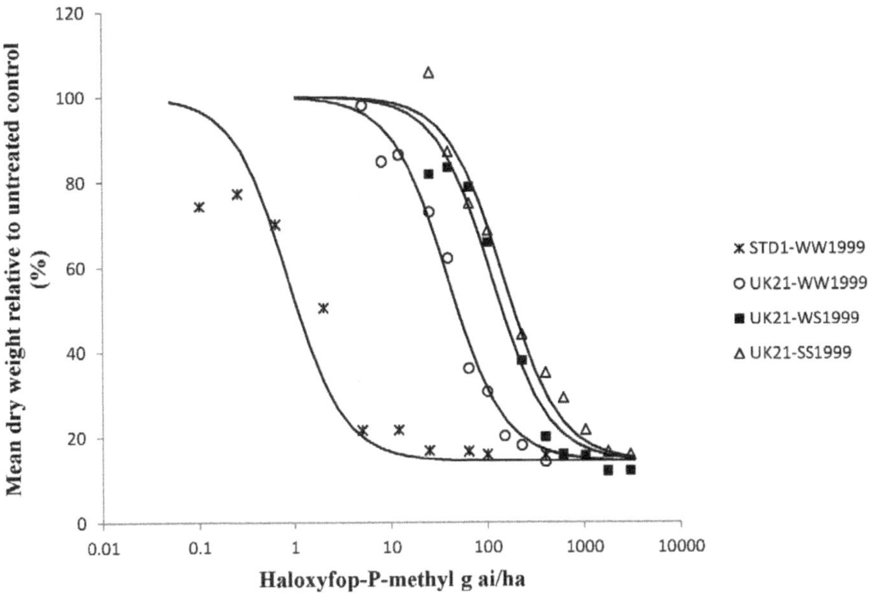

Figure 4. Haloxyfop-P-methyl dose response tests on four plant groups: homozygous wild type WW1999, heterozygous WS1999 and homozygous mutant SS1999 from the mixed resistant population UK21 and standard sensitive STD1 plants (WW1999) for comparison. Observed values represent dry weight relative to untreated (%) averaged across biological and technical replicates.

resistant weed populations being compared with a standard sensitive population having a very different genetic background. Our approach of creating wild type, heterozygous and homozygous 1999 mutant ACCase sub-populations from the same parental population UK21 and comparison with a standard sensitive population has allowed for a better estimation of the level of resistance conferred by target site and non-target site resistance.

Differential impact of the W1999S mutation on ACCase herbicides

The novel target site resistance mutation identified in UK21 consists of a change from a tryptophan 1999 to serine (W1999S) amino acid residue in the carboxyltransferase domain of the ACCase enzyme. At ACCase codon position 1999 two other mutations have been identified in grass weeds that have evolved resistance to ACCase herbicides. These include a tryptophan to cysteine change in *Avena sterilis* [14], *Avena fatua* [31] and *Lolium multiflorum* [33], and a tryptophan to leucine change (W1999L) in

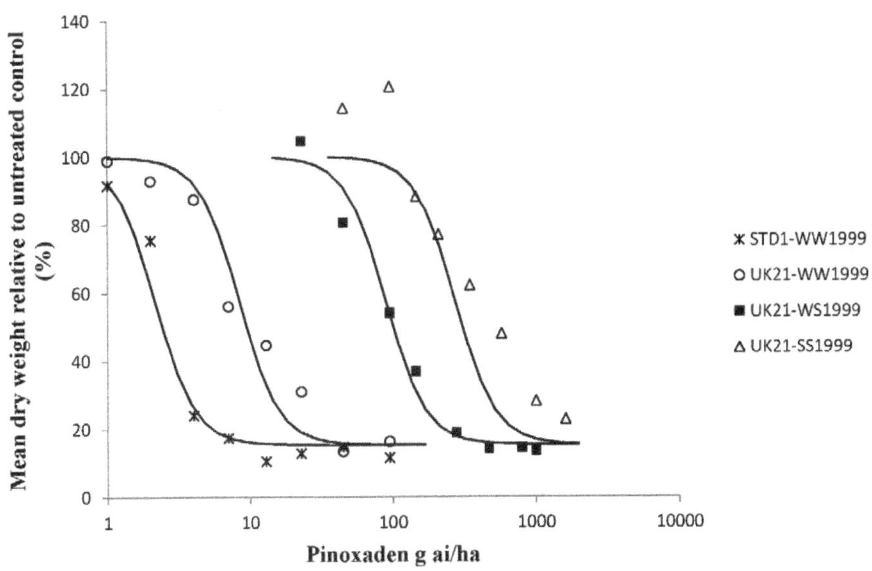

Figure 5. Pinoxaden dose response tests on four plant groups: homozygous wild type WW1999, heterozygous WS1999 and homozygous mutant SS1999 from the mixed resistant population UK21 and standard sensitive STD1 plants (WW1999) for comparison. Observed values represent dry weight relative to untreated (%) averaged across biological and technical replicates.

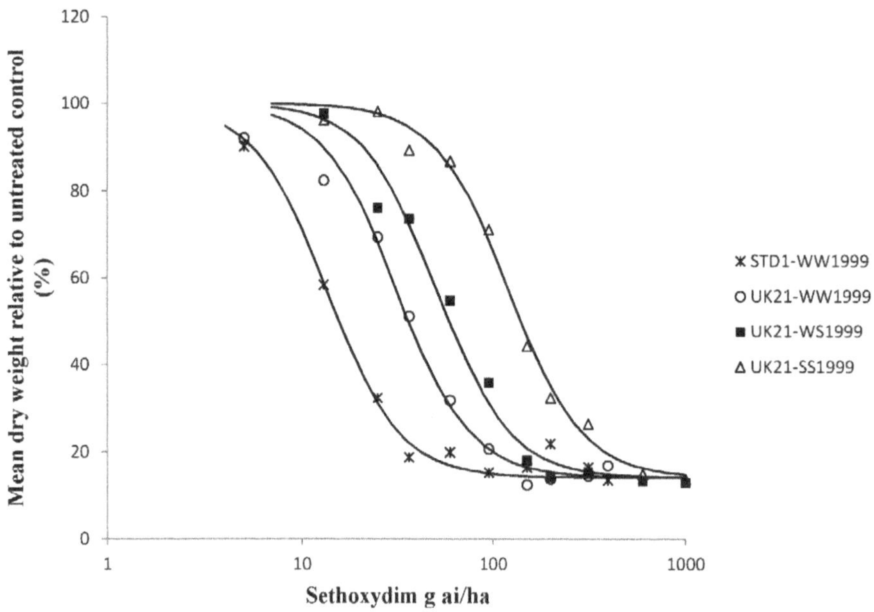

Figure 6. Sethoxydim dose response tests on four plant groups: homozygous wild type WW1999, heterozygous WS1999 and homozygous mutant SS1999 from the mixed resistant population UK21 and standard sensitive STD1 plants (WW1999) for comparison. Observed values represent dry weight relative to untreated (%) averaged across mother plants. The curves are constructed based on the average GR50 across biological and technical replicates.

Alopecurus myosuroides [34] and *Lolium multiflorum* [16] populations. The tryptophan 1999 to cysteine (W1999C) mutation was found to confer resistance to fenoxaprop-P-ethyl but not to two other FOP and DIM herbicides in the *Avena sterilis* population [14]. The role of the W1999C mutation on the five different ACCase herbicides was adequately investigated via comparison of plant phenotypes as established with glasshouse herbicide assays and PCR based genotyping at ACCase codon position 1999. As some plants that did not contain the 1999 mutation were also resistant to fenoxaprop-P-ethyl, the importance of the W1999C mutation on ACCase herbicides was further confirmed via yeast-gene replacement assays with strains differing at the 1999 codon position only [14]. On the other hand, the impact of the W1999L remains to be assessed in the blackgrass and ryegrass populations as well as W1999C mutation in the ryegrass and *Avena fatua* populations. In this study the W1999S mutation was found to confer variable

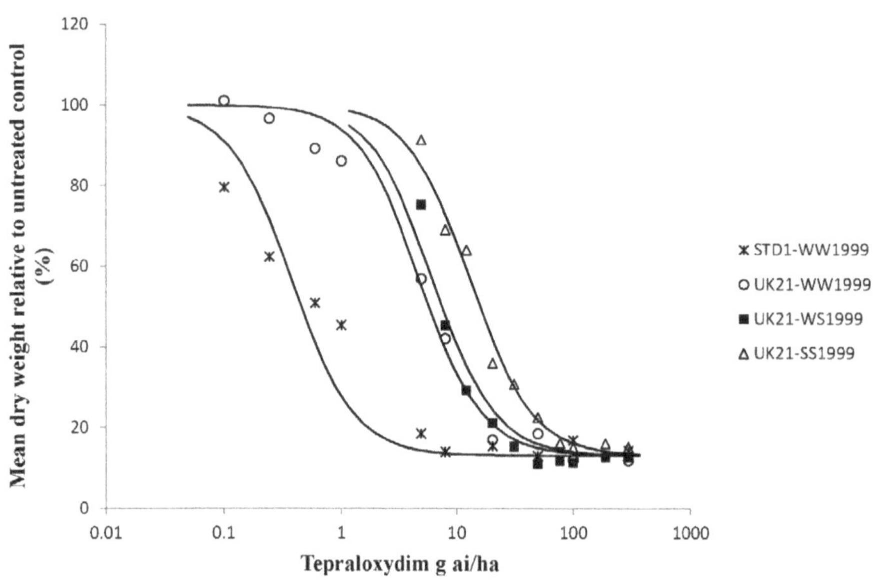

Figure 7. Tepraloxydim dose response tests on four plant groups: homozygous wild type WW1999, heterozygous WS1999 and homozygous mutant SS1999 from the mixed resistant population UK21 and standard sensitive STD1 plants (WW1999) for comparison. Observed values represent dry weight relative to untreated (%) averaged across mother plants. The curves are constructed based on the average GR50 across biological and technical replicates.

Table 3. Estimated GR50 values for five ACCase herbicides and different 1999 ACCase genotypes.

	STD1-WW1999	UK21-WW1999	UK21-WS1999	UK21-SS1999
Pinoxaden	2.1 (0.7–6.7)	8.4 (2.7–26.1)	89.2 (28.6–278.5)	272.9 (87.4–851.9)
Diclofop-methyl	132.1 (1.5–11290.8)	>6082	>15000	>15000
Haloxyfop-P-methyl	0.9 (0.3–2.3)	39.1 (14.4–105.9)	112.7 (41.6–305.0)	148.7 (54.9–402.5)
Sethoxydim	13.5 (9.2–19.7)	31.6 (21.7–46.1)	51.2 (35.1–74.6)	121.4 (83.3–177.0)
Tepraloxydim	0.4 (0.2–0.7)	4.8 (2.4–9.5)	6.2 (3.2–12.3)	13.0 (6.6–25.7)

95% confidence limits in brackets.

levels of resistance to ACCase herbicides within and between the different herbicide subclasses. Clethodim and tepraloxydim controlled all plants containing the serine mutant allele either in homozygous or heterozygous states while the W1999S mutation conferred only partial resistance to the herbicides cycloxydim and sethoxydim. In contrast, the W1999S mutation conferred dominant resistance to haloxyfop-P-methyl and pinoxaden and very probably to the other FOP herbicides tested here.

The difference in the resistance profiles between the *Avena sterilis* [14] and *Lolium multiflorum* populations characterised by the two different W1999C and W1999S mutations is at first sight surprising given that cysteine shares a very similar structure to serine. However, minor differences in amino acid residues such as the isoleucine to leucine change at codon 1781 have previously been shown to have very dramatic effects on the efficacy of ACCase herbicides [3,18]. The differential impact of the W1999C and W1999S mutations on ACCase inhibiting herbicides could also be due the difference in ploidy levels between these two grass weed species. *Avena sterilis* and *Lolium multiflorum* are hexaploid and diploid species respectively. Thus for *Avena* species, for which all homeologous ACCase genes are expressed [14], the dilution factor is three times more accentuated compared to *Lolium* species when only one of the loci is mutated. While this may have little bearing on plant survivorship for mutations that confer very high levels of resistance to ACCase herbicides such as the I1781L, D2078G, C2088R mutations [13], the number of ACCase gene copies becomes important for amino acid changes that confer lower levels of resistance such as the ones at codon positions 1999. A similar observation was made for the proline 106 mutations in the 3-phosphoshikimate 1-carboxyvinyltransferase glyphosate target. The impact of the P106 mutations, which generally confer low levels of resistance to glyphosate, was greater in *Eleusine indica* characterised by a single *EPSPS* copy compared to *Lolium* species for which the corresponding gene exists as a small family with at least three copies identified. [35]. Therefore the conclusion regarding the impact of the W1999C or W1999L mutations in the ryegrass and blackgrass populations cannot be based on the wild oat study [14] and should be properly evaluated, potentially by using a similar approach to the one employed in this study.

Occurrence and spread of the W1999S mutation in grass weed species

The mechanism of target site resistance to ACCase herbicides has been extensively studied over the past ten years, thanks to growing accessibility of molecular biology techniques and publications on the ACCase target gene and relevant DNA markers for the most commercially significant grass weed species [36,37]. Mutations at seven ACCase codon positions have been shown to confer resistance to at least one ACCase inhibiting herbicide [13]. The first target site mutation identified in most grass weeds is the I1781L mutation probably due to the high level of resistance it confers and the absence of a fitness cost associated with this mutation [38,39]. In contrast, mutations at codon position 1999 were the last to be identified and the least frequently encountered with only a handful of grass weed populations identified to date [14,16,31,33,34]. In a comprehensive survey of 54 ryegrass populations and 384 plants from the UK, a single plant was identified that contained the W1999C mutation [33]. The rarity of the 1999 mutations could imply that they are characterised by a fitness penalty. Indeed, in yeast gene replacement experiments, strains carrying the wild type W1999 allele grew two times faster than the equivalent strain with the mutant C1999 allele [14]. That

Table 4. Estimated resistance factors for five ACCase herbicides for different 1999 ACCase plant groups.

Genotypes	Pinoxaden	Diclofop-methyl	Haloxyfop-P-methyl	Sethoxydim	Tepraloxydim
UK21-WW1999 v STD1-WW1999	3.9 (1.3–12.2)	>46.0	45.5 (16.8–123.2)	2.3 (1.6–3.4)	12.9 (6.5–25.5)
UK21-WS1999 v STD1-WW1999	41.7 (13.3–130.1)	>113.6	131.2 (48.5–355.1)	3.8 (2.6–5.5)	16.7 (8.4–33)
UK21-SS1999 v STD1-WW1999	127.4 (40.8–397.9)	>113.6	173.1 (63.9–468.6)	9.0 (6.2–13.1)	34.8 (17.6–68.8)
UK21-WS1999 v UK21-WW1999	10.7 (3.4–33.3)	-	2.9 (1.1–7.8)	1.6 (1.1–2.4)	1.3 (0.7–2.6)
UK21-SS1999 v UK21-WW1999	32.6 (10.4–101.8)	-	3.8 (1.4–10.3)	3.8 (2.6–5.6)	2.7 (1.4–5.3)
UK21-SS1999 v UK21-WS1999	3.1 (1.0–9.6)	-	1.3 (0.5–3.6)	2.4 (1.6–3.5)	2.1 (1.1–4.1)

UK21- WW1999, UK21-WS1999 andUK21- SS1999 originates from the same population and thus shares the same genetic background except for their amino acids at ACCase position 1999.
STD1 is the standard sensitive population used for comparison.
Resistance factors could not be estimated for diclofop-methyl when no dose responses were observed for both genotypes being compared.
95% confidence limits in brackets.

a fitness cost can be associated with the 1999 ACCase mutations is plausible given that tryptophan is very conserved in nature and that a substitution with either serine or cysteine is a rather rare event [40]. Proper fitness trials using wild and mutant 1999 subpopulations created as part of this study and tested under competition as outlined by Vila-Aiub et al. [41] would shed light on the potential fitness penalty associated with the W1999S mutation in *Lolium multiflorum.* The infrequent occurrence of the W1999S mutations could also be due to the fact that it is partially recessive to herbicides such as sethoxydim and cycloxydim, and sensitive to tepraloxydim and clethodim used in rotational dicotyledonous crops grown in the same field. Thus, plants containing the 1999 ACCase mutations are likely to take longer to be selected in the field in particular in an out-breeder species such as *L. multiflorum.* This contrasts with other mutations that confer broader resistance to ACCase herbicides such as the I1781L, D2078G and C2088R mutations. Analysis of a large number of *Lolium* spp. populations from the UK [33] and Australia [42], and *Alopecurus myosuroides* from Europe [30] showed that the D2078G, I2041N and I1781L mutations were the most prevalent respectively. As populations containing the D2078G mutations have been shown to have a fitness cost [39], herbicide use patterns rather that a fitness penalty appears to be a more important determinant regarding the selection of specific mutations in grass weeds subjected to herbicide pressure.

Detection of mutations at ACCase codon position 1999

Upon discovery of the W1999C mutation which was found to confer resistance to fenoxaprop-P-ethyl only in an *Avena sterilis* resistant population, an allele specific PCR based assay was developed for the high throughput identification of either the tryptophan or mutant cysteine allele in this species [14]. In addition to intrinsic issues associated with allele specific assays [37], the method developed by Liu et al [14] will not allow detection of other nucleotide changes recently found at ACCase codon position 1999. Subsequently, two more robust and universal dCAPS assays were established taking into account the newly identified leucine allele at codon position 1999 [34]. One of the assays uses the same enzyme *Xcm*I employed here to detect the wild type TGG triplet characteristic of the tryptophan amino acid residue. The second 'cysteine' (TGC/T) assay uses the *Xcm*I dCAPS primer but with an extension of an additional two bases at the 3′ end of the primer. An unrestricted PCR fragment would indicate the absence of a guanine on the third nucleotide of the triplet or presence of cysteine at ACCase codon position 1999. However these two assays alone will not resolve the ambiguity between leucine (TTG) and serine (TCG) residues. The dCAPS assay developed in this study allows positive identification of the wild type tryptophan and mutant serine alleles. This assay is original in the sense that it uses only one PCR product but two different enzyme reactions to detect the tryptophan and serine amino acid residues at ACCase codon position 1999. This amounts to a significant cost reduction with respect to PCR and further demonstrates the flexibility offered by the dCAPS method [27]. When used in conjunction with the second 'cysteine' assay developed by Delye et al [34], the dCAPS method will allow inference and high throughput screening of all possible amino acid changes identified at ACCase codon position 1999 to date.

Specificity and high levels of resistance conferred by non-target mechanisms

Non target site resistance and in particular metabolism is often believed to confer lower levels of resistance compared to target site resistance [43]. At the field level, this type of resistance mechanism is habitually suspected when the weeds are damaged but grow out of the herbicide treatment. In this study, additional underlying resistance, very likely non-target site based, were detected to diclofop-methyl, potentially due to the accumulation of several minor genes over time. Diclofop-methyl is the least potent of all ACCase herbicides and significant levels of non-target site resistance can be selected in an originally sensitive *Lolium* sp. population in only a few generations [22]. Very high levels of non-target site resistance to diclofop-methyl were also found in a different UK *Lolium multiflorum* population characterised by the D2078G mutation [26]. A clear shift in dose responses was also observed for haloxyfop-P-methyl with target site resistance and non-target site resistance conferring a 3-fold and 45-fold resistance respectively. This is interesting given that haloxyfop-P-methyl is a poorly metabolisable herbicide. Though not significant at practical field rates, non-target site resistance had a bigger impact compared to target site resistance on sethoxydim and tepraloxydim, two other non-selective ACCase herbicides on cereal crops. Additionally non target site resistance is often seen as a threat for current and future herbicides in development since it is increasingly being identified as the most prevalent resistance mechanism and because selection with one herbicide can confer resistance to herbicides belonging to a different mode of action [44,45]. Here we found that non target site resistance can also be specific even among the cereal selective herbicides. Indeed, non-target site mechanisms conferred resistance to practical field rates of diclofop-methyl and clodinafop-propragyl but not pinoxaden, even though the latter two herbicides use the same safener for crop selectivity.

Conclusion

This detailed molecular and biological study demonstrates the need for a rigorous methodological approach for identifying and quantifying target site and non-target site resistance in weed populations that have evolved resistance to herbicides. It also underlines the unpredictability of resistance conferred by target site and non-target site mechanisms within and between subclasses of ACCase herbicides. As resistance to ACCase herbicides can be quite specific, rotation between ACCase compounds and ideally with other herbicide modes of action and inclusion of non-chemical weed control approaches will ensure their prolonged sustainability for controlling grass weed species.

Acknowledgments

The authors are very grateful to the plant production and screening teams from Jealott's Hill UK for their help in producing and spraying the plants as part of this study. The authors also express their gratitude to the Syngenta technical managers for their encouragement for carrying out this study.

Author Contributions

Conceived and designed the experiments: SSK. Performed the experiments: GCB SJH RD. Analyzed the data: EM. Contributed reagents/materials/analysis tools: SSK GCB SJH RD. Wrote the paper: SSK.

References

1. Shorrosh BS, Dixon RA, Ohlrogge JB (1994) Molecular cloning, characterization, and elicitation of acetyl-CoA carboxylase from alfalfa. Proc Natl Acad Sci USA 91: 4323–4327.
2. Konishi T, Shinohara K, Yamada K, Sasaki Y (1996) Acetyl-CoA carboxylase in higher plants: most plants other than Gramineae have both the prokaryotic and the eukaryotic forms of this enzyme. Plant and Cell Physiology 37: 2.
3. Délye C (2005) Weed resistance to acetyl coenzyme A carboxylase inhibitors: an update. Weed Sci 53: 728–746.
4. Nikolau BJ, Ohlrogge JB, Wurtele ES (2003) Plant biotin-containing carboxylases. Arch of Biochem and Biophys 414: 211–222.
5. Incledon BJ, Hall CJ (1997) Acetyl-coenzyme A carboxylase: quaternary structure and inhibition by graminicidal herbicides. Pestic Biochem Physiol 57: 255–271.
6. Sasaki Y, Nagano Y (2004) Plant acetyl-CoA carboxylase: structure, biosynthesis, regulation, and gene manipulation for plant breeding. Biosci Biotechnol Biochem 68: 1175–1184.
7. Herbert D, Walker K, Price L, Cole D, Pallett K, et al. (1997) Acetyl-CoA carboxylase: a, graminicide target site. Pestic Sci 50: 67–71.
8. Hofer U, Muehlebach M, Hole S, Zoschke A (2006) Pinoxaden – for broad spectrum grass weed management in cereal crops. J Plant Dis Prot 113: 989–995.
9. Konishi T, Sasaki Y (1994) Compartmentalization of two forms of acetyl-CoA carboxylase in plants and the origin of their tolerance toward herbicides. Proc Nati Acad Sci USA 91: 3598–3601.
10. Heap IM (2012) International Survey of Herbicide-Resistant Weeds. http://www.weedscience.com.
11. Heap I, Knight R (1982) A population of ryegrass tolerabnt to the herbicide diclofop-methyl. Journal of the Australian Institute of Agricultural Science 48: 156–157.
12. Devine MD (1997) Mechanisms of resistance to acetyl-coenzyme A carboxylase inhibitors: a review. Pestic Sci 51: 259–264.
13. Beckie H, Tardif FJ (2012) Herbicide cross resistance in weeds. Crop prot 35: 15–28.
14. Liu W, Harrison DK, Chalupska D, Gornicki P, O'Donnell CC, et al. (2007) Single-site mutations in the carboxyltransferase domain of plastid acetyl-CoA carboxylase confer resistance to grass-specific herbicides. Proc Natl Acad Sci USA 104: 3627–3632.
15. Délye C, Zhang XQ, Michel S, Matéjicek A, Powles SB (2005) Molecular bases for sensitivity to acetyl-coenzyme A carboxylase inhibitors in black-grass. Plant Physiol 137: 794–806.
16. Scarabel L, Panozzo S, Varotto S, Sattin M (2011) Allelic variation of the ACCase gene and response to ACCase-inhibiting herbicides in pinoxaden-resistant Lolium spp. 67: 932–941.
17. Yu Q, Collavo A, Zheng M-Q, Owen M, Sattin M, et al. (2007) Diversity of Acetyl-Coenzyme A Carboxylase Mutations in Resistant Lolium Populations: Evaluation Using Clethodim. Plant Physiol 145: 547–558.
18. Powles S, Yu Q (2010) Evolution in action: plants reistant to herbicides. Annu Rev Plant Biol 61: 317–347.
19. Alarcon-Reverte R, Moss SR (2008) Resistance to ACCase inhibiting herbicides in the weed *Lolium multiflorum* (Italian rye-grass). Commun Agric Appl Biol Sci 73: 899–902.
20. Petit C, Bay G, Pernin F, Délye C (2010) Prevalence of cross- or multiple resistance to the acetyl-coenzyme A carboxylase inhibitors fenoxaprop, clodinafop and pinoxaden in black-grass (*Alopecurus myosuroides* Huds.) in France. Pest Manag Sci 66: 168–177.
21. Preston C, Tardif FJ, Christopher JT, Powles SB (1996) Multiple Resistance to Dissimilar Herbicide Chemistries in a Biotype of *Lolium rigidum* Due to Enhanced Activity of Several Herbicide Degrading Enzymes. Pesticide Biochemistry and Physiology 54: 123–134.
22. Neve P, Powles SB (2005) Recurrent selection with reduced herbicide rates results in the rapid evolution of herbicide resistance in *Lolium rigidum*. Theor Appl Genet 110: 1154–1166.
23. Preston C (2003) Inheritance and linkage of metabolism-based herbicide cross-resistance in rigid ryegrass (*Lolium rigidum*). Weed Sci 51: 4–12.
24. Petit C, Duhieu B, Boucansaud K, Délye D (2010) Complex genetic control of non-target-site-based resistance to herbicides inhibiting acetyl-coenzyme A carboxylase and acetolactate-synthase in Alopecurus myosuroides Huds. Plant Science 178: 501–509.
25. Busi R, Vila-Aiub MM, Powles SB (2011) Genetic control of a cytochrome P450 metabolism-based herbicide resistance mechanism in Lolium rigidum. Heredity 106: 817–824.
26. Kaundun SS (2010) An aspartate to glycine change in the carboxyl transferase domain of acetyl-CoA carboxylase is in part associated with resistance to ACCase inhibitor herbicides in a Lolium multiflorum population. Pest Manag Sci 66: 1249–1256.
27. Neff MM, Neff JD, Chory J, Pepper AE (1998) dCAPS, a simple technique for the genetic analysis of single nucleotide polymorphisms: Experimental applications in Arabidposis thaliana genetics. Plant Journal 14: 387–392.
28. Neff MM, Turk E, Kalishman M (2002) Web-based primer design for single nucleotide polymorphism analysis. Trends Genet 18: 613–615.
29. Kaundun SS. Comparison of pinoxaden's resistance profile with other ACCase inhibitors; 2008; Vancouver, Canada 23–27 June.
30. Délye C, Michel S, Berard A, Chauvel B, Brunel B, et al. (2010) Geographical variation in resistance to acetyl-coenzyme A carboxylase-inhibiting herbicides across the range of the arable weed *Alopecurus myosuroides* (black-grass). New Phytologist 186: 1005–1017.
31. Beckie H, Warwick S, Sauder C (2012) Basis of resistance in Canadian populations of wild oat (*Avena fatua*). Weed sci 60: 10–18.
32. Kaundun S, Hutchings S-J, Dale R, McIndoe E (2012) Broad resistance to ACCase inhibiting herbicides in a ryegrass population Is due only to a cysteine to arginine mutation in the target enzyme. PLoS ONE 7(6): e39759. doi:10.1371/journal.pone.0039759.
33. Alarcon-Reverte R, Shanley S, Kaundun S, Karp A, Moss S (2012) A SNaPshot assay for the rapid and simple detection of known point mutations conferring resistance to ACCase-inhibiting herbicides in Lolium spp. Weed Res In press.
34. Délye C, Pernin F, Michel S (2011) Universal' PCR assays detecting mutations in acetyl-coenzyme A carboxylase or acetolactate synthase that endow herbicide resistance in grass weeds. Weed Res 51: 353–362.
35. Kaundun SS, Dale R, Zelaya, Dinelli G, Maroti I, et al. (2012) Importance of a novel P106L mutation in conferring resistance to glyphosate in a rye grass population; February 5th–9th, Honolulu, Hawai, USA.
36. Délye C, Michel S (2005) 'Universal' primers for PCR-sequencing of grass chloroplastic acetyl-CoA carboxylase domains involved in resistance to herbicides. Weed Res 45: 323–330.
37. Kaundun SS, Windass JD (2006) Derived cleaved amplified polymorphic sequence, a simple method to detect a key point mutation conferring acetyl CoA carboxylase inhibitor herbicide resistance in grass weeds. Weed Res 46: 34–39.
38. Vila-Aiub M, Neve P, Powles S (2005) Resistance cost of a cytochrome P450 herbicide metabolism mechanism but not an ACCase target site mutation in a multiple resistant Lolium rigidum population. New Phytologist 167: 787–796.
39. Menchari Y, Chauvel B, Darmency H, Delye C (2008) Fitness costs associated with three mutant acetyl-coenzyme A carboxylase alleles endowing herbicide resistance in black-grass *Alopecurus myosuroides*. J appl ecol 45: 939–947.
40. Dayhoff MO, Schwartz RM, Orcutt BC (1978) A model of evolutionary change in proteins. In: Dayhoff MO, editor. Atlas of Protein Sequence and Structure: National Biomedical Research Foundation, Washington. 345–352.
41. Vila-Aiub M, Neve P, Powles S (2009) Fitness costs associated with evolved herbicide resistance alleles in plants. New Phytologist 184: 751–767.
42. Malone J, Boutsalis P, Preston C (2010) Spread of herbicide resistance alleles in Lolium rigidum Gaud. (annual ryegrass). In: Zydenbos S, editor; Christchurch, New Zealand. New Zealand Plant Protection Society. 258–261.
43. Moss SR (2002) Experience with herbicide resistance in Europe: global implications. In: Spafford Jacob H, Dodd J, Moore J, editors; Perth, Western Australia, 8–13 September 2002. Plant Protection Society of WA. 660–665.
44. Christopher JT, Powles SB, Holtum JAM (1992) Resistance to acetolactate synthase-inhibiting herbicides in annual ryegrass (*Lolium rigidum*) involves at least two mechanisms. Plant Physiol 100: 1909–1913.
45. Délye C, Gardin JAC, Boucansaud K, Chauvel B, Petit C (2011) Non-target-site-based resistance should be the centre of attention for herbicide resistance research: Alopecurus myosuroides as an illustration. Weed Research 51: 433–437.

3

Assessment of Genetic Diversity among Barley Cultivars and Breeding Lines Adapted to the US Pacific Northwest, and Its Implications in Breeding Barley for Imidazolinone-Resistance

Sachin Rustgi[1*¶], Janet Matanguihan[1¶], Jaime H. Mejías[1,2], Richa Gemini[1], Rhoda A. T. Brew-Appiah[1], Nuan Wen[1], Claudia Osorio[1], Nii Ankrah[1], Kevin M. Murphy[1], Diter von Wettstein[1,3,4*]

1 Department of Crop & Soil Sciences, Washington State University, Pullman, Washington, United States of America, **2** Instituto de Investigaciones Agropecuarias INIA, Vilcún, Chile, **3** School of Molecular Biosciences, Washington State University, Pullman, Washington, United States of America, **4** Centre for Reproductive Biology, Washington State University, Pullman, Washington, United States of America

Abstract

Extensive application of imidazolinone (IMI) herbicides had a significant impact on barley productivity contributing to a continuous decline in its acreage over the last two decades. A possible solution to this problem is to transfer IMI-resistance from a recently characterized mutation in the 'Bob' barley *AHAS* (*acetohydroxy acid synthase*) gene to other food, feed and malting barley cultivars. We focused our efforts on transferring IMI-resistance to barley varieties adapted to the US Pacific Northwest (PNW), since it comprises ~23% (335,000 ha) of the US agricultural land under barley production. To effectively breed for IMI-resistance, we studied the genetic diversity among 13 two-rowed spring barley cultivars/breeding-lines from the PNW using 61 microsatellite markers, and selected six barley genotypes that showed medium to high genetic dissimilarity with the 'Bob' *AHAS* mutant. The six selected genotypes were used to make 29–53 crosses with the *AHAS* mutant and a range of 358–471 F_1 seeds were obtained. To make informed selection for the recovery of the recipient parent genome, the genetic location of the *AHAS* gene was determined and its genetic nature assessed. Large F_2 populations ranging in size from 2158–2846 individuals were evaluated for herbicide resistance and seedling vigor. Based on the results, F_3 lines from the six most vigorous F_2 genotypes per cross combination were evaluated for their genetic background. A range of 20%–90% recovery of the recipient parent genome for the carrier chromosome was observed. An effort was made to determine the critical dose of herbicide to distinguish between heterozygotes and homozygotes for the mutant allele. Results suggested that the mutant can survive up to the 10× field recommended dose of herbicide, and the 8× and 10× herbicide doses can distinguish between the two *AHAS* mutant genotypes. Finally, implications of this research in sustaining barley productivity in the PNW are discussed.

Editor: Tianzhen Zhang, Nanjing Agricultural University, China

Funding: This work was supported by the Washington Grain Commission Grant #13C-3019-3590. The funders had no role in study design, data collection and analysis, decision to publish, or preparation of the manuscript.

Competing Interests: The authors have declared that no competing interests exist.

* Email: rustgi@wsu.edu (SR); diter@wsu.edu (DvW)

¶ These authors contributed equally to this work.

Introduction

Barley is a short-season, early maturing annual grain crop with some degree of tolerance to drought and salinity, which allows its production in a wide range of climatic zones including both irrigated and dryland production areas [1]. Barley is the third major feed grain crop produced in the United States, after corn and sorghum [2]. Spring barley is a preferred rotational crop in the US Pacific Northwest (PNW) for two- or three-year rotations with winter wheat (*Triticum aestivum* L.), pea (*Pisum sativum* L.), lentil (*Lens culinaris* L.), or fallow [1,3]. A cropping system like spring wheat-fallow or winter wheat-fallow is generally practiced in the PNW, which encourages populations of summer and winter annual-grassy weeds, respectively [4]. These weed cycles can be broken with a winter wheat-barley-fallow rotation [6]. Depending upon the management practices followed in an area, this cropping system results in a buildup of crown and root rot pathogens including *Fusarium*, *Rhizoctonia* and *Phythium* species, which frequently result in significant yield losses [5]. Similarly, in an eight-year dryland no-till cropping systems experiment conducted near Ritzville, Washington, a significant drop in the incidence of bare patches caused by *Rhizoctonia* was observed by adaptation of a two-year spring wheat rotation with spring barley. A significant gain in average yield of spring wheat was also documented with this change [5]. Likewise, in continuous cropping systems, spring barley fits well after winter wheat because the time interval between harvesting the barley crop and planting winter wheat is usually sufficient to allow soil moisture recharge to support an optimum winter wheat stand [6,7]. In addition to its agronomical relevance and commercial value as a feed or malt grain crop, barley is regaining popularity as human food due to the antioxidant and β-glucan (dietary fiber) rich grains [8,9]. Despite its agronomical importance and rising market value, barley acreage in the US has declined from 8.94 million acres in 1991 to 3.48 million acres in 2013 [10]. In Washington State alone the acreage has dropped significantly from 500,000 acres planted in 1999 to 180,000 acres in 2013 [10].

The significant drop in barley acreage during the last two decades can be partly attributed to the wide scale application of imidazolinone herbicides in combination with the introduction of imidazolinone (IMI)-resistant crops, and the residual activity of the herbicides of this family [1]. The decline in acreage can also be explained by the overlapping distribution of regions under barley cultivation in the PNW and the regions under extensive application of Imazamox (Beyond) and/or Imazethapyr (Pursuit) [11]. Collectively, the major reason for the decline in barley acreage is its sensitivity to commonly used herbicides. Many of the widely used herbicides, which impose barley plant-back restrictions, belong to the group B herbicides [12]. Thus, identification of IMI-resistant mutant(s) in barley and its transfer to relevant feed, food and/or malting barley cultivars adapted to the PNW is of extreme importance to sustain barley productivity in this region and elsewhere.

The group of herbicides belonging to the imidazolinone family targets acetohydroxyacid synthase (AHAS) or acetolactate synthase (ALS), an octameric enzyme with four catalytic and four regulatory subunits [13]. The enzyme AHAS catalyses two parallel reactions in the synthesis of branched chain amino acids. The first reaction is condensation of two pyruvate molecules to yield acetolactate leading to the production of valine and leucine, and the other reaction is the condensation of pyruvate and α-ketobutyrate that give rise to acetohydroxybutyrate, which subsequently results in the synthesis of isoleucine [14]. The AHAS-inhibiting herbicides are known to bind at the substrate access channel, blocking the path of substrate to the active site. When AHAS is inhibited, deficiency of the amino acids (valine, leucine and isoleucine) causes a decrease in protein synthesis, which in turn slows down the rate of cell division. This process eventually kills the plant, with symptoms observed in meristematic tissues where biosynthesis of amino acids primarily takes place [12]. In most cases, resistant plants have a reduced sensitivity to these herbicides due to amino acid substitution(s) in AHAS that give rise to catalytically active isoforms of the enzyme. Most AHAS isoenzymes resistant to the herbicides carry substitutions for the amino acid residues Ala122, Pro197, Ala205, Asp376, Trp574 or Ser653 (amino acid numbering refers to the sequence in *Arabidopsis thaliana*) [13]. Amino acid substitutions at Ala122 and Ser653 confer high levels of resistance to imidazolinone herbicides, whereas substitutions at Pro197 endow high level of resistance against sulfonylureas and provide low-level resistance against imidazolinone and triazolopyrimidine herbicides. Likewise, substitutions at Trp574 provide high levels of resistance to imidazolinones, sulfonylureas and triazolopyrimidines, while substitutions at Ala205 confer resistance against all AHAS-inhibiting herbicides [15].

In the case of barley, there is no IMI-resistance reported for any of the varieties cultivated in the PNW. Thus, introduction of a barley variety with IMI-resistance will provide greater flexibility to barley as a rotational crop after winter wheat [11]. An IMI-resistant mutant was earlier isolated by our group from an extensive screening of two million seeds of 'Bob' treated with sodium azide. Molecular characterization of the mutant revealed an amino acid substitution in the substrate access channel of the catalytic subunit of the AHAS enzyme, changing a serine to asparagine at amino acid location 653 [16]. This mutation in the substrate access channel does not allow imazamox to block the path of the substrate to the active site, thus allowing the plant to survive with no obvious effects on plant fitness even when exposed to field recommended dose of herbicide used on the IMI-tolerant winter wheat (i.e., 0.118 L/Acre Beyond with 1% non-ionic surfactant).

In view of the agronomical importance of this trait and the great demand for IMI-resistant barley cultivars in the PNW, this study was undertaken with the following objectives: i) estimation of genetic diversity among the 13 two-rowed spring barley cultivars/breeding-lines adapted to the US PNW using 61 microsatellite markers to select for lines showing sufficient genotypic differences with the 'Bob' *AHAS* mutant, to be used in the crossing program; and (ii) transfer the IMI-resistance to selected food, feed and malting barley cultivars using marker-assisted foreground and background selections.

Materials and Methods

Plant material

Seeds of the 13 two-rowed spring barley cultivars or breeding lines were procured from the variety testing program at the Washington State University (WSU), Pullman. Of the 13 genotypes selected for genetic analysis, eight are feed barleys, three are food barleys and the remaining two are malting barleys (Table 1).

Crossing scheme

To transfer IMI-resistance from the 'Bob' *AHAS* mutant, crosses were made between the mutant and each of the six barley genotypes, selected on the basis of genetic diversity analysis performed using microsatellite markers specific to chromosome 6H (see later for details). Twenty nine to fifty three crosses were made per genotype combination during the summers of 2012 at the Spillman Agronomy Farm (WSU, Pullman) and a range of 358 to 471 F_1 grains were harvested. The F_1 plants were propagated in 48-well flats in the glasshouse to obtain F_2 seeds. Subsequently, a range 2158 to 2846 F_2 plants per cross combination were evaluated for herbicide resistance by spraying two-week-old seedlings with 0.236 L/Acer Beyond (twice the field recommended dose applied to the IMI-tolerant winter wheat) with 1% methylated seed oil (MSO). A month after herbicide spray, the survivors (i.e., resistant plants) were evaluated for plant height as an indicator of early vigor and the 250 top ranking lines per cross combination were raised to maturity for seeds. Later, one to three F_3 plants each from the six most vigorous F_2 lines per cross combination were evaluated for the genotype at the *AHAS* locus by DNA sequencing, and the percent recovery of the recipient parent genome using chromosome 6H-specific SSR markers.

DNA extraction and PCR amplification

DNA was extracted from the one-month-old seedlings of each of the 13 barley genotypes, and the two-week-old seedlings of the F_3 progeny of selected F_2 lines, using the modified CTAB (Cetyl Trimethyl Ammonium Bromide) method [17]. DNA was treated by RNAse and purified by phenol extraction (25 phenol: 24 chloroform: 1 isoamyl alcohol, v/v/v) followed by ethanol precipitation [18]. Concentration of DNA samples was adjusted

Table 1. List of two-rowed spring barley varieties/breeding lines used in the study.

Genotype	Pedigree	Class
Baronesse	([(Mentor×Minerva)×mutant of Vada]×[(Carlsberg×Union)×(Opavsky×Salle)×Ricardo]) ×(Oriol×6153 P40)	hulled, feed barley (originally released as malting barley)
Bob	(Lewis somaclonal line)/Baronesse	hulled, feed barley
Champion	Baronesse/Camas	hulled, feed barley
Clearwater	Baronesse*2/pmut882//HB317 (CDC Dawn sib)	hulless, low phytate, food barley
Lenetah	94Ab12981/Criton	hulled, feed barley
Conrad	B1215/B88-5336	hulled, malting barley
Radiant	*ant29-667* (an induced mutant in Harrington)/Baronesse	hulled, malting barley, pro-anthocyanidine-free
Spaulding	Vanguard/Imber//Zephyr/3/ Heavyweight/4/VD403582	hulled, feed barley
WAS4	01WA-13862.3/Radiant	hulless, food barley
05WA-316.99	Baronesse/Spaulding	hulled, feed barley
Lyon	Baronesse/Spaulding	hulled, feed barley
07WA-682.1	WA 10701-99/AC Metcalfe	hulled, feed barley
Meresse	Merlin/Baronesse	hulless, food barley

to 50 ng μl^{-1} using *Hind* III digested λ DNA as a marker. DNA amplification was carried out on a C1000 thermal cycler (Bio-Rad Laboratories). The PCR reactions were performed in 20 μl reaction mixtures, each containing 50 ng template DNA, 0.25 μM primers, 200 μM dNTPs, 1.5 mM $MgCl_2$, 1×PCR buffer and 0.5 U Ex *Taq* DNA polymerase (TAKARA, Bio Inc.) using the following PCR profile: initial denaturation at 95°C for 3 min followed by 40 cycles at 95°C for 30 sec, 53–61°C (depending upon the primer pair used) for 30 sec (for primer details, cf. [19]), 72°C for 45 sec, and a final extension at 72°C for 5 min. The amplification products were resolved on 10% polyacrylamide denaturing gels followed by silver staining [20]. A hundred base pair ladder was used as a size marker (New England BioLabs, Inc., Beverly, USA). The amplified product/allele sizes were determined using Fragment Size Calculator available at http://www.basic.northwestern.edu/biotools/SizeCalc.html.

DNA sequencing and sequence analysis

To determine the genotype at the *AHAS* locus, genomic DNA extracted from the F_3 progeny of selected F_2 lines was amplified using the *AHAS* gene-specific sequence tagged site (STS) primers that flank the point mutation responsible for the IMI-resistance (for primer details, cf. [16]). The amplification product was resolved on 1% agarose gel. A 100-bp ladder was used as a size marker (New England BioLabs). The band of expected size was excised from the gel, and DNA was eluted from the band using the Geneclean kit following the manufacturer's instructions (MP Biomedicals). The eluted DNA was used as a template for the sequencing reaction using either forward or reverse primers in separate reactions. The sequencing reactions were carried out at the DNA Sequence Core, WSU, Pullman. Alignment of the DNA sequences was performed using the Vector NTI AdvanceTM 9.1 (Invitrogen).

Determination of the polymorphic information content (PIC) and genetic diversity

For each microsatellite or simple sequence repeat (SSR) locus, PIC was calculated using the following equation: $PIC = 1-\Sigma(Pi)^2$, where Pi is the proportion of genotypes carrying the ith allele [21]. For dissimilarity analysis, null alleles were scored as zero (0) and other microsatellite alleles (length variants) were each scored in the form of single bands of expected sizes, which were later converted into the number of repeat units as allele codes (all modalities were given equal weight during the analysis). The numerical data thus obtained was used to calculate Sokal and Michener dissimilarity indices ($di\text{-}j$) [22]. The dissimilarity indices between pairs of accessions using genotypic data were calculated on the basis of the following equation: $di\text{-}j = (n11+n00)/(n11+n01+n10+n00)$, where n11 is the number of fragments present in both *i* and *j*, n01 and n10 is the number of fragments present in one accession but absent in the other, and n00 is the number of fragments absent in both *i* and *j*. From the obtained distance matrix, an un-weighted Neighbor-Joining tree [23] was computed using the Darwin 5.0 software [24] and branch robustness was tested using 1000 bootstraps.

Enzyme extraction

Soluble proteins from 'Bob' and 'Bob' *AHAS* mutant were extracted following Singh et al. [25], with minor modifications. Briefly, two batches of 500 mg of the fresh leaf tissue were pulverized each with 5 mL of the protein extraction buffer [consisting of 100 mM potassium phosphate buffer (pH 7.5), 10 mM sodium pyruvate, 5 mM $MgCl_2$, 5 mM EDTA, 100 μM flavin adenine dinucleotide (FAD) and 10% Glycerol], using a polypropylene mesh bag (supplied with the P-PER Plant Protein Extraction Kit, Thermo Scientific). After adding the extraction buffer to the leaf tissue, the bag was rubbed from the outside with a ceramic pestle until a homogeneous mixture of the tissue was obtained. Later, the lysate was suctioned from the bag using a pipette and placed into a 15 mL conical tube and centrifuged at 22,000×g for 20 min at 4°C. The supernatant was transferred to a new tube and mixed with an equal volume of saturated $(NH_4)_2SO_4$. The mixture was incubated on ice for 30 min, and then centrifuged at 4°C for 20 min at 22,000×g. The supernatant was discarded and the pellet containing protein was re-suspended in 700 μL of the buffer solution containing 50 mM potassium phosphate (pH 7.5), 100 mM sodium pyruvate, 10 mM $MgCl_2$, 1 mM EDTA, 10 μM FAD, 100 mM NaCl and 1 mM thiamine pyrophosphate (TPP).

After extraction, protein concentration was determined using Bradford colorimetric micro-assay by mixing 80 μL of protein extract with 20 μL of the Bradford reagent (containing 1 mL of concentrated Bradford solution in 4 mL of deionized water), and measuring absorbance at 590 nm wavelength. The presence of the enzyme in the extract was also confirmed by loading protein extracts on 10% sodium dodecyl sulfate (SDS) polyacrylamide gel. For this purpose 15 μL of protein extract was mixed with 3 μL of the loading buffer, and electrophoresed on polyacrylamide gel for 2 h at 120 volts. After electrophoresis, the gel was stained with Coomassie brilliant blue reagent (80% Coomassie and 20% methanol, v/v) for 24 h. A protein band of ~65 kDa was observed, which corresponds with the size of AHAS enzyme monomers, confirming its presence in the extract.

Colorimetric enzyme activity assay

Enzyme activity was tested by using five different doses of Beyond (i.e., 1×, 4×, 6×, 8× and 10× the field recommended dose applied on IMI-tolerant winter wheat) with 0.25% (v/v) nonionic surfactant (NIS). Initial reaction was performed in 1.5 mL microfuge tube by adding 52 μL of enzyme (in extraction buffer containing the substrate and co-factors, see above for the buffer composition) to equal volume of herbicide and incubating the mixture at 37°C for 1 h to facilitate acetolactate production. Later, the reaction was stopped by adding 21 μL of 5% H_2SO_4, and incubating at 60°C for 15 min. After incubation, tubes were spiked with 175 μL of color change solution containing 0.32 g of NaOH, 0.12 g of 1-naphtol and 0.01 g of creatine in 4 mL of deionized water, and the mixture was re-incubated at 60°C for 15 min. After incubation, 200 μL sub-samples of the reaction mixture were added to a 96-well microtiter plate (Falcon cat#353077) to determine the enzyme activity by studying color change using a microplate reader spectrophotometer (Spectra Max, M2, Molecular Devices) at 520 nm wavelength.

Results and Discussion

Chromosomal assignment of the gene encoding catalytic subunit of barley AHAS enzyme

The AHAS holoenzyme (~548 kDa) consists of two halves where one half, known as the large or catalytic subunit, is comprised of a homotetramer of ~65 kDa polypeptides, and the second half, known as the small or regulatory subunit, consists of homo-tetramer/-pentamer of polypeptides of ~52 kDa each [16,26,27]. The regulatory subunit stimulates enzyme activity and is required for the feedback regulation of the branched-chain amino acid biosynthesis, whereas the catalytic subunit is solely responsible for the enzyme activity and is also the site of point mutation(s) that confers resistance against IMI-herbicides [16]. Due to the importance of the catalytic subunit in providing IMI-resistance, the genes encoding it have been studied in common wheat and assigned to group 6 chromosomes [6A (*imi3*), 6B (*imi2*) and 6D (*imi1*)], using nulli-tetrasomic lines [28]. Later, the genetic location of *imi1* gene on the long arm of chromosome 6D was determined using three mapping populations, namely Cashup/cv. 9804, Madsen/cv. 9804 and Opata 85/W7984 [28]. However, the genetic location of the *AHAS* gene in barley remains unknown. Therefore, we used the map location of the *AHAS* gene in wheat to decipher its location in barley, which is possible in this particular case due to the shared ancestry of the two genera, and high levels of synteny as well as colinearity between them [29]. The availability of common markers between wheat and barley maps allowed an approximation of the barley *AHAS* gene location on chromosome 6H (Fig. S1). Moreover, we used the complete *AHAS* gene sequence we had previously obtained to blast against the barley genomic DNA sequences available in the public domain (http://webblast.ipk-gatersleben.de/barley/). The BLASTn search (score = 2834 and E-value = 0.0) allowed unambiguous assignment of the gene to genetically anchored 'Morex' BAC contig numbered 40275 on chromosome 6H at 67.917 cM (Fig. S1). In addition, the initial genotyping of the F_3 progeny of selected F_2 lines (carrying the *AHAS* mutant allele in hetero-/homozygous state) from all six cross combinations with chromosome 6H specific microsatellite markers showed higher recovery rate (50–72%) of recipient parent alleles for markers mapping to the non-proximal long arm in comparison with the short arm and the centromeric region (37–58%) (see next section for details). This is an indication of suppressed recombination, likely due to selection for the trait of interest. Collectively, the *in silico* and experimental data strongly indicate that the gene encoding the catalytic subunit of the AHAS enzyme maps to the sub-centromeric region of the barley chromosome arm 6HL.

Polymorphism survey using chromosome 6H-specific microsatellite markers

The level of genetic diversity among 13 two-rowed spring barley cultivars/breeding lines adapted to the PNW was assessed using microsatellite or simple-sequence repeat (SSR) markers specific to the barley chromosome 6H. Out of the 13 genotypes selected for the analysis, eight are feed barleys, three are food barleys and two are malting barleys (Table 1). The 61 SSR markers selected for the analysis are evenly distributed along the entire length of chromosome 6H (Table 2) [19]. The major reason behind selecting markers from chromosome 6H lies in the fact that this chromosome carries the gene encoding for the catalytic subunit of acetohydroxyacid synthase (AHAS) enzyme and the mutation providing IMI-resistance (see above). It is known through trait-introgression studies that due to linkage-drag, it always takes longer (several backcrossing and selfing generations) to recover the recipient parent genotype for the carrier chromosome in comparison with non-carrier chromosomes, which assort independently [30]. Thus, to identify the rare recombinant(s) carrying the precise gene introgression in the early generation, it is important to screen large segregating populations with the markers derived from the carrier chromosome.

Of the 61 markers used for analysis, two markers (*HvWaxy4* and *GBM1319*) were non-functional (no amplification observed in any of the genotypes), three markers (*HVM22*, *GBM1215* and *GMS6*) were monomorphic, and 56 markers were polymorphic. These polymorphic markers allowed us to detect 62 loci. Of the 56 polymorphic markers, one marker detected three loci, another marker detected two loci, while the remaining 52 markers each detected a single locus. (Fig. S2). These 56 markers amplified 1 to 12 alleles from the 13 barley genotypes (Table 2). The number of alleles detected by each marker and their frequencies were used to calculate the polymorphic information content (PIC) of the marker. The PIC value, which depends on the number of detectable alleles and the distribution of their frequency, indicates the marker's utility in detecting polymorphism within a population

Table 2. List of chromosome 6H specific microsatellite markers used for the genetic diversity analysis and marker-assisted background selection, their repeat elements, respective locations in the genetic-linkage map [17], number of alleles detected and their polymorphic information content (PIC).

Marker/loci	Repeat element	Position (cM)	PIC	Allele#
Af166121	$(A)_{10}$	0.00	0.38	3
84c21j33	$(T)_{10}$	0.00	0.14	2
Bmac0316	$(AC)_{19}$	7.16	0.80	7
scssr09398	$(CTT)_9$	7.16	0.43	2
Bmag0500	$(AG)_6CG(AG)_{29}(AGAGGG)_3(AG)_6$	31.65	0.72	6
GBM1270	$(GCC)_8$	36.52	0.56	4
GBM1355	$(GCA)_7$	40.43	0.14	2
GBM1212	$(AGG)_5$	55.10	0.14	2
Bmag0807	$(TC)_{18}$	56.11	0.39	4
Bmag0173	$(CT)_{29}$	57.79	0.86	9
GBM1423	$(CGGCTC)_5$	58.46	0.36	2
HVM31	$(AC)_9$	60.90	0.57	3
Bmac0040	$(AC)_{20}$	61.07	0.77	7
Bmag0174	$(AG)_9$	61.40	0.72	6
EBmac0560	$(AC)_7$	61.70	0.77	5
GBM1267	$(TTG)_9$	61.70	0.69	4
Bmac0018	$(AC)_{11}$	61.79	0.77	7
Bmac0144	$(AT)_4(AC)_{20}$	61.79	0.91	12
Bmac0175	$(AC)_{12}$	61.79	0.57	3
GBM5012	$(ACG)_7$	61.95	0.49	2
Ebmac0674	$(TG)_{18}(AG)_9$	61.96	0.46	3
EBmac0874.1	$(CA)_8AA(CA)_4CG(CA)_8AA(CA)_7AA(CA)_9(TA)_8$	61.96	0.67	4
EBmac0874.2	$(CA)_8AA(CA)_4CG(CA)_8AA(CA)_7AA(CA)_9(TA)_8$	61.96	0.80	7
HVM65	$(GA)_{10}$	62.11	0.71	6
Bmag0009	$(AG)_{13}$	62.21	0.57	3
Ebmac0639	$(TG)_5(TG)_8$	62.21	0.57	4
EBmatc0028.1	$(ATC)_3N3(ATC)_6$	62.21	0.50	2
EBmatc0028.2	$(ATC)_3N3(ATC)_6$	62.21	0.58	4
EBmatc0028.3	$(ATC)_3N3(ATC)_6$	62.21	0.67	5
Bmac0297.1	$(AC)_9(AC)_{10}$	62.23	0.58	4
Bmac0297.2	$(AC)_9(AC)_{10}$	62.23	0.77	5
Bmac0297.3	$(AC)_9(AC)_{10}$	62.23	0.46	3
Bmac0047	$(AC)_{16}$	62.27	0.47	2
Bmac0127	$(AC)_{26}$	62.27	0.47	2
GBM1389	$(GCCT)_5$	62.27	0.26	2
HVM14	$(CA)_{11}$	62.28	0.50	2
Bmag210	$(AG)_7T(AG)_{13}$	62.28	0.57	3
HVM34	$(GA)_{10}$	62.43	0.36	2
HVM74	$(GA)_{13}$	62.66	0.56	3
Bmag0003.1	$(AG)_{28}$	63.49	0.78	6
Bmag0003.2	$(AG)_{28}$	63.49	0.59	6
Bmag0004	$(AG)_{14}$	64.71	0.91	12
BMG001	$(G)_{10}$	64.71	0.52	3
Bmgt0001	$(GTTTTT)_5$	64.71	0.27	3
scssr02093	$(GA)_{18}$	67.20	0.49	2
Bmag0344	$(CT)_{10}GT(CT)_{16}$	67.80	0.66	6
GBM1400	$(CACG)_5$	67.80	0.27	3
Bmac0251	$(AC)_{12}A(AC)_{13}$	69.25	0.38	3

Table 2. Cont.

Marker/loci	Repeat element	Position (cM)	PIC	Allele#
Bmag0613	$(GA)_{17}$	69.82	0.66	6
Bmac0218	$(AC)_{14}$	71.99	0.79	6
Bmac602	$(AC)_9AT(AC)_7(AG)_9$	75.42	0.49	4
GBM1256	$(GA)_8$	75.46	0.52	3
HVM11	$(GGA)_3(GGA)(GAA)_2$	88.47	0.56	3
scssr05599	$(AAG)_4$	96.34	0.63	4
GBM1140	$(ATC)_5$	97.31	0.52	3
GBM1356	$(GTG)_7$	98.38	0.57	3
scssr00103	$(GT)_{10}$	105.26	0.59	3
GBM1274	$(TCG)_7$	123.45	0.27	3
GBM1275	$(TGC)_7$	124.29	0.15	2
GBM1276	$(TGC)_7$	124.29	0.46	3
GBM1087	$(AGG)_5$	127.70	0.54	3
GBM1404	$(TATG)_5$	129.76	0.46	3

[21]. The PIC values ranged from 0.14 (*84c21j33, GBM1355, GBM1212*) to 0.91 (*Bmac0144, Bmag0004*) (Table 2). When the PIC value for each marker was plotted against its location on the genetic-linkage map, it showed a multimodal distribution, with low levels of PIC values observed at the sub-telomeric and centromeric regions of the chromosome (Fig. 1). This distribution shows the level of nucleotide diversity along the entire length of the chromosome and suggests the possibility of identifying a polymorphic marker from a specific region of the chromosome. The type of repeat element, chromosomal location, number of repeat units, and sequence of repeat element can influence the level of nucleotide diversity. Thus, we classified the SSR markers according to the type of repeat element into simple and compound repeats. Whenever two or more repeat runs were present adjacent to each other or microsatellite array of same repeat was interrupted by non-repeat base(s) the repeat was classified as compound repeat. We further classified simple repeats into mono-, di-, tri-, tetra-, penta- and hexa-nucleotide repeats and reported their mean PIC values. Compound repeats in general showed higher PIC values in comparison with simple repeats, whereas, among simple repeats the di-nucleotide repeats showed highest PIC values (Table 3). To distinguish the effect of chromosomal location from the microsatellite element type, the PIC values obtained for different microsatellite types (i.e, mono-, di-, tri, tetra-, hexa-nucleotide repeats and compound repeats) were individually plotted against their respective location on the genetic-linkage map. The analysis revealed reduced levels of nucleotide diversity in the peri-centromeric region for di-nucleotide repeats and in sub-telomeric regions for the tri-nucleotide repeats (Fig. S3). However, it was apparent from the analysis that the number of repeat units does not have any influence on the number of alleles detected per locus.

Preferential association of different SSR elements of variable sequences and lengths (i.e., total number of repeat units) with physical chromosome landmarks like the centromere, telomere, heterochromatin and euchromatin, and their relevance in determining chromosome function, has been extensively documented in literature [31–33]. Thus, the influence of the genomic locations of these markers on their evolvability and/or divergence is plausible. For instance, a low level of nucleotide diversity was observed in the proximal chromosomal regions of both *Triticum aestivum* and wild emmer (*Triticum turgidum* ssp. *dicoccoides*) [34]. Moreover, the effect of direct or indirect selection on genomic diversity is also a likely cause of observed fluctuations in genetic diversity along the chromosome length. Similar regions of low diversity associated with sites of domestication loci and genomic regions under selection in later breeding efforts were reported in maize [35]. Since barley genotypes selected in this study were bred in the PNW, they share some common ancestry. Thus, the regions of low diversity observed in the present study are likely to represent the genomic regions providing adaptive advantage to these genotypes. However, this aspect needs further investigation.

Assessment of genetic diversity among barley genotypes

The genetic relationships among the barley genotypes were evaluated based on the combined profiles of 62 SSR loci. The genetic dissimilarity coefficient (GD) values were calculated for all possible 78 pairs of genotypes, and ranged from 0.339 (between Bob and Baronesse) to 0.806 (between WAS4 and Conrad) with a mean of 0.601 (Fig. S4). All 13 genotypes were grouped into three clusters (Fig. 2). Two clusters were further subdivided into two sub-clusters each. As expected on the basis of pedigree information (Table 1), Bob, Baronesse, Meresse, 05WA-316.99 and Clearwater formed a single cluster (middle), where the first three genotypes grouped into one sub-cluster and the latter two genotypes grouped into the other sub-cluster. Clustering of these genotypes in a single group can be explained by the presence of Baronesse in their lineages. The cultivars Radiant, Champion, Lenetah, 07WA-682.1, WAS4 and Conrad formed another cluster (top), where the first five genotypes formed a sub-cluster and Conrad alone formed a sub-cluster. The first sub-cluster was further divided into two sub-sub clusters, the former containing Radiant, Champion and Lenetah, and the later containing 07WA-682.1 and WAS4. The remaining two genotypes Spaulding and Lyon formed a separate cluster (bottom), which is well justified due to the Spaulding lineage of Lyon. The above diversity analysis proved useful in selecting lines to cross with the Bob *AHAS* mutant to transfer IMI-resistance, and will also prove useful in future breeding efforts where these lines will be used. Nevertheless, Baronesse has been extensively used in barley breeding programs in the PNW; the

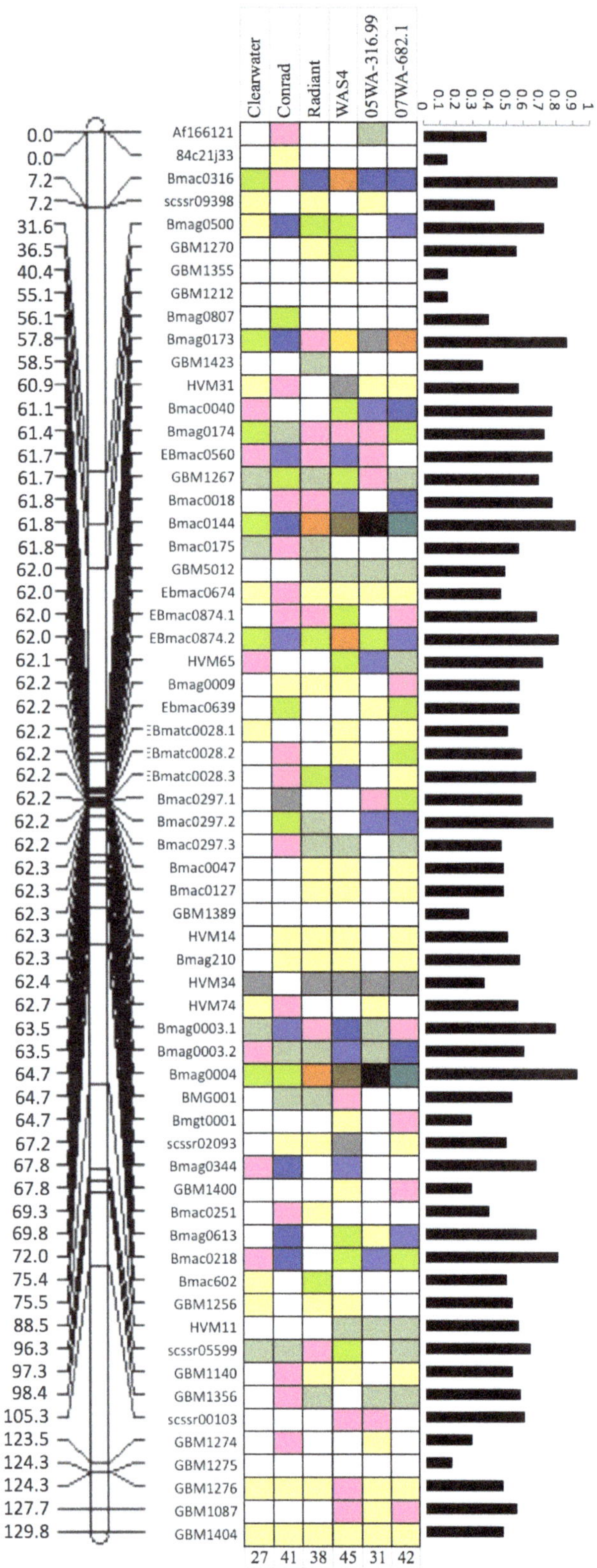

Figure 1. Genetic linkage map of chromosome 6H showing the respective locations of 61 microsatellite markers used in the present study (left). Various alleles detected from six barley genotypes used for crossing with the Bob *AHAS* mutant are indicated by colored boxes (middle), where each color represents a unique allele and the white color represents the 'Bob'-type allele. The total number of polymorphic markers identified per genotype pair with the Bob mutant is shown below. The PIC value calculated for each marker was plotted against its location on the genetic linkage map (right) to indicate the level of nucleotide diversity observed using 13 barley genotypes, and its distribution along the entire length of chromosome 6H.

results clearly demonstrated high level of genetic diversity among studied genotypes, which is very important for the success of any breeding program. Thus, this study uniquely provides information about the genetic makeup of cultivars/breeding lines developed in the US PNW.

In summary, the polymorphism survey and diversity analysis i) allowed determination of genetic relationships of barley genotypes adapted to the US PNW; ii) provided data to make informed selection of barley genotypes used for crossing with the Bob *AHAS* mutant; iii) allowed identification of the most divergent pair of genotypes with the Bob mutant to be used for the genetic mapping of the *AHAS* gene; iv) allowed identification of the polymorphic markers for each pair of genotypes with Bob mutant to uniquely track and reconstitute the genetic-background of the recipient genotype; and v) allowed determination of the level of nucleotide diversity along the entire length of the barley chromosome 6H. This information not only proved useful during the present study but will also prove useful in later studies.

Determination of the critical dose of herbicide

From previous experience we know that the 0.118 L/acre dose of Beyond is sufficient to distinguish the susceptible barley genotypes from the resistant ones [16]. However, a critical herbicide dose, which could discriminate between the heterozygous and homozygous states of the *AHAS* mutation, remains unknown. Thus, in the present study, an attempt was made to determine the critical herbicide dose by spraying 0.118, 0.236 and 0.295 L/acre doses of Beyond on the segregating F_2 population derived from WAS4×Bob mutant cross. A non-significant deviation from the 2:1 segregation ratio (at $p<0.05$) of surviving vs dead plants was observed at each herbicide dose, which indicates the semi-dominant nature or dominant transmission of this mutation with incomplete penetrance (see next section for details). Subsequently, an effort was made to determine the maximum dose of herbicide, which can be tolerated by the IMI-resistant AHAS isoform. In order to achieve this objective, crude enzyme extracted from the leaf tissues of the Bob *AHAS* mutant was fed with the substrate (pyruvate) in presence of the increasing concentrations of the herbicide (see Materials and Methods). The assay suggested that the mutant enzyme can survive up to 1.18 L/acre Beyond that is 10 times field recommended dose applied on the IMI-tolerant winter wheat (Fig. 3). The assay also allowed discrimination of homozygotes from heterozygotes at 8× and 10× field recommended doses of the herbicide, displayed in the test by the intensity of red color as determined by the spectrophotometer. The heterozygotes took longer to produce same intensity of color that homozygotes produced in shorter duration of time (data not shown). However, these high doses of herbicide are impractical for use in glasshouse and field trials. In actual field conditions, the plant only receives a maximum of 0.236 L/acre dose, especially in the overlapping areas. Thus, for rest of the analyses, we used 0.236 L/acre herbicide dose.

Collectively these results suggested that the mutant AHAS enzyme can survive up to 10× field recommended dose of herbicide, which makes it unlikely to find a critical herbicide dose that can discriminate homozygotes from heterozygotes at the *AHAS* locus.

Table 3. Microsatellite markers classified according to repeat element type.

Repeat type	SSR markers used	Mean PIC	Number of repeats
Simple	45	0.41	-
Mononucleotide	3	0.35	10
Dinucleotide	24	0.63	7 to 29
Trinucleotide	13	0.43	4 to 9
Trtranucleotide	3	0.33	5
Hexanucleotide	2	0.31	5
Compound	17	0.61	-

Number of simple sequence repeats (SSRs) or microsatellites falling in each category is listed and the range of alleles detected by SSRs in these categories and their average PIC (polymorphic information content) values are shown.

Transfer of the IMI-resistance to other barley cultivars

A large collection of recombinants was screened in order to transfer IMI-resistance to selected genotypes in a single generation, and to identify rare recombinants carrying a small chromosomal segment with the gene of interest introgressed in the desired genetic background (Table 4). This will alleviate the need of backcrossing and avoid overriding the 'Breeder's Code of Ethics'. As mentioned in the Materials and Methods, the F_1s were grown to obtain F_2 seeds and a range of 2158 to 2846 F_2 lines per cross combination were evaluated for the presence of the mutant allele. This has been achieved by spraying the F_2 populations with 2× equivalent to the field recommended dose of Beyond used on the IMI-tolerant winter wheat (i.e., 0.236 L/Acre Beyond with 1% methylated seed oil), and by phenotyping the resistant plants for early vigor a month after spraying with the herbicide. The expected 3:1 ratio of resistant vs susceptible plants, an indicative of the dominant nature of the mutation was not observed with any of the six segregating populations. Instead, the crosses between WAS4, Radiant, and Clearwater with the Bob mutant showed a 2:1 segregation ratio of resistant vs susceptible plants (>0.05 probability). Collectively, the observed segregation ratios obtained from the greenhouse herbicide tests of the six segregating populations, at the best suggested a semi-dominant nature of the mutation or incomplete penetrance of the trait (Table 4). This low trait penetrance could be explained due to the cumulative effect of a number of factors like genetic differences for leaf and/or culm wax coating in the parental genotypes of a population, though this possibility needs further investigation.

The semidominant nature of the mutant prompted us to determine the genotype at the *AHAS* locus (the foreground selection) by DNA sequencing of the *AHAS* gene fragment from 1 to 3 F_3 lines each from the six most vigorous F_2 plants selected per cross combination (Figs. 4 and 5). Although, an allele-specific agarose based assay exists for genotyping of segregating populations for the *AHAS* mutant allele, it is unsuitable for use in this situation due to its dominant nature (i.e., incapability of distinguishing between a heterozygote and a mutant type homozygote) [16]. Later, the six F_3 plants showing the *AHAS* mutant allele in homo- or heterozygous state were selected to

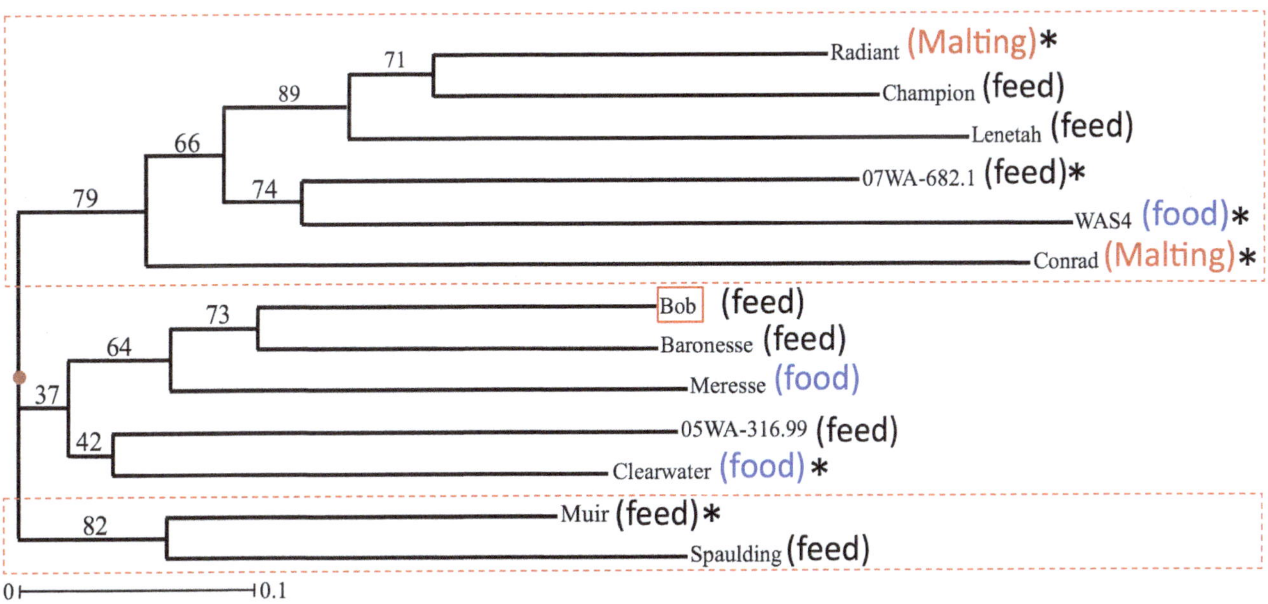

Figure 2. A dendrogram showing the clustering pattern of the 13 barley genotypes (see Table 1 for genotype details). Genetic distances for the dendrogram were estimated using SSR (simple sequence repeat) polymorphism data. Bootstrap values are indicated at each node. The genotypes selected for crossing with the Bob *AHAS* mutant are marked with asterisk.

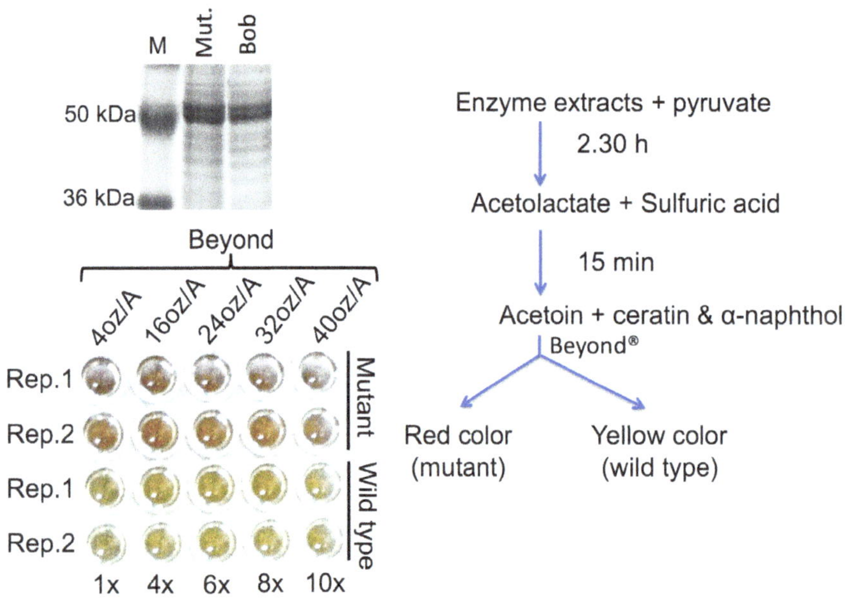

Figure 3. Results of the *in vitro* colorimetric enzyme activity assay performed for AHAS enzyme in the presence of inhibiting concentrations of an imidazolinone herbicide Beyond. Upper left, the crude enzyme was extracted from the leaves of 'Bob' and 'Bob' *AHAS* mutant and loaded on 10% SDS-polyacrylamide gel to check for the presence of the AHAS enzyme in the extracts. The presence of the enzyme in the extract was confirmed by a ~65 kDa protein band on the gel. Lower left, the enzyme activity assay further confirms the presence of enzyme in the extracts from wild type and mutant. In the assay, the enzyme was fed with pyruvate (substrate) in presence of increasing concentrations of Beyond. The assay clearly showed that the mutant AHAS enzyme can survive up to 1.18 L/acre dose of herbicide and show equal activity if measured 15 min after addition of the color change solution (for more details, see Materials and Methods). The assay was performed as summarized in the line-diagram on the right.

check for carrier chromosome recovery using 10–12 SSR markers specific to barley chromosome 6H. A range of 20 to 90% recovery of the recipient parent genome for the carrier chromosome was observed in the different cross combinations (Fig. 5). Collectively, this pilot study clearly demonstrates the feasibility of transferring IMI-resistance to desired barley genotypes in a single generation with the possibility of finding lines showing good recovery of the recipient parent genome.

Conclusion

Results of the study are of high significance not only to growers in the Pacific Northwest but also to growers in other parts of the US and the world, wherever IMI-herbicides are applied and IMI-resistant crops are cultivated. In this study we determined the genetic diversity among 13 barley cultivars/breeding lines, which benefitted the present study and is expected to prove useful in future breeding efforts. Chromosomal localization of the gene encoding the catalytic subunit of the barley AHAS enzyme will

Table 4. List of the number of crosses made, the F_1 seeds obtained per cross combination and the F_2 lines screened for imidazolinone (IMI)-resistance.

Female	Male	Crosses	F_1 harvested	F_2 sampled	F_2 screened against herbicide	
					Susceptible	Resistant
Feed						
05WA-316.99	Bob mutant	53	471	2815	856	1959
07WA-682.1	Bob mutant	29	394	2251	813	1438
Malting						
Radiant*	Bob mutant	37	358	2130	669	1461
Conrad	Bob mutant	35	445	2671	806	1865
Food						
Clearwater*	Bob mutant	38	394	2336	790	1546
WAS4*	Bob mutant	38	403	2342	736	1606
Total		230	2465	14545	4670	9875

*Fitted in 2:1 (resistant vs susceptible) segregation ratio at 0.05 significance level.

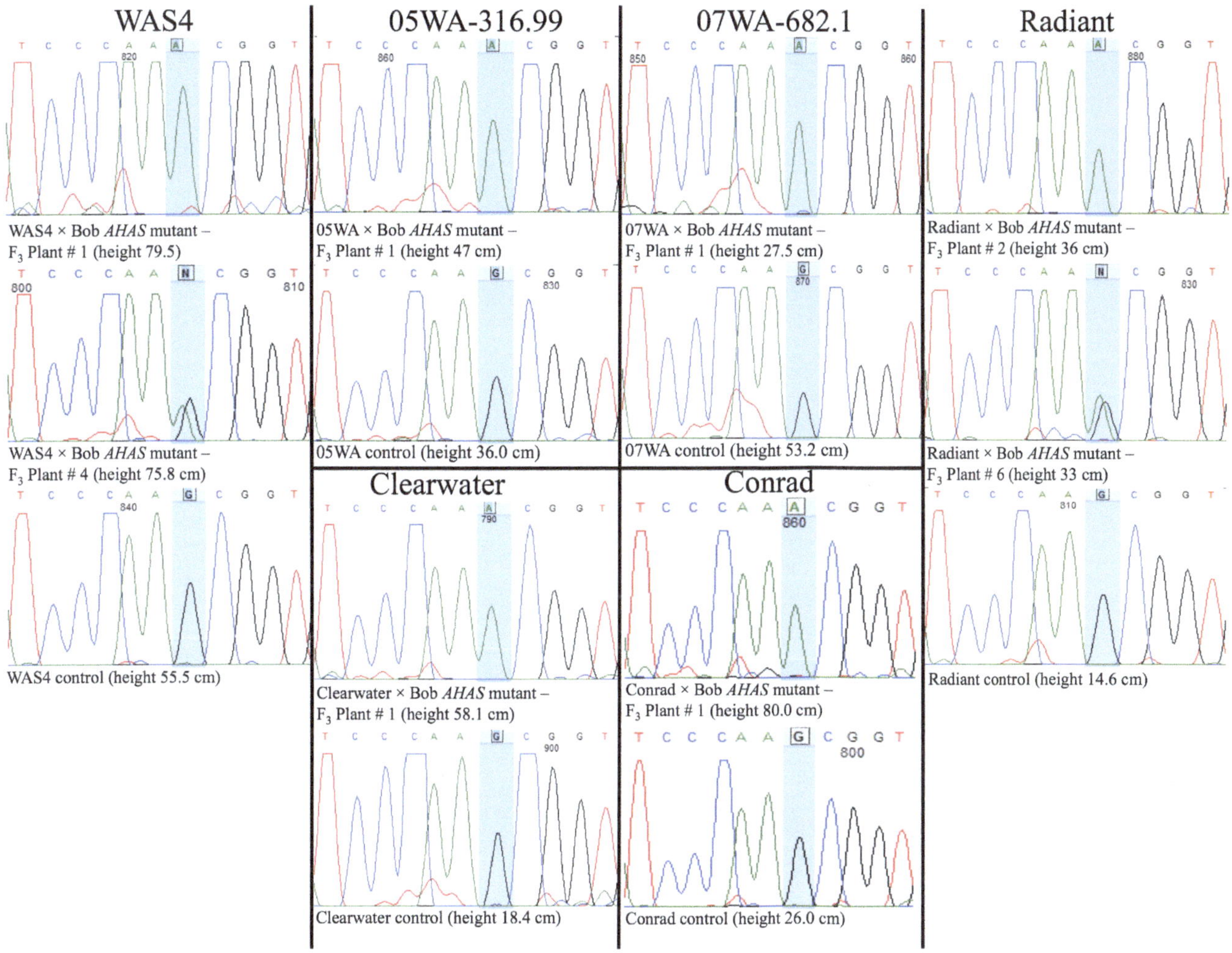

Figure 4. A part of the DNA sequence of the *AHAS* gene showing the point mutation responsible for IMI-resistance (highlighted in blue). The DNA sequencing results clearly demonstrated the transfer of IMI-resistance to two feed barleys, 05WA-316.99 and 07WA-682.1, two food barleys, Clearwater and WAS4, and two malting barleys, Radiant and Conrad.

also prove useful in future gene-transfer studies leading to the development of herbicide-resistant cultivars with other agronomically important traits. Determination of the working dose of herbicide used for phenotypic screening of this trait will be used in future breeding efforts to transfer IMI-resistance. This pilot study with a limited number of selected F_2 lines shows that it is possible to identify genotypes showing good recovery of the recipient parent genome by screening large $F_{2:3}$ populations and following a strategic selection scheme (Fig. 5).

Our future objective is to take the recently developed IMI-resistant food, feed and malting barley genotypes from the glasshouse to the field by i) screening large numbers of F_3 families, representing the 250 top ranking F_2 lines selected per cross combination, based on their vigor a month after herbicide spray, for their genetic backgrounds using DNA markers; ii) fixing heterozygosity (which confounds phenotypic evaluations) in selected lines by doubled haploid (DH) production; iii) field evaluation of the DH lines for their performance on herbicide residue and under spray trials. This will allow identification of barley lines showing more genetic proximity to their respective recipient parents.

For the first objective, F_3 seeds belonging to the 250 F_2 lines which survived the herbicide spray (at the rate of 0.236 L/acre Beyond with methylated seed oil) and showed early vigor a month after spray are currently being propagated in herbicide treated soil in the glasshouse. Cultivating plants on herbicide treated soil will allow elimination of susceptible individuals, which are expected in a segregating population at a proportion of one in four individuals. Genotype of the survivors will be determined at the AHAS locus by DNA sequencing following the procedure described above. It is of considerable importance to differentiate homozygotes from heterozygotes at the AHAS locus, as the two genotypic states at this locus are undistinguishable from each other using herbicide treatment alone. This is due to the semi-dominant nature of the AHAS mutation. The lines possessing the mutant allele(s) at the AHAS locus either in homo- or heterozygous state will be evaluated for their genetic background in a stepwise fashion first using 10 carrier chromosome (6H) specific microsatellite markers followed by 4 DNA markers per non-carrier chromosomes (2 markers per arm). The second step of background selection will be performed on the F_3 plants showing good carrier chromosome recovery in the first step. The lines showing good recovery of recipient parent genome will be converted to doubled haploids via

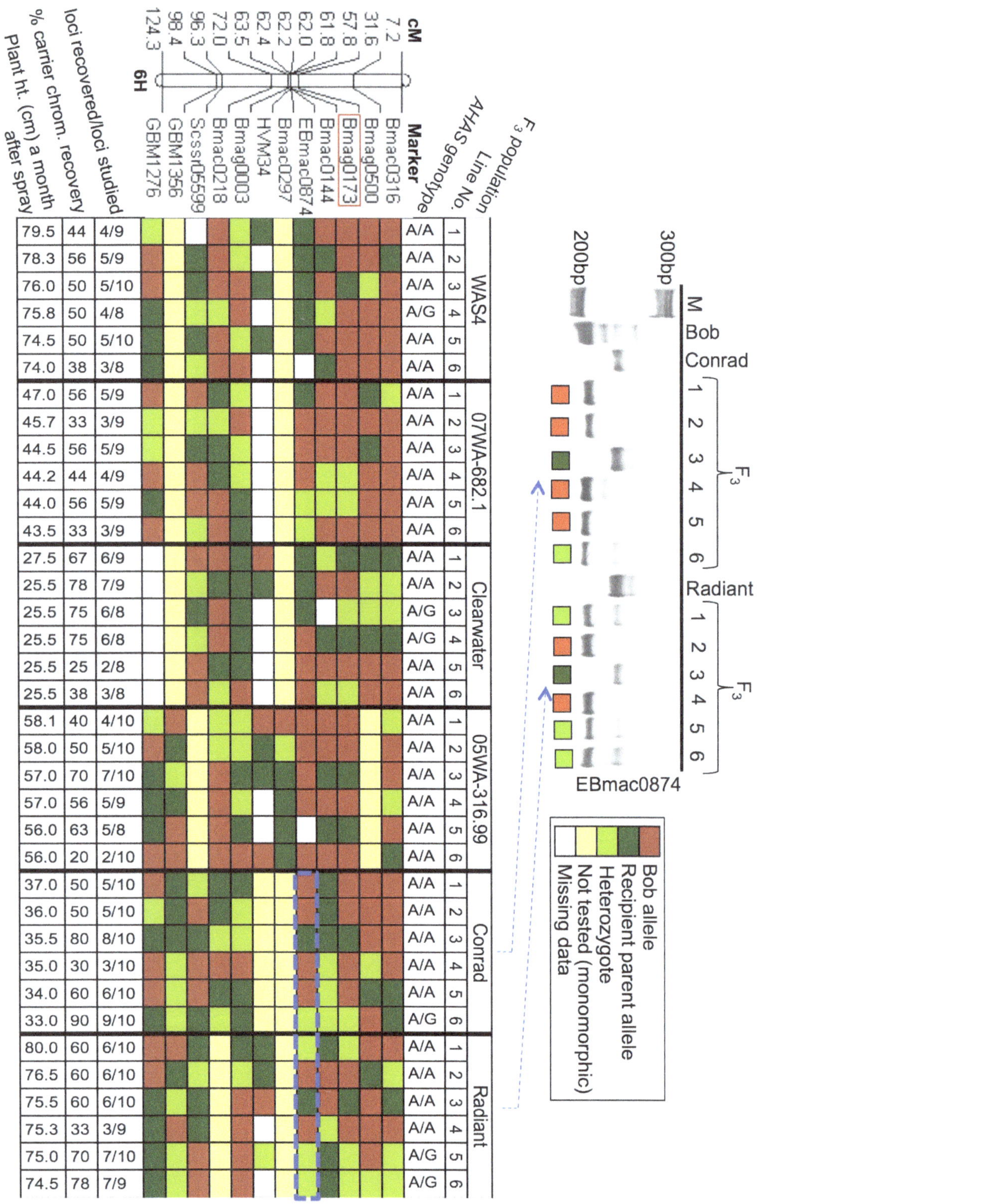

Figure 5. Diagrammatic representation of the results of the marker assisted background selection on the F_3 progeny of the selected F_2 lines. After foreground selection, the six F_3 lines per cross combination were evaluated for the recovery of recipient parent background by genotyping each line with 10 to 12 chromosome 6H-specific microsatellite markers. The markers were selected on the basis of polymorphism data obtained earlier during diversity analysis and their respective location on chromosome 6H. Map locations of selected markers are shown on the left. Each column in the picture represents a F_3 line and each row represents a DNA marker, whereas each cell represents the marker genotype in an individual. The marker genotype is represented by a color code: a) light green color denotes heterozygotes carrying marker alleles from both parents; b) dark green color denotes a marker allele similar to the recipient parent; and c) red color denotes the marker allele of the donor parent. Thus, a column with more dark and light green cells represent a genotype showing high percentage of the recipient parent genome, as observed for the 6th F_3 individual in the Conrad ×Bob mutant cross, which had 90% recovery of the Conrad alleles. In contrast, a column showing more red cells represents

a donor-like carrier chromosome as examplified by the 6th F_3 individual in the 05WA-316.99×Bob mutant cross, which had only 20% recovery of the 05WA-316.99 parent genome.

a microspore culture based method following Kasha et al. [36]. The resultant doubled haploids will be evaluated for their performance in the field on herbicide residue and herbicide spray trials.

The major outcome of this project will be the development of IMI-resistant barley varieties and germplasm with a combination of beneficial traits including resistance for various biotic and abiotic stresses, higher grain yield and better quality. Moreover, adding imidazolinone resistance to barley cultivars adapted to the PNW will certainly improve the sustainability of barley, which is one of the best rotational crops for this region.

Supporting Information

Figure S1 Comparative mapping of wheat chromosome 6D and barley chromosome 6H to determine approximate location of the *AHAS* gene on chromosome 6H. (a) Genetic linkage map of wheat chromosome 6D. (b) Physical map of wheat chromosome 6D. Short arm is at the top, and the black circle indicates the centromere. Deletion-line breakpoints and fraction lengths (FLs) are indicated by the horizontal line to the left. Breakpoint positions are drawn approximately to scale. Darkened areas within chromosome arms are C-bands (cf. Endo and Gill. 1996. Journal of Heredity 87:295). (c) Microsatellite consensus map of barley chromosome 6H (modified from Varshney et al. 2007. Theoretical and Applied Genetics 114:1091). (d) Genetic location of the *AHAS* gene determined on the basis of *in silico* analysis. The gene was assigned to the 'Morex' BAC-contig #40275 anchored to the consensus genetic linkage map at 67.917 cM (cf. Close et al. 2009. BMC Genomics 10:582).

Figure S2 Consensus map of barley chromosome 6H (left; Varshney et al. 2007 Theoretical and Applied Genetics 114:1091) used to select simple sequence repeat (SSR)-markers for diversity analysis of two-rowed spring barley genotypes. Amplification profile of a few SSR markers used for analysis of barley genotypes are shown on left, and their locations on the genetic-linkage map are highlighted by red rectangles. Different SSR alleles are coded by different numbers and shown on the bottom of each SSR profile.

Figure S3 Polymorphic information content (PIC) values for different SSRs (classified according to repeat element type) are plotted against their respective location (in cM) on the genetic linkage map, showing variation in nucleotide diversity observed along the entire length of chromosome 6H.

Figure S4 The dissimilarity coefficient (GD) values calculated for 78 pairs of genotypes. High to low dissimilarity coefficient values with 'Bob' are shown on a red to white scale, with the highest value (0.726) shaded with the darkest red color, and the lowest value (0.435) in white.

Acknowledgments

The authors would like to thank Patrick Reisenauer, Dennis Pittmann, Shanshan Wen, Mingming Yang and Jayanthi Mohan for assistance in the glasshouse and laboratory experiments, and Dr. Ian Burke for useful discussions.

Author Contributions

Conceived and designed the experiments: SR KMM DvW. Performed the experiments: SR JM JHM RATB-A NW CO NA. Analyzed the data: SR RG JM JHM. Contributed to the writing of the manuscript: SR JM KMM DvW.

References

1. Garstang JR, Spink JH, Suleimenov M, Schillinger WF, McKenzie RH, et al. (2011) Cultural practices: Focus on major barley-producing regions. In: Ullrich SE, editor. Barley: Production, improvement, and uses, Wiley-Blackwell, Oxford, UK. 221–281.
2. Brester G (2012) Barley profile. Available: http://www.agmrc.org/commodities__products/grains __oilseeds/barley-profile/#. Accessed 2014 Mar 1.
3. Smiley RW, Ogg AG, Cook RJ (1992) Influence of glyphosate on Rhizoctonia root rot, growth, and yield of barley. Plant Disease 76: 937–942.
4. Janosky J, Young D, Schillinger W (2002) Economics of conservation tillage in a wheat-fallow rotation. In: Pacific Northwest Conservation Tillage Handbook Series No. 16, Chapter 10 - New Technology Access, Adaptation and Economics, February 2002. Available: http://pnwsteep.wsu.edu/tillagehandbook/chapter10/101602.htm. Accessed 2014 Mar 1.
5. Schillinger WF, Paulitz TC (2006) Reduction of Rhizoctonia bare patch in wheat with barley rotations. Plant Disease 90: 302–306.
6. Black AL, Bauer A (1986) Optimizing winter wheat management in the Northern Great Plains region. In: Proc. A Hands-On Workshop for Implementing Maximum Economic Yield (MEY) Systems, North Dakota Coop. Ext. Serv., Kirkwood Inn, Bismarck, ND, July 8–10, 1986.
7. Turner WJ, Yenish J, Line RF, Chen X (2001) Crop profile for barley in Washington. Available: http://www.ipmcenters.org/cropprofiles/docs/wabarley.html. Accessed 2014 Mar 1.
8. Baik BK, Newman CW, Newman RK (2011) Food uses of barley. In: Ullrich SE, editor. Barley: Production, improvement, and uses, Wiley-Blackwell, Oxford, UK. 532–562.
9. Newman RK, Newman CW (2008) Barley for food and health: Science, technology and products. Wiley Publishers, New York.
10. United States Department of Agriculture, Economic Research Service (2013) Available: http://www.ers.usda.gov/data-products/feed-grains-database/feed-grains-yearbook-tables.aspx.Accessed 2014 Mar 1.
11. Rustgi S (2013) Bringing barley back in crop rotation by breeding for imidazolinone resistance. Medicinal & Aromatic Plants 2: e148.
12. Shaner DL, Singh BK (1997) Acetohydroxyacid synthase inhibitors. In: Roe RM, Burton JD, Kuhr RJ, editors. Herbicide activity: Toxicity, biochemistry and molecular biology, IOS Press, Washington DC. 69–110.
13. Duggleby RG, McCourt JA, Guddat LW (2008) Structure and mechanism of inhibition of plant acetohydroxyacid synthase. Plant Physiology and Biochemistry 46: 309–324.
14. Duggleby RG, Pang SS (2000) Acetohydroxyacid Synthase. Journal of Biochemistry and Molecular Biology 33: 1–36.
15. Tan S, Evans RR, Dahmer ML, Singh BK, Shaner DL (2005) Imidazolinone-tolerant crops: history, current status, and future. Pest Management Science 61: 246–257.
16. Lee H, Rustgi S, Kumar N, Burke IC, Yenish JP, et al. (2011) Single nucleotide mutation in the barley *acetohydroxy acid synthase* (*AHAS*) gene confers resistance to imidazolinone herbicides. Proceedings of the National Academy of Sciences of the United States of America 108: 8909–8913.
17. Saghai-Maroof MA, Soliman KM, Jorgensen RA, Allard RW (1984) Ribosomal DNA spacer length polymorphisms in barley: Mendelian inheritance, chromosomal location, and population dynamics. Proceedings of the National Academy of Sciences of the United States of America 81: 8014–8018.
18. Sambrook J, Fritsch EF, Maniatis T (1989) Molecular Cloning: a laboratory manual, 2nd edition. Cold Spring Harbor Laboratory Press, Plainview, New York, USA.
19. Varshney RK, Marcel TC, Ramsay L, Russell J, Röder MS, et al. (2007) A high density barley microsatellite consensus map with 775 SSR loci. Theoretical and Applied Genetics 114: 1091–1103.

20. Tegelstrom H (1992) Detection of mitochondrial DNA fragments. In: Hoelzel AR, editor. Molecular genetic analysis of populations: a practical approach. Oxford: IRL Press. 89–114.
21. Botstein D, White RL, Skolnick M, Davis RW (1980) Construction of a genetic linkage map in man using restriction fragment length polymorphisms. American Journal of Human Genetics 32: 314–331.
22. Sokal RR, Michener CD (1958) A statistical method for evaluating systematic relationships. University of Kansas Science Bulletin 28: 1409–1438.
23. Saitou N, Nei M (1987) The neighbor-joining method: a new method for reconstructing phylogenetic trees. Molecular Biology and Evolution 4: 406–425.
24. Perrier X, Flori A, Bonnot F (2003) Data analysis methods. In: Hamon P, Seguin M, Perrier X, Glaszmann JC, editors. Genetic diversity of cultivated tropical plants. Montpellier (France): Enfield, Science Publishers. 43–76.
25. Singh BK, Stidham MA, Shaner DL (1988) Assay of acetohydroxyacid synthase. Analytical Biochemistry 171: 173–179.
26. McCourt JA, Pang SS, King-Scott J, Guddat LW, Duggleby RG (2006) Herbicide-binding sites revealed in the structure of plant acetohydroxyacid synthase. Proceedings of the National Academy of Sciences of the United States of America 103: 569–573.
27. Lee YT, Duggleby RG (2001) Identification of the regulatory subunit of *Arabidopsis thaliana* acetohydroxyacid synthase and reconstitution with its catalytic subunit, Biochemistry 40: 6836–6844.
28. Anderson JA, Matthiesen L, Hegstad J (2004) Resistance to an imidazolinone herbicide is conferred by a gene on chromosome 6DL in the wheat line cv. 9804. Weed Science 52: 83–90.
29. Mayer KF, Martis M, Hedley PE, Simková H, Liu H, et al. (2011) Unlocking the barley genome by chromosomal and comparative genomics. Plant Cell 23: 1249–1263.
30. Randhawa HS, Mutti JS, Kidwell K, Morris CF, Chen X, et al. (2009) Rapid and targeted introgression of genes into popular wheat cultivars using marker-assisted background selection. PLoS ONE 4(6): e5752.
31. Cuadrado A, Cardoso M, Jouve N (2008) Physical organization of simple sequence repeats (SSRs) in Triticeae: structural, functional and evolutionary implications. Cytogenetics and Genome Research 120: 210–219.
32. Molnár I, Cifuentes M, Schneider A, Benavente E, Molnár-Láng M (2011) Association between simple sequence repeat-rich chromosome regions and intergenomic translocation breakpoints in natural populations of allopolyploid wild wheats. Annals of Botany 107: 65–76.
33. Bandopadhyay R, Rustgi S, Chaudhuri RK, Khurana P, Khurana JP, et al. (2011) Use of methylation filtration and C_0t fractionation for analysis of genome composition and comparative genomics in bread wheat. Journal of Genetics and Genomics 38: 315–325.
34. Akhunov ED, Akhunova AR, Anderson OD, Anderson JA, Blake N, et al. (2010) Nucleotide diversity maps reveal variation in diversity among wheat genomes and chromosomes. BMC Genomics. 11: 702.
35. Hufford MB, Xu X, van Heerwaarden J, Pyhäjärvi T, Chia JM, et al. (2012) Comparative population genomics of maize domestication and improvement. Nature Genetics 44: 808–811.
36. Kasha KJ, Simion E, Oro R, Yao QA, Hu TC, et al. (2001) An improved in vitro technique for isolated microspore culture of barley. Euphytica 120: 379–385.

Cyclodextrin Inclusion Complex to Improve Physicochemical Properties of Herbicide Bentazon: Exploring Better Formulations

Claudia Yáñez[1]*, Paulina Cañete-Rosales[1], Juan Pablo Castillo[1], Nicole Catalán[1], Tomás Undabeytia[2], Esmeralda Morillo[2]

1 Departamento de Química Orgánica y Fisicoquímica, Facultad de Ciencias Químicas y Farmacéuticas, Universidad de Chile, Santiago, Chile, **2** Instituto de Recursos Naturales y Agrobiología, Consejo Superior de Investigaciones Científicas, Sevilla, España

Abstract

The knowledge of the host-guest complexes using cyclodextrins (CDs) has prompted an increase in the development of new formulations. The capacity of these organic host structures of including guest within their hydrophobic cavities, improves physicochemical properties of the guest. In the case of pesticides, several inclusion complexes with cyclodextrins have been reported. However, in order to explore rationally new pesticide formulations, it is essential to know the effect of cyclodextrins on the properties of guest molecules. In this study, the inclusion complexes of bentazon (Btz) with native βCD and two derivatives, 2-hydroxypropyl-β-cyclodextrin (HPCD) and sulfobutylether-β-cyclodextrin (SBECD), were prepared by two methods: kneading and freeze-drying, and their characterization was investigated with different analytical techniques including Fourier transform infrared spectroscopy (FT-IR), differential thermal analysis (DTA), X-ray diffractometry (XRD) and differential pulse voltammetry (DPV). All these approaches indicate that Btz forms inclusion complexes with CDs in solution and in solid state, with a stoichiometry of 1:1, although some of them are obtained in mixtures with free Btz. The calculated association constant of the Btz/HPCD complex by DPV was 244±19 M^{-1} being an intermediate value compared with those obtained with βCD and SBECD. The use of CDs significantly increases Btz photostability, and depending on the CDs, decreases the surface tension. The results indicated that bentazon forms inclusion complexes with CDs showing improved physicochemical properties compared to free bentazon indicating that CDs may serve as excipient in herbicide formulations.

Editor: Piero Andrea Temussi, Università di Napoli Federico II, Italy

Funding: This research was supported by Fondecyt, grant 1090254 (http://www.fondecyt.cl). The funders had no role in study design, data collection and analysis, decision to publish, or preparation of the manuscript.

Competing Interests: The authors have declared that no competing interests exist.

* E-mail: cyanez@ciq.uchile.cl

Introduction

Cyclodextrins (CDs) are cyclic oligosaccharides which provide an interesting organic host system, since they have a hydrophobic inner cavity available to form non-covalent host-guest inclusion complexes with a wide variety of organic molecules of appropriate shape and size [1–3]. This property has attracted considerable attention in the field of the encapsulation of molecules because some physicochemical properties of the guest can be notably changed by its inclusion. In this context, the pharmaceutical industry has successfully used this distinctive aspect of CDs in different pharmaceutical applications [4]. Some commercial formulations using CD are mentioned by Szejtli [5].

Nowadays, thanks to large improvements in cyclodextrin production and purification, the application of this organic host in another important area such as agriculture, is becoming a real possibility [6]. In agricultural production, the application of pesticides is of vital importance. However, excessive amount of pesticide is sometimes used to compensate that part of them is lost by different mechanism (evaporation, hydrolysis, inactivation by wetness or by contact with soil, degradation by light, etc). For this reason, an optimization in the herbicides formulations is necessary.

Molecular encapsulation of pesticides by CDs can enhance the efficacy and facilitate formulation of these agrochemicals in a variety of ways, for example, by increasing their flowability, wettability, dissolution rate, chemical and thermal stability, and reducing their volatility [7]. CDs have the capacity to increase the apparent solubility of several hydrophobic pesticides and thus, their bioavailability for biodegradation [8–10]. Besides, CDs present several advantages over solvents and nonionic surfactants including their lower toxicity and higher biodegradability [11].

Herbicides are applied to control or kill undesired plants, enabling greater crop productivity and promoting the mechanization of the agricultural production [12]. Bentazon (Btz) 3-(1-methylethyl)-1H-2,1,3-benzothiadiazin-4(3H)-one-2,2-dioxide (Figure 1A) provides effective control of broadleaf weeds and sedges in many crops such as beans, corn, rice, peanuts, peas, asparagus, peppers and peppermint [13]. Btz is a contact post-emergence herbicide that inhibits photosynthesis [14] and is photo-degraded in water with a half-life lower than 24 h [15].

A

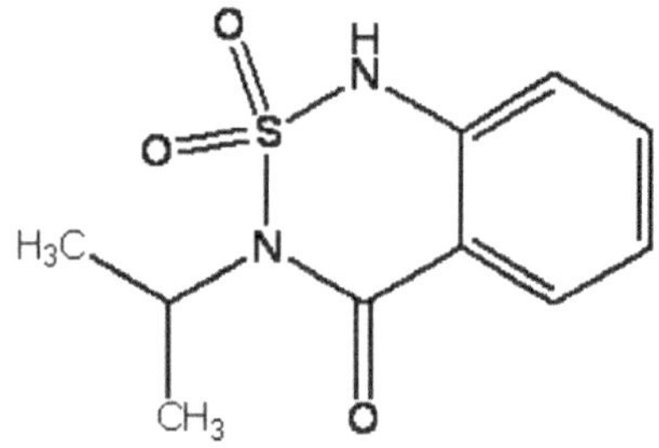

B

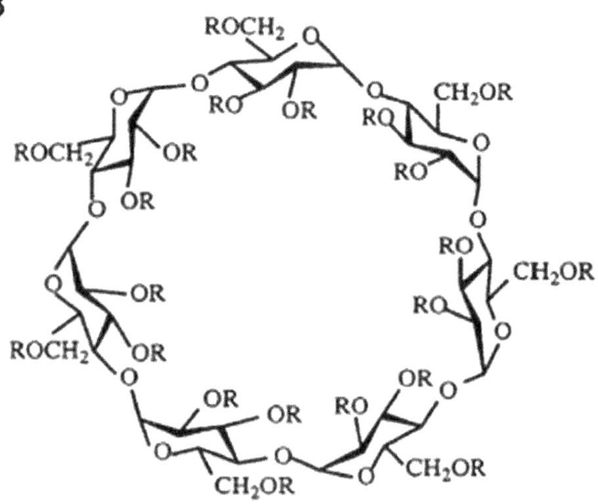

Figure 1. Molecular structures. Bentazon (A) and CD derivative (B), where R is H, 2-hydroxypropyl ($CH_2CHOHCH_3$) or sulfobutylether ($(CH_2)_4SO_3Na$) for βCD, HPCD and SBECD, respectively.

The effectiveness of Btz is often improved by adding adjuvants that increase herbicide activity but also its retention and adsorption; therefore, adjuvants increase phytotoxicity of Btz on many weeds and enhance crop damage [16].

Recently, Ge et al. [17] stated that the herbicide chloropropham increased its solubility and thermal stability when it was in complex with beta-cyclodextrin, βCD, suggesting the use of the latter for herbicide formulations. The utilization of hydroxypropylcyclodextrin (HPCD) in an insecticide formulation has been proposed [18]. In addition, we reported previously the inclusion complex of Btz with βCD and sulfobutylether-βCD (SBECD) in solution, which showed dependence on the functional groups present in the cyclodextrin [19]. General uses of cyclodextrins and their patents were revised by Del Valle [20]. Several commercial products based on CDs are available, as reviewed by Hashimoto [21]. Although an increase of work has been lead towards agrochemical-CD formulations, no large scale commercial applications can be found. Publications and patents have been reviewed by Morillo [22].

In this research we aim to report the preparation and characterization of some inclusion complexes formed by bentazon with native βCD and two derivatives hydroxypropyl-βCD and sulfobutylether-βCD (HPCD and SBECD; Figure 1B). Our main interest is to explore some physicochemical properties of these inclusion complexes with the aim to contribute to the future applications of CDs to solid samples in agriculture field.

Results and Discussion

Characterization of inclusion complexes in solution

The electrochemical behavior of Btz at glassy carbon electrode in aqueous solution in presence of two different cyclodextrin has been previously studied by our research group [19]. The presence of electroactive functionality permits us to use this technique to study the incorporation of Btz into the cyclodextrin cavity. In the case of HPCD, the effect observed is similar to previous results informed, showing a decrease in the current intensity with increasing HPCD concentration (Figure 2A). This decrease of the current (Figure 2B) is associated to the change in apparent diffusion coefficient of the complex compared with the bentazon-free form [19]. The association constant (K_a) for the complex Btz/HPCD was obtained using the equation described for current titration (Equation 2, Materials and Methods). A linear relationship was obtained in 1/CD vs $1/(1\text{-}i/i_0)$ plot (Figure 2B, inset). The value for the association constant (K_a) obtained by DPV was $244\pm19\ M^{-1}$. This value is between those obtained for βCD and SBECD (118 ± 20 and $317\pm25\ M^{-1}$, respectively [19]). The K_a value for Btz/HPCD inclusion complex differs from that obtained by Porini and Escandar using fluorescence emission [23]. However, the difference is not very significant taking into account that two different methods were used, and also that the association constants are practically of the same order (log K_a are 2.3 and 2.1 obtained by DPV and fluorescence, respectively). An association constant value using HPCD higher than that obtained using βCD was expected considering that the polarity of the substituents of the host could contribute to the complex formation. Thus a less polar group like 2-hydroxypropyl can be more favourable to the formation of the complex [23]. In spite of electrochemical results would indicate that a 1:1 inclusion complex was formed, we have verified the stoichiometry of the inclusion complex obtaining the Job plot (Figure 3). As shown, the maximum value is reached when the mole fraction is 0.5, indicating a 1:1 stoichiometry.

It is interesting to note that no change in peak potential is observed with HPCD. Previously, we have found different effect of CD on peak potential which could indicate a different insertion of the guest molecules (a change close to 45 mV was observed for a nitro compound indicating that nitro group probably was located inside the cavity) [24]. Thus, no change in peak potential of oxidation of Btz suggests that the electroactive moiety of the molecule is located outside the cavity, consequently the benzene ring would be inside the cavity of cyclodextrin. This possible structure is in agreement with the structure recently proposed by Porini and Escandar [23]. We also have informed a slight shift in a positive direction using βCD and SBECD (a change lower than 20 mV was observed) [19]. This slight shift observed with other CDs, is not enough to propose a different structure. Additional information would be neccesary to verify the structures of the complexes.

Characterization of solid inclusion complexes

Solid inclusion complexes (1:1) were prepared by kneading (KN) and freeze-drying (FD) method. Native βCD and two derivatives were used. FT-IR analysis was made to verify the inclusion of the pesticide in the CDs. We have found that this technique is effective in these cases because the principal bands of Btz were not masked by bands from the CDs For the inclusion complexes and the physical mixtures (PM) the bands of Btz are overlapped by the

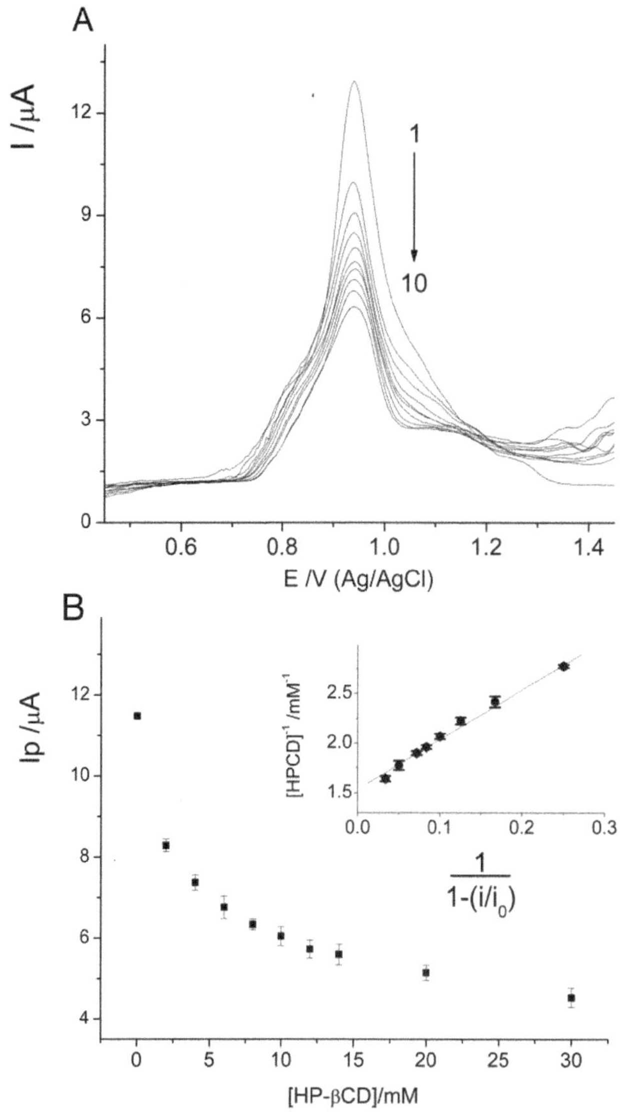

Figure 2. Electrochemical oxidation of bentazon. (A) Differential pulse voltammetry curves for 5.0×10^{-4} M Btz in buffer phosphate pH 6.0 in the absence and presence of different HPCD concentrations. Curves 1–10: 0, 2, 4, 6, 8, 10, 12, 14, 20 and 30 mM HPCD; (B) Current dependence on the concentration of HPCD for bentazon in buffer phosphate pH 6.0. Current values obtained from DPV measurements. Inset: Plot of 1/[CD] versus $1/(1-I/I_0)$ for bentazon in buffer phosphate pH 6.0.

bands of CDs in the range of 3800–2500 cm^{-1}, thus these bands are not considered. The results of FTIR spectroscopy are shown in figures 4, 5 and 6. The spectrum of Btz is shown in all of the figures for comparison, and the band assignments are shown in Table 1. The peaks in 1934 and 1809 cm^{-1}, corresponding to the bending vibration in the aromatic ring of pure Btz, are not present in the inclusion complexes and physical mixtures. On the other hand, two bands of Btz [25] (1478 and 1356 cm^{-1} corresponding to the C-H bending of the methyl group and S = O stretching vibration in sulfone group, respectively) are visible in the inclusion complexes as well as the physical mixtures, which confirm its presence in the sample. The IR spectra of βCD, HPCD and SBECD showed their characteristic bands in agreement with the previously reported literature [2,26,27].

As can be seen, in the spectrum of the inclusion complex Btz/βCD prepared by kneading (Figure 4d) and Btz/HPCD prepared by kneading (Figure 5d) or freeze-drying (Figure 5e), a peak in 1681 cm^{-1} originally positioned in 1646 cm^{-1} in the spectrum of Btz was observed. This shift shows the disruption of the interaction between molecules of Btz (N-H—O = C), indicating that the herbicide is entrapped into the host cavities as a monomer [25]. The band associated to the bending of N-H (1616 cm^{-1}) in the spectrum of Btz, shifts to a lower frequency (1609 cm^{-1}) due to the formation of a hydrogen bond. These bands are not present in the physical mixture confirming the formation of an inclusion complex in solid state. Although a small and broad signal in 1648 cm^{-1} is observed for all the CDs, it is clear that there is no overlapping with the bands attributed to the inclusion complexes.

In the other cases (Figure 4e, 6d and 6e) a small band about 1646 cm^{-1} is also present indicating that a mixture of free molecules of bentazon and inclusion complex was obtained. Although Btz/SBECD showed the highest association constant, the solid product was a mixture, even if another ratio (1:2, Btz:SBECD) was tested (Figure 6f).

The formation of inclusion complex has been also checked using both, X-ray diffraction (XRD) and differential thermal analysis. The results of the complex Btz/HPCD are shown in figures 7 and 8. The pattern for pure Btz shows its high crystallinity (Figure 7a), with strong and well defined peaks at 10.95 Å and 5.47 Å. The pattern corresponding to HPCD shows its amorphous character (Figure 7b), with two wide bands at approximately 9.95 Å and 4.52 Å. The kneaded complex exhibits no Btz diffraction peaks (Figure 7c), but the new pattern shows the 9.95 Å diffraction of HPCD as a small shoulder over a new band with a maximum at 13.03 Å, suggesting that a new complex has probably been formed but with an amorphous state. The above results support that Btz and HPCD form true inclusion complexes in solid state, with amorphous character when kneading both components. Similar results were obtained for the complex prepared by freeze-drying method (not shown). Finally, differential thermal analysis (DTA) allowed us to determined differences between the thermal stability of the pesticide in a free-form and in the form of an inclusion complex. The thermo-analysis curves obtained for the isolated host, guest and the complex can be seen in figure 8 and data are summarized in Table 2. HPCD shows a broad endothermic peak between 50–100°C corresponding to the process of dehydration. Otherwise, Btz shows an endothermic peak between 130–140°C which corresponds to its melting point and a broad exothermic peak between 170–200°C related with its decomposition [28]. After the inclusion complex is formed, a minor endothermic peak around 60°C is associated to the dehydration of the cyclodextrin. This peak is smaller than the observed for HPCD due to the exchange of Btz for the water molecules located into the cavity of the cyclodextrin. On the other hand, the peak of the melting point of pure Btz disappears, which implies that Btz is included into the cyclodextrin cavity and decompose when inclusion complex does. The exothermic peak around 180°C is associated to the decomposition of the inclusion complex and its enthalpy value is lower than pure Btz. Moreover, the decomposition of the inclusion complex occurs at lower temperatures (about 10°C less), indicating that the complex is slightly less stable at high temperatures.

Photodegradation studies

Previous to the photochemical experiments carried out with Btz in solution, photodegradation of solid Btz was also studied for a further study of its photodegradation in the solid complexes with cyclodextrins, but it reached only about 3% after 40 h of exposition. It indicates that the possibility of Btz photodegradation

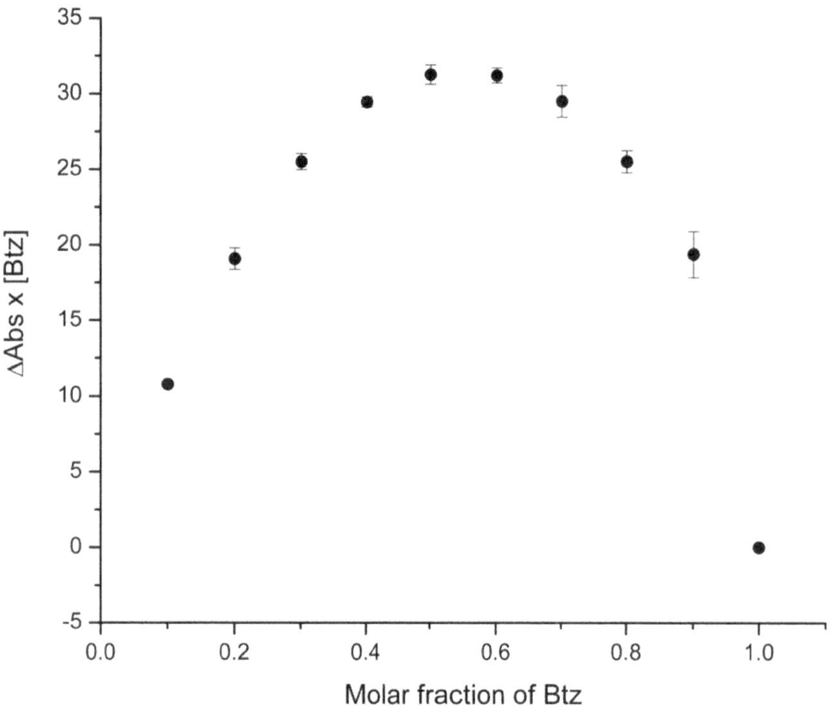

Figure 3. Determination of the stoichiometry of the complex. Continuous variation plot (Job plot) of the absorption change in dependence of the bentazon mole fraction.

when applied as formulation will be higher in the soil solution than in solid state. According to the formation constants of its complexes in solution with βCD, HPCD and SBECD (118, 244 and 317 M^{-1}, respectively), more than 90% of Btz was forming complexes with the cyclodextrins; therefore we decided to study the photodegradation of Btz and their complexes in solution. The photodegradation of Btz in aqueous solution is shown in figure 9. The data were plotted as the concentration of remaining herbicide versus irradiation time. Btz photodegradation profile followed an exponential decay showing that it was clearly affected by photoirradiation in solution, remaining only about 2 mg L^{-1} of Btz after 60 hours. On the contrary, no disappearance of Btz was detected in the dark experiments (not shown), demonstrating that the disappearance of the herbicide was only due to photodegradation, permitting to exclude other phenomena such as volatilization or thermal degradation. Simple first-order kinetics (Eq. 1) was found to be the best descriptor for the experimental data.

$$C_t = C_0 e^{-kt} \quad (1)$$

where C_t is the concentration of pesticide remaining in solution (mg L^{-1}) at time t (hours), C_0 is the initial concentration of pesticide (mg L^{-1}) and k is the rate of degradation ($hours^{-1}$). The linearized form of Eq. 1 was used to calculate the degradation constants (k) and the time required for 50% disappearance (DT_{50}) of Btz, with values of 3.84×10^{-2} h^{-1} and 17.9 h, respectively.

There is a relative controversy among the values of DT_{50} for Btz photodegradation in distilled water in the literature, and also with the value obtained in our paper. Peschka et al. [29], using experimental conditions similar to ours, observed complete photodegradation and formation of SO_3-bentazon after 13.4 h in pure water, with a DT_{50} about 4 h. Davezza et al. [30] obtained the complete degradation of Btz after 60 min, and its DT_{50} at about 15 min, when using a Xenon lamp of 1500 W m^{-2} and pyrex glass containers instead of quartz ones.

The irradiation of Btz in aqueous solution in the presence of the three cyclodextrins showed no disappearance of the herbicide along the course of irradiation (Figure 9), demonstrating the high photoprotective effect of these cyclodextrins. However, a loss of the herbicide of 32% is observed after 60 h in the case of the complex Btz-SBECD. This loss was not due to Btz photodegradation, because the same loss was also observed in the corresponding dark control, while those corresponding to the other Btz complexes did not show any loss. It indicates that the presence of SBECD facilitated its decomposition after many hours of irradiation and could be probably related to the increased temperature inside the quartz flasks during photochemical experiments (about 60°C).

Although few studies about the photocatalytic effect of CDs on pesticides in aqueous solution have been reported previously, pesticides can exhibit an increase or decrease in the intensity of light absorption when included in the cavity of a cyclodextrin. Zeng et al. [31], concluded that several cyclodextrins acted like photocatalysts in the degradation of the pesticide methyl-parathion, and the same effect occurred when the herbicide norflurazon was irradiated in aqueous solutions in the presence of five different cyclodextrins [32]. However, Kamiya and Nakamura [33] observed that some cyclodextrins promoted the photodegradation of parathion and paraoxon, while others inhibited it. Hanna et al. [34] observed the inhibition effect of β-CD on the photocatalytic degradation of pentachlorophenol in distilled water, and also Orgovãnyi et al. [35] reported an inhibition of the photodegradation of the insecticide cypermethrin by complexation with two methylated CDs (RAMEB and DIMEB).

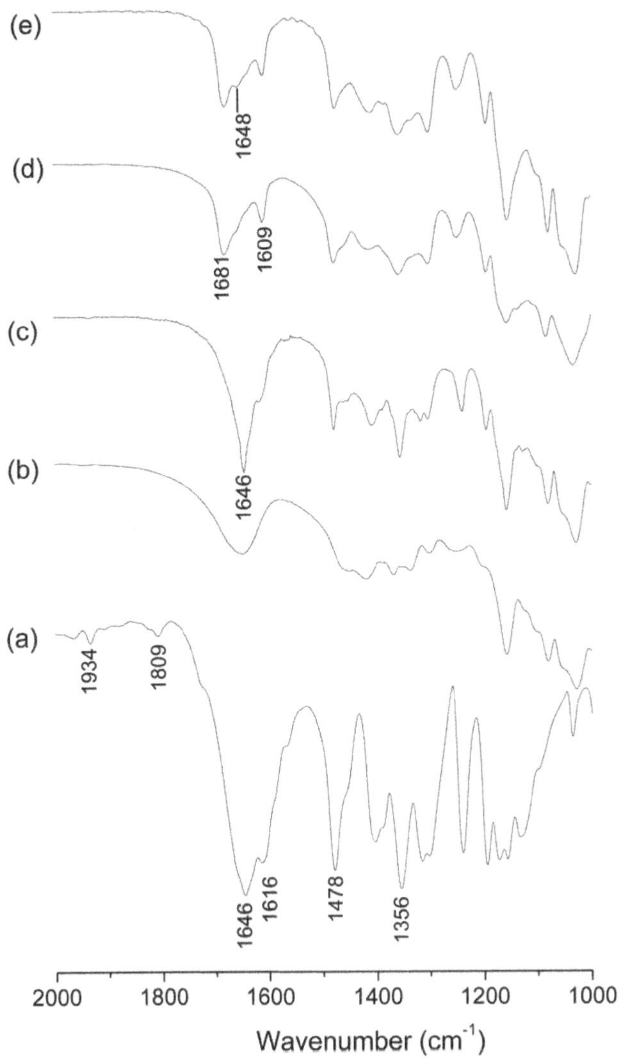

Figure 4. FT-IR spectra. (a) Bentazon, (b) βCD, (c) Btz-βCD (PM), (d) Btz/βCD (KN), (e) Btz/βCD (FD).

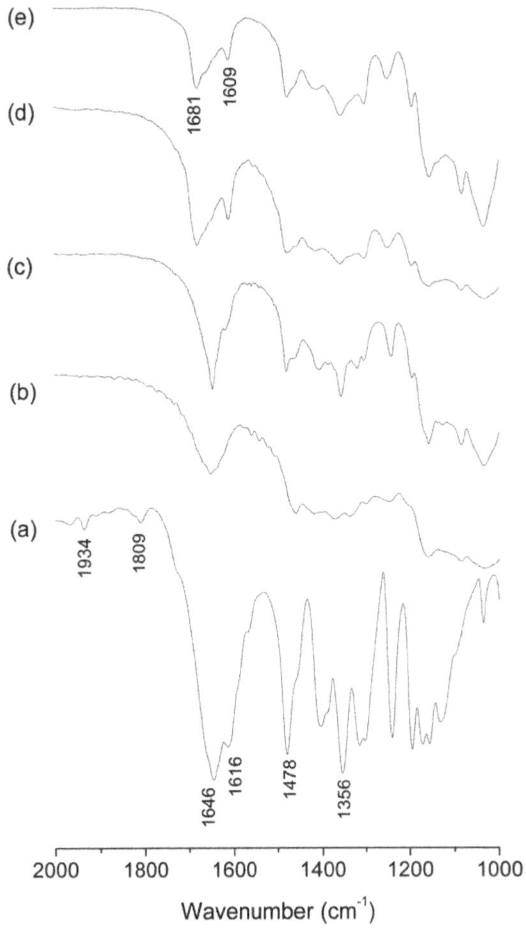

Figure 5. FT-IR spectra. (a) Bentazon, (b) HPCD, (c) Btz-HPCD (PM), (d) Btz/HPCD (KN), (e) Btz/HPCD (FD).

Studies of inclusion complexes in solution

The phase solubility diagram of Btz in the presence of HPCD is shown in figure 10. Btz showed a Bs type diagram where the solubility of Btz increased linearly with HPCD concentrations up to 22 mM. Then, there is no additional increase in solubility of Btz. Under the experimental conditions used in this experiment, the solubilization efficiency of HPCD, defined here as the increment of the apparent solubility of Btz at the maximum solubility reached with respect to Btz solubility in the absence of HPCD, was about 3.75. HPCD solubilization efficiency is higher than that obtained in presence of βCD although lower than SBECD [19]. This result correlates well with the tendency observed for the association constants of the inclusion complexes under study.

The dissolution studies are shown in figure 11. A faster dissolution rate for inclusion complex with HPCD with respect to the free Btz can be seen, reaching a plateau in a very short time. Btz is completely dissolved before 5 min compared to only 28% of free Btz for the same time and the dissolution of the herbicide (95% dissolved) is reached after 1.5 h. It is shown that no better results were obtained using βCD and SBECD. Although a higher solubility of Btz is reached using SBECD [19], the result obtained in the dissolution test is lower than HPCD probably because the inclusion complex of Btz/SBECD is actually a mixture of both, inclusion complex and free Btz.

The increase in the dissolution of the herbicide is due to the formation of the inclusion complex with CDs. According to Dua et al. [36], the increase in dissolution rate is caused by the effect of CD on lowering the surface tension. On this way, we have measured the surface tension of free Btz, CDs and inclusion complexes solutions. Also, physical mixtures of them were tested for comparison. Results are shown in table 3. No changes in surface tension of water are observed in presence of Btz, βCD or SBECD. However, the surface tension of water decreased to 65.3±0.2 mN/m in presence of HPCD. This value is in agreement to those reported by Leclercq [37] and Nogueiras-Nieto [38], which pointed out that some modified CDs provoke a decrease in surface tension in a different way compared from typical surfactants. We have found that although there was no change in surface tension of Btz or βCD solutions, the inclusion complex showed a lower value. Also, a slight decrease is observed in physical mixtures indicating that some of the complex is being formed.

The lowest value was obtained with Btz/HPCD inclusion complex. The change observed would indicate that the Btz molecules are interacting with CDs to form inclusion complexes that have surface activity, therefore are adsorbed at the interface

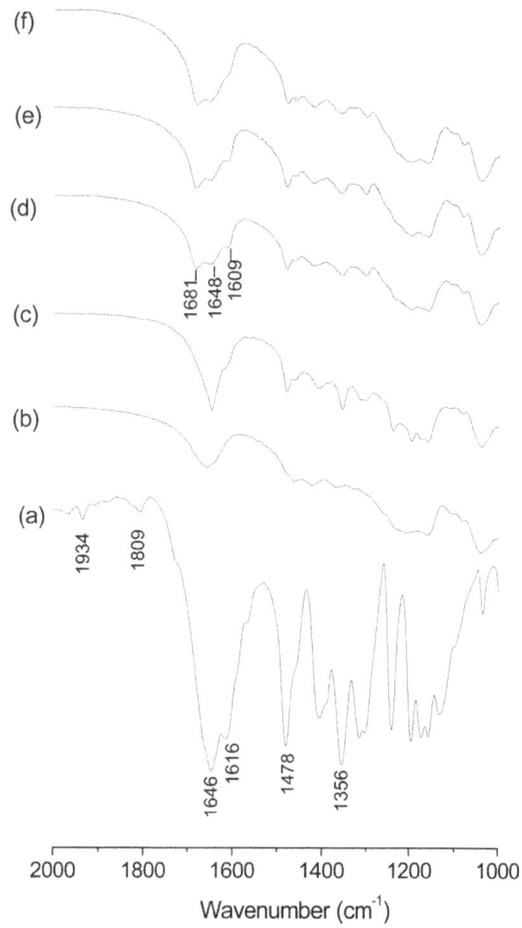

Figure 6. FT-IR spectra. (a) Bentazon, (b) SBECD, (c) Btz-SBECD (PM), (d) Btz/SBECD (KN), (e) Btz/SBECD (FD, 1:1), (f) Btz/SBECD (FD, 1:2).

which leads a surface excess slightly higher than Btz or even CDs themselves. It is expected that this effect permit an increasing of the contact of the herbicide on the leaf surface after application. A slightly lower surface tension of the Btz/HPCD complex prepared by freeze drying method compared to that prepared by the kneading method is seen in table 3. It is probable that the solid produced by both two different methods exhibits different physical characteristics in solid state, where particle size and crystalline structure of the product, among other factors, could be affecting the wettability and dissolution, and therefore, surface tension. Thus, for example, a reduction in particle size could provoke an increase in wettability and dissolution and, hence lower effect on surface tension. Although the lowest value was obtained for complex prepared by freeze-drying method, the time necessary to obtain such solid inclusion complex is higher than kneading method.

Conclusions

In this work, the inclusion complexes of Btz with three different CDs were prepared. Two different methods were carried out: kneading and freeze-drying. Host guest ratio 1:1 was determined by electrochemical technique and Job Plot. FTIR was used in all the cases to verify the formation of the solid inclusion complexes. DTA and XRD also were used to confirm the formation of an inclusion complex. Solid complexes were obtained for Btz/HPCD

Table 1. FT-IR spectrum assignments for Btz.

Wavenumber (cm^{-1})	Assignment
3180	N-H stretching vibration in secondary amine
2981, 2875	C-H stretching vibration in methyl groups
1934, 1809	C-H bending vibration for aromatic ring
1646	C=O stretching vibration in tertiary amide
1616	N-H bending vibration in secondary amine
1478, 1450	C-H bending vibration in methyl groups
1356, 1158	S=O stretching vibration in sulfone group

independently of the processing method. Besides, solid inclusion complex of Btz/βCD was achieved by kneading; In other cases, mixtures where free bentazon was present together with complex were obtained for Btz/βCD prepared by freeze-drying and complexes using SBECD. We showed that the formation of inclusion complexes with the three CDs used greatly increased the photostability of Btz in aqueous solutions. Besides, CDs improved slightly the solubility of Btz, increased the dissolution rate and decreased the surface tension. Thus, solid complexes with CDs can be obtained by kneading or freeze-drying (depending on the CD) and improved physicochemical properties are observed as compared with Btz. In this work, better properties are obtained for Btz/HPCD complex. Therefore, it is expected that the formation of inclusion complexes with CDs may to improve the properties of Btz formulations, increasing the herbicide effectiveness for future applications.

Materials and Methods

Materials

All reagents were of analytical grade and used without prior purification. Bentazon (Btz) was supplied by SIGMA-Aldrich (PS-1011, SUPELCO, Bellefonte, PA, USA). βCD was purchased from Calbiochem (A Brand of EMD Biosciences, Inc. La Jolla, CA. An Affiliate of Merck KGaA, Darmstadt, Germany) 2-Hydroxypropyl-beta-cyclodextrin (HPCD) was supplied by SIGMA-Aldrich (St.Louis, USA). SBECD was obtained from CyDex, Inc. (Lenexa, Kansas). The solutions were prepared with ultrapure water (18.2 MΩ cm) from a Millipore Milli-Q system.

Preparation and analysis of Btz/HPCD inclusion complexes in solution

An aqueous solution of Btz 5.0×10^{-4} M was divided in flasks of 20 mL, and then HPCD was added in each one in order to obtain samples of different concentrations in the range between 2 and 30 mM. The samples were stirred for 12 h at 25°C and kept in repose for three hours before the measurements.

Differential pulse voltammetry experiments were performed with a potentiostat CHI 440 (CH Instruments). A 10 mL home made cell with a glassy carbon electrode (GCE) (3 mm diam., CHI 104, CH Instruments Inc.) as working electrode was used. A platinum wire and Ag/AgCl (Bioanalytical System) were used as counter and reference electrodes, respectively. The cleanliness of the working electrode was carried out by polishing it with 0.3 and 0.05 μm alumina slurries (Buehler) and then was profusely rinsed with water. Parameters selected: potential scan rate: 20 mV s^{-1}, pulse amplitude: 50 mV and pulse width 50 ms.

Voltammetric experiments were carried out by keeping constant the concentration of Btz in 5.0×10^{-4} M, prepared in 0.1 M buffer

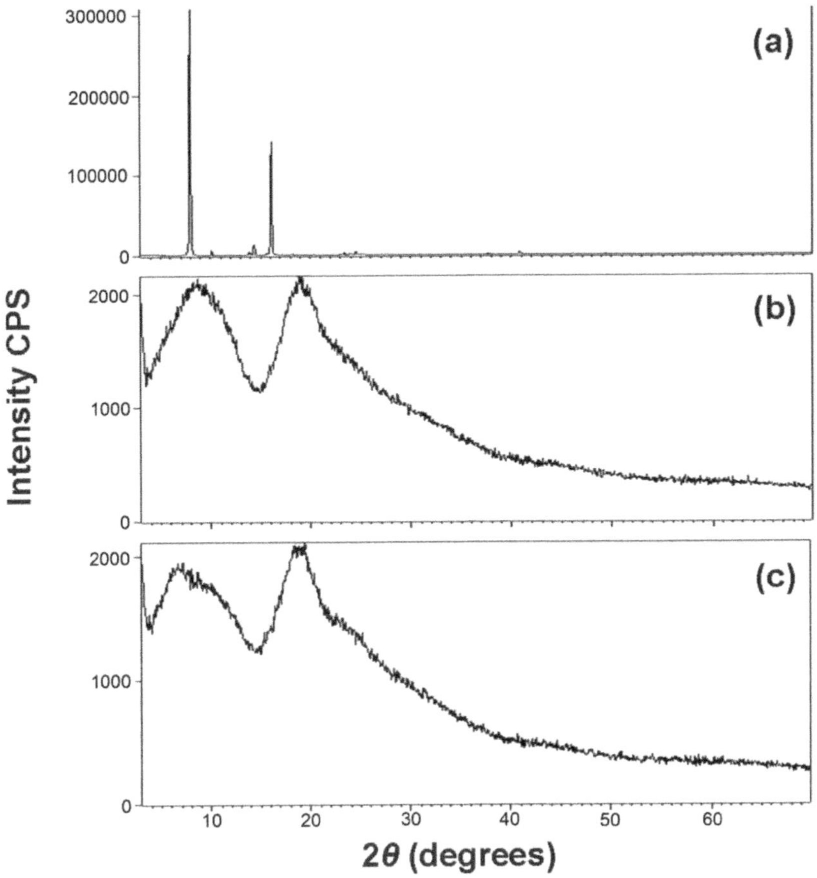

Figure 7. XRD patterns. (A) Bentazon, (B) HPCD and (C) Btz/HPCD inclusion complex (KN).

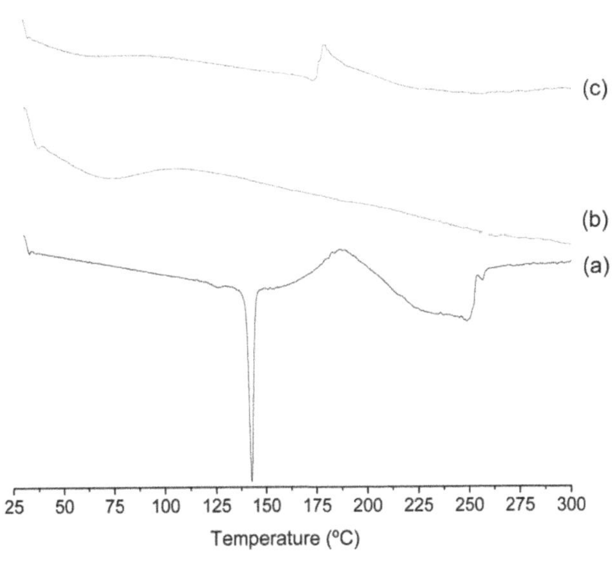

Figure 8. Thermal stability of free molecules and the inclusion complex. DTA thermogram of (a) bentazon, (b) HPCD and (c) Btz/HPCD inclusion complex (KN).

phosphate (pH 6.0); while the concentration of CD was varied. The current titration equation has been described as follows [39–41]:

$$\frac{1}{[CD]} = K_a \frac{(1-A)}{1-I/I_0} - K_a \tag{2}$$

where K_a is the complex association constant, I_0 and I are the peak currents without and with CD, respectively; [CD] is the molar concentration of CD and A is a constant. The condition for using this equation is that a 1:1 association complex is formed and CD concentrations are much larger than the total concentration of the guest. All voltammetric experiments were performed after flushing the cell with N_2 for ten minutes before each run. The temperature was kept constant at 25°C in all experiments.

The stoichiometry of the Btz/HPCD inclusion complex was determined by the equimolar variation method (Job plot) [42] based on the difference in UV-vis absorbance, ΔA ($\Delta A = A_0 - A$), of Btz in presence (A) and absence (A_0) of HPCD. A series of solutions in which the total concentration of host and guest species was maintained constant and the mole fraction of the guest varied between 0 and 1 were prepared. The concentration used for Btz and HPCD was 2.0×10^{-4} M. The using 1 cm quartz cell and the UV-Vis range used were from 190 to 400 nm.

Preparation and characterization of solid complexes

Inclusion complexes of Btz with βCD, HPCD and SBECD were prepared by kneading (KN) and freeze-drying (FD) procedures.

Table 2. Thermal decomposition temperature obtained by DTA.

Sample	Temperature range (°C)	Temperature Peak (°C)	Enthalpy (J g^{-1})	Process
HPCD	40–110	70.4	95.0	Dehydration
Btz	130–140	139.5	184.0	Melting point
	170–200	184.9	−530.0	Decomposition
Inclusion Complex	40–110	65.0	64.5	Dehydration
	170–200	176.9	−156.0	Decomposition

The 1:1 stoichiometric ratio used to prepare solid complexes was obtained from electrochemical results and Job Plot obtained in the present paper for HPCD complex and in a previous paper for BCD and SBECD complexes [19]. Physical mixtures (PM) have also been prepared for comparative purpose, where the Btz and the different CDs (1:1) were put together without kneading or freezing them. The two different methods employed for the preparation of solid complexes are described:

Kneading method (KN)

Btz and CDs (1:1) were mixed in a mortar and kneaded for 45 min. During this process 3 mL of ethanol was added to the mixture to maintain a suitable consistency. The final product was dried at 36°C for 48 h and then was ground to powder and characterized.

Freeze-drying method (FD)

1:1 molar ratio of bentazon and CD were dissolved in 60 mL of deionized water and magnetic stirred at 25°C for 24 hours. The solution was freeze-dried in a FreeZone 1.0 Labconco Benchtop Freeze Dry System (Temperature Collector −50°C; Vaccum 0.080 mbar). The time required to dry completely the sample was three days.

FT-IR analyses were carried out on a Bruker IFS 56 FT spectrometer. Spectra were measured with a KBr disk.

Differential thermal analyses were obtained using a differential temperature analyzer (Mettler, FP90 DTA). Samples of about 2–5 mg were encapsulated in aluminum pans and covered with lids which were pierced to permit the gas release during the heat process. The parameters used were a heating rate of 5°C/min over the temperature range from 25 to 300°C.

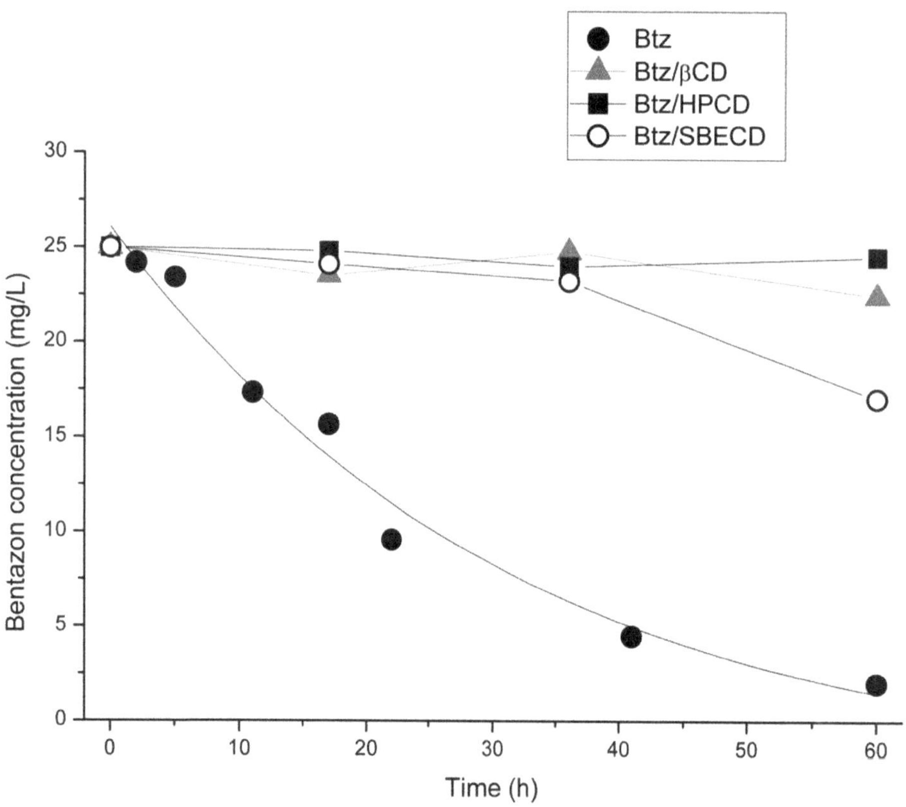

Figure 9. Photodegradation of bentazon. Btz photodegradation profiles in distilled water and in the presence of different cyclodextrins. Bentazon (●), Btz/βCD (▲), Btz/SBECD (○), and Btz/HPCD (■).

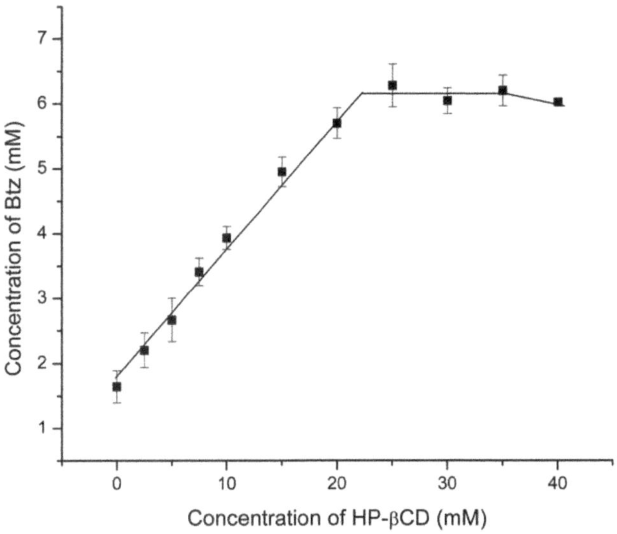

Figure 10. Phase solubility diagram. Concentration of bentazon against increasing concentrations of HPCD.

Table 3. Surface tension (γ) of water and aqueous solutions of systems under study.

	γ/mN m^{-1}
Water	70.8±0.1
Btz	70.8±0.1
βCD	70.8±0.1
SBECD	70.8±0.1
HPCD	65.3±0.2
Btz- βCD (PM)	66.5±0.4
Btz/βCD (KN)	61.8±0.1
Btz/βCD (FD)*	66.1±0.2
Btz-SBECD (PM)	68.2±0.2
Btz/SBED (KN)*	68.0±0.1
Btz/SBED (FD)*	68.0±0.1
Btz-HPCD (PM)	64.8±0.2
Btz/HPCD (KN)	63.1±0.2
Btz/HPCD (FD)	59.9±0.1

KN (kneading) and FD (freeze drying) indicates methods of preparation of inclusion complexes; PM (physical mixtures) of bentazon (Btz) with different CDs (βCD, SBECD, HPCD).
*: inclusion complex and free bentazon mixture.

X-Ray diffraction patterns were obtained using an X-ray diffractometer Siemens model Kristallofex D-5000. The samples were irradiated with Ni-filtered Cu Kα radiation (36 kV, 26 mA), at a scanning rate of 1°/min; chart speed, 1 cm min^{-1} and adequate sensibility, usually 4×10^4 counts per second.

Photodegradation studies were carried out with a Suntest CPS photoreactor (Heraeus, Germany) equipped with a Xe lamp (500 W/m^2) with a permanent filter, which selects wavelengths ≥ 290 nm. An aqueous Btz solution (100 mL, 25 mg/L) was irradiated in quartz flasks during 60 hours. The same experiment was carried out in the presence β-CD, HPCD and SBECD at a relation Btz:CD 1:1 (mol/mol). Dark controls of all experiments were conducted in a similar manner except that the flasks were kept within the reactor covered with aluminum foil. All experiments were carried out in duplicate. Samples were collected from the flasks at different time intervals and the concentration of Btz determined by HPLC under the following conditions: mobile

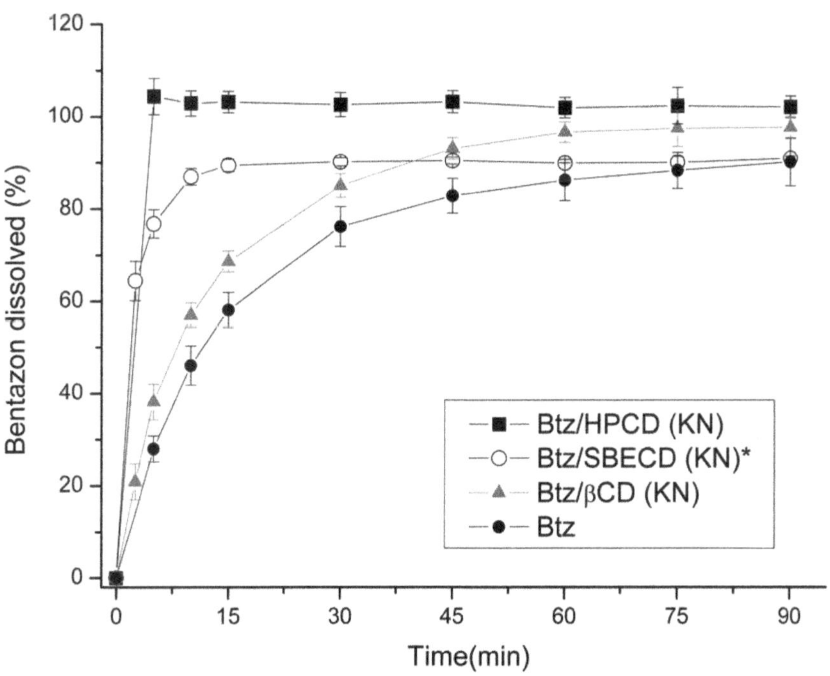

Figure 11. Dissolution test. Dissolution profiles in water of bentazon (●), Btz/βCD (KN) (▲), Btz/SBECD (KN)* (○), and Btz/HPCD (KN) (■). (*: inclusion complex and free bentazon mixture).

phase, 40:60 acetonitrile/water containing 0.1% H_3PO_4; flow 1 mL min^{-1}; chromatographic column, Kromasil C18 (150 mm length×4 mm i.d.) (Teknokroma, Spain); diode array detector (Shimadzu SPD-M10A) at a selected λ of 225 nm. The retention time for Btz under these conditions was 5.2 min.

Phase solubility diagrams were performed in triplicate according to the method carried out previously [19]. The dissolution rate studies were carried out separately in 500 mL deionized water with paddle method using Varian VK 7010 dissolution testing instrument with a Varian VK 750D heater-circulator. Samples containing 30 mg of Btz were sprinkled over the surface of the dissolution media. The stirring speed was 50 rpm and the temperature was maintained at 25±0.1°C. 5 mL aliquots were withdrawn at different time intervals up to 1.5 hours, then were filtered and analyzed by spectrophotometry UV-Visible. Same volume of fresh deionized water was added to maintain the total volume. The amount of Btz was estimated by measuring the absorbance at 333.5 nm. The dissolution experiments were done in triplicate.

Surface tension measurements were performed preparing Btz, HPCD, physical mixture and Btz/HPCD inclusion complex solutions (5×10^{-4} M) in ultrapure water. Surface tension measurements were made in quintuplicate at 25±0.1°C by the Du Noüy Ring method using a tensiometer EasyDyne Krüss.

Author Contributions

Conceived and designed the experiments: CY PC-R EM. Performed the experiments: PC-R JPC NC TU. Analyzed the data: CY PC-R EM TU. Contributed reagents/materials/analysis tools: CY EM. Wrote the paper: CY PC-R EM TU.

References

1. Szejtli J (1998) Introduction and General Overview of Cyclodextrin Chemistry. Chemical Reviews 98: 1743–1754.
2. Misiuk W, Zalewska M (2009) Investigation of inclusion complex of trazodone hydrochloride with hydroxypropyl-β-cyclodextrin. Carbohydrate Polymers 77: 482–488.
3. Kawasaki N, Araki M, Nakamura T, Tanada S (2001) Inclusion Behavior of 4-Nonylphenol into Cyclodextrin Derivatives. Journal of Colloid and Interface Science 238: 215–218.
4. Loftsson T, Brewster ME (1996) Pharmaceutical applications of cyclodextrins. 1. Drug solubilization and stabilization. Journal of Pharmaceutical Sciences 85: 1017–1025.
5. Szejtli J (2004) Past, present and futute of cyclodextrin research. Pure and Applied Chemistry 76: 1825–1845.
6. Dodziuk H (2006) Cyclodextrins and Their Complexes: Chemistry, Analytical Methods, Applications: John Wiley & Sons.
7. Pérez-Martínez JI, Morillo E, Ginés JM (1999) β-CD effect on 2,4-D soil adsorption. Chemosphere 39: 2047–2056.
8. Villaverde J, Morillo E, Pérez-Martínez JI, Ginés JM, Maqueda C (2004) Preparation and Characterization of Inclusion Complex of Norflurazon and β-Cyclodextrin To Improve Herbicide Formulations. Journal of Agricultural and Food Chemistry 52: 864–869.
9. Villaverde J, Pérez-Martínez JI, Maqueda C, Ginés JM, Morillo E (2005) Inclusion complexes of α- and γ-cyclodextrins and the herbicide norflurazon: I. Preparation and characterisation. II. Enhanced solubilisation and removal from soils. Chemosphere 60: 656–664.
10. Villaverde J (2007) Time-dependent sorption of norflurazon in four different soils: Use of β-cyclodextrin solutions for remediation of pesticide-contaminated soils. Journal of Hazardous materials 142: 184–190.
11. Ribeiro AB, Mateus EP, Rodríguez-Maroto J-M (2011) Removal of organic contaminants from soils by an electrokinetic process: The case of molinate and bentazone. Experimental and modeling. Separation and Purification Technology 79: 193–203.
12. Pourata R, Khataee AR, Aber S, Daneshvar N (2009) Removal of the herbicide Bentazon from contaminated water in the presence of synthesized nanocrystalline TiO2 powders under irradiation of UV-C light. Desalination 249: 301–307.
13. International Program on Chemical S, International Labour O, World Health O, United Nations Environment P (1990) Bentazone health and safety guide. Geneva: World Health Organization.
14. Galhano V, Peixoto F, Gomes-Laranjo J, Fernández-Valiente E (2010) Comparative toxicity of bentazon and molinate on growth, photosynthetic pigments, photosynthesis, and respiration of the Portuguese ricefield cyanobacterium Nostoc muscorum. Environmental Toxicology 25: 147–156.
15. Agency USEP (1985) Chemical fact sheet for bentazon and sodium bentazon. Fact sheet no. 64.. In: Programs OoP, editor. Washington, DC.
16. Al-Khatib K, Kadir S., Libbey C. (1995) Effect of Adjuvants on Bentazon Efficacy in Green Pea (Pisum sativum. Weed Technology 9: 426–431.
17. Ge X, He J, Qi F, Yang Y, Huang Z, et al. (2011) Inclusion complexation of chloropropham with β-cyclodextrin: Preparation, characterization and molecular modeling. Spectrochimica Acta Part A: Molecular and Biomolecular Spectroscopy 81: 397–403.
18. Mount D (2005) Insecticide-impregnated fabric and method of production. 6,896,892
19. Yañez C, Araya M, Bollo S (2010) Complexation of herbicide bentazon with native and modified β-cyclodextrin. Journal of Inclusion Phenomena and Macrocyclic Chemistry 68: 237–241.
20. Del Valle EMM (2004) Cyclodextrins and their uses: a review. Process Biochemistry 39: 1033–1046.
21. Hashimoto H (2002) Present Status of Industrial Application of Cyclodextrins in Japan. Journal of Inclusion Phenomena and Macrocyclic Chemistry 44: 57–62.
22. Morillo E (2006) Application of Cyclodextrins in Agrochemistry. In: Dodziuk H, editor. Cyclodextrins and Their Complexes: Chemistry, Analytical Methods, Applications: John Wiley & Sons. pp. 459–467.
23. Porini JA, Escandar GM (2011) Spectrofluorimetric study of the herbicide bentazone in organized media: analytical applications. Analytical Methods 3: 1494–1500.
24. Yáñez C, Núñez-Vergara LJ, Squella JA (2003) Differential Pulse Polarographic and UV-Vis Spectrophotometric Study of Inclusion Complexes Formed by 1,4-Dihydropyridine Calcium Antagonists, Nifedipine and Nicardipine with β-Cyclodextrin. Electroanalysis 15: 1771–1777.
25. Carrizosa MJ, Koskinen WC, del Carmen Hermosín M (2004) Interactions of acidic herbicides bentazon and dicamba with organoclays. Soil Science Society of America Journal 68: 1863–1866.
26. De Araújo MVG, Vieira EKB, Lázaro GS, de Souza Conegero L, Ferreira OP, et al. (2007) Inclusion complexes of pyrimethamine in 2-hydroxypropyl-β-cyclodextrin: Characterization, phase solubility and molecular modelling. Bioorganic & Medicinal Chemistry 15: 5752–5759.
27. Mayo DW, Miller FA, Hannah RW (2004) Course notes on the interpretation of infrared and Raman spectra: Wiley-Interscience.
28. Food and Agriculture Organization Of The United Nations F (1999).Specifi-Specifications and Evaluations For Plant Protection Products: BENTAZONE. Available: http://www.fao.org/agriculture/crops/core-themes/theme/pests/pm/jmps/ps/ps-new/en/#B. Accessed: 11 May 2012
29. Peschka M, Petrovic M, Knepper T, Barceló D (2007) Determination of two phototransformation products of bentazone using quadrupole time-of-flight mass spectrometry. Analytical and Bioanalytical Chemistry 388: 1227–1234.
30. Davezza M, Fabbri D, Pramauro E, Bianco Prevot A (2012) Photocatalytic degradation of bentazone in soil washing wastes containing alkylpolyoxyethylene surfactants. Chemosphere 86: 335–340.
31. Zeng Q, Liao B, Luo Y, Liu C, Tang H (2003) Inclusion Effects of Highly Water-Soluble Cyclodextrins on the Solubility, Photodegradation, and Acute Toxicity of Methyl Parathion. Bulletin of Environmental Contamination and Toxicology 71: 0668–0674.
32. Villaverde J, Maqueda C, Undabeytia T, Morillo E (2007) Effect of various cyclodextrins on photodegradation of a hydrophobic herbicide in aqueous suspensions of different soil colloidal components. Chemosphere 69: 575–584.
33. Kamiya M, Nakamura K (1995) Cyclodextrin inclusion effects on photodegradation rates of organophosphorus pesticides. Environment International 21: 299–304.
34. Hanna K, de Brauer C, Germain P, Chovelon JM, Ferronato C (2004) Degradation of pentachlorophenol in cyclodextrin extraction effluent using a photocatalytic process. Science of the Total Environment 332: 51–60.
35. Orgoványi J, Otta KH, Pöppl L, Fenyvesi É, Záray G (2005) Spectrophotometric and thermoanalytical study of cypermethrin/cyclodextrin complexes. Microchemical Journal 79: 77–82.
36. Dua K, Ramana MV, Singh Sara UV, Himaja M, Agrawal A, et al. (2007) Investigation of Enhancement of Solubility of Norfloxacin -Cyclodextrin in Presence of Acidic Solubilizing Additives. Current Drug Delivery 4: 21–25.
37. Leclercq L, Bricout H, Tilloy S, Monflier E (2007) Biphasic aqueous organometallic catalysis promoted by cyclodextrins: Can surface tension measurements explain the efficiency of chemically modified cyclodextrins? Journal of Colloid and Interface Science 307: 481–487.
38. Nogueiras-Nieto L, Alvarez-Lorenzo C, Sandez-Macho I, Concheiro A, Otero-Espinar FJ (2009) Hydrosoluble Cyclodextrin/Poloxamer Polypseudorotaxanes at the Air/Water Interface, in Bulk Solution, and in the Gel State. The Journal of Physical Chemistry B 113: 2773–2782.
39. Yañez C, Basualdo J, Jara-Ulloa P, Squella JA (2007) Inclusion complexes of estrone and estradiol with β-cyclodextrin: Voltammetric and HPLC studies. Journal of Physical Organic Chemistry 20: 499–505.

40. Ibrahim MS, Shehatta IS, Al-Nayeli AA (2002) Voltammetric studies of the interaction of lumazine with cyclodextrins and DNA. Journal of Pharmaceutical and Biomedical Analysis 28: 217–225.
41. Ureta-Zañartu MS, Yáñez C, Gutiérrez C (2004) Cyclic voltammetric, EQCM and impedance study of the 1-octanol-β-cyclodextrin-electrodeposited platinum system in perchloric medium. Journal of Electroanalytical Chemistry 569: 275–286.
42. Job P (1928) Formation and stability of inorganic complexes in solution. Annali di Chimica Applicata 9: 113–203.

5

Enantioselective Induction of a Glutathione-*S*-Transferase, a Glutathione Transporter and an ABC Transporter in Maize by Metolachlor and Its (*S*)-Isomer

Sen Pang[❡], **Zhaojin Ran**[❡], **Zhiqian Liu**[¤], **Xiaoyu Song, Liusheng Duan, Xuefeng Li*, Chengju Wang***

College of Sciences, China Agricultural University & Engineering Research Center of Plant Growth Regulators, Ministry of Education, China Agricultural University, Beijing, People's Republic of China

Abstract

The metabolism of chiral herbicides in plants remains poorly understood. Glutathione conjugation reactions are one of the principal mechanisms that plants utilize to detoxify xenobiotics. The induction by *rac*- and *S*-metolachlor of the expression of three genes, *ZmGST27*, *ZmGT1* and *ZmMRP1*, encoding respectively a glutathione-*S*-transferase, a glutathione transporter and an ATP-binding cassette (ABC) transporter was studied in maize. The results demonstrate that the inducing effect of *rac*- and *S*-metolachlor on the expression of *ZmGST27* and *ZmGT1* is comparable. However, the inducing effect of *rac*-metolachlor on *ZmMRP1* expression is more pronounced than that of *S*-metolachlor. Furthermore, vanadate, an ABC transporter inhibitor, could greatly reduce the difference in herbicidal activity between *rac*- and *S*-metolachlor. These results suggest that the ABC transporters may preferentially transport conjugates of *rac*-metolachlor, leading to a faster metabolism of the latter. Through comparing the expression of *ZmGST27*, *ZmMRP1* and *ZmGT1* after treatment by *rac*- and *S*-metolachlor, we provide novel insights into the metabolic processes of chiral herbicides in plants.

Editor: Vladimir N. Uversky, University of South Florida College of Medicine, United States of America

Funding: This study was supported by Chinese Universities Scientific Fund (Project No. 2011JS074). The funders had no role in study design, data collection and analysis, decision to publish, or preparation of the manuscript.

Competing Interests: The authors have declared that no competing interests exist.

* E-mail: Wangchengju@cau.edu.cn (CW); lixuefeng2007@163.com (XL)

❡ These authors contributed equally to this work.

¤ Current address: Department of Primary Industries, Biosciences Research Division, Victorian AgriBiosciences Centre, Bundoora, Victoria, Australia

Introduction

Chirality is a common phenomenon in life sciences. More than 30% of currently used pesticides are chiral compounds, including synthetic pyrethroids, organophosphate insecticides, imidazolinone and chloracetanilide herbicides [1,2]. The percentage of chiral pesticides is increasing with the introduction of more complex structures [3]. Enantiomers are defined as molecules that contain chiral structures and mirror images of each other [4]. Enantiomers of chiral pesticides have identical physicochemical properties, but exhibit quite different biochemical activities because biochemical processes usually show high stereo- or enantioselectivity [5]. Over the past two decades, studies on enantimers differing in biological properties have been widely reported [6], including their biodegradation [7,8], toxicities to non-target organisms [4,9–14], endocrine-disrupting activities [15–17], etc. However, the enantioselective physiological effects and toxicities of chiral herbicides in plants have not received as much attention as insecticides in animals [1]. Although the enantioselective phytotoxicity of several chiral herbicides in plants has been evaluated [2,4,6,18,19], the metabolic processes of chiral herbicides in plants remain poorly understood.

Metabolism of herbicides in plants can generally be divided into three phases. In phase I, the herbicide may be oxidized, reduced or hydrolyzed to introduce or reveal a functional group. In phase II, the herbicide is conjugated to glutathione, glucose or malonate by the respective transferase to form a water-soluble conjugate. In phase III, herbicide conjugates are transported from the cytosol to the vacuole for further degradation. Glutathione conjugation plays a major role in the resistance of plants to herbicides. There have been numerous reports concerning the role of glutathione S-transferases (GSTs) in conjugating xenobiotics and the role of tonoplast ATP-binding cassette (ABC) transporters in transporting glutathione conjugates (GS conjugates) into the vacuole [20–22]. In addition to ABC transporters, recent studies showed that glutathione transporters, located in the plasma membrane, are able to mediate the transport of both reduced glutathione (GSH) and GS conjugates [23,24]. Our earlier work found that the expression of a glutathione transporter gene isolated from maize, named *ZmGT1*, was inducible by herbicides atrazine and metolachlor, and the inducing effect of metolachlor in different maize cultivars was correlated to their tolerance to this herbicide, suggesting an involvement of *ZmGT1* in the detoxification of xenobiotics by plants [25,26]. Co-induction of *ZmGST27*, *ZmMRP1* and *ZmGT1* in maize by xenobiotics further suggests that, in addition to GSH, GSTs and ABC transporters, glutathione transporters located in the plasma membrane are also an important component in the glutathione conjugation-related plant detoxification system [27].

Metolachlor is a widely used herbicide that inhibits the synthesis of fatty acids in broadleaf weeds [28]. It was first introduced into the market as a racemic product, which contains both *R*- and *S*-enantiomers present in an equal ratio [19]. Racemic metolachlor is currently being replaced by *S*-metolachlor, which contains approximately 90% *S*-isomers and has the same herbicidal effect as the former when used at 65% of its dosage [29]. The sorption and desorption of metolachlor in the soil have been studied, along with its dissipation properties and effects on non-target species [30–33]. However, only limited information is available on the enantioselective behavior of *rac*- and *S*-metolachlor in plants [2,19]. Our previous work tested the effect of metolachlor on the expression of a glutathione-*S*-transferase (*ZmGST27*), an ABC transporter (*ZmMRP1*) and a glutathione transporter (*ZmGT1*) in maize leaves, but the effect of *S*-metolachlor on the expression of these genes was not investigated. Therefore, in the present study we have compared the expression pattern of *ZmGST27*, *ZmMRP1* and *ZmGT1* in maize leaves after treatment by *rac* and *S*-metolachlor. The aim was to provide insights into the metabolism of chiral herbicides in plants.

Results

Induction of *ZmGST27*, *ZmMRP1* and *ZmGT1* Expression by *rac*- and *S*-metolachlor

The semi-quantitative RT-PCR results showed that the expression of *ZmGST27*, *ZmMRP1* and *ZmGT1* was promoted by both *rac*- and *S*-metolachlor after 48 h treatment. While no significant difference was found between *rac*- and *S*-metolachlor in increasing the transcript level of *ZmGST27* and *ZmGT1* (Fig. 1A and 1C), *rac*-metolachlor was found to be a stronger inducer for *ZmMRP1* expression as compared to *S*-metolachlor (Fig. 1B).

A time course study was designed to further verify the differential up-regulation of *ZmMRP1* expression by *rac*- and *S*-metolachlor in maize leaves. The expression level of *ZmMRP1* analyzed at 4, 8, 24, 48, 72 and 96 h after treatment by *rac*- and *S*-metolachlor confirmed the initial finding. As shown in Fig. 2, while a consistent but moderate increase in *ZmMRP1* transcript level was afforded by *S*-metolachlor during the entire treatment period (from 4 to 96 h), the induction of *ZmMRP1* expression by *rac*-metolachlor was moderate during the first hours but became dramatic from 24 h of treatment onwards.

Effects of *rac*- and *S*-metolachlor on the Growth of Maize Shoot

Dose-response experiments in the presence of *rac*- and *S*-metolachlor allowed determination of their respective EC_{50} values on maize growth. The EC_{50} value of *rac*-metolachlor is about 2.1-fold that of *S*-metolachlor, indicating *S*-metolachlor is more efficient in inhibiting the growth of maize shoot than *rac*-metolachlor (Table 1). When used alone, vanadate, an inhibitor of ABC transporters, had no significant effect on the growth of maize shoots (results not shown). However, this same compound significantly enhanced the inhibitory effect of *rac*-metolachlor on maize growth, as judged by the decrease of its EC_{50} from 375 to 184 μM. By contrast, only a small reduction in the EC_{50} of *S*-metolachlor was found when vanadate was applied (Table 1). As a result, the ratio of EC_{50} values of *rac*-metolachlor and *S*-metolachlor decreased from 2.1 to 1.3 in the presence of the ABC transporter inhibitor.

Discussion

Studies on chiral pesticides started to appear in the early 1990s [4]. However, there is still a severe lack of knowledge about the metabolism of chiral herbicides in plants [28]. Metolachlor was introduced into the market in 1976 as a racemic product. It was later found that the two 1′*S*-isomers of metolachlor afforded most of its biological activity. Our earlier work has provided evidence for the involvement of a glutathione-*S*-transferase (*ZmGST27*), a glutathione transporter (*ZmGT1*) and an ABC transporter (*ZmMRP1*) in the detoxification of metolachlor in maize [25,26]. However, a comparative analysis of the expression patterns of GSTs, ABC transporters and glutathione transporters in the same plant tissue after exposure to metolachlor and its *S*-isomer was lacking.

In this work, we investigated the enantioselective induction of *ZmGST27*, *ZmGT1* and *ZmMRP1* in maize by *rac*- and *S*-metolachlor. The role of GSTs in the detoxification of certain herbicides has been known for many years. The activation of GST genes in response to biotic and abiotic stresses has also been reported previously [34,35]. The non-selective up-regulation of *ZmGST27* expression by *rac*- and *S*-metolachlor is likely to be a general response of this gene to chemical stress.

Our earlier work has shown that *ZmGT1* is able to mediate the uptake of both GSH and GS conjugates by yeast cells, suggesting a potential role of this gene in xenobiotic detoxification [25]. The fact that *S*-metolachlor, which is enantiomerically enriched with the biologically active 1′*S*-isomer, gives rise to similar level of up-regulation of *ZmGT1* expression in comparison to *rac*-metolachlor implies that this glutathione transporter has no preference in transporting GS conjugates of *R*- and *S*-isomer of metolachlor.

It is known that ABC transporters of plants have a wider spectrum of substrates as compared to glutathione transporters [27] and they are able to transport GS conjugates as well as glucose conjugates of certain herbicides [36–38]. A smaller increase in *ZmMRP1* transcripts by *S*-metolachlor as compared to *rac*-metolachlor suggests that ABC transporters may preferentially transport GS conjugates of *R*-isomers. This may cause accelerated degradation of *R*-isomers, leading to reduced herbi-

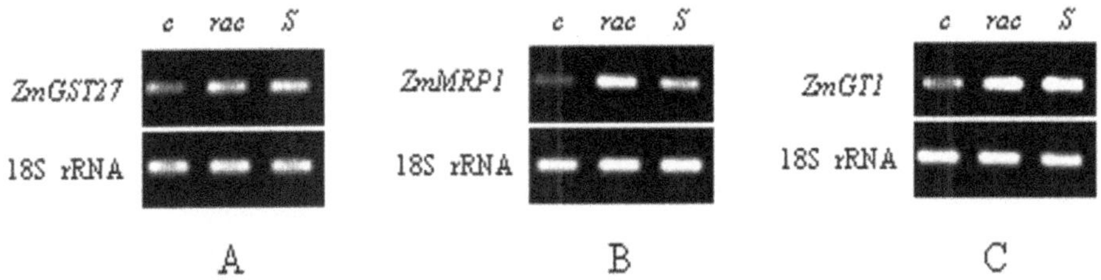

Figure 1. Semi-quantitative RT-PCR analysis of *ZmGST27, ZmMRP1* and *ZmGT1* expression in maize leaves at 48 h after treatment by *rac*- and *S*-metolachlor. A: *ZmGST27*; B: *ZmMRP1*; C: *ZmGT1*. *c*: non-treated control; *rac*: *rac*-metolachlor; *S*: *S*-metolachlor. The experiment was repeated twice with similar results.

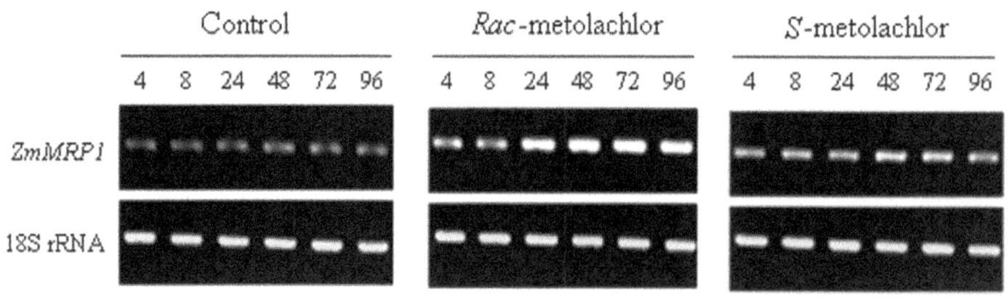

Figure 2. Semi-quantitative RT-PCR analysis of *ZmMRP1* expression in maize leaves after treatment by *rac*- and *S*-metolachlor. The numbers (4, 8, 24, 48, 72 and 96) above each lane indicate the time intervals (h) after treatment. The experiment was repeated twice with similar results.

cidal activity of *rac*-metolachlor as compared to *S*-metolachlor. The dramatic decrease of EC_{50} value of *rac*-metolachlor and the quasi-equal growth inhibition effect between *rac*- and *S*-metolachlor in the presence of an ABC transporter inhibitor (vanadate) further support this hypothesis. However, further investigation is needed to obtain direct evidence for faster degradation of *R*-isomers. More work is also needed to find whether the stronger herbicidal activity of *S*-isomer is related solely to its slower metabolism in plants.

In summary, racemic metolachlor is currently being replaced by *S*-metolachlor in application around the world. Through comparing the expression of *ZmGST27*, *ZmMRP1* and *ZmGT1* after treatment by *rac*- and *S*-metolachlor, we have found that *rac*- and *S*-metolachlor display differential activation on one of the detoxifying genes, suggesting a possible link between ABC transporter activity and differential plant sensitivity to chiral herbicides.

Materials and Methods

Chemicals

Rac-metolachlor (95% purity) and *S*-metolachlor (96% purity) were kind gifts from Institute of Plant Protection, Chinese Academy of Agricultural Sciences. Sodium orthovanadate was purchased from Sinopharm Chemical Reagent Beijing Co., Ltd.

Plant Material

Maize (*Zea mays*, cv Bainuo No.2) were grown on sand in the glasshouse as described previously [25]. Plants grown to the 3-leaf stage were transferred to a hydroponic culture (three plants per pot). Each pot contained 50 mL of nutrient solution. Plants were adapted for 7 d and the nutrient solution was changed every 2 d. After the 7 d adaptation period, *rac*- and *S*-metolachlor were added into the nutrient solution to a final concentration of 100 μM and the nutrient solution was changed daily. At the end of the treatment, leaves were harvested, immediately frozen in liquid nitrogen, and stored at −80°C.

Table 1. Effects of various chemicals on the growth of maize shoots (4 d after treatment).

Herbicides	Regression equation	EC_{50} (μM)	R^2
Rac-metolachlor	$y=0.58\ln x-3.25$	374.51	0.91
S-metolachlor	$y=0.60\ln x-3.11$	175.41	0.90
Rac-metolachlor + vanadate	$y=0.45\ln x-2.34$	184.40	0.97
S-metolachlor + vanadate	$y=0.68\ln x-3.35$	138.58	0.90

*Vanadate was applied at a constant concentration of 100 μM.

RNA Isolation

Frozen leaf samples were ground into a fine powder in liquid nitrogen. Total RNA was isolated using TRIzol reagent (Invitrogen, USA), and then digested by DNase I (TakaRa, China). First strand cDNA was synthesized with AMV reverse transcriptase and Oligo$(dT)_{15}$, according to the manufacturer's protocol of Reverse Transcription System (Promega, USA).

Semi-quantitative RT-PCR Analysis of Transcripts

Three primer pairs (*ZmGST27-F* 5′- GAC CTG CTC CTC GCC TCC AA -3′and *ZmGST27-R* 5′- CCT CCA GCG TGT CCA TAG CG -3′; *ZmGT1-F* 5′- GTG CCG CAG TGG TGG TTC -3′ and *ZmGT1* 5′- GTG ACG ACG AAG GCG AGC -3′; *ZmMRP1-F* 5′- CTA GAA TAT GAA ACA CCA GCC AAG - 3′and *ZmMRP1-R* 5′- CTG CAA TAA TGG TAG ATC ATG TTG -3′) were designed from the ORF region of *ZmGST27* (accession number AF244692), *ZmGT1* (accession number FJ573212), and *ZmMRP1* (accession number AY186244) respectively to amplify a single fragment for each gene. Another pair of primers (P3 5′- GCT CTT TCT TGA TTC TAT GGG TGG -3′ and P4 5′- GTT AGC AGG CTG AGG TCT CGT TC -3′) was used to amplify a fragment from maize 18S ribosomal RNA (accession number M82386), chosen as an internal reference. PCR amplification conditions were: 30 s at 95°C, 30 s at 62°C, 30 s at 72°C (30 cycles).

Growth Inhibition Tests

The effect of *rac*- and *S*-metolachlor on the growth of maize shoots was assessed using the dose-response test. *Rac*- and *S*-metolachlor were dissolved in methanol to make a stock solution (10,000 mg L^{-1}). The stock solution was then diluted to five serial concentrations (25, 50, 100, 200 and 400 μM) using distilled water. Triton X-100 was added at the rate of 0.5 mL L^{-1} as a wetting agent. Maize seeds were sown in plastic pots (9 cm diameter, 10 seeds per pot) filled with sand (250 g). After applying herbicide solutions, the pots (three replicates for each concentration) were placed in a growth chamber (25±2°C, 80±5% relative humidity) and watered as needed. Shoot lengths were measured 4 d after sowing. A 2nd dose-response test of *rac*-metolachlor and *S*-metolachlor was performed in parallel in the presence of an ABC transporter inhibitor, vanadate (sodium orthovanadate), which was added to the herbicide dilutions to a final concentration of 100 μM. Controls were treated with vanadate only. The effective concentration of herbicide causing 50% reduction (EC_{50}) in shoot length was determined from the dose-response curve by Probit Analysis using the SPSS (version 16.0).

Acknowledgments

The authors acknowledge the technical support of Dr Zhonghua Yang of the College of Sciences, China Agricultural University.

Author Contributions

Conceived and designed the experiments: SP LD XL CW. Performed the experiments: SP ZR XS. Analyzed the data: SP ZL. Contributed reagents/materials/analysis tools: SP CW. Wrote the paper: SP ZL.

References

1. Sekhon BS (2009) Chiral pesticides. J Pestic Sci 34: 1–12.
2. Liu HJ, Huang RN, Xie F, Zhang SX, Shi J (2012) Enantioselective phytotoxicity of metolachlor against maize and rice roots. J Hazard Mater 217–218: 330–337.
3. Liu WP, Gan J, Schlenk D, Jury WA (2005) Enantioselectivity in environmental safety of current chiral insecticides. Proc Natl Acad Sci USA 102: 701–706.
4. Qian HF, Hu HJ, Mao YY, Ma J, Zhang AP, et al. (2009) Enantioselective phytotoxicity of the heibicide imazethapyr in rice. Chemosphere 76: 885–892.
5. Muller TA, Kohler HP (2004) Chirality of pollutants–effects on metabolism and fate. Appl Microbiol Biotechnol 64: 300–316.
6. Qian HF, Lu T, Peng XF, Han X, Fu ZW, et al. (2011) Enantioselective phytotoxicity of the herbicide imazethapyr on the response of the antioxidant system and starch metabolism in *Arabidopsis thaliana*. PLoS ONE 6: 1–12.
7. Falconer RL, Bidleman TF, Szeto SY (1997) Chiral pesticides in soils of the Fraser Valley, British Columbia. J Agr Food Chem 45: 1946–1951.
8. Kurt-Karakus PB, Bidleman TF, Jones KC (2005) Chiral organochlorine pesticide signatures in global background soils. Environ Sci Technol 39: 8671–8677.
9. Wang LM, Ye WH, Zhou SS, Lin KD, Zhao MR, et al. (2009) Acute and chronic toxicity of organophosphate monocrotophos to *Daphnia magna*. J Environ Sci Heal B 44: 38–43.
10. Xu C, Zhao MR, Liu WP, Wang JJ, Chen SW, et al. (2008) Enantioselective separation and zebrafish embryo toxicity of insecticide acetofenate. Chem Res Toxicol 47: 4236–4242.
11. Liu HJ, Xiong MY (2009) Comparative toxicity of racemic metolachlor and *S*-metolachlor to *Chlorella pyrenoidosa*. Aquat Toxicol 93: 100–106.
12. Vallotton N, Moser D, Eggen RIL, Junghans M, Chevre N (2008) *S*-metolachlor pulse exposure on the alga *Scenedesmus vacuolatus*: effects during exposure and the subsequent recovery. Chemophere 73: 395–400.
13. Zhan XM, Liu HJ, Miao YG, Liu WP (2006) A comparative study of rac and *S*-metolachlor on some activities and metabolism of silkworm, *Bombyx mori L.* Pestic Biochem Phys 85: 133–138.
14. Liu HJ, Ye WH, Zhan XM, Liu WP (2006) A comparative study of rac and *S*-metolachlor toxicity to *Daphnia magna*. Ecotox Environ Safe 63: 451–455.
15. Miyashita M, Shimada T, Nakagami S, Kurihara N, Miyagawa H, et al. (2004) Enantioselective recognition of mono-demethylated methoxychlor metabolites by the estrogen receptor. Chemosphere 54: 1273–1276.
16. Hoekstra PF, Burnison BK, Garrison AW, Neheli T, Muir DC (2006) Estrogenic activity of dicofol with the human estrogen receptor: isomer- and enantiomer specific implications. Chemosphere 64: 174–177.
17. Jin YX, Chen RJ, Sun LW, Wang WY, Zhou L, et al. (2009) Enantioselective induction of estrogen-responsive gene expression by permethrin enantiomers in embryo-larval zebrafish. Chemosphere 74: 1238–1244.
18. Liu W, Ye J, Jin MQ (2009) Enantioselective phytoeffects of chiral pesticides. J Agric Food Chem 57: 2087–2095.
19. Xie F, Liu HJ, Cai WD (2010) Enantioselectivity of racemic metolachlor and S-metolachlor in maize seedlings. J Environ Sci Health B 45: 808–816.
20. Sheehan D, Meade G, Foley VM, Dowd CA (2001) Structure, function and evolution of glutathione transferases: implications for classification of non-mammalian members of an ancient enzyme superfamily. Biochem J 360: 1–16.
21. Theodoulou FL, Clark IM, He X, Pallett KE, Hallahan DL (2003) Co-induction of glutathione-S-transferases and multidrug resistance associated protein by xenobiotics. Pest Manag Sci 59: 202–214.
22. Swarbreck D, Ripoll PJ, Brown DA, Edwards KJ, Theodoulou F (2003) Isolation and characterization of two multidrug resistance associated protein genes from maize. Gene 315: 153–164.
23. Zhang MY, Bourbouloux A, Cagnac O, Shrikanth CV, Rentsch D, et al. (2004) A novel family of transporters mediating the transport of glutathione derivatives in plants. Plant Physiol 134: 482–491.
24. Cagnac O, Bourbouloux A, Chakrabarty D, Zhang MY, Delrot S (2004) AtOPT6 transports glutathione derivatives and is induced by primisulfuron. Plant Physiol 135: 1378–1387.
25. Pang S, Li XF, Liu ZQ, Wang CJ (2010) *ZmGT1* transports glutathione conjugates and its expression is induced by herbicide atrazine. Prog Biochem Biophys 37: 1120–1127.
26. Pang S, Duan LS, Liu ZQ, Song XY, Li XF, et al (2012) Metolachlor-induced *ZmGT1* expression in maize cultivars is correlated with their tolerance to the herbicide. J Food Agric Environ 10: 621–623.
27. Pang S, Duan LS, Liu ZQ, Song XY, Li XF, et al (2012) Co-induction of a Glutathione-*S*-transferase, a Glutathione Transporter and an ABC Transporter in Maize by Xenobiotics. PLoS ONE 7(7): e40712. doi:10.1371/journal.-pone.0040712.
28. Liu WP, Tang ML (2011) Enantioselective Activity and Toxicity of Chiral Herbicides, Herbicides - Mechanisms and Mode of Action, Dr. Mohammed Nagib Hasaneen (Ed.), ISBN: 978-953-307-744-4. InTech website. Available: http://www.intechopen.com/books/herbicides-mechanisms-and-mode-of-action/enantioselective-activity-and-toxicity-of-chrial-herbicides. Accessed 2010.
29. Buser HR, Poiger T, Muller MD (2000) Changed enantiomer composition of metolachlor in surface water following the introduction of the enantiomerically enriched product to the market. Environ Sci Technol 34: 2690–2696.
30. Ding GW, Novak JM, Herbert S, Xing BS (2002) Long-term tillage effects on soil metolachlor sorption and desorption behavior. Chemosphere 48: 897–904.
31. Cao PY, Wang XY, Liu FM, Zhao E, Han LJ (2008) Dissipation and residue of S-metolachlor in maize and soil. Bull Environ Contam Toxicol 80: 391–394.
32. Cook ME, Moore PA (2008) The effects of the herbicide metolachlor on agonistic behavior in the Crayfish, *Orconectes rusticus*. Arch Environ Contant Toxicol 55: 94–102.
33. Pereira SP, Fernandes MAS, Martins JD, Santos MS, Moreno AJM, et al. (2009) Toxicity assessment of the herbicide metolachlor compatative effects on bacterial and mitochondrial model systems. Toxicol In Vitro 23: 1585–1590.
34. Edwards R, Dixon DP, Walbot V (2000) Plant glutathione S-transferases: enzymes with multiple functions in sickness and in health. Trends Plant Sci 5: 193–198.
35. Basantani M, Srivastava A, Sen S (2011) Elevated antioxidant response and induction of tau-class glutathione *S*-transferase after glyphosate treatment in *Vigna* radiate (L.) Wilczek. Pestic Biochem Physiol 99: 111–117.
36. Yuan JS, Tranel PJ, Steward CN (2007) Non-target-site herbicide resistance: a family business. Trends Plant Sci 12: 6–13.
37. Klein M, Burla B, Martinoia E (2006) The multidrug resistance-associated protein (MRP/ABCC) subfamily of ATP-binding cassette transporters in plants. FEBS Lett 580: 1112–1122.
38. Gaillard C, Dufaud A., Tommasini R, Kreuz K, Amrhein N, et al (1994) A herbicide antidote (safener) induces the activity of both the herbicide detoxifying enzymes and a vacuolar transporter for the detoxified herbicide. FEBS Lett 352: 219–221.

6

Swimming with Predators and Pesticides: How Environmental Stressors Affect the Thermal Physiology of Tadpoles

Marco Katzenberger[1]*, John Hammond[2], Helder Duarte[1], Miguel Tejedo[1], Cecilia Calabuig[3], Rick A. Relyea[4]

1 Department of Evolutionary Ecology, Doñana Biological Station - Spanish Council for Scientific Research, Sevilla, Spain, **2** Department of Biology, University of New Mexico, Albuquerque, New Mexico, United States of America, **3** Department of Animal Sciences, Federal Rural University of the Semiarid Region, Mossoró, Rio Grande do Norte, Brazil, **4** Department of Biological Sciences, University of Pittsburgh, Pittsburgh, Pennsylvania, United States of America

Abstract

To forecast biological responses to changing environments, we need to understand how a species's physiology varies through space and time and assess how changes in physiological function due to environmental changes may interact with phenotypic changes caused by other types of environmental variation. Amphibian larvae are well known for expressing environmentally induced phenotypes, but relatively little is known about how these responses might interact with changing temperatures and their thermal physiology. To address this question, we studied the thermal physiology of grey treefrog tadpoles (*Hyla versicolor*) by determining whether exposures to predator cues and an herbicide (Roundup) can alter their critical maximum temperature (CT_{max}) and their swimming speed across a range of temperatures, which provides estimates of optimal temperature (T_{opt}) for swimming speed and the shape of the thermal performance curve (TPC). We discovered that predator cues induced a 0.4°C higher CT_{max} value, whereas the herbicide had no effect. Tadpoles exposed to predator cues or the herbicide swam faster than control tadpoles and the increase in burst speed was higher near T_{opt}. In regard to the shape of the TPC, exposure to predator cues increased T_{opt} by 1.5°C, while exposure to the herbicide marginally lowered T_{opt} by 0.4°C. Combining predator cues and the herbicide produced an intermediate T_{opt} that was 0.5°C higher than the control. To our knowledge this is the first study to demonstrate a predator altering the thermal physiology of amphibian larvae (prey) by increasing CT_{max}, increasing the optimum temperature, and producing changes in the thermal performance curves. Furthermore, these plastic responses of CT_{max} and TPC to different inducing environments should be considered when forecasting biological responses to global warming.

Editor: Michael Sears, Clemson University, United States of America

Funding: This work was funded by a Fundação para a Ciência e Tecnologia (FCT; www.fct.pt) PhD fellowship (SFRH/BD/60271/2009) to MK and a U.S. National Science Foundation (NSF; www.nsf.gov) grant to RAR. The funders had no role in study design, data collection and analysis, decision to publish, or preparation of the manuscript.

Competing Interests: The authors have declared that no competing interests exist.

* E-mail: katzenberger@ebd.csic.es

Introduction

Biological mechanisms underlying a response to environmental changes can be quite complex. To forecast these biological responses, we need to understand how a species' physiology varies through space and time [1,2] and assess how changes in physiological function induced by environmental changes (e.g., increasing environmental temperatures) may interact with phenotypic changes induced by other types of environmental variation [3,4,5,6].

Species can possess the ability to respond to new or altered environments with flexible phenotypes that are environmentally induced and can potentially contribute to adaptive evolution [7]. Stressful environments can induce non adaptive plasticity, increasing the variance around the mean phenotypic response or distancing it from the favored optimum. Nevertheless, if plasticity is adaptive and promotes establishment and persistence in a new environment, by placing populations close enough to a new phenotypic optimum for directional selection to act, it can predictably enhance fitness and is most likely to facilitate adaptive evolution on ecological timescales [7].

The presence of predators in the environment can induce behavioral and morphological changes in prey that result in the prey being less susceptible to the predator (e.g., [8,9,10,11]). Furthermore, pesticides can also induce behavioral and morphological changes in organisms. Sublethal exposure to pesticides early in life can make the individuals more tolerant of the pesticide later in life [12,13] and they can induce phenotypic changes that resemble predator-induced phenotypes [14,15,16,17]. In other cases, pesticides impede the induction of predator-induced morphology [18,19,20,21].

In the current scenario of climate change, there has been a renewed interest in the thermal physiology of organisms and the estimation of thermal tolerance and sensitivity, using physiological traits such as the critical thermal maximum (CT_{max}; e.g., the temperature at which animals become immobile [22,23]), the optimum temperature (T_{opt}) for performing some function, or the shape of the thermal performance curve (TPC), which describes how an animal's performance changes across a range of temperatures. Although some pesticides are known to affect CT_{max} and burst speed, usually in a negative way (e.g., [24]), there is limited information on how pesticides affect optimum temperature and performance over a range of temperatures (i.e. how pesticides affect TPCs), especially for amphibians. Likewise, much is known about predator-induced changes in organisms, including some interactions with pesticides [17]. Predators also influence thermoregulation and thermal preferences of prey, resulting in behavioral changes and coevolution of thermal optima between species [25]. Other than these behavioral responses that indirectly affect physiology, little is known about whether predator cues can directly affect the thermal physiology of prey.

We addressed these issues by studying the thermal physiology of grey treefrog tadpoles (*Hyla versicolor* LeConte 1825) that were exposed to predator cues and pesticides. Tadpoles are excellent model organisms for this study because they are practically isothermal with their aquatic environment [23] and their thermal physiology traits (CT_{max} and T_{opt}) are not influenced by confounding processes such as dehydration. Tadpoles are also well known for expressing predator-induced changes in behavior and morphology (e.g., [9,26,27]. Furthermore, at least two species of tadpoles can alter their morphology when exposed to the herbicide Roundup and exhibit morphological changes that closely resemble predator-induced changes in tadpoles [17].

Given that pollutants and predators can both affect many aspects of tadpole biology, including development and metamorphosis (e.g., [28,29]), and the interaction of pollutants with other stressors are often negative to the organism (e.g., glyphosate, [30]), we expect the impact of these stressors on the thermal physiology of tadpoles to be mainly negative. Therefore, we hypothesized that tadpoles exposed a sublethal concentration of an herbicide will have reduced tolerance to higher temperatures (CT_{max}) and exhibit a lower optimal temperature (T_{opt}) compared to tadpoles not exposed to the herbicide. Furthermore, because predator cues and the herbicide can induce deeper tails in tadpoles, we hypothesized that tadpoles exposed to either stressor will suffer a vertical shift upward in their TPC across a range of temperatures [31], and have increased swimming performance (e.g., [32]). However, it is also possible that the herbicide will have a negative effect on swimming performance (e.g., [33]) if induced morphological changes are countered by other phenotypic changes that impair swimming ability.

Methods

Inducing the tadpoles

The induction experiment was conducted at the University of Pittsburgh's Pymatuning Laboratory of Ecology in northwest Pennsylvania, USA. The experiment used a completely randomized, 2×2 factorial design comprised of the presence or absence of predator cues crossed with the presence or absence of an herbicide (nominal concentrations of 0 or 2 mg active ingredient per liter (a.e./L). Based on past studies, this herbicide concentration should remain sublethal to gray treefrog tadpoles while inducing morphology changes (e.g., [34,35]).

The four treatment combinations were replicated four times for a total of 16 mesocosms, which consisted of 120-L wading pools, set outdoors (air temperature ranged from 9°C to 28°C), that we filled with 100 L of well water on 11 June 2011. We then added 100 g of dry leaves (*Quercus* spp.) and 5 g of rabbit chow to serve as habitat structure and an initial nutrient source, respectively. We also added an aliquot of zooplankton and phytoplankton that was a mixture from 5 local ponds. Each mesocosm was equipped with a predator cage constructed of 10×10 cm well pipe covered with window screen at each end. These cages allow the chemical cues emitted during predation to diffuse through the water while preventing the predators from killing the target tadpoles [36,37,38]. Mesocosms were covered with a 60% shade cloth, for the duration of the outdoor experiment.

To obtain tadpoles for the experiment, we collected >20 amplecting pairs of grey treefrogs from a nearby wetland (41° 34′ 9.55" N, 80° 27′ 22.29" W) on 18, 21 and 22 May 2011, and allowed them to lay eggs in tubs containing aged well water. Once the eggs hatched, the tadpoles were held in outdoor pools and fed rabbit pellets *ad libitum* until used in the experiment.

On 15 June 2011, which we defined as day 0 of the experiment, we added 40 tadpoles to each mesocosm from a mixture of the clutches with an initial mass (±SE) of 37.5±2.1 mg per tadpole (subsample, N = 20). On day 1, we applied the herbicide treatment. To achieve nominal concentrations of 2 mg a.e./L, we prepared 8 equal mixtures containing 372 μL of stock solution (Roundup Power Max; concentration = 540 g a.e./L) and 250 ml of water. For the eight mesocosms assigned the herbicide treatment, we drizzled one mixture into each mesocosm. For the eight mesocosms assigned the no-herbicide treatment, we drizzled 250 mL of water into each mesocosm. Approximately 1 hr after dosing, we collected water samples from each tank to confirm the concentration of the herbicide. An independent analysis found that the concentrations in the water were 0 and 1.55 mg a.e./L (Mississippi State Chemical Laboratory, Mississippi State, MS). Observing lower actual concentrations is a common phenomenon in mesocosm experiments (reviewed in Brock et al. 2000), likely as the result of binding to surfaces in the mesocosm and degradation of the samples before the testing is conducted. Jones et al. [39] measured little herbicide breakdown for a similar time period, so we assumed there was little change in herbicide concentration during the induction experiment.

After sampling the water, we manipulated the predator environment. For mesocosms assigned the no-predator treatment, the cages remained empty. For mesocosms assigned the predator-cue treatment, we placed a single dragonfly nymph (*Anax junius*) inside the predator cage. Each dragonfly was fed ~300 mg of grey treefrog tadpole biomass every 2 d (see [38]). Prior to each feeding, we observed no tadpoles left in the predator cage, which indicates that the dragonfly nymphs consumed the tadpoles in the cages. The feeding continued until day 10 to allow tadpole growth and induction by the herbicide and predator cues.

Determining the critical thermal maximum of the tadpoles

On day 10, we brought sets of tadpoles into the laboratory to allow them to acclimate at a temperature of 20°C (approximately the average temperature experienced in the mesocosms), with a 12L:12D photoperiod, for 4 to 5 d before testing them for CT_{max} and T_{opt} [22,40]. During acclimation, tadpoles were fed rabbit pellets *ad libitum* and we maintained the predator and herbicide environments to help prevent the loss of any phenotype induction [41]. All tested larvae were below Gosner stage 38 [42]. This is important because tadpoles close to metamorphic climax exhibit a significant decline in thermal tolerance [43].

We obtained upper critical thermal tolerances (CT_{max}) by using a slightly modified version of Hutchison's dynamic method [23]. We exposed tadpoles to a constant heating rate of 0.05°C min^{-1} (3°C h^{-1}), which simulates a natural rate of temperature increase in ponds (H. Duarte, M. Tejedo, J. Hammond, M. Katzenberger, R.A. Relyea, unpublished data from dataloggers; see also [44]) until we observed complete immobility, which signaled the endpoint of the experiment. After reaching CT_{max}, we transferred tadpoles to cooler water (~20°C) to allow recovery. After complete recovery, the tadpoles were weighed and we found that the mass of the tadpoles had increased by 13- to 15-fold since day 0. We tested 3 to 4 tadpoles from each mesocosm, for a total of 56 tadpoles from the 16 mesocosms, as seen in Table 1.

We performed an analysis of variance (ANOVA) that used CT_{max} as the dependent variable, predator cues and herbicide as categorical factors (including the interaction of these factors), and mesocosm nested within the interaction of predator cues and herbicide (i.e. mesocosm nested within treatment). Given that tadpole mass was not correlated with CT_{max} (see results), we did not include it as a covariate. No data transformations were required for this analysis.

Determining the thermal performance curves for tadpole burst speed

Locomotor performance, measured as a TPC, is considered to be a proxy of maximum physiological performance and has been used to estimate optimum temperatures in amphibians [45,46]. We obtained TPCs by measuring each tadpole's maximal burst swimming speed (i.e. burst speed) across a range of temperatures. To determine burst speed, tadpoles were placed individually in a portable thermal bath (patent license ES 2372085), which consists of an opened cross section methacrylate tube (1 m long ×6 cm wide ×3 cm deep) filled with water of a given temperature. We then gently prodded the tadpole with a thin stick to stimulate swimming. Each trial was recorded using a digital camera (30 frames/s) positioned above the tube (JVC Everio GZ-MG505). TPCs were defined using a set of six temperatures (20°, 24°, 28°, 32°, 35° and 38°C). This set includes temperatures tadpoles experienced in the mesocosms (20°-32°C) and two more (35° and 38°C) which they might be exposed to in a scenario of increasing environmental temperatures (but lower than their critical thermal maximum). Temperatures were tested in a random order and, for each temperature, tadpoles from the four treatments were tested in the same session; therefore, all treatments had the same temperature order. Prior to swimming, tadpoles were held individually in 250-ml containers at the test temperature for approximately 1 hr. A different set of tadpoles (total N = 570) was used for each temperature (Table 2) and each wading pool was represented equally in each set.

After the tadpole started to move, we used the software Measurement in Motion [47] to estimate burst speed over three frames (0.1 s) by measuring the distance the center of mass moved between frames [48,49]. After conducting at least three bouts, we used the fastest speed measured for a given tadpole as our measure of that individual's burst speed. Since maximal swimming speed may scale with body size [45] and body size may confound the effect of speed on escape success [50], we used size-corrected burst speed (using tadpole total length) when constructing TPCs.

To describe the TPCs for burst speed, we used the Template Mode of Variation method (TMV, [51]) which employs a polynomial function to decompose variation among TPCs into three predetermined modes of variation with biological connotation: vertical shift (faster-slower), horizontal shift (hotter-colder), and specialist-generalist trade-offs ([31]; see [51] and supporting information for details on calculations). Since we tested tadpole performance at six temperatures, we assumed that the common template curve was a fourth-degree polynomial, as in previous studies (e.g., [46]). Making this assumption avoids inadequately describing TPCs, which can happen when using a lower-order polynomial [51,52].

In addition to using the TMV method, we also calculated maximum performance (z_{max}) to evaluate changes in maximum swimming speed at the optimum temperature and a more traditional measurement of performance breadth to confirm specialist-generalist trade-offs (using B_{95}, which is the range of temperatures at which performance values exceed 95% of the maximum;[53]). We used B_{95} instead of the traditional B_{80} because the lower limit of B_{80} would fall below 20°C, which is outside the tested range of temperatures. All computations regarding the TPCs, except for B_{95}, were made using the Matlab code by R. Izem (available online in the appendix of [51]). We also confirmed the fit of each treatment's curve and calculated standard error (SE) of each curve's parameters using nlinfit and nlparci functions, respectively, in Matlab [54].

We conducted an ANOVA that used burst speed as the dependent variable, temperature, predator cues and the herbicide (including the interaction of these factors) as categorical factors and, mesocosm nested within the interaction of predator cues and herbicide (i.e. mesocosm nested within treatment). ANOVA analysis was followed by a Tukey post-hoc test.

Assessing the morphology of the tadpoles

After the swimming trials, we determined the mass and developmental stage of each tested tadpole. We then took lateral photos of each tadpole and digitized the images for morphometric measurements. We captured the shape of tadpoles by digitizing 10 landmarks and 15 semi-landmarks (see supporting information; see also [49,55]) on each tadpole using tpsDig2 software [56]. We then extracted partial warps and the uniform component with tpsRelw software [57], which we used as our shape variables in a subsequent analysis. We visualized variation in landmark positions using the thin-plate spline approach (transformation grids, [58] in MorphJ [59]. As an alternative approach to quantify tadpole morphology, we also took the following linear measurements of each tadpole: total tadpole length (TTL, distance between snout and tip of tail fin), body length (BL, distance between snout and point where bottom edge of tail muscle meets body), body depth

Table 1. Critical thermal maximum (CT_{max}), sample size (N) and body mass (Mass) of *Hyla versicolor* tadpoles, in four treatments.

Treatment	N	CT_{max} (°C±SE)	Mass (mg±SE)
Control	13	41.78±0.1	483.7±22.9
Predator	13	42.14±0.1	520.4±29.3
Roundup	15	41.76±0.1	545.4±28.0
Predator + Roundup	15	42.17±0.1	489.8±34.2

Tested tadpoles are representative of the four mesocosms used for each treatment.

Table 2. Experimental temperatures, sample size (N), total tadpole length in mm (TTL±SE), and maximum swimming speeds in cm/s (mean ±SE) for gray treefrog tadpoles that were exposed to predator cues and the herbicide Roundup.

Temp.	Control			Predator			Roundup			Predator + Roundup		
	N	TTL	Speed	N	TTL	Speed	N	TTL	Speed	N	TTL	Speed
20°C	22	37.3±0.6	39.2±1.0	24	38.2±0.6	41.1±1.1	23	38.8±0.6	41.0±1.4	24	37.9±0.7	40.2±1.3
24°C	24	38.1±0.5	41.3±1.1	24	41.0±0.5	46.9±1.5	23	38.8±0.5	44.0±1.3	23	38.0±0.6	44.0±1.3
28°C	24	39.9±0.6	45.4±1.5	24	42.1±0.5	52.3±1.7	24	41.8±0.5	50.7±1.8	24	39.7±0.6	50.3±1.7
32°C	24	39.4±0.6	46.7±1.3	24	39.7±0.5	52.5±1.2	24	40.2±0.6	50.2±1.4	24	39.1±0.6	52.5±1.1
35°C	24	39.6±0.6	45.8±1.7	24	40.8±0.5	51.5±1.6	24	40.6±0.5	47.1±1.7	24	40.4±0.6	50.1±1.8
38°C	24	37.9±0.5	40.2±1.8	24	37.6±0.6	44.6±1.6	24	39.1±0.6	41.5±2.1	23	36.2±0.6	42.3±2.1

(BD, deepest point of the body), tail length (TL, distance between point where bottom edge of tail muscle meets body and tip of tail fin), muscle depth (MD, deepest point of the muscle) and tail depth (TD, maximum depth of the tail fin).

We conducted canonical correlation analysis as a dimension-reducing procedure to obtain two morphological indices (i.e. a linear combination of shape variables); one was for the linear measurements (MI_{lin}) and the other was for the partial warps and uniform component (MI_{geo}). We then examined these two indices for correlations with burst speed (across all treatments; see [55]). To determine if predator cues, herbicide, and their interaction influenced tadpole size (i.e. centroid) or shape (MI_{lin} or MI_{geo}), we performed three ANOVAs followed by Tukey HSD post-hoc tests; mesocosms were nested within the interaction of predator cues and herbicide (i.e. mesocosm nested within treatment). Shape variables (MI_{lin} and MI_{geo}) and tadpole size (centroid) were then used as continuous predictors, along with temperature, predator cues and herbicide as a categorical predictors, in two ANCOVA analysis (testing either MI_{lin} or MI_{geo} separately), to evaluate their effects on burst speed. We performed all analyses using Matlab [54], except when mentioned otherwise, and used a significance level of $\alpha = 0.05$.

All experiments were approved by the University of Pittsburgh's Institutional Animal Care and Use Committee (Protocol #12050451).

Results

Critical thermal maxima of the tadpoles

In our analysis of CT_{max}, there were no differences among mesocosms within a given treatment. We found an effect of predator cues but no effect of the herbicide or the interaction of both (Table 3). Averaged across herbicide treatments, tadpoles exposed to predators had a CT_{max} that was 0.4°C higher than tadpoles not exposed to predators (Table 1). CT_{max} was not correlated with tadpole mass (Pearson's $R = -0.17$, $p = 0.22$).

Thermal performance curves for tadpole burst speed

When we test tadpole swimming ability across different water temperatures, we found that swimming burst speed varied with temperature (Table 2). When we used the TMV method on size-corrected performance data, we obtained both a common template curve, which provided a good approximation of the common shape of each treatment's curve (Fig. 1), and a three-parameter shape-invariant model (with the use of a fourth-degree polynomial), which explained over 99% of the variation for swimming speed. Decomposition of the total variation into the three pre-determined directions of variation reveals that TPCs for swimming speed vary mostly in the specialist-generalist (53.27%) direction and the vertical (45.98%) direction, but very little in the horizontal (0.59%) direction. This indicates that tadpoles in the control treatment had a wider swimming TPC than tadpoles exposed to predator cues or the herbicide, even when comparing more traditional measures of curve width (B_{95}; Table 4, Fig. 2). Thus, most of the variation in the TPCs is due to specialist-generalist trade-offs and differences in overall performance (faster-slower), rather than changes in T_{opt} (hotter-colder). Indeed, tadpoles raised in the herbicide treatment exhibited only a small decrease in T_{opt} (−0.4°C) while tadpoles raised with predator cues exhibited an increase in T_{opt} (1.5°C). Tadpoles raised with both predators and herbicide exhibited a T_{opt} that was intermediate in magnitude between the latter two treatments but still higher (0.5°C) than tadpoles raised in the control treatment. The only significant difference in T_{opt} was between tadpoles exposed only to

Table 3. ANOVA using CT_{max} as dependent variable, predator cues and Roundup as categorical factors (including the interaction of these factors) and, mesocosm nested within the interaction of predator cues and Roundup, for *Hyla versicolor.*

	SS	d.f.	MS	F	p
Predator	1.993	1	1.993	14.9	<0.001
Roundup	0.006	1	0.006	0.04	0.834
Predator*Roundup	0.009	1	0.009	0.06	0.801
Mesocosm (Predator*Roundup)	1.329	12	0.111	0.83	0.622
Error	5.350	40	0.134		

Univariate tests of significance for CT_{max}. In this model, we used Sigma-restricted parameterization and Type III sum of squares.

herbicide and those exposed only to predator cues (1.8°C; 2-tailed *t*-test, $p<0.05$). Maximal performance (z_{max}) was marginally correlated with performance breadth (Pearson's $R=-0.95$, $p=0.051$).

Temperature and predator cues both influenced burst speed (Table 5). There was also a significant interaction between predator cues and herbicide. Tadpoles in the control treatment had slower burst speeds accross all temperatures than tadpoles in the other three treatments (all $p<0.05$). Tadpoles raised in the predator treatment were also faster than those from herbicide treatment ($p<0.05$). Furthermore, tadpoles in all treatments containing predator cues or herbicide had higher maximum performance (z_{max}) than tadpoles in the control treatment, so that their burst speed at the optimum temperature was higher than the burst speed of tadpoles raised without any cues. These differences in the parameters of the TPCs can be seen as changes in the overall shape of the curves (Fig. 2). Our analysis of burst speed also revealed a significant effect of mesocosms (nested within treatment), however the magnitude of this effect was much smaller than in other effects, such as the interaction of predator cues and herbicide (Table 5). Nevertheless, we checked for burst speed differences among tanks of the same treatment and temperature and we found no significant effect of mesocosm on burst speed, in any of the treatment-temperature combinations (all $p>0.05$).

Induced morphology of the tadpoles

We observed size and shape changes in tadpoles exposed to the herbicide and predator cue treatments (Fig. 3). Predator cues and herbicide had no main effects on tadpole centroid size (Table 6a) but they did have a significant interaction; tadpoles exposed to predator cues + herbicide were smaller than those exposed only to the herbicide or only to the predator cues (both $p<0.05$). Similarly, tadpoles in the control treatment were smaller than those exposed only to the herbicide or only to the predator cues (both $p<0.05$). For geometric morphometric measurements, both predator cues and herbicide influenced tadpole shape (Table 6b) and there was a significant interaction between the two factors. Tadpoles raised in the control treatment differed from those raised in the other three treatments (all $p<0.05$), however these did not differ amongst themselves. For linear measurements, only predator cues significantly influenced shape of tadpoles (Table 6c). Tadpoles raised in predator or predator + herbicide treatment differed from those raised in herbicide or control treatments (all $p<0.05$). Mesocosm effect on either centroid or shape (MI_{lin} or MI_{geo}) was non-significant (Table 6). Overall, compared to tadpoles in the control,

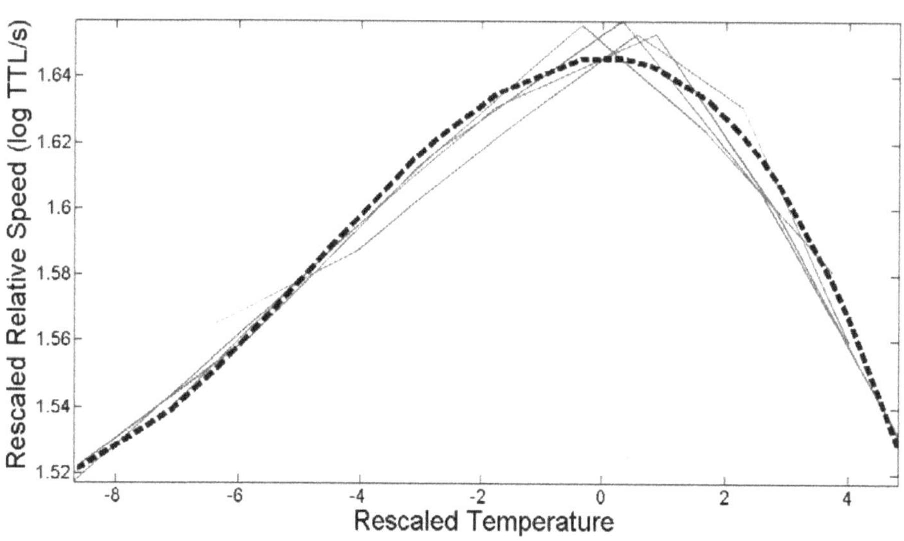

Figure 1. Rescaled thermal performance curves for swimming speed in each treatment with fitted common template shape. Common template shape z(T) is represented by a dashed line nad the treatments by solid lines. Each thermal performance curve of a treatment (i) and temperature were standardized with respect to the estimates of height (h), location (m; T_{opt}), and width (w) parameters from the fit to model. Rescaled optimum temperature $T_{opt}=0$. (see [46,51]). Swimming $z(T) = 1.6458-0.004T^2-0.00023982T^3+0.000003493T^4$.

Table 4. Parameters of thermal performance curves for maximum swimming speed in four treatments, for *Hyla versicolor*, estimated with TMV method (Izem and Kingsolver, 2005) and nlinfit/nlparci functions in Matlab (Mathworks, 2009).

Treatment	TMV parameters						nlinfit/nlparci		
	$h^{¥}$	T_{opt}	w	z_{max} 1	z_{max} 2	B_{95}	h±SE	T_{opt}±SE	w±SE
Control	0.12	31.05	1.74	1.07	11.65	18.36	0.12±0.10	31.07±0.76	1.73±0.19
Predator	−0.01	32.52	1.47	1.11	12.96	14.29	−0.01±0.09	32.53±0.52	1.47±0.13
Roundup	0.00	30.70	1.52	1.09	12.27	14.97	0.00±0.09	30.71±0.65	1.52±0.14
Predator + Roundup	−0.11	31.58	1.34	1.12	13.13	12.35	−0.11±0.09	31.58±0.51	1.33±0.11

$h^{¥}$, height (log TTL/s); T_{opt}, optimal temperature (°C); w, width (dimensionless); z_{max} 1 (TPC), maximum performance (log TTL/s); z_{max} 2 (TTL/s), maximum performance (TTL/s); B_{95}, thermal performance breadth (°C).

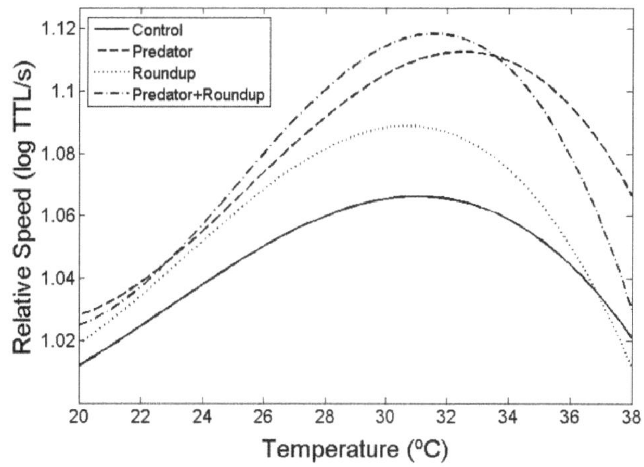

Figure 2. Overall shape of the thermal performance curves for each of the four induction treatments. Each treatment is represented by a thermal performance curve for tadpole swimming speed: control - solid line, predator - dashed line, Roundup - dotted line and predator+Roundup - dash-dot line.

tadpoles in the other three treatments exhibited relatively shorter bodies. Furthermore, in the two treatments containing predator cues, tadpoles exhibited an increase in their relative tail length and tail depth (Fig. 3). Apart from temperature and predator cues, burst speed was also influenced by tadpole's size, either when using morphometric geometric data (Table 7a) or linear measurements (Table 7b). We also found a significant effect of shape on burst speed when using geometric morphometric data (Table 7a).

Discussion

We discovered that predator cues and the herbicide Roundup can affect the thermal physiology of *Hyla versicolor* tadpoles. Predator cues induced tadpoles to have CT_{max} values that were 0.4°C higher whereas the herbicide had no effect. Predator cues and Roundup also influenced the shape of the thermal performance curves, resulting in changes in optimum temperature, performance breadth and maximal performance (Fig. 2). Furthermore, predator cues also induced morphological changes that increased the tadpoles' burst speed.

Roundup, a glyphosate based broad-spectrum systemic herbicide, did not have any effect on CT_{max} estimates of tadpoles. However there have been reports of other contaminants affecting the thermal physiology of vertebrates. Among insecticides, for example, endosulfan (an organochlorine insecticide that affects the central nervous system) and chlorpyrifos (an organophosphate insecticide that inhibits acetylcholinesterase) are known to decrease CT_{max} in fishes [60]. Other environmental contaminants, such as cadmium and copper, can adversely affect the ability of fish to withstand high temperature stress [61,62]. Whether all of these observations in fishes can be extrapolated to other species of aquatic organisms, such as tadpoles, is yet to be determined. Based on these studies and our own results, it seems that the effects of pesticides on CT_{max} may depend on the type of pesticide, the concentration of the pesticide, and how it affects the organism (i.e. its mode of action). There is the possibility that using higher concentrations of the herbicide might induce a decrease in CT_{max}, but higher concentrations will cause tadpole death [63]. Furthermore, the herbicide also did not interfere with the increase in CT_{max} induced by predator cues; tadpoles exposed to predator

Table 5. ANOVA using burst speed as dependent variable, and temperature, mesocosm, predator cues and Roundup as categorical predictors, with mesocosm nested within the interaction of predator cues and Roundup, for *Hyla versicolor*.

	SS	d.f.	MS	F	p
Temperature	0.891	5	0.178	32.17	<0.001
Predator	0.106	1	0.106	19.16	<0.001
Roundup	0.002	1	0.002	0.38	0.537
Predator*Roundup	0.070	1	0.070	12.65	<0.001
Mesocosm (Predator*Roundup)	0.127	12	0.010	1.92	0.03
Predator*Temperature	0.023	5	0.005	0.83	0.528
Roundup*Temperature	0.017	5	0.003	0.62	0.683
Predator*Roundup*Temperature	0.009	5	0.002	0.33	0.903
Error	3.085	546	0.006		

Univariate tests of significance for burst speed. We used Sigma-restricted parameterization and Type III sum of squares.

cues + herbicide had similar CT_{max} values to those exposed only to predator cues.

Different methodological protocols and biological sources can affect estimates of upper thermal tolerances (see [44,64]). For example, the ramping rate used [65,66,67,68], the selection of end-point [23], variations in previous thermal acclimation [40], ontogenetic stage [43], time of day, and photoperiod [69] all may promote shifts in amphibian upper thermal tolerances. We discovered that predatory cues can also affect CT_{max} estimates of prey. An increase in thermal tolerance of predator-induced tadpoles would cause an increase in their warming tolerance, which is the difference between CT_{max} and maximum temperature of the environment to which an ectotherm is exposed [70,71]. This means that tadpoles exposed to predator cues would be less susceptible to acute thermal stress than tadpoles that were not exposed to predator cues. In contrast, an exposure to the herbicide, at least at the concentration used in our study, would not affect the warming tolerance of tadpoles.

An exposure to predator cues and the herbicide had interactive effects on tadpole burst speed. The interaction occurred because the herbicide alone and predator cues alone each increased burst speed compared to the control, but the combination of the

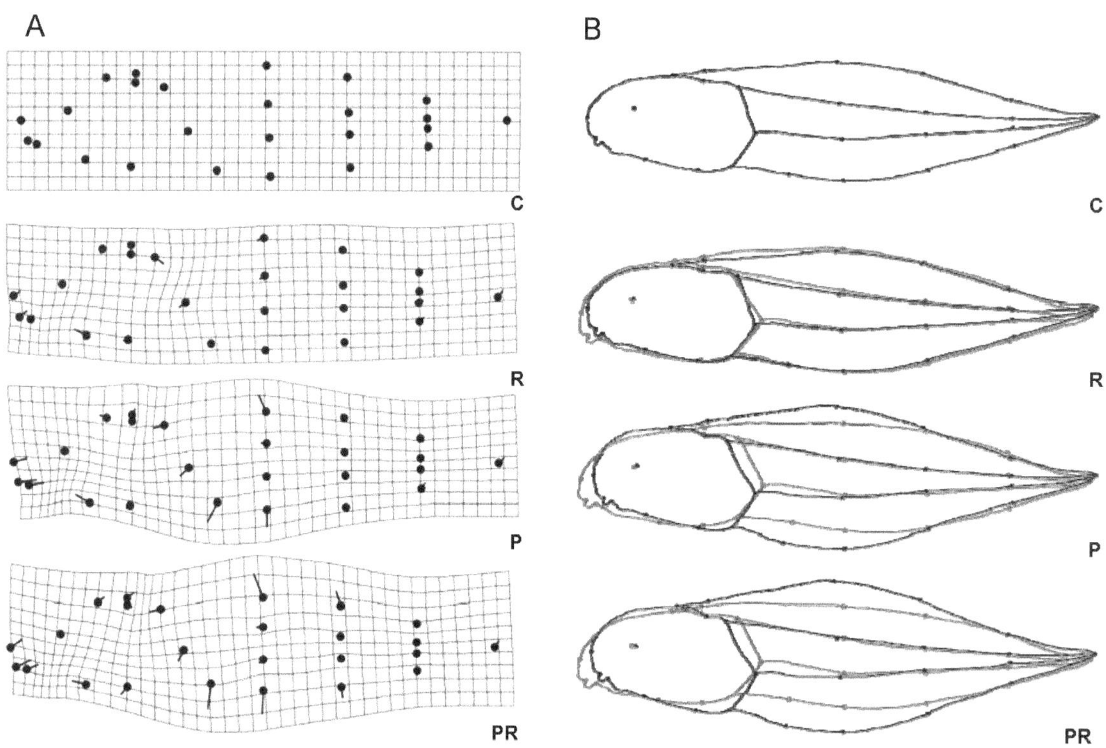

Figure 3. Transformation grids with landmarks and warped outline drawings for each treatment's tadpole shape. a) Transformation grids with landmarks (black dots) and vectors showing direction of variation; b) comparison of warped outline drawings for each treatment shape (black) and control shape (grey). Transformation grids and warped outline drawings were magnified (x5) to better illustrate the differences. C – Control, R – Roundup, P – Predator and PR – Predator + Roundup.

Table 6. ANOVAs to determine if predator cues and Roundup (including their interaction) influenced size (a; centroid), or shape (b and c) of tadpoles (MI_{geo}, for geometric morphometric measurements, or MI_{lin}, for linear measurements, respectively) with mesocosm nested within the interaction of predator cues and Roundup (i.e. mesocosm nested within treatment).

a) Centroid (size)	SS	d.f.	MS	F	p
Predator	19.9	1	19.91	0.97	0.326
Roundup	4.3	1	4.32	0.21	0.647
Predator*Roundup	521.7	1	521.7	25.38	<0.001
Mesocosm (Predator*Roundup)	423.0	12	35.25	1.72	0.06
Error	11386.2	554	20.55		
b) MI_{geo} (shape)	**SS**	**d.f.**	**MS**	**F**	**p**
Predator	11.77	1	11.766	12.21	<0.001
Roundup	5.17	1	5.172	5.37	0.021
Predator*Roundup	5.68	1	5.684	5.90	0.016
Mesocosm (Predator*Roundup)	12.37	12	1.031	1.07	0.383
Error	533.95	554	0.964		
c) MI_{lin} (shape)	**SS**	**d.f.**	**MS**	**F**	**p**
Predator	29.35	1	29.348	30.95	<0.001
Roundup	3.32	1	3.317	3.50	0.062
Predator*Roundup	2.46	1	2.463	2.60	0.108
Mesocosm (Predator*Roundup)	8.67	12	0.723	0.76	0.690
Error	525.32	554	0.948		

We used Sigma-restricted parameterization and Type III (Effective hypothesis) sum of squares.

herbicide and predator cues induced an increase that was not larger than predator cues alone.Therefore, since the combination of the herbicide and predators cues was not additive, in the presence of predator cues, exposure to the herbicide caused no change in burst speed.

The presence of either predator cues or the herbicide narrowed the performance breadth of the TPC while increasing maximal performance. As performance breadth is negatively correlated with maximal performance, we would expect a generalist-specialist trade-off. Tadpoles from a treatment which induced a more

Table 7. ANCOVA analysis using burst speed as dependent variable, shape variables MI_{geo} (a) or MI_{lin} (b) and tadpole size (centroid) as continuous predictors, alongside temperature, predator cues and Roundup as categorical predictors.

a)	SS	d.f.	MS	F	p
Predator	0.068	1	0.068	15.01	<0.001
Roundup	0.000	1	0.000	0.01	0.909
Temperature	0.197	5	0.039	8.66	<0.001
Size (Centroid)	0.167	1	0.167	36.70	<0.001
Shape (MI_{geo})	0.129	1	0.129	28.27	<0.001
Predator*Roundup	0.011	1	0.011	2.34	0.127
Error	2.544	559	0.005		
b)	**SS**	**d.f.**	**MS**	**F**	**p**
Predator	0.101	1	0.101	21.24	<0.001
Roundup	0.003	1	0.003	0.59	0.443
Temperature	0.507	5	0.101	21.30	<0.001
Size (Centroid)	0.410	1	0.410	86.09	<0.001
Shape (MI_{lin})	0.011	1	0.011	2.37	0.124
Predator*Roundup	0.012	1	0.012	2.45	0.118
Error	2.661	559	0.005		

Univariate tests of significance for burst speed. In both models, we used Sigma-restricted parameterization and Type III (Effective hypothesis) sum of squares.

specialist curve (as demonstrated by predator cues + herbicide) would perform better at the optimum temperature but gradually decrease in performance, as moving away from the optimum temperature, until reaching a point were tadpoles from a treatment which induced a more generalist curve (as demonstrated by control) would outperform them (see [31,51]; Table 4). However, we do not see a decline in performance at the extremes of the thermal performance curve, at the tested temperatures, as a result of this trade-off. This observation is confirmed by thermal tolerance data where none of the tadpoles raised in any of the treatments with predator cues or the herbicide had lower CT_{max} than those from the control treatment. Instead, it appears the expected decline in sub-optimal performance resulting from a generalist-specialist trade-off is compensated by the increase in overall performance, so that tadpoles raised in the control treatment always perform, on average, worse than herbicide- or predator-induced tadpoles, at least at the tested temperatures. Therefore, when comparing thermal performance curves, the resulting increase in overall performance was asymmetric, being greater around the optimum temperature and lower at the extreme temperatures.

Surprisingly, predator cues and the herbicide also produced changes in the optimum temperature, but in opposite directions. Of course, the small decrease in optimum temperature caused by the herbicide (0.4°C) may have little or no biological relevance. In contrast, the increase in optimum temperature promoted by predator cues (approximately 1.5°C) may be important, especially when new assessments suggested that environmental impacts will require smaller degrees of global warming than previously thought [72]. Since predator cues increase optimum temperature, the difference between optimum temperature and the environmental temperature should also increase (i.e. thermal safety margins (TSM); see [70]), which would be beneficial to the tadpoles in the current scenario of increasing global temperatures.

Previous studies have demonstrated that changes in the shape or position of thermal performance curves can occur due to acclimation (e.g., [73,74,75]) or that thermal performance curves of different locomotor strategies for the same organism can have different shapes (e.g., [46,76]). In the present study, we demonstrate that the presence of sublethal concentrations of an herbicide and cues from predators can also produce changes in the thermal performance curves and therefore affect how tadpoles respond to environmental temperature changes.

Although it has been documented that predators can affect the behavioral thermoregulation of their prey (e.g., [25]), to our knowledge this is the first study to demonstrate a predator altering the thermal physiology of their prey by increasing CT_{max}, increasing the optimum temperature, and producing changes in the thermal performance curves. It has also been demonstrated that Roundup's lethality increases with competition stress [35] and that predator cues can improve tadpole survival when tadpoles are exposed to the herbicide under stratified water conditions [17]. Therefore, one could make the argument that acclimation to predator cues might be beneficial under warmer temperatures. However, we should also keep in mind that predation simultaneously has a negative effect on tadpole populations and can select for particular phenotypes (see [9]). To display a predator-induced phenotype, tadpoles need to detect chemical cues that are released when other tadpoles (particularly conspecifics) are consumed. So, the possible positive effects of predator cues on the thermal physiology, in a global warming scenario, would only be beneficial for those phenotypes that survive predation.

Predator cues in our study induced morphology changes (relative smaller bodies, deeper tails and deeper tail muscle) that were similar to those observed in previous studies (e.g., [77]). These morphological changes likely explain why tadpoles exposed to predator cues swam faster than control tadpoles. Exposure to the herbicide (see figure 3) induced relative smaller bodies, and the observed changes partially resembled the predator-induced phenotype (see also [17]). The induction of relatively deeper tadpole tails by the herbicide was less evident in the current work than in the study of Relyea [17]. However, this may be due to a number of differences in the experimental protocol including the duration of exposure and a substantially different experimental venue.

Predator cues and the herbicide caused interactive effects on tadpole size. Tadpoles exposed to predator cues + herbicide were smaller than those exposed only to the herbicide or only to predator cues. Tadpoles raised in the control treatment also tended to be smaller than those exposed only to the herbicide or only to predator cues. This may explain why tadpoles from the herbicide treatment also swam faster than tadpoles from the control treatment. As a result, all three treatments had better overall swimming performance than in control, with increase in burst speed related to the magnitude of morphology change (more induction, higher performance) and size. Furthermore, predator-induced morphology changes can be reversed if cues are removed [41]. As a result, some of the changes in the thermal performance curve may also be reversible. If so, in the absence of cues, the predator- and herbicide-induced TPC shapes would revert back to the original curve (i.e. the control curve).

The mechanism underlying the ability of the herbicide to induce morphological changes in tadpoles is still unknown. It has been suggested that the herbicide may be interfering with the stress hormones that induce anti-predator defenses [78] or that herbicides and predator cues activate shared endocrinological pathways [17]. We have demonstrated that predator cues and the herbicide can affect the thermal physiology of tadpoles, although not all changes occur in the same direction. However, the mechanisms behind these thermal physiology changes are also unknown, with possible scenarios arising from our results: a) herbicide interferes only with the stress hormones that induce anti-predator defenses; b) they do not share the same physiological pathways, or at least not all of them; c) they both activate shared endocrinological pathways but predator cues also indirectly activate temperature-stress response mechanisms; or d) stress response mechanisms are more general than previous thought and predator-induced stress produces similar physiological responses as temperature-induced stress.

Conclusions

Apart from inducing morphology changes, predator cues promoted an increase in CT_{max} and optimum temperature of *Hyla versicolor* tadpoles. As such, in the presence of predators, we can expect tadpoles to have greater warming tolerance and broader thermal safety margins. These changes might indirectly help tadpoles cope with increasing environmental temperatures. The herbicide Roundup is not only toxic to amphibians (and lethal over certain concentrations), but it also produces changes in morphology [17]. With this work, we now know that it also interferes, to some extent, with the thermal physiology of tadpoles (in particular in the thermal performance curves), although the effect on warming tolerance and thermal safety margins appears to be marginal. However, Roundup is just one of hundreds of chemicals currently used in anthropogenic activities (e.g., agriculture) and tadpoles can face predation by a wide variety of predator species. Because combinations of pesticides, which are a common situation in natural environments, can have greater impacts than each pesticide alone [79], future studies should test whether combinations of pesticides and predators could have different effects on the thermal physiology of organisms.

In the current scenario of climate change, it is important that we understand the physiological mechanisms underlying tolerance to abiotic stress [80,81] and the sensitivity of organisms to changes in the environment [80,82]. However, it also is important that we understand the indirect effects of physiological responses (in particular thermal physiology) on species interactions, such as predation, competition and disease transmission [2]. Therefore, understanding the plasticity of thermal performance curves and thermal limits (CT_{max} and CT_{min}) and how these parameters are altered by environmental stressors may be critical to understanding how physiological variation can influence a species' response to climate change [83].

Supporting Information

Methods S1 Appendices 1–4. Appendix 1, Detailed information on laboratory conditions for rearing tadpoles during acclimation for the experiments. Appendix 2, Description of method and apparatus used for measuring CT_{max}. Appendix 3, TMV method equation for calculating thermal performance curve's parameters. Appendix 4, Description of the side-view landmarks and semi-landmarks, and linear measurements in a hypothetical tadpole.

Raw Data S1 CT_{max} and swimming performance database.

TPS Data S1 Geometric morphometry database.

Acknowledgments

Our thanks to Aaron Stoler, Jessica Hua, Will Brogan, R.J. Bendis, Keri Simonette, Zach Zbinden, Lindsey Freed, Patrick Noyes, Beverly French, Chris Hensley, Elena Valdés, Sara Tripodi and Casilda Maldonado for their assistance with the experiment. We also thank the anonymous reviewers for comments on previous versions of this manuscript.

Author Contributions

Conceived and designed the experiments: MK JH RAR. Performed the experiments: MK JH. Analyzed the data: MK CC. Contributed reagents/materials/analysis tools: MK MT RAR. Wrote the paper: MK JH HD MT CC RAR.

References

1. Kearney M, Porter W (2009) Mechanistic niche modelling: combining physiological and spatial data to predict species' ranges. Ecol Lett 12: 334–350.
2. Helmuth B, Broitman BR, Yamane L, Gilman SE, Mach K, et al. (2010) Organismal climatology: analyzing environmental variability at scales relevant to physiological stress. J Exp Biol 213: 995–1003.
3. Chown SL, Terblanche JS (2007) Physiological diversity in insects: ecological and evolutionary contexts. Adv In Insect Phys 33: 50–152.
4. Pörtner HO, Farrell AP (2008) Physiology and climate change. Science 322: 690–692.
5. Hoffmann AA (2010) Physiological climatic limits in Drosophila: patterns and implications. J Exp Biol 213: 870–880.
6. Chown SL, Hoffmann AA, Kristensen TN, Angilletta Jr MJ, Stenseth NC, et al. (2010) Adapting to climate change: a perspective from evolutionary physiology. Clim Res 43: 3–15.
7. Ghalambor CK, McKay JK, Carroll SP, Reznick DN (2007) Adaptive versus non-adaptive phenotypic plasticity and the potential for contemporary adaptation in new environments. Funct Ecol 21: 394–407.
8. Arnqvist G, Johansson F (1998) Ontogenic reaction norms of predator- induced defensive morphology in dragonfly larvae. Ecology 79: 1847–1858.
9. Relyea RA (2002) The many faces of predation: how induction, selection, and thinning combine to alter prey phenotypes. Ecology 83: 1953–1964.
10. Domenici P, Turesson H, Brodersen J, Brönmark C (2008) Predator-induced morphology enhances escape locomotion in crucian carp. Proc R Soc B Biol Sci 275: 195–201.
11. Brönmark C, Lakowitz T, Hollander J (2011) Predator-induced morphological plasticity across local populations of a freshwater snail. PLoS One 6: e21773.
12. Poupardin R, Reynaud S, Strode C, Ranson H, Vontas J, et al. (2008) Cross-induction of detoxification genes by environmental xenobiotics and insecticides in the mosquito Aedes aegypti: impact on larval tolerance to chemical insecticides. Insect Biochem Mol Biol 38: 540–551.
13. Hua J, Morehouse N, Relyea RA (2013) Pesticide resistance in amphibians: Induced resistance in susceptible populations, constitutive tolerance in tolerant populations. Evol Appl 6: 1028–1040.
14. Hanazato T (1991) Pesticides as chemical agents inducing helmet formation in Daphnia ambigua. Freshw Biol 26: 419–424.
15. Barry MJ (1998) Endosulfan-enhanced crest induction in Daphnia longicephala: evidence for cholinergic innervation of kairomone receptors. J Plankton Res 20: 1219–1231.
16. Oda S, Kato Y, Watanabe H, Tatarazako N, Iguchi T (2011) Morphological changes in Daphnia galeata induced by a crustacean terpenoid hormone and its analog. Environ Toxicol Chem 30: 232–238.
17. Relyea RA (2012) New effects of Roundup on amphibians: predators reduce herbicide mortality; herbicides induce antipredator morphology. Ecol Appl 22: 634–647.
18. Hanazato T (1999) Anthropogenic chemicals (insecticides) disturb natural organic chemical communication in the plankton community. Environ Pollut 105: 137–142.
19. Barry MJ (1999) The effects of a pesticide on inducible phenotypic plasticity in Daphnia. Environ Pollut 104: 217–224.
20. Barry MJ (2000) Effects of endosulfan on Chaoborus-induced life-history shifts and morphological defenses in Daphnia pulex. J Plankton Res 22: 1705–1718.
21. Sakamoto M, Chang K-H, Hanazato T (2006) Inhibition of development of anti-predator morphology in the small cladoceran Bosmina by an insecticide: impact of an anthropogenic chemical on prey-predator interactions. Freshw Biol 51: 1974–1983.
22. Hutchison VH (1961) Critical thermal maxima in salamanders. Physiol Zool 34: 92–125.
23. Lutterschmidt WI, Hutchison VH (1997) The critical thermal maximum: history and critique. Can J Zool 75: 1561–1574.
24. Heath AG, Cech JJ, Brink L, Moberg P, Zinkl JG (1997) Physiological responses of fathead minnow larvae to rice pesticides. Ecotoxicol Environ Saf 37: 280–288.
25. Angilletta Jr MJ (2009) Thermal Adaptation: A Theoretical and Empirical Synthesis. Oxford: Oxford University Press, USA. 320 p.
26. Van Buskirk J (2002) A comparative test of the adaptive plasticity hypothesis: relationships between habitat and phenotype in anuran larvae. Am Nat 160: 87–102.
27. Miner BG, Sultan SE, Morgan SG, Padilla DK, Relyea RA (2005) Ecological consequences of phenotypic plasticity. Trends Ecol Evol 20: 685–692.
28. Relyea RA (2007) Getting out alive: how predators affect the decision to metamorphose. Oecologia 152: 389–400.
29. Hayes TB, Falso P, Gallipeau S, Stice M (2010) The cause of global amphibian declines: a developmental endocrinologist's perspective. J Exp Biol 213: 921–933.
30. Wagner N, Reichenbecher W, Teichmann H, Tappeser B, Lötters S (2013) Questions concerning the potential impact of glyphosate-based herbicides on amphibians. Environ Toxicol Chem 32: 1688–1700.
31. Huey RB, Kingsolver JG (1989) Evolution of thermal sensitivity of ectotherm performance. Trends Ecol Evol 4: 131–135.
32. Van Buskirk J, McCollum SA (2000) Influence of tail shape on tadpole swimming performance. J Exp Biol 203: 2149–2158.
33. Hopkins WA, Winne CT, DuRant SE (2005) Differential swimming performance of two natricine snakes exposed to a cholinesterase-inhibiting pesticide. Environ Pollut 133: 531–540.
34. Relyea RA (2005) The impact of insecticides and herbicides on the biodiversity and productivity of aquatic communities. Ecol Appl 15: 618–627.
35. Jones DK, Hammond JI, Relyea RA (2011) Competitive stress can make the herbicide Roundup more deadly to larval amphibians. Environ Toxicol Chem 30: 446–454.
36. Petranka JW, Kats LB, Sih A (1987) Predator-prey interactions among fish and larval amphibians: use of chemical cues to detect predatory fish. Anim Behav 35: 420–425.
37. Kats LB, Petranka JW, Sih A (1988) Antipredator defenses and the persistence of amphibian larvae with fishes. Ecology 69: 1865–1870.
38. Relyea RA, Auld JR (2005) Predator- and competitor-induced plasticity: how changes in foraging morphology affect phenotypic trade-offs. Ecology 86: 1723–1729.

39. Jones DK, Hammond JI, Relyea RA (2010) Roundup and amphibians: the importance of concentration, application time, and stratification. Environ Toxicol Chem 29: 2016–2025.
40. Brattstrom BH (1968) Thermal acclimation in anuran amphibians as a function of latitude and altitude. Comp Biochem Physiol 24: 93–111.
41. Relyea RA (2003) Predators come and predators go: the reversibility of predator-induced traits. Ecology 84: 1840–1848.
42. Gosner KL (1960) A simplified table for staging anuran embryos and larvae with notes on identification. Herpetologica 16: 183–190.
43. Sherman E (1980) Ontogenetic change in thermal tolerance of the toad Bufo woodhousii fowleri. Comp Biochem Physiol Part A 65A: 227–230.
44. Terblanche JS, Hoffmann AA, Mitchell KA, Rako L, le Roux PC, et al. (2011) Ecologically relevant measures of tolerance to potentially lethal temperatures. J Exp Biol 214: 3713–3725.
45. Gvoždík L, Van Damme R (2006) Triturus newts defy the running-swimming dilemma. Evolution 60: 2110–2121.
46. Gvoždík L, Van Damme R (2008) The evolution of thermal performance curves in semi-aquatic newts: Thermal specialists on land and thermal generalists in water? J Therm Biol 33: 395–403.
47. Learning in Motion, Inc (2004) Measurement in Motion software, v3.0.
48. Arendt JD (2009) Influence of sprint speed and body size on predator avoidance in New Mexican spadefoot toads (Spea multiplicata). Oecologia 159: 455–461.
49. Arendt J (2010) Morphological correlates of sprint swimming speed in five species of spadefoot toad tadpoles: Comparison of morphometric methods. J Morphol 271: 1044–1052.
50. Van Damme R, Van Dooren TJ (1999) Absolute versus per unit body length speed of prey as an estimator of vulnerability to predation. Anim Behav 57: 347–352.
51. Izem R, Kingsolver JG (2005) Variation in continuous reaction norms: quantifying directions of biological interest. Am Nat 166: 277–289.
52. David JR, Gibert P, Gravot E, Petavy G, Morin J-P, et al. (1997) Phenotypic plasticity and developmental temperature in Drosophila: analysis and significance of reaction norms of morphometrical traits. J Therm Biol 22: 441–451.
53. Huey RB, Stevenson RD (1979) Integrating thermal physiology and ecology of ectotherms: a discussion of approaches. Am Zool 366: 357–366.
54. The MathWorks, Inc.. (2009) MATLAB v7.8 and Statistics Toolbox Release 2009a. Natick, Massachusetts, United States.
55. Dayton GH, Saenz D, Baum KA, Langerhans RB, DeWitt TJ (2005) Body shape, burst speed and escape behavior of larval anurans. Oikos 111: 582–591.
56. Rohlf FJ (2010) tpsDig2, digitize landmarks and outlines, version 2.16. Department of Ecology and Evolution, State University of New York at Stony Brook.
57. Rohlf FJ (2010) tpsRelw, relative warps analysis, version 1.49. Department of Ecology and Evolution, State University of New York at Stony Brook.
58. Bookstein FL (1991) Morphometric tools for landmark data. Cambridge/New York/Port Chester/Melbourne/Sydney: Cambridge University Press.
59. Klingenberg CP (2011) MorphoJ: an integrated software package for geometric morphometrics. Mol Ecol Resour 11: 353–357.
60. Patra RW, Chapman JC, Lim RP, Gehrke PC (2007) The effects of three organic chemicals on the upper thermal tolerances of four freshwater fishes. Environ Toxicol Chem 26: 1454–1459.
61. Carrier R, Beitinger TL (1988) Reduction in thermal tolerance of Notropis lutrensis and Pimephales promelas exposed to cadmium. Water Res 22: 511–515.
62. Lydy MJ, Wissing TE (1988) Effect of sublethal concentrations of copper on the critical thermal maxima (CTMax) of the fantail (Etheostoma flabellare) and johnny (E. nigrum) darters. Aquat Toxicol 12: 311–321.
63. Relyea RA, Jones DK (2009) The toxicity of Roundup Original Max to 13 species of larval amphibians. Environ Toxicol Chem 28: 2004–2008.
64. Navas CA, Gomes FR, Carvalho JE (2008) Thermal relationships and exercise physiology in anuran amphibians: integration and evolutionary implications. Comp Biochem Physiol Part A 151: 344–362.
65. Terblanche JS, Deere JA, Clusella-Trullas S, Janion C, Chown SL (2007) Critical thermal limits depend on methodological context. Proc R Soc London B 274: 2935–2942.
66. Chown SL, Jumbam KR, Sørensen JG, Terblanche JS (2009) Phenotypic variance, plasticity and heritability estimates of critical thermal limits depend on methodological context. Funct Ecol 23: 133–140.
67. Mitchell KA, Hoffmann AA (2010) Thermal ramping rate influences evolutionary potential and species differences for upper thermal limits in Drosophila. Funct Ecol 24: 694–700.
68. Rezende EL, Tejedo M, Santos M (2011) Estimating the adaptive potential of critical thermal limits: methodological problems and evolutionary implications. Funct Ecol 25: 111–121.
69. Mahoney JJ, Hutchison VH (1969) Photoperiod acclimation and 24-hour variations in the critical thermal maxima of a tropical and a temperate frog. Oecologia 2: 143–161.
70. Deutsch CA, Tewksbury JJ, Huey RB, Sheldon KS, Ghalambor CK, et al. (2008) Impacts of climate warming on terrestrial ectotherms across latitude. Proc Natl Acad Sci 105: 6668–6672.
71. Duarte H, Tejedo M, Katzenberger M, Marangoni F, Baldo D, et al. (2012) Can amphibians take the heat? Vulnerability to climate warming in subtropical and temperate larval amphibian communities. Glob Chang Biol 18: 412–421.
72. Smith JB, Schneider SH, Oppenheimer M, Yohe GW, Hare W, et al. (2009) Assessing dangerous climate change through an update of the Intergovernmental Panel on Climate Change (IPCC) "reasons for concern." Proc Natl Acad Sci 106: 4133.
73. Kingsolver JG, Huey RB (1998) Evolutionary analyses of morphological and physiological plasticity in thermally variable environments. Am Zool 545: 1–16.
74. Condon CH, Chenoweth SF, Wilson RS (2010) Zebrafish take their cue from temperature but not photoperiod for the seasonal plasticity of thermal performance. J Exp Biol 213: 3705–3709.
75. Lachenicht MW, Clusella-Trullas S, Boardman L, Le Roux C, Terblanche JS (2010) Effects of acclimation temperature on thermal tolerance, locomotion performance and respiratory metabolism in Acheta domesticus L. (Orthoptera: Gryllidae). J Insect Physiol 56: 822–830.
76. Gvoždík L, Puky M, Šugerková M (2007) Acclimation is beneficial at extreme test temperatures in the Danube crested newt, Triturus dobrogicus (Caudata, Salamandridae). Biol J Linn Soc 90: 627–636.
77. Relyea RA (2001) Morphological and behavioral plasticity of larval anurans in response to different predators. Ecology 82: 523–540.
78. Glennemeier KA, Denver RJ (2002) Small changes in whole-body corticosterone content affect larval Rana pipiens fitness components. Gen Comp Endocrinol 127: 16–25.
79. Relyea RA (2009) A cocktail of contaminants: how mixtures of pesticides at low concentrations affect aquatic communities. Oecologia 159: 363–376.
80. Gilman SE, Wethey DS, Helmuth B (2006) Variation in the sensitivity of organismal body temperature to climate change over local and geographic scales. Proc Natl Acad Sci 103: 9560–9565.
81. Tewksbury JJ, Huey RB, Deutsch CA (2008) Putting the Heat on Tropical Animals. Science (80-) 320: 1296–1297.
82. Porter WP, Gates DM (1969) Thermodynamic equilibria of animals with environment. Ecol Monogr 39: 227–244.
83. Buckley LB (2008) Linking traits to energetics and population dynamics to predict lizard ranges in changing environments. Am Nat 171: E1–E19.

7

Factors Affecting the Design of Slow Release Formulations of Herbicides Based on Clay-Surfactant Systems. A Methodological Approach

María del Carmen Galán-Jiménez[1], Yael-Golda Mishael[2], Shlomo Nir[2], Esmeralda Morillo[1], Tomás Undabeytia[1]*

1 Agrochemistry and Soil Conservation Department, Spanish Research Council-Institute of Natural Resources and Agrobiology, Seville, Spain, **2** The Seagram Center for Soil and Water Sciences, The Robert H. Smith Faculty of Agriculture, Food and Environment, Hebrew University of Jerusalem, Rehovot, Israel

Abstract

A search for clay-surfactant based formulations with high percentage of the active ingredient, which can yield slow release of active molecules is described. The active ingredients were the herbicides metribuzin (MZ), mesotrione (MS) and flurtamone (FL), whose solubilities were examined in the presence of four commercial surfactants; (i) neutral: two berols (B048, B266) and an alkylpolyglucoside (AG6202); (ii) cationic: an ethoxylated amine (ET/15). Significant percent of active ingredient (*a.i.*) in the clay/surfactant/herbicide formulations could be achieved only when most of the surfactant was added as micelles. MZ and FL were well solubilized by berols, whereas MS by ET/15. Sorption of surfactants on the clay mineral sepiolite occurred mostly by sorption of micelles, and the loadings exceeded the CEC. Higher loadings were determined for B266 and ET/15. The sorption of surfactants was modeled by using the Langmuir-Scatchard equation which permitted the determination of binding coefficients that could be used for further predictions of the sorbed amounts of surfactants under a wide range of clay/surfactant ratios. A possibility was tested of designing clay-surfactant based formulations of certain herbicides by assuming the same ratio between herbicides and surfactants in the formulations as for herbicides incorporated in micelles in solution. Calculations indicated that satisfactory FL formulations could not be synthesized. The experimental fractions of herbicides in the formulations were in agreement with the predicted ones for MS and MZ. The validity of this approach was confirmed in *in vitro* release tests that showed a slowing down of the release of *a.i.* from the designed formulations relative to the technical products. Soil dissipation studies with MS formulations also showed improved bioactivity of the clay-surfactant formulation relative to the commercial one. This methodological approach can be extended to other clay-surfactant systems for encapsulation and slow release of target molecules of interest.

Editor: John Vontas, University of Crete, Greece

Funding: This research was supported by the projects CTM2009-07425 and PRI-PIBAR-1393 from the Spanish Ministry of Economy and Competitiveness, and Junta de Andalucía Project P09-RNM4581. These projects received funding by the European Social Fund. M.C. Galán-Jiménez acknowledges a research contract from Junta de Andalucía in the frame of the project P09-RNM4581. The funders had no role in study design, data collection and analysis, decision to publish, or preparation of the manuscript.

Competing Interests: The authors have the following interest. Campi y Jove S.A. (Barcelona, Spain) supplied the commercial surfactants for this study. Syngenta Spain provided the authors with some samples of the commercial formulation of MS (Callisto, 10% w/V). There are no further patents, products in development or marketed products to declare. This does not alter the authors' adherence to all the PLOS ONE policies on sharing data and materials, as detailed online in the guide for authors.

* E-mail: undabeyt@irnase.csic.es

Introduction

Clay minerals are widely used in numerous industrial applications as fillers in production of paper, plastics; heat insulators in building industry; nacreous pigments in cosmetics; corrosion proof agents in coatings; anti-caking agents in chemical industry; UV-, heat-stable and under-water agents in paintings, etc. [1]

Recently there is a growing use of clays in delivery technology in order to confer slow release properties of the active ingredient, which is of particular interest in pharmaceutical and agrochemical industry. Levis and Deasy [2] showed prolonged release of the antihyperthensive propranolol from loaded halloysite tubes. Park et al. [3] intercalated donepezil into several layered clays and the formulated products showed a biphasic release pattern with an initial burst effect followed by a slow release of the drug. Clay-polymer complexes were also designed for slowing down the release pattern of the drug docetaxel [4]. The polymer improved the mechanical and rheological properties of the formulations [5].

Several clay-based approaches for slow release have been used for pesticides, e.g., modification of the clay surface from hydrophilic to hydrophobic by preadsorbing organic cations and, therefore, enhancing the affinity for hydropobic herbicides [6–8]. Other strategies involve trapping of the herbicide by coagulation of delaminated clay particles [9]; sorption on thermally treated-clay [10]; intercalation of copolymers of the pesticide [11]; encapsulation into a polymer-clay [12] and clay-gel [13] matrix, etc.

In general, clay minerals are incorporated in crop protection products as rheological modifiers to give large stability over the products lifespan, thus avoiding syneresis. It has been previously reported that incorporation of active ingredients in the micelles and vesicles formed by surfactant molecules which adsorbed on clays, such as montmorillonite, provided a slow release system [14–16]. The fraction of the pesticide encapsulated in surfactant-clay complexes prolonged their action. Clearly the encapsulated fraction and the released pattern depend on the chemical properties of both the surfactant and the pesticide, of the type of clay mineral and of the range of concentrations used.

In the current work, we test the applicability of a methodological approach for the design of slow release formulations based on clay-surfactant. The following elements were considered for formulation optimization: (i) determination of the solubility of the active ingredient in surfactant systems; (ii) determination of the binding affinity of the surfactants to the clay surface; (iii) optimization of surfactant sorption at several clay/surfactant ratios; and (iv) estimating the amount of active ingredient sorbed for that clay/surfactant ratio. We tested the approximation that the same proportionality between the active ingredient included in micelles and the surfactant in solution, could be retained in the presence of clay. The applicability of this approach was examined for three herbicides (MS, MZ and FL) and four surfactants employed in commercial formulations (neutral: an alkypolyglucoside, two ethoxylated alcohols and one cationic ethoxylated amine). The clay mineral used was sepiolite, which can form weak gels at very low concentrations yielding good viscosity properties [17].

Materials and Methods

Materials

Metribuzin [4-amino-6-ter-butyl-3-methylthio-1,2,4-triazin-5(4H)-one] (MZ), mesotrione [2-(4-mesyl-2-nitrobenzoyl)cyclohexane-1,3-dione] (MS) and flurtamone [(2RS)-5-methylamino-2-phenyl-4-(α,α,α-trifluoro-m-tolyl)furan-3(2H)-one] (FL) were purchased from Sigma-Aldrich Co., St Louis, MO). The commercial formulation of MS (Callisto, 10% w/V) was a gift from Syngenta Spain. The surfactants AG6202, Ethomeen T/15 (ET/15), Berol 266 (B266) and Berol 048 (B048) were kindly provided by Akzo Nobel Co. (Amsterdam, The Netherlands). HPLC grade-methanol was purchased from Sigma-Aldrich (Sigma Chemical Co., St Louis, MO); H_3PO_4 (85% w:v) and HPLC grade-acetonitrile was obtained from Teknokroma S.A. (Barcelona, Spain). Sepiolite (Pangel S9) was obtained from Tolsa S.A. (Madrid, Spain). Sepiolite is a hydrated Mg phylosilicate with fibrous morphology formed by blocks and cavities (tunnels) growing in the direction of the fibers [18]. Each structural block is composed of a central Mg sheet sandwiched between two tetrahedral silica sheets. Silanol groups are covering the external surface which is accessible to reagents and acting as neutral sorption sites. In addition, certain isomorphic substitutions in the tetrahedral sheets are responsible for the cationic exchange capacity of this clay (0.15 $cmol_c\ kg^{-1}$).

Figure 1 shows the sepiolite structure and the structural formulas of the herbicides and surfactants. Tables 1 and 2 list properties of the surfactants and herbicides, respectively.

The upper horizon of a clay soil from Lebrija (Sevilla, Spain) with a 1.1% organic matter content and a basic pH (7.6) was collected and passed through a 2 mm sieve before use.

Surface tension

Surface tension (γ) measurements were performed with a LAUDA TD 3 tensiometer using Du Nuoy ring detachment method. The ring was cleaned with ethanol and flamed after every measurement. For each measurement at least five readings were taken and the mean γ value was recorded. Before each experiment the instrument was calibrated and checked by measuring the surface tension of distilled water. Surface tension measurements of ET/15 solutions were performed in the presence of 0.01 M NaCl as a background electrolyte. The minimum area occupied by a surfactant molecule at the air/solution interface, a_{min}, was estimated from its maximum surface excess concentration (Γ_{max}) as determined from the Gibbs adsorption equation.

Solubility studies

10 mL of surfactant solutions at concentrations ranging up to 20 g/L were added in duplicate to a suspension of each herbicide beyond its solubility limit, and the suspensions were shaken for 1 week at 25°C. Then the suspensions were let to settle down; the supernatant was removed and filtered through 0.20 μm PTFE membranes, and the herbicide analyzed. The solubility curves were constructed by plotting the amount of herbicide solubilized versus the amount of surfactant employed.

A solubility enhancement factor of the herbicides by the use of surfactants was determined from S_w*/S_w, where S_w* is the apparent solubility at a 20 g/L surfactant concentration and S_w is the intrinsic herbicide solubility in water. S_w is 0.41 mM for MS, 5.13 mM for MZ and 0.02 mM for FL.

Sorption studies

The adsorption of surfactants on the clay was carried out in glass vials in duplicate by mixing 15 ml of surfactant solutions whose concentrations were up to 12 g/l with 24 mg of sepiolite. The clay concentration in the vials was 1.6 g/L. After shaking for 24 h at 20°C, the suspensions were centrifuged at 12000 g for 10 min and the supernatants removed. Kinetic experiments showed that equilibrium was reached within 24 h shaking of clay-surfactant suspensions. Preliminary experiments with only surfactant solutions were performed to determine the operating conditions for centrifugation by avoiding sedimentation of the surfactant molecules. The pellets were dry-frozen and surfactant concentration in the pellets was determined by elemental C analysis.

Surfactant sorption was modeled by using the Langmuir-Scatchard equation:

$$\frac{Lo - L}{L} = \frac{R_0 * K}{1 + K * L}$$

Where $L_0$and L denote the molar concentration of total and free surfactant, respectively, $R_0$is the molar concentration of sorption sites and K is the binding coefficient.

Sorption of the herbicides was analogously performed on sepiolite. Clay concentration was again 1.6 g/L. The added herbicide concentrations were up to 20 mg L^{-1} for MZ and MS, and up to 5 mg L^{-1} for FL.

Sorption-desorption isotherms were fitted to the logarithmic form of the Freundlich equation:

$$\log C_s = \log K_f + n \log C_e$$

were C_s (mg g^{-1}) is the amount of adsorbed herbicide, C_e (mg mL^{-1}) is the equilibrium concentration of herbicide, and K_f and n are constants that characterize the relative sorption capacity and sorption intensity, respectively.

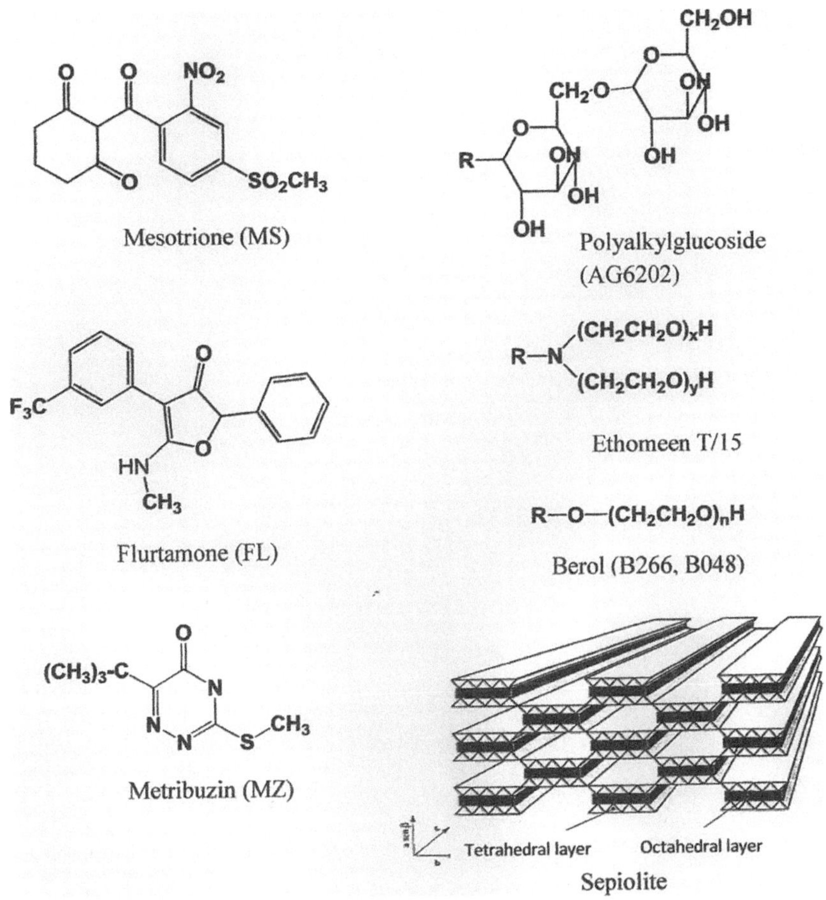

Figure 1. Sepiolite structure and structural formulas of the herbicides and surfactants.

Zeta potential measurements

The surfactant-clay composites were redispersed at a concentration of 1.6 g/l. The samples were allowed to equilibrate for 1 h and few milliliters of dispersion were measured. The temperature of the samples was 25±1°C, ξ was measured using a Zetasizer Nanosystem (Malvern Instruments, Southborough, MA). The ξ value was deduced from the mobility of the particles using the Smoluchowski equation.

Table 1. Surface and chemical properties of the surfactants.

	B048	B266	ET/15	AG6202
M.W.	640	400	480	370
R	C_{13}	C_{10}	C_{17}	C_8
n. ethoxyl units	10	5.5	5	-
HLB[§]	13.5	12.1	9.2	11.2
cmc (mM)[§]	0.16	0.075	0.041	37.83
a_{min} (nm^2)[f]	0.64	0.50	0.19	0.20

[§]HLB, hydrophilic-lipophilic balance; cmc, critical micellar concentration.
[f]Determined from surface tension measurements.

Preparation of herbicide formulations

Herbicide-clay-surfactant formulations were prepared by dissolving several amounts of the herbicides in surfactant solutions,

Table 2. Physico-chemical and environmental properties of the herbicides.[a]

	FL	MZ	MS
M.W.	333.30	214.29	339.32
S_w (at 20°C) (mg L^{-1})	10.7	1165	160
Log Kow[b]	3.2	1.65	0.11
pKa	–	0.99	3.12
Freundlich parameters			
K_f	0.66–7.19	0.018–1.9	0.33–4.48
1/n	0.9	0.89–1.52	–
DT_{50} (days)	48–211	5.3–17.7	32
GUS leaching index	2.59	2.57	3.43
Mobility	Transition state	Transition state	High leachability

a. From the IUPAC-Pesticide properties data base (http://sitem.herts.ac.uk/aeru/iupac/).
b. Kow, octanol-water partition coefficient at pH 7 and 20°C.

which were later added to sepiolite. The concentrations of herbicides and surfactant were determined from solubility diagrams and the corresponding surfactant sorption on sepiolite. Optimization of herbicide in the formulation considered both the enhanced solubility of the herbicide and sorption of the complexes formed in solution on the clay surface. The employed concentrations are listed in Table 3. The calculations of *a.i.* (w/w%) in Table 3 were as follows. As an approximation to the amount of herbicides in micelles of the surfactant, the solution concentrations of herbicides in the absence of surfactants were subtracted from their enhanced amounts in the presence of surfactants. The adsorbed amounts of surfactants were calculated by means of the Langmuir equation. The amount of a herbicide in the formulations surfactant/herbicide/clay was calculated by multiplying its amount in the micelles by the fraction of adsorbed micelles. After shaking for 24 h the suspensions were centrifuged at 12000 g for 10 min, and the pellets were dry-frozen yielding the herbicide formulations. 10 mg of these formulations were extracted with 20 mL of methanol to determine their active ingredient (*a.i.*) content. A nomenclature for herbicide-clay-surfactant formulations was introduced (see Table 3) where the first two letters indicate the herbicide (MS: mesotrione; MZ: metribuzin) followed by letters indicating the surfactant used, the first number denotes the surfactant concentration, the second one gives the initial herbicide concentration and the third one the clay concentration used.

Studies of release in water

Release tests of technical grade herbicide and selected clay-based formulations were performed in duplicate with a rotating paddle apparatus (Sotax). For each formulation 5 mg of herbicide were added to 1 L of deionized water at 25°C and stirred at 50 rpm. At appropriate time intervals from 0 to 96 h, samples were taken, immediately passed through PTFE filters and the herbicide was analyzed.

Soil dissipation

MS dissipation was carried out in pot experiments with the commercial formulation and one of the clay-surfactant-herbicide formulations (MSET/15 12/23/5). Seeds of sunflower (*Helianthus annuus*) were cultivated under hydroponic conditions. Then, 10 seedlings were transplanted after appearing the first true leaf to pots of 10 cm diameter filled with 0.32 kg of the clay soil. The formulations were sprayed on the soil at the recommended dose ($100\ g\ ha^{-1}$), and at ½- and 2-fold of that. After 10 days in a climatic chamber, the bleaching intensity was obtained by measuring the chlorophyll content and determining the inhibition percentage in relation to controls (no herbicide was added). The chlorophyll content was determined by cutting the fresh shoot of plants from the starting of the first true leaf and extracted with 12 mL of *N,N*-dimethylformamide. The solutions were incubated for 48 h, and the chlorophyll (*a* and *b*) content was measured by vis spectroscopy at 664 and 647 nm and related to the weight of the extracted amount. The experiments were done in triplicate for each formulation and dose.

Herbicide analysis

Herbicides were analyzed by HPLC (Shimadzu Model 10A) equipped with a PDA detector. The reverse phase column was a 15 cm Kromasil 100 C18. The flow rate was $1.0\ mL\ min^{-1}$. The mobile phase was 40% acetonitrile and 60% water containing 0.1% H_3PO_4. The wavelengths were set at 220 nm for FL, 254 nm for MS and 230 nm for MZ. The retention times were 15.19, 3.05 and 2.86 min for FL, MS and MZ, respectively.

Results and Discussion

Herbicide solubility

Solubility enhancement of the herbicides (Table 4, Figure 2) was strongly influenced by their polarity and chemical structure as well as by their compatibility with the surfactant molecules (Figure 1). The data in Table 4 showed that MS and FL were appreciably solubilized by ET/15, whereas in the case of berols (B048, B266), the largest solubility enhancement factors were obtained for FL. In contrast, the solubility factors of MZ were low in all surfactants used and ranged from 1.09 to 2.02.

B266 solubilized the more hydrophobic herbicide FL relative to B048 and ET/15, which did not follow the HLB values of the surfactants (Table 1) and revealing other factors are influencing the solubilizing capacity of the surfactants. Probably, interactions of the herbicide molecules with the polar groups of the surfactants play an important role. Ethoxyl units are strongly hydrated and may form water bridges with polar moieties of the herbicide molecules, resulting in an enhanced solubilization. This mechanism is of less importance for very hydrophobic herbicides. Therefore, the enlarged solubilization of FL by B266 was due to its lower number of ethoxyl units, facilitating FL penetration and interaction with the micellar core.

Table 3. Clay-surfactant formulations.

Herbicide	Surfactant, conc. (g/L)	Herbicide added (mM)	Clay conc. (g/L)	Notation of formulation	Calculated *a. i.* (w/w%)	Experimental *a. i.* (w/w%)
Mesotrione	Berol 048, 0.07	0.69	0.5	MSB048 0.07/0.69/0.5	3.9	6.0±0.4
	Berol 266, 1	0.48	1	MSB266 1/0.48/1	1.9	2.7±0.2
	ET/15, 12	23.00	5	MSET 12/23/5	19.6	15.8±0.8
	ET/15, 12	23.00	15	MSET 12/23/15	18.8	12.5±0.4
	ET/15, 12	23.00	20	MSET 12/23/20	17.7	11.1±0.3
Metribuzin	Berol 048, 3	6.44	1.6	MZB048 3/6.44/1.6	2.4	2.8±0.1
	Berol 266, 5	7.55	1.6	MZB266 5/7.55/1.6	3.2	4.6±0.1
	ET/15, 3	6.29	3	MZET 3/6.29/3	2.3	3.9±0.1
	ET/15, 5	6.95	5	MZET 5/6.95/5	3.8	3.2±0.1
	ET15/20	10	20	MZET 20/10/20	1.8	4.5±0.1

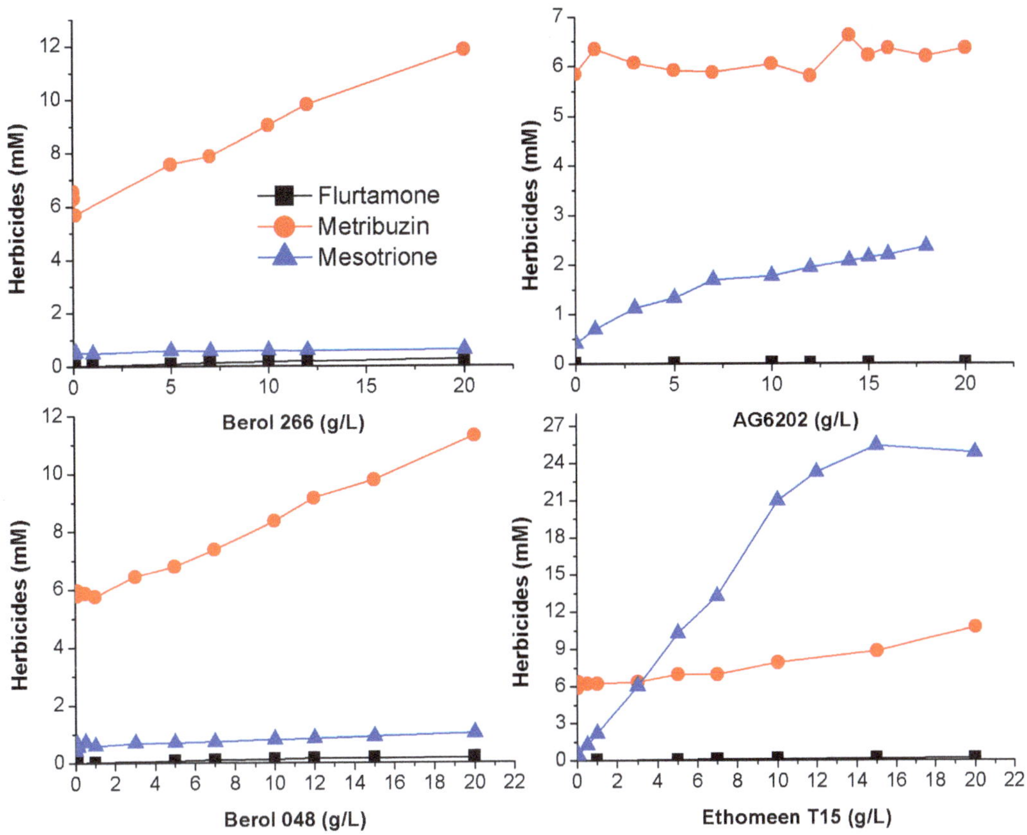

Figure 2. Phase solubility diagrams of the herbicides in the presence of surfactants.

When the phase solubility diagrams of the herbicides in the presence of surfactants were examined (Figure 2), MZ solubility increased with increasing concentration of the two berols from 0 to 20 g/L, the total amount dissolved was respectively increased from 6 up to 12 mM. MS solubility was greatly enhanced in the presence of ET/15 from 1 to 25 mM when increasing concentration of ET/15 from 0 to 20 g/L. This large solubilization was mainly driven by electrostatic interactions. MS is a weak acid (pKa = 3.12) remaining mostly as anionic species at the equilibrium pH (5.4) in the solubility studies, whereas the surfactant molecules of ET/15 were positively charged (pKa = 8.5). Finally, phase solubility of FL was almot not affected by any surfactant, the total amount dissolved was very small. These values seem to be contradictory to those in Table 4. For example, the solubility enhancement factors for FL with berols were larger than those for MZ. However, the solubility enhancement factors were calculated relative to the intrinsic solubility of the herbicides. FL has a very small solubility (10.7 mg/L), therefore the total amount of herbicide in solution dissolved for a fixed concentration of berols was always smaller than with MZ, which has a high value of water solubility (1165 mg/L). In the design of clay-based formulations, it is desirable to have (i) a large ratio between the amount of herbicide dissolved and surfactant used, (ii) a high probability of sorption of the herbicide-surfactant complexes to the clay and, (iii) a large amount of the active ingredient (*a.i.*) in the formulations. Consequently, the amount of MZ in solution dissolved for the same amount of surfactant will be much larger than with FL, and also the probability of sorption of the surfactant-herbicide complexes, yielding formulations with larger *a.i.*

Table 4. Solubility enhancement factor for the herbicide-surfactant systems.

	MS	MZ	FL
B048	2.40	1.93	18.9
B266	1.29	2.02	22.5
ET/15	59.0	1.84	18.1
AG6202	5.90	1.09	1.15

In general, enhanced solubility of the herbicides by the surfactants occurred through incorporation of the herbicide molecules into the micelles formed in solution, as no solubilization was noticed for surfactant concentrations below their cmc, with the exception of AG6202, which showed a monotonic increase in the solubilized amount of mesotrione at concentrations below its high cmc (14 g/L). This enhanced MS solubilization may be due to the formation of large premicellar aggregates that have been often observed [19,20], with which the herbicide would be interacting.

Surfactant sorption

The sorption capacity of surfactants by the clay mineral was affected by the type of surfactant (Figure 3). A larger sorption was observed with ET/15, followed by berol surfactants, the lowest adsorption was obtained for polyalkylglucoside AG6202. Higher loading of the surfactant AG6202 on sepiolite was not possible by increasing the total amount since its adsorption was limited due to precipitation during centrifuging at higher concentrations. Consequently, this surfactant was discarded for preparation of herbicide formulations because of its low sorption on the clay and its generally poor efficacy for enhanced solubility of herbicides. The sorption of the surfactants on the sepiolite was well-fitted to Freunlich equation with the exception of B266 (Table 5). The comparison between K_f values was only allowed for B048 and ET/15 due to its similar n values, indicating a higher loading and thus affinity for ET/15 on sepiolite over B048.

ET/15 molecules were protonated at equilibrium pH, so that they were first sorbed on sepiolite by ion exchange with Mg as reported for quaternary and primary amines. For higher loads adsorption was followed by combination of this electrostatic mechanism with chain-chain interactions [21,22]. Li et al [23] showed formation of admicelles in the sorption of hexadecyltrimethylammonium on the external surface of fibrous clay minerals, resulting in a positively charged surface. Zeta potential was examined for the ET/15-sepiolite complexes formed during ET/15 sorption onto sepiolite (Figure 4). Charge reversal was observed for added concentrations of 1.25 mol/kg; the zeta potential was close to the p.z.c. and slightly positive, and the sorbed amount of ET/15 was 1.2-fold larger than the CEC of the sepiolite. These results indicate an initial sorption of the surfactant by cationic exchange and then, the sorbed monomers of the surfactant serve as nuclei for the development of surface aggregates (admicelles), in an arrangement where a large fraction of the positively charged headgroups is not located at external surfaces.

In the case of non-ionic surfactants, their sorption on clay minerals has been reported to occur through hydrogen bonding with the oxygen atoms of the silicate surface, bonding to the interlayer cations through a water bridge, direct coordination or ion-dipole interaction with the exchangeable cations, and Van der Waals interactions between the alkyl chain of the surfactant and clay surface areas with low charge density [24–26]. Shen [27] observed a high sorption of an alkylphenol exthoxylated surfactant on silicon oxide over aluminium and iron oxides, which was due to hydrogen bonding with the strongest Bronsted acid sites of SiO_2. Because of the fibrous structure of sepiolite, sorption of the non-ionic surfactants would be expected to occur through H-bonding to silanol groups covering the surface, similarly to sorption on silicon oxide.

Table 5. Parameters of the Freundlich equation describing the adsoption isotherms of the herbicides and surfactants on sepiolite.

	K_f (mg$^{1-(1/n)}$ mL$^{1/n}$ g^{-1})	1/n	R^2
MS	43.65	0.75	0.95
FL	6.33	0.49	0.98
MZ	4.73	1.12	0.97
B048	346.73	0.10	0.86
B266	295.12	0.20	0.74
ET/15	467.33	0.12	0.99
AG6202	34.67	0.67	0.98

In Figure 3, sorption of B048 on sepiolite reached 0.74 mol kg^{-1} which was close to the amount of neutral sites in sepiolite, 0.6 mol kg^{-1} as determined from grafting reaction of organoclorosilanes on the silanol groups [28]. This value is somewhat larger than that of 0.51 mol kg^{-1} estimated from the analysis of the sorption of Triton X100 on sepiolite by Rytwo et al. [29]. These authors pointed out that geometrical considerations marginally allowed the entry of Triton X100 molecules into the channels of the sepiolite, and there was no cooperativity in the binding of the non-ionic molecules. Albeit B048 poses the same number of ethoxyl units as Triton X100, its larger hydrophobic moiety will modify its sorption pattern as compared to Triton X100. It has been reported that sorption of non-ionic surfactants on hydrophilic surfaces will result in a monolayer from monomer sorption on the surface. In the case of a relatively strong interaction between the surfactant hydrophilic moieties and those of the surface as in silica or oxides, the monomers will adopt a conformation with the headgroup anchored to the surface and the alkyl chain displaced from the surface and protruded into the solution. An increase in the sorption will cause a vertical orientation of the alkyl chains with the further development of aggregate structures (hemimicelles) by interaction between surfactant molecules [30,31]. A calculation of the packing area per molecule of B048 at the sorption plateau yielded 0.60 nm^2, which is quite close to

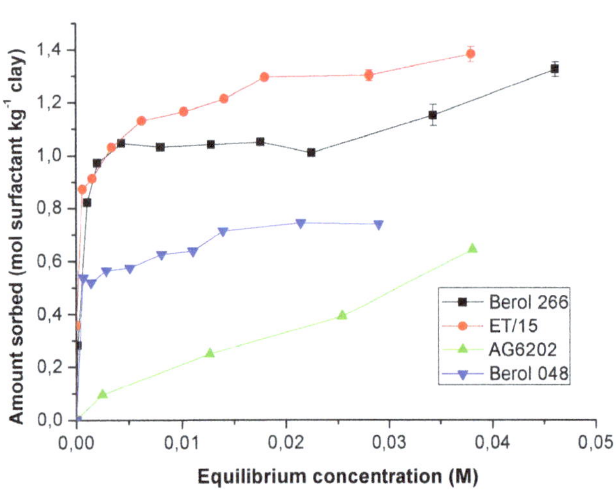

Figure 3. Sorption isotherms of surfactants on sepiolite.

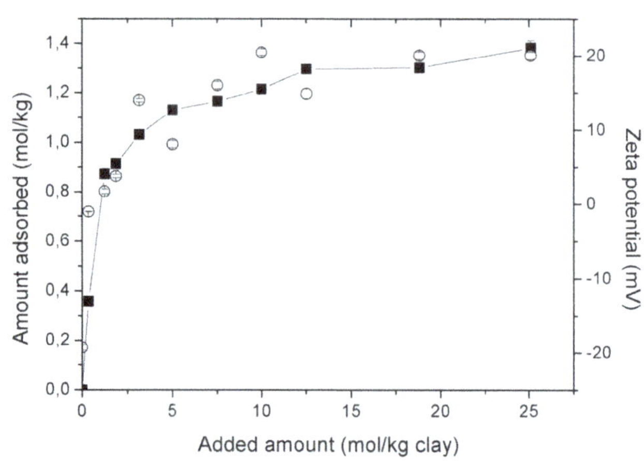

Figure 4. Zeta potential (ζ) of clay-ET/15 complexes as a function of ET/15 sorption on sepiolite. Open and closed symbols denote the zeta potential and sorbed amount, respectively.

a_{min}(Table 1). indicating the development of a monolayer with the molecules standing up with no further aggregation process.

With B266 (Figure 3), a plateau is observed at 1.04 mol kg^{-1}, and the calculation of the packing area in the plateau yielded a similar value (0.45 nm^2) to a_{min}. However, unlike B048, a further increase in the sorption after the plateau is noticed, revealing the formation of surfactant aggregates on the sepiolite surface. Formation of surface aggregates due to lateral interactions between surfactant molecules on the sepiolite surface is energetically more favoured for B266 than for B048, as reflected by its larger hydrophobicity (lower HLB value). Correspondingly, sorption of B266 on sepiolite was larger than that of B048, despite its lower content of ethoxyl groups, and thus, its reduced capability to form H-bonding to surface groups of sepiolite. A more dense packing enhances the energy of interaction between the alkyl chains, which apparently exceeds the interaction energy between the surfactant molecules and the clay surface.

In order to elucidate the mechanisms involved in the sorption of the surfactants on the clay, their sorption was also performed at added concentrations of the surfactant below the cmc for the same clay concentration (1.6 g L^{-1}); the amount of surfactant in solution was determined by surface tension measurements. Whereas in the case of ET/15, all the molecules were sorbed on the clay, exhibiting a H-type isotherm (data not shown), an adsorption plateau was observed both for B266 and B048 (Figure 5), which was several-fold lower than their higher sorption values recorded previously from sorption as micelles (Figure 3). This indicated that the sorption of non-ionic surfactants at relevant concentrations as in Figure 3, was mainly occurring as micelles, which decomposed after interaction with the clay surface, yielding a monolayer of sorbed surfactant. These results were in contrast to those from previous studies by Levitz [32] and Cases and Villieras [33], who reported that the adsorption layer was built from isolated molecules of the surfactant arriving from aqueous solution and not by direct adsorption of micelles. Surfactant sorption as micelles was also observed for the sorption of cationic surfactants on the negatively charged clay mineral montmorillonite [34,35]. These authors showed that at high surfactant/clay ratios micelles were adsorbed without undergoing premature decomposition, whereas in the presence of an excess of clay particles over the surfactant concentration, the rapid sorption of monomers on the clay particles resulted in decomposition of the micelles. Likewise, surfactant loading from monomer sorption was lower than sorption as micelles. The effect of the surfactant/clay ratio in the sorption of surfactants on clays is a critical factor.

Preparation of slow release herbicide formulations

The design of slow release herbicide formulations requires an optimization of two processes: (i) increased solubilization of the herbicide in surfactant solutions; and (ii) high sorption on the clay of the surfactants encapsulating the herbicide molecules. The sorption of surfactants was modeled by using the Langmuir-Scatchard equation. This provided a binding coefficient for each surfactant, that allowed to predict their sorption for several surfactant/clay ratios. The validity of this approach has been extensively proven for other surfactants [16,36]. In a second stage, once the amount of sorbing surfactant was determined for particular surfactant/clay ratios, the maximal amount of herbicide which could be sorbed for the employed surfactant concentration was calculated from the solubility studies by considering the same ratio between the solubilized (included in micelles) herbicide and the surfactant in solution as in the formulation with the clay. This assumption guided the choice of herbicide/clay/surfactant concentrations to be used for the preparation of the formulations.

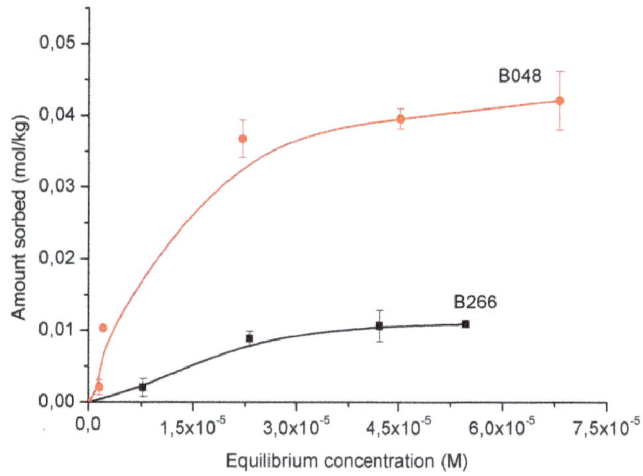

Figure 5. Sorption isotherms of surfactants on sepiolite at added concentrations below their cmc.

As in Nir et al. [37], only one fitting parameter was used in the modeling of the adsorption isotherms by using the Langmuir-Scatchard equation, the binding coefficient *K*In the determination of the binding coefficients, the concentration of adsorption sites of sepiolite (R_0) was fixed for each surfactant to be the largest sorbed amount from Fig. 3. The binding coefficients were 900 M^{-1} for Berol 048, 680 M^{-1} for Berol 266 and 950 M^{-1} for ET/15. The fit was reasonably good, yielding low values of the root mean square error (RMSE), 0.027 mol kg^{-1} for Berol 048, 0.084 mol kg^{-1} for Berol 266 and 0.063 mol kg^{-1} for ET/15.

Table 3 shows an estimation of the *a.i.* content by employing several herbicide/clay/surfactant ratios, which were calculated from the binding coefficients as well as from the solubility diagrams. Formulations were prepared for MS and MZ, but not for FL, since in this case the calculations yielded formulations with very low *a.i.* content (<1% w:w). The low herbicide content of FL formulations arises from its very low dissolved amount versus the amount of surfactant used.

The herbicides' affinity on the clay mineral was very little (Figure 6). The sorption isotherms of the herbicides were also fitted to Freundlich equation (Table 5), and the fitting was pretty good (R^2~0.95–0.98). Again, with the exception of MZ, no comparison in K_f values may be established with reported values because of the large differences in *n* values. MZ was classified as a mobile herbicide with K_f values ranging from 0.018–1.9 mL g^{-1} (Table 2). The experimental K_f value (4.73 mL g^{-1}) on sepiolite was higher but still indicative of poor retention. This value was quite similar for those found in topsoils presenting small sorption, great risk of leaching and fast degradation of MZ: 3.17–3.26 mL g^{-1} [38], 1.49–3.33 mL g^{-1} [39].

Consequently, sorption of the herbicides on the clay occurred mostly through their inclusion in micelles. Overall, the predicted amounts of MS and MZ corroborated reasonably well with the experimental ones (Table 3). The formulations of MS based on ET/15 reached high *a.i.* contents, as much as 15%, whereas smaller contents were obtained for its formulations based on berols or for MZ formulations.

The experimental *a.i.* content of MZ for the clay-surfactant formulations was several-fold lower compared to MS. The lower *a.i.* levels which could be achieved with MZ formulations are due to its relatively large solubility, which implies that a relatively smaller fraction could be included in micelles. Furthermore, table 4

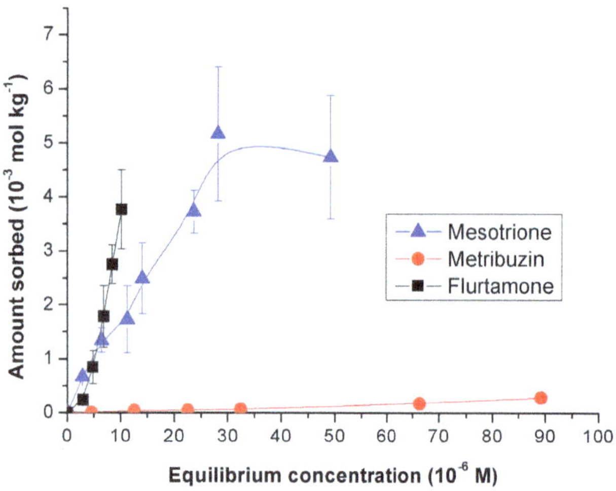

Figure 6. Sorption isotherms of the herbicides on clay.

shows that unlike MS, solubility enhancement of MZ was not large in the presence of the surfactant ET/15.

Overall, MS formulations can be designed for several surfactants at optimal ratios between the concentrations of herbicide, surfactant and clay, thus yielding complexes with high *a.i.* contents that are relevant for agricultural applications. In contrast, clay-based formulations of MZ yielded very low *a.i.* contents for all the surfactants selected in the current study.

Herbicide release in water

Figure 7 shows the cumulative release of MS and MZ from the clay-surfactant formulations as well as for the technical products. The release of herbicides presented an initial burst followed by a more gradual increase in the released amounts. This release pattern was strongly dependent on the herbicide, yielding larger released amounts for MZ in accord with its higher water solubility, compared to MS. Whereas the amount released from the technical product in the first five hours was 90% for MZ, the release for MS was 20% of the total amount. In addition, MS was not completely released during the course of the experiment, unlike MZ.

Clay-surfactant formulations showed a slower release relative to the technical products. The release of *a.i.* from clay-surfactant formulations can be modulated by the choice of the surfactant. The approach used can be extended to other clay-surfactant systems for encapsulation and slow release of target molecules of particular technological interest in agrochemical and pharmaceutical applications.

Soil dissipation

The clay-surfactant formulation MSET/15 12/23/5 was chosen from the clay-surfactant formulation for bioassay tests because of its higher *a.i.* content and slower *in vitro* release. This formulation proved to be more bioactive in pot experiments relative to the commercial formulation (Table 6). This was due to the slower release of the active ingredient from the clay-surfactant formulation that prolonged MS bioactivity and improved weed control. At the field rate, the inhibition percent of the clay-based formulation was 68.9±2.2% amounting approximately to a 20% enhanced bioactivity relative to the commercial formulation, and as bioactive as this one when applied at two-fold the field rate (72.7±5.7%). At two-fold the field rate, the difference in the

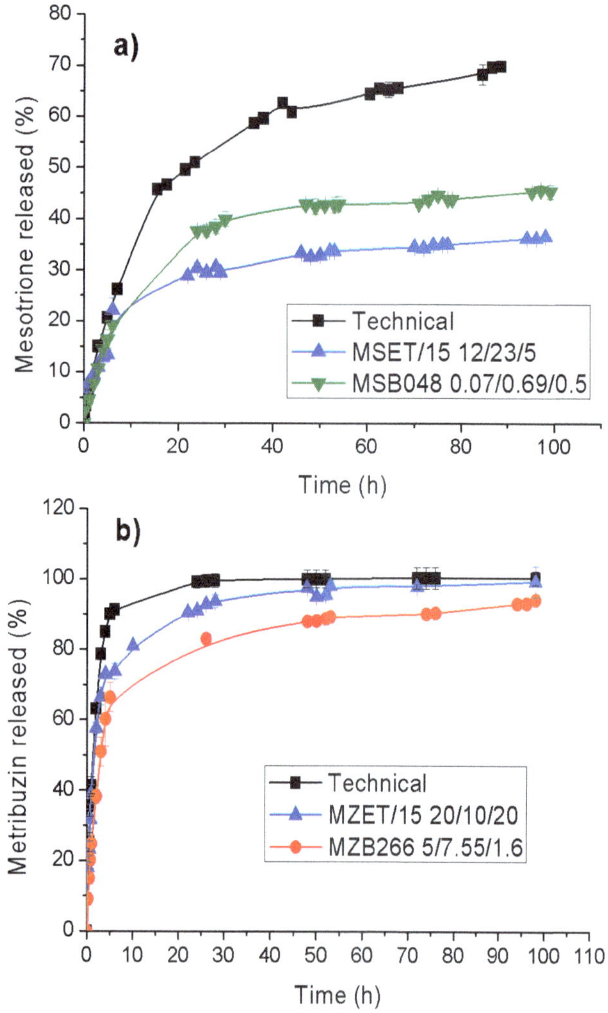

Figure 7. *In-vitro* water release of MS (a) and MZ (b) formulations versus their technical products.

bioactivity of both formulations was reduced, but still higher for the clay-surfactant formulation.

The commercial formulation was inactive at the lowest rate. On the contrary, the clay-surfactant formulation showed 21.9% inhibition. All these results indicate that lower doses may be applied from the clay-surfactant formulations to achieve a good weed control relative to the commercial formulation, with the subsequent reduction in environmental and economic costs.

Table 6. Inhibition percents (as measured by reduction in chlorophyll content relative to a control) of MS formulations applied at several rates in pot experiments.[a]

Rate applied	Commercial formulation	MSET/15 12/23/5
½-field rate	0	21.9±1.1
Field rate	55.3±2.2	68.9±2.2
2-field rate	72.7±5.8	83.0±1.6

The field rate was 100 g ha^{-1}.

Acknowledgments

The authors are grateful to Campi y Jove S.A (Barcelona, Spain) for the supply of commercial surfactants. The authors also acknowledge the Analytical Services of Seville University (CITIUS) for zeta potential (ζ) measurements and elemental analysis.

Author Contributions

Conceived and designed the experiments: TU MCG EM. Performed the experiments: MCG. Analyzed the data: MCG TU SN YM EM. Contributed reagents/materials/analysis tools: TU SN. Wrote the paper: MCG TU SN YM EM.

References

1. Harvey CC, Lagaly G (2006) Chapter 10.1. Conventional applications. *In:* Handbook of Clay Science; Bergaya, Theng & Lagaly (eds.); Elsevier, Amsterdam, The Netherlands; pp. 501–540.
2. Levis SR, Deasy PB (2003) Use of coated microtubular halloysite for the sustained release of diltiazem hydrochloride and propranolol hydrochloride. International Journal of Pharmaceutics 253: 145–157.
3. Park JK, Choy YB, Oh JM, Kima JY, Hwanga SJ, et al. (2008) Controlled release of donepezil intercalated in smectite clays. International Journal of Pharmaceutics 359: 198–204.
4. Si-Shen F, Lin M, Panneerselvan A, Chee-Wee G, Zhou W (2009) Poly(lactide)-vitamin E derivative/montmorillonite nanoparticle formulations for the oral delivery of Docetaxel. Biomaterials 30: 3297–3306.
5. Viseras C, Cerezo P, Sánchez R, Salcedo I, Aguzzi C (2010) Current challenges in clay minerals for drug delivery. Applied Clay Science 48: 291–295.
6. Undabeytia T, Nir S, Rubin B (2000a) Organo-clay formulations of the hydrophobic herbicide norflurazon yield reduced leaching. Journal of Agricultural and Food Chemistry 48: 4767–4773.
7. Undabeytia T, Nir S, Tel Or E, Rubin B (2000b) Photostabilization of the herbicide norflurazon by using organo-clays. Journal of Agricultural and Food Chemistry 48: 4774–4779.
8. El-Nahhal Y, Nir S, Serban C, Rabinovitz O, Rubin B (2001). Organo-clay formulations of acetochlor for reduced movement in soil. Journal of Agricultural and Food Chemistry 49: 5364–5371.
9. Nennemann A, Kulbach S, Lagaly G (2001) Entrapping pesticides by coagulating smectites. Applied Clay Science 18: 285–298.
10. Bojemueller E, Nennemann A, Lagaly G (2001) Enhanced pesticide adsorption by thermally modified bentonites. Applied Clay Science 18: 277–284.
11. Rehab A, Akelah A, El-Gamal MM (2002) Controlled-release systems based on the intercalation of polymeric metribuzin onto montmorillonite. Journal of Polymer Science Part A-Polymer Chemistry 40: 2513–2525.
12. Radian A, Mishael YG (2008) Characterizing and designing polycation-clay nanocomposites as a basis for imazapyr controlled release formulations. Environmental Science and Technology 42: 1511–1516.
13. Maqueda C, Villaverde J, Sopeña F, Undabeytia T, Morillo E (2008) Novel system for reducing leaching of the herbicide metribuzin using clay-gel-based formulations. Journal of Agricultural and Food Chemistry 56: 11941–11946.
14. Mishael YG, Undabeytia T, Rabinovitz O, Rubin B, Nir S (2003) Sulfosulfuron incorporated in micelles adsorbed on montmorillonite for slow release formulations. Journal of Agricultural and Food Chemistry 51: 2253–2259.
15. Undabeytia T, Nir S, Gomara MJ (2004) Clay-vesicle interactions: Fluorescence measurements and structural implications for slow release formulations of herbicides. Langmuir 20: 6605–6610.
16. Sánchez-Verdejo T, Undabeytia T, Nir S, Maqueda C, Morillo E (2008) Environmentally friendly slow release formulations of alachlor based on clay-phosphatidylcholine. Environmental Science and Technology 42: 5779–5784.
17. Maqueda C, Partal P, Villaverde J, Pérez-Rodríguez JL (2009) Characterization of sepiolite-gel-based formulations for controlled release of pesticides. Applied Clay Science 46: 289–295.
18. Ruiz-Hitzky E (2001) Molecular access to intracrystalline tunnels of sepiolite. Journal of Materials Chemistry 11: 86–91.
19. Hou Y, Han Y, Deng M, Xiang J, Wang Y (2010) Aggregation behavior of a tetrameric cationic surfactant in aqueous solution. Langmuir 26: 28–33.
20. Toerne K, Rogers R, Von Wandruska R (2001) Thermal stability of nonionic surfactant aggregates. Langmuir 17: 6119–6121.
21. Sabah E, Turan M, Çelik MS (2002) Adsorption mechanism of cationic surfactants onto acid- and heat-activated sepiolites. Water Research 36: 3957–3964.
22. Akçay G, Akçay M, Yurdakoç K (2005) Removal of 2,4-dichlorophenoxyacetic acid from aqueous solutions by partially characterized organophilic sepiolite: thermodynamic and kinetic calculations. Journal of Colloidal Interface Science 281: 27–32.
23. Li Z, Willms CA, Kniola K (2003) Removal of anionic contaminants using surfactant-modified palygorskite and sepiolite. Clays and Clay Minerals 51: 445–451.
24. Backhaus WK, Klumpp E, Narres H-D, Schwuger MJ (2001) Adsorption of 2,4-dichlorophenol on montmorillonite and silica: Influence of nonionic surfactants. Journal of Colloidal Interface Science 242: 6–13.
25. Sonon LS, Thompson ML (2005) Sorption of a nonionic polyoxyethylene lauryl ether surfactant by 2:1 layer silicates. Clays and Clay Minerals 53: 45–54.
26. Deng Y, Dixon JB, White GN (2006) Bonding mechanisms and conformation of poly(ethylene oxide)-based surfactants in interlayer of smectite. Colloid and Polymer Science 284: 347–356.
27. Shen Y-H (2000) Sorption of non ionic surfactants to soil: the role of soil mineral composition. Chemosphere 41: 711–716.
28. Ruiz-Hitzky E, Fripiat JJ (1976) Organomineral derivatives obtained by reacting organochlorosilanes with the surfaces of silicates in organic solvents. Clays and Clay Minerals 24: 25–30.
29. Rytwo G, Nir S, Margulies L, Casal B, Merino J, et al. (1998) Adsorption of monovalent organic cations on sepiolite: experimental results and model calculations. Clays and Clay Minerals 46: 340–348.
30. Denoyel R, Rouquerol J (1991) Thermodynamic (including microcalorimetry) study of the adsorption of nonionic and anionic surfactants onto silica, kaolin and alumina. Journal of Colloidal Interface Science 143: 555–572.
31. Paria S, Khilar KC (2004) A review on experimental studies of surfactant adsorption at the hydrophilic solid-water interface. Advances in Colloid and Interface Science 110: 75–95.
32. Levitz P (1991) Aggregative adsorption of nonionic surfactants onto hydrophilic solid/water interface. Relation with bulk micellization. Langmuir 7: 1595–1608.
33. Cases JM, Villieras F (1992) Thermodynamic model of ionic and nonionic surfactant adsorption-abstraction on heterogeneous surfaces. Langmuir 8: 1251–1264.
34. Mishael Y-G, Undabeytia T, Rytwo G, Papahadjopoulos-Sternberg B, Rubin B, et al. (2002) Sulfometuron incorporation in cationic micelles adsorbed on montmorillonite. Journal of Agricultural and Food Chemistry 50: 2856–2863.
35. Undabeytia T, Mishael Y, Nir S, Papahadjopoulos-Sternberg B, Rubin B, et al. (2003) A novel system for reducing leaching from formulations of anionic herbicides. Environmental Science and Technology 37: 4475–4480.
36. Undabeytia T, Nir S, Sánchez-Verdejo T, Villaverde J, Maqueda C, et al. (2008) A clay-vesicle system for water purification from organic pollutants. Water Research 42: 1211–1219.
37. Nir S, Undabeytia T, Yaron-Marcovich D, El-Nahhal Y, Polubesova T, et al. (2000) Optimization of adsorption of hydrophobic herbicides on montmorillonite preadsorbed by monovalent organic cations: Interaction between phenyl rings. Environmental Science and Technology 33: 1269–1274.
38. Daniel PE, Bedmar F, Costa JL, Aparicio VC (2002) Atrazine and metribuzin sorption in soils of the Argentinean humic pampas. Environmental Toxicology and Chemistry 21: 2567–2572.
39. Henriksen T, Svensmark B, Juhler RK (2004) Degradation and sorption of metribuzin and primary metabolites in a sandy soil. Journal of Environmental Quality 33: 619–627.

8

Avian Species Richness in Relation to Intensive Forest Management Practices in Early Seral Tree Plantations

Jay E. Jones[1]*, Andrew J. Kroll[2], Jack Giovanini[1], Steven D. Duke[1], Tana M. Ellis[3], Matthew G. Betts[3]

1 Statistics, Mathematics, and Operations Research, Weyerhaeuser NR, Federal Way, Washington, United States of America, **2** Timberlands Technology, Weyerhaeuser NR, Federal Way, Washington, United States of America, **3** Department of Forest Ecosystems and Society, Oregon State University, Corvallis, Oregon, United States of America

Abstract

Background: Managers of landscapes dedicated to forest commodity production require information about how practices influence biological diversity. Individual species and communities may be threatened if management practices truncate or simplify forest age classes that are essential for reproduction and survival. For instance, the degradation and loss of complex diverse forest in young age classes have been associated with declines in forest-associated Neotropical migrant bird populations in the Pacific Northwest, USA. These declines may be exacerbated by intensive forest management practices that reduce hardwood and broadleaf shrub cover in order to promote growth of economically valuable tree species in plantations.

Methodology and Principal Findings: We used a Bayesian hierarchical model to evaluate relationships between avian species richness and vegetation variables that reflect stand management intensity (primarily via herbicide application) on 212 tree plantations in the Coast Range, Oregon, USA. Specifically, we estimated the influence of broadleaf hardwood vegetation cover, which is reduced through herbicide applications, on bird species richness and individual species occupancy. Our model accounted for imperfect detection. We used average predictive comparisons to quantify the degree of association between vegetation variables and species richness. Both conifer and hardwood cover were positively associated with total species richness, suggesting that these components of forest stand composition may be important predictors of alpha diversity. Estimates of species richness were 35–80% lower when imperfect detection was ignored (depending on covariate values), a result that has critical implications for previous efforts that have examined relationships between forest composition and species richness.

Conclusion and Significance: Our results revealed that individual and community responses were positively associated with both conifer and hardwood cover. In our system, patterns of bird community assembly appear to be associated with stand management strategies that retain or increase hardwood vegetation while simultaneously regenerating the conifer cover in commercial tree plantations.

Editor: Han Y.H. Chen, Lakehead University, Canada

Funding: This work was partially supported by grants from the National Science Foundation (NSF-ARC-0941748), the United States Department of Agriculture Food and Research Initiative (AFRI-2009-04457), and the National Council for Air and Stream Improvement to M.G. Betts. The funders had no role in study design, data collection and analysis, decision to publish, or preparation of the manuscript.

Competing Interests: The authors have declared that no competing interests exist.

* E-mail: jay.jones@weyerhaeuser.com

Introduction

Landscapes dedicated to timber commodity production are often managed for multiple objectives, including retention of populations and communities of native organisms, maintenance of ecosystem services, and the sustainable flow of commodities [1,2]. However, relatively few studies quantify the relationship between specific practices that are employed over large areas and species and community responses, despite the fact that even modest changes to current practices may provide substantial ecological benefits [3,4,5].

Rising global demand for wood and pulp products has led to intensification of forest management practices and a commensurate increase in concern about how these practices influence biological diversity [6,7]. Intensive forest management practices typically include clearcutting, rapid regeneration of single-species conifer stands, and chemical control of competing vegetation, resulting in truncated successional stages [8]. In particular, herbicide applications are designed to suppress naturally regenerating vegetation, including hardwood and deciduous broadleaf plants that are important components of biological diversity [9,10]. As a result, species that rely on early seral conditions may be vulnerable, as their preferred habitat is reduced in quality and available for only short periods of time [11,12].

The potential link between stand management practices and population declines of broadleaf-associated avian species has been observed around the globe where forest management has favored planted conifers over naturally regenerated broadleaf species [13,14,15]. For example, forest practices on private lands in the Pacific Northwest (PNW), USA, coupled with reduced harvest rates and the promotion of late seral forest on federal lands, have resulted in a decline in the amount of high quality, early seral habitat available for Neotropical migrant bird species [16,17,18]. Breeding Bird Survey results from 1966 to 2007 in the South Pacific Rainforest Bird Conservation Region indicate declining trends for several Neotropical migrant bird species that breed in early seral forest [19,20]. Species more strongly associated with broadleaf forest are declining at the greatest rates [18].

We evaluated relationships between stand management intensity and avian species responses in the Oregon Coast Range, USA. Specifically, we examined how differences in species-level occupancy and community richness varied across gradients in four important measures of vegetation composition: coniferous, broadleaf, deciduous broadleaf, and hardwood vegetation (see Table S1 for dominant species). Herbicide control is directed at broadleaf, deciduous broadleaf, and hardwood vegetation (Figure 1). By examining avian association with these three stand characteristics, we quantify, albeit indirectly, evidence for the biological impact of herbicide control. We expected that leaf-gleaning species would respond more strongly to reductions in hardwood and deciduous broadleaf vegetation, given their reliance on these vegetation features for foraging and breeding [12,21]. As a result, we summarized leaf-gleaner responses both separately and as part of the overall avian community.

Results

Population-level mean occupancy was positively associated with cover of both coniferous and hardwood vegetation (Figure 2). This result indicates that on average, occupancy probabilities across all species in this study tended to be higher in stands with greater percentages of conifer and hardwood cover. In contrast, posterior intervals for the community hyper-parameters for elevation and deciduous broadleaf cover were centered near zero, indicating little overall population-level effect of these covariates.

We estimated species richness in each of the 212 stands using the community model (circles, Figure 3). Congruent with previous results (12), the response to hardwood cover was stronger than for broadleaf or deciduous broadleaf cover, suggesting a greater biological value of this vegetation feature to breeding birds. Importantly, estimates of total species and leaf-gleaner richness were 35–80% lower if we ignored variation associated with detection probability (crosses, Figure 3).

Conifer cover showed the largest contribution to species richness in our model, with an average predictive comparison (APC) of approximately 3.7 (SE = 0.9) (Figure 4). This result indicates that for two sites with otherwise similar levels of other covariates, we expect ~3.7 additional species will occur on the site

Figure 1. The range of vegetation cover resulting from herbicide control of competing vegetation in conifer plantations. A 10-year old Douglas-fir plantation with <5% hardwood cover (A) and a 7-year old Douglas-fir plantation with ~50% hardwood cover (B). Our results indicate that species richness was positively associated with both hardwood and conifer cover.

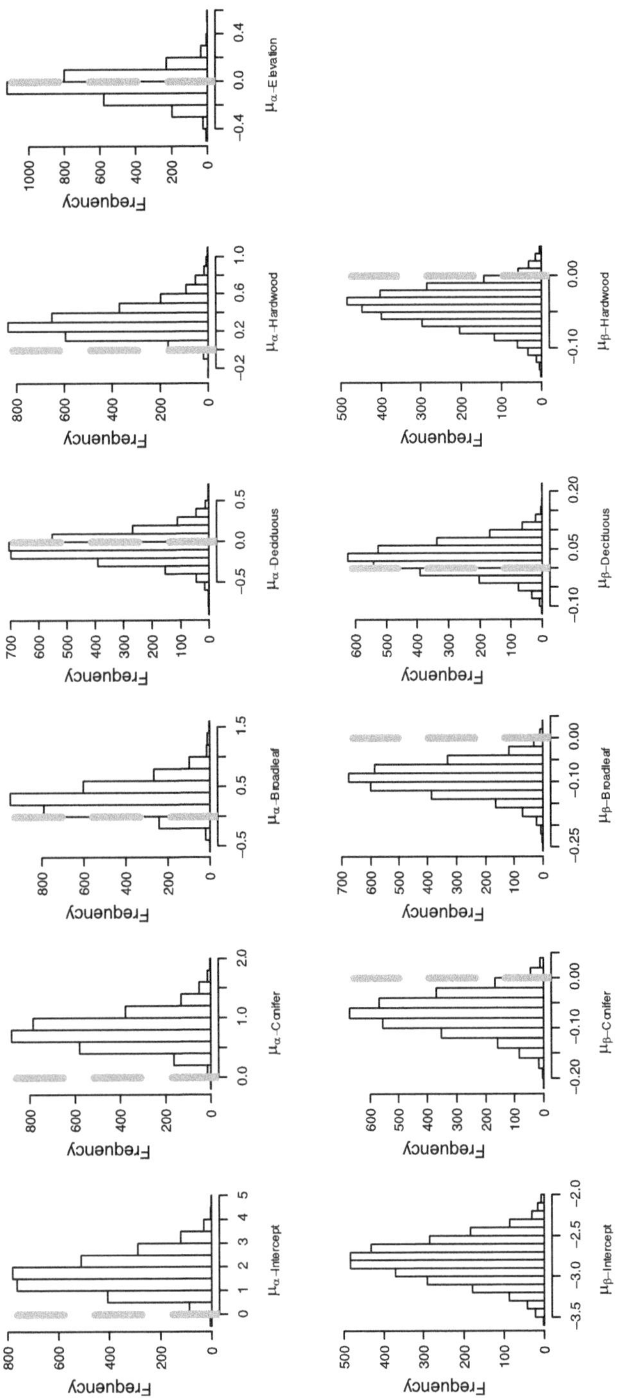

Figure 2. Posterior distributions of the population level hyper-parameter means for occupancy (top row) and detection (bottom row) covariates, Oregon Coast Range, USA, 2008–2009. Dashed gray line indicates zero. Covariates were centered and scaled for analysis.

with 10% greater conifer cover (hereafter, all predictive comparisons for vegetation variables correspond to a 10% difference in cover). The APC for hardwood and broadleaf cover were 2.5 and 2.2, respectively, although the estimate for broadleaf cover indicated substantial uncertainty in this estimate. The APC for deciduous broadleaf cover suggested a negative association of avian richness with this covariate although the estimate was small and one SE included 0.

The difference in estimate magnitude for each cover type suggests that vegetation composition, rather than total cover, may drive species richness. For example, the difference in richness associated with a 10% difference in conifer cover is substantially larger than that associated with a similar difference in deciduous cover. If total cover were the primary driver of richness, we would expect all of the cover APC values (i.e., broadleaf, deciduous broadleaf, hardwood, conifer) to be similar, and to be the same as the APC for total cover – i.e., the association between richness and total cover would be independent of how cover was obtained. The APC for a 10% difference in total cover (not shown in Figure 4), as calculated from our model, was 2.3 additional species (SE = 0.4), a value that is less than our estimate for conifer cover, but greater than for deciduous cover.

Leaf gleaners showed similar, but reduced, trends for each of the model inputs (Figure 4) with APCs of 1.6, 1.1 and 1.0 for conifer, hardwood, and broadleaf cover, respectively. For leaf gleaner total richness, the APC for total cover was 1.0 (SE = 0.2).

Estimates of individual species' occupancy probabilities across all study sites (at median covariate levels) ranged from 0.23–0.99. For many species, detection probabilities were low, ranging from 0.01–0.65. We include posterior summaries of occupancy and detection, as well as estimates and 95% credible intervals (the Bayesian equivalent of a confidence interval) [22] for occupancy and detection covariates, in Table S2.

Both conifer and hardwood cover were more strongly associated with species occupancy than broadleaf or deciduous broadleaf cover (Table S2). For example, the occupancy of 56 of 64 species increased positively with conifer cover (i.e., >200% increase in odds of occurrence across the sampled values of conifer cover), 4 species showed a decrease in occupancy probability (>200% decrease in odds of occurrence across the sampled values of conifer cover), and 4 species showed no change (less than 200% change in odds of occurrence in either direction). Among species with >10 detections, white-crowned sparrow *Zonotrichia leucophrys* and willow flycatcher *Empidonax traillii* showed the strongest negative and positive responses, respectively, to conifer cover.

The occupancy of fifty-six of 64 species showed a strong positive response to increases in hardwood cover (i.e., >200% increase in odds of occurrence across the sampled values of hardwood cover), 1 species showed a decrease in occupancy probability (>200% decrease in odds of occurrence across the sampled values of conifer cover), and 7 species showed no change (less than 200% change in odds of occurrence in either direction). Among species with >10 detections, White-crowned sparrow *Zonotrichia leucophrys* and black-throated gray warbler *Setophaga nigrescens* showed the strongest negative and positive responses, respectively, to hardwood cover.

Mean posterior occupancy probability for each of the 23 leaf-gleaning species was positively associated with both conifer and hardwood cover (Figure 5 and Table S2). However, some species that showed positive associations with hardwood cover were negatively associated with deciduous broadleaf cover. For example, bushtit *Psaltriparus minimus*, common yellowthroat *Geothlypis trichas*, ruby-crowned kinglet *Regulus calendula*, and western tanager *Piranga ludoviciana* occupancy probability estimates were negatively associated with broadleaf cover. Similarly, Bewick's wren *Thryomanes bewickii*, black-throated gray warbler *Setophaga nigrescens*, golden-crowned kinglet *Regulus satrapa*, hermit warbler *Setophaga occidentalis*, Hutton's vireo *Vireo huttoni*, MacGillivray's warbler *Geothlypis tolmiei*, Townsend's warbler *Setophaga townsendi*, warbling vireo *Vireo gilvus*, western tanager, and wrentit *Chamaea fasciata* were negatively associated with deciduous broadleaf cover (Figure 4 and

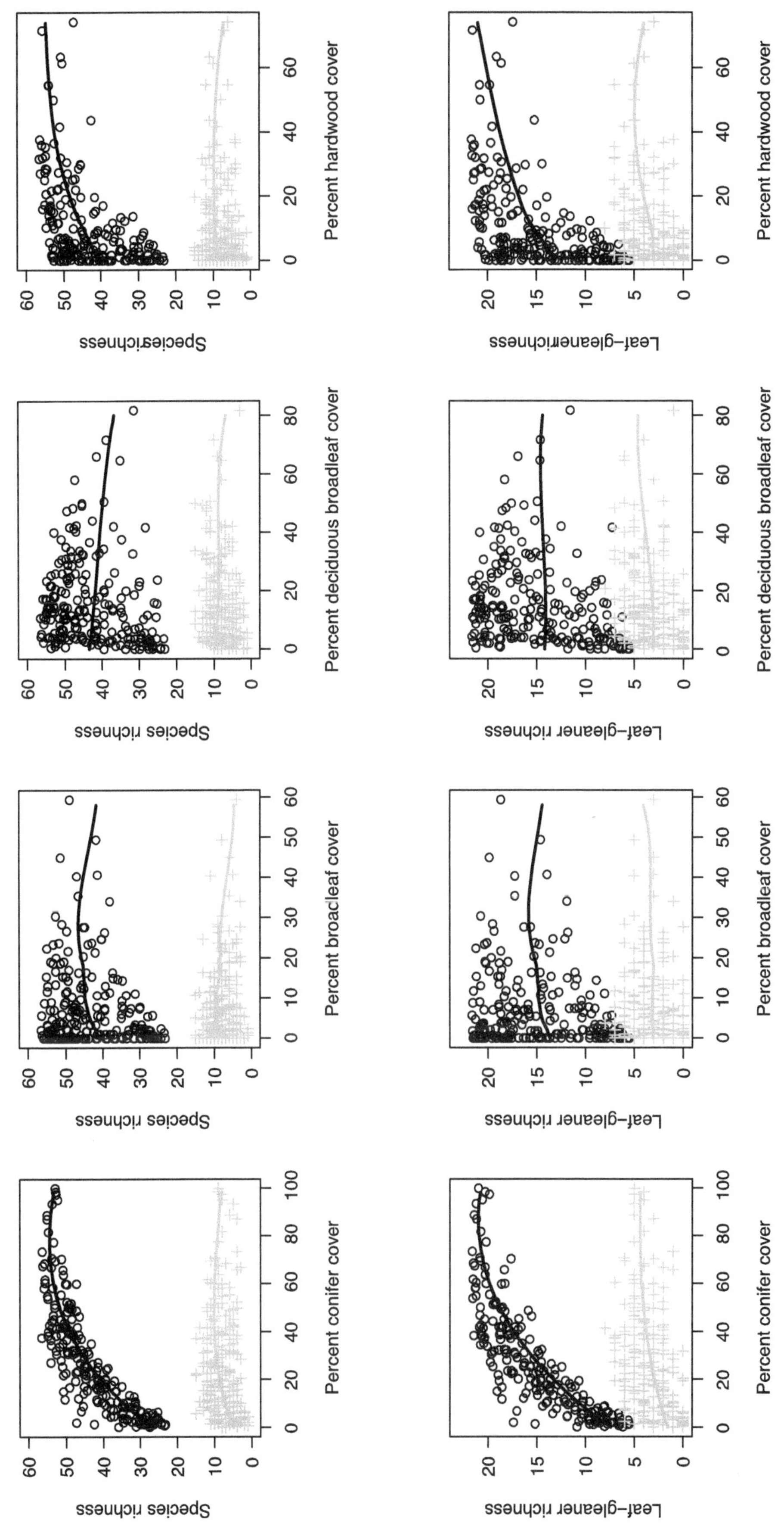

Species richness
Leaf−gleaner richness
Percent conifer cover
Percent broadleaf cover
Percent deciduous broadleaf cover
Percent hardwood cover

Figure 3. Total estimated species richness (top row) and leaf-gleaner richness (bottom row), based on elevation (not shown) and conifer, broadleaf, deciduous broadleaf, and hardwood vegetation cover, on 212 early seral forest stands, Oregon Coast Range, USA, 2008–2009. Estimates are from a model that incorporates variation in species detection (circles) and a model that ignores variation in species detection (crosses). Solid lines show smoothed trends in expected richness while holding all other covariates at their median values.

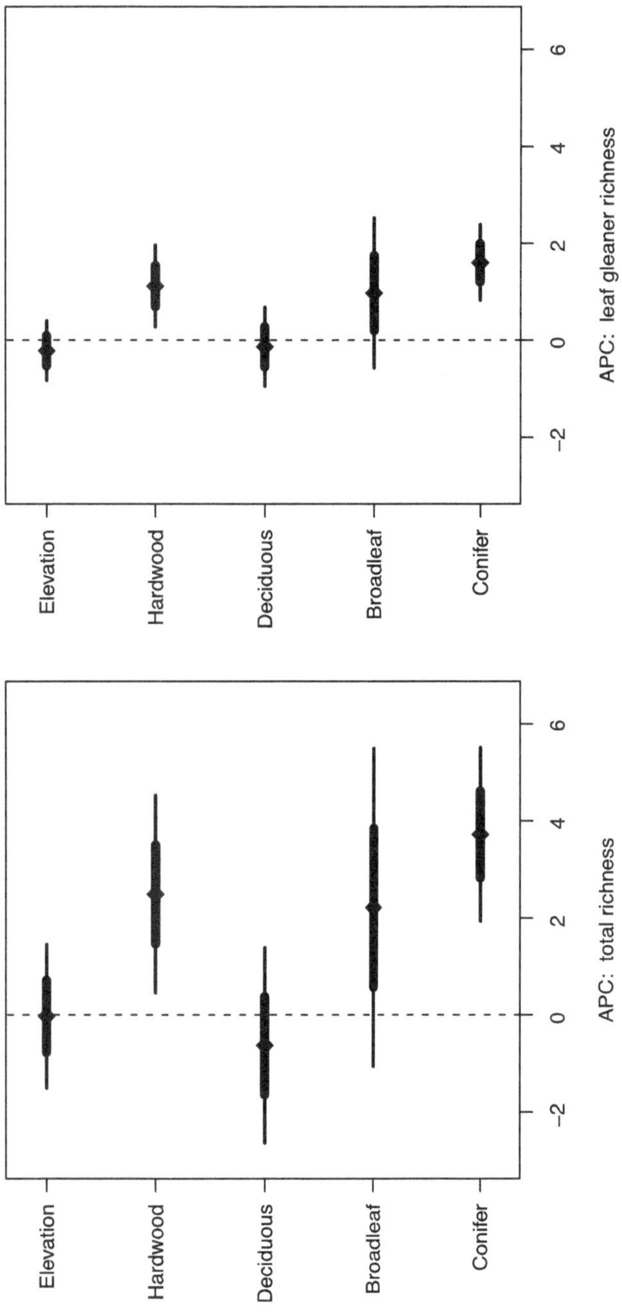

Figure 4. Average predictive comparisons (+/−1 and 2 standard errors) for the association between total species richness and leaf-gleaner richness and elevation (m), percent cover of hardwoods, deciduous, deciduous broadleaf, and conifer vegetation, Oregon Coast Range, USA, 2008–2009. Comparisons were calculated based on a modification of methodology described in Gelman and Pardoe [66].

Table S2). Although the number of detections of these species varied substantially (2–197), our use of the community model allowed us to estimate the responses of the individual species to the vegetation covariates, which would have been challenging with more conventional single-species models.

Discussion

Stand regeneration practices that limit plant community composition and structural complexity may reduce habitat quality and amount for early seral forest species [23]. However, our results indicate that efforts may be undertaken in intensive forest management to mitigate impacts of plantation forestry. Bird species richness was strongly and positively related to the total amount of hardwood cover. Though these results are correlative, they suggest that maintaining non-coniferous vegetation within a stand should have useful conservation benefits. These results concur with similar work from the same study region that examined abundance of birds caught in mistnets (12) and productivity of individual species [24]. Interestingly, species richness also showed a substantial positive association with conifer cover that was larger than the estimated association with total cover. Given that high rates of seedling survival are often positively associated with conifer cover in plantations in our region [25], these results suggest that successful stand regeneration is not necessarily incompatible with conservation goals across the stand ages we observed.

We expect that food availability and predation avoidance were the primary reasons for the large effects of both hardwood and conifer cover on bird species richness. That is, stands with higher vegetation cover will likely support larger invertebrate numbers per unit area than stands with reduced vegetation cover and, consequently, support enhanced reproductive success and survival [26,27,28]. Also, increased amounts of vegetation cover may provide high quality nesting sites and reduce efficiency of predators that target nests and adults [29,30].

Given that we evaluated occupancy as the primary response, we note two important caveats from our study. First, although we did not find strong or consistent effects of deciduous broadleaf and broadleaf cover on avian species richness, the abundance of individual bird species may have been reduced if they relied on specific plant species that were in turn reduced by regeneration practices [10]. Second, our sampling method counted singing males and we do not know how regeneration practices may have influenced measures of demographic performance, including nesting success, productivity, or individual survival. Given the intensification of forest management both regionally and globally [6,31], evaluation of demographic responses across a gradient of regeneration intensity could provide managers with powerful tools to integrate conservation of biological diversity with commodity production [32,33].

Our inferences are confined to the stand ages (0–18 years old) evaluated in our study, and we cannot address changes in avian species richness beyond that time frame. Indeed, both theory and empirical evidence indicate that plantations with closed conifer canopies will have simplified vertical structure and plant diversity, leading to reductions in biological diversity [8]. As a result, the positive relationship between diversity and conifer canopy cover

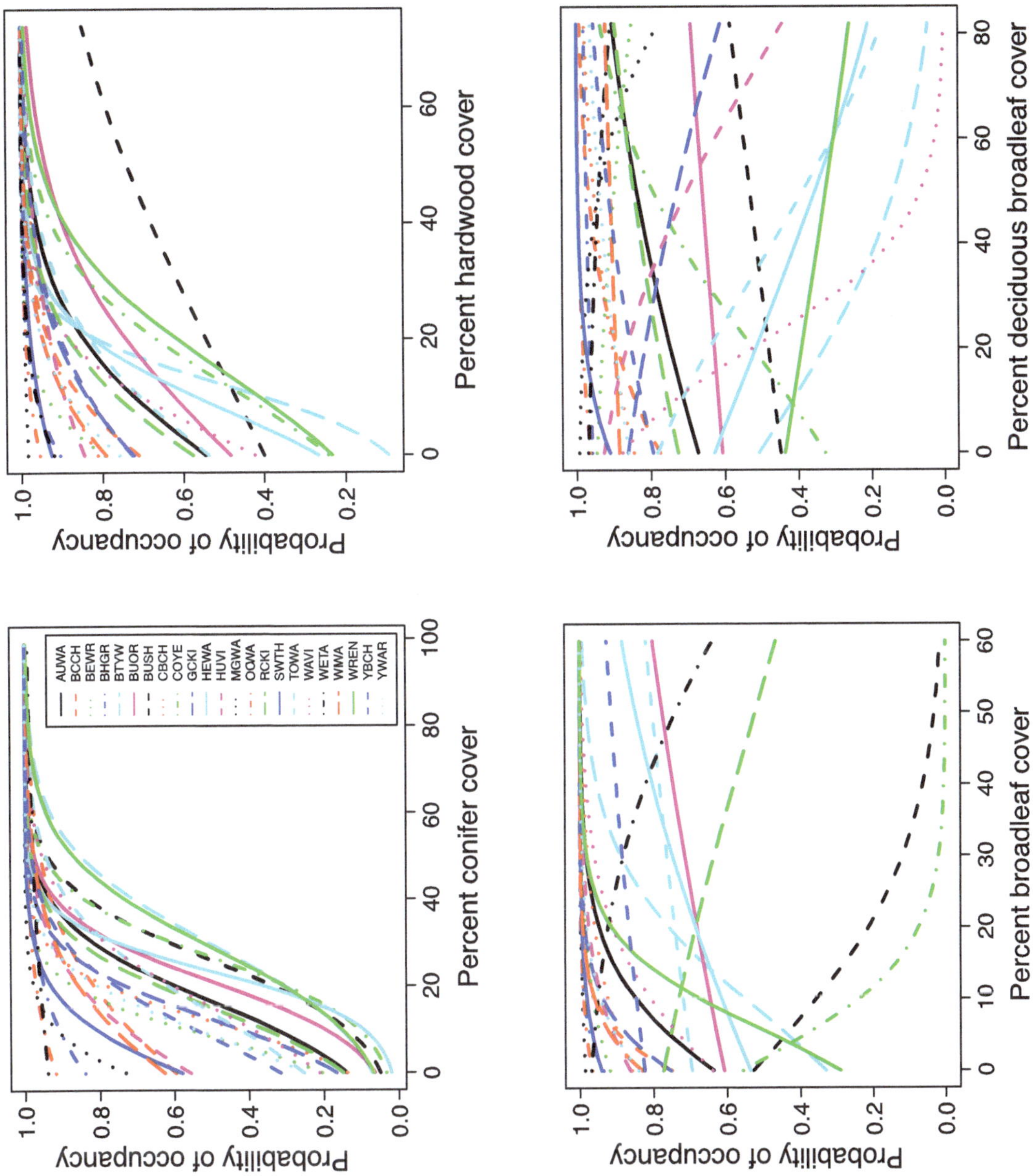

Figure 5. Mean occupancy probabilities of 23 species of leaf-gleaning birds based on percent cover of conifers, broadleaf, deciduous broadleaf, and hardwood vegetation, Oregon Coast Range, USA, 2008–2009. Species codes are in Table S2. Estimates in each panel were calculated while holding all other covariates constant at their median values.

that we observed in our study may not be maintained as plantations age. Importantly, the correlation between hardwood and conifer cover across our study sites is not consistent over time. In the initial years of forest succession, these variables are positively correlated. However, after eight years, they become negatively correlated (See Methods and Figure S1). Thus, in the initial years of plantation development, the two key factors influencing diversity coincide. Later, they become decoupled, likely due to competition between conifers and hardwoods as the canopy closes [34,35]. This indicates a potential trade-off in the later years of stand development between the relative contributions to richness of conifers versus hardwoods.

Our study was a natural experiment rather than a manipulative experiment, thus, our ability to sample the full gradient in hardwood cover was constrained by the characteristics of existing early seral stands within our study area. Stands with high hardwood cover are rare (current state policy requires reduction of these competing hardwood species), so we were not able to predict avian diversity in stands with >60% hardwood cover.

Previous studies have found strong correlations between vegetation structure (i.e., vertical distribution of foliage) and avian species richness [36], but less support exists for a relationship with vegetation composition [37,38,39,40,41,42]. Our results raise the questions of whether vegetation composition did not exert an independent effect on richness in these studies, or whether variation associated with the detection process masked an effect? For example, if we had not modeled the detection process, our estimates of species richness would have been much lower (Figure 3) and substantially less correlated with the variables we examined. Vegetation structure is known to interfere with the detection process during avian sampling [43,44], in which case making unbiased inferences about how vegetation structure influences ecological responses (e.g., occupancy or abundance) will be challenging if the detection process is ignored. In general, we found very low detection probabilities for many species (Table S2), a result that is consistent with findings from recent studies of forest bird communities [45,46]. Although a diverse suite of statistical tools now exist to incorporate variation from the detection process into ecological modeling [47,48], we encourage investigators to consider this issue in the design stage of research projects [49].

We did not evaluate the effect of plant species richness on avian species richness, as we assumed that total hardwood/broadleaf cover was associated with hardwood and broadleaf species richness in our study area [12]. The response of higher-level taxa to plant species richness has a long-standing theoretical basis [50], as well as recent empirical support [51], although this response could be a function of both greater resource availability and more complex vegetation structure [52]. For instance, although we did not predict the strong association between conifer cover and avian species richness, conifer species richness clearly did not account for this relationship as study stands were dominated by a single conifer species.

We could not determine if species richness on our study sites was in response to vegetation composition per se (i.e., growth of hardwood and conifer cover) or to community assembly following disturbance (i.e., forest harvesting) [53]. Tree regeneration in our study area is generally very rapid, and older plantations are likely to have high conifer cover. We think that among stand variation in trajectories of plant succession is the most likely explanation for our observations, although we cannot preclude the possibility that intrinsic biological processes influenced the responses that we observed. For instance, bird richness might be higher in older stands because of the time required for species to re-colonize following disturbance (i.e., timber harvest) rather than conifer growth per se. However, given the high vagility of most bird species in our system, it seems unlikely that dispersal limitation is the primary driver of our results.

Land use intensification plays a critical role in provisioning rapidly growing human populations and has potentially severe consequences for the conservation of native biological diversity [54,55]. Species richness is frequently measured in research studies and management programs to assess community responses to anthropogenic disturbances [56,57], but reasons exist for considering species richness as only a preliminary, and potentially not very informative, assessment. First, species occupancy (e.g., at the stand level) may remain unchanged even if demographic measures such as survival and reproduction are changing, a critical result for management of individual populations. Second, and more importantly, species richness can remain constant despite substantial changes in community membership. For example, Harvey and Villalobos [58] reported bird assemblages that were equally abundant, speciose, and diverse in agro-forestry systems compared to unmanaged forests. However, the species composition of these assemblages was highly modified, with fewer forest-dependent species, more open area species, and different dominant species. In these cases, the critical question is not what species are present, but the roles they play in ecosystem functioning [55].

Methods

Study Area & Bird Observations

We collected our data over a 2-year period (May–July in 2008 and 2009) in 212 forested stands located in the western hemlock zone in western Oregon, USA. We selected stands using a stratified sampling design that represented available gradients in stand age 0–8 years) and proportion of hardwood tree cover (estimated visually upon initial encounter; 0–75%). We did not use a stratified-random design because stands with >10% hardwood cover were relatively rare in our study area. We sampled all stands with >10% cover that we could locate on state or private land within Benton and Polk Counties, Oregon, USA. We received written or verbal permission to sample sites from all private and public landowners involved in the study. No formal permits were required. This approach allowed us to sample across a broad gradient of hardwood cover (0–60%) in existing Douglas-fir plantations, which we considered to be a proxy for intensive forest management. Sites were not broadcast-burned prior to planting of conifer seedlings and treatment with herbicides.

The avian community was sampled using a single fixed-radius point count station in each stand [59]. Point count stations were located >50 m from clearly identifiable forest edges (e.g., roads, forest of different age classes). The average distance between points was 685 m (SE = 41). We conducted two, 5-minute counts on separate occasions, spaced >10 days apart, between 0530 and 1000 hours. Counts were not conducted in the rain or when wind exceeded 20 km/h. We recorded all male birds seen or heard within a 50-m radius as present. At each point count station, we estimated the total cover of coniferous, broadleaf, deciduous broadleaf, and hardwood vegetation (Table S1) in three 3 m-radius circles distributed throughout the 50-m count area (the point center, and 50 m from the point in two random directions). In order to evaluate major broadleaf types separately, we defined deciduous broadleaf cover exclusive of hardwood cover (i.e., total deciduous broadleaf cover minus total hardwood cover) and broadleaf cover exclusive of deciduous broadleaf cover.

Due to the strong correlation between stand age and conifer cover in our study (Pearson correlation coefficient = 0.71; 90% confidence interval = 0.62, 0.80), we were unable to separate effects of stand regeneration (i.e., increasing vegetation cover) and time since disturbance (i.e., forest harvesting) [53]. In general, older stands with low conifer cover are not a deliberate management objective and are rare in our study region. As a result, we present the association of conifer cover and species richness rather than stand age per se. In addition, we evaluated the association between conifer and hardwood cover across the stand ages that we sampled (Figure S1). We found that the correlation between stand conifer and hardwood cover differed by stand age, with a positive correlation for stands less than 8 years of age (0.27; 90% confidence interval = 0.12, 0.41), and a negative correlation for stands greater than 8 years of age (−0.20; 90% confidence interval = −0.38, −0.02).

Analysis

We used the Dorazio-Royle community occupancy model [48] to examine the relationships between stand management intensity and avian species responses in the Oregon Coast Range, USA. We used the model to estimate species level covariate effects, as well as population level measures of occupancy, including species richness [45,48]. Following Russell et al. [60], we do not account for the contribution of unobserved species in our population estimates, instead conditioning on the set of observed breeding species in our study.

We let $z_{i,j}$ denote true occupancy status, in which $z_{i,j} = 1$ if species i occupies site j for the study interval, or $z_{i,j} = 0$ otherwise. The occupancy state is taken to be a Bernoulli random variable, $z_{i,j} \sim Bern(\psi_{i,j})$, where $\psi_{i,j}$ is the probability that species i occupies site j. We take species detection to again follow a Bernoulli distribution: $y_{i,j,k} \sim Bern(p_{i,j,k} \cdot z_{i,j})$, where $y_{i,j,k}$ is 1 if the species i is detected at site j during visit k, or 0 otherwise. Note that under this parameterization, the probability of detecting the species i at site j will be zero if the species does not occupy site j, since $z_{i,j} = 0$.

We modeled species-specific occupancy probabilities as a function of the 4 vegetation covariates, plus site-level elevation, using a logit link function. The effect of elevation was not a focus of our study, but prior studies have found associations between bird species richness and elevation and we wanted to control for this source of variation [61,62]:

$$\begin{aligned} logit(\psi_{i,j}) &= a_0 + a_{1i} \cdot Conifer_j + a_{2i} \cdot Broadleaf_j \\ &+ a_{3i} \cdot Deciduous\ broadleaf_j + a_{4i} \cdot Hardwood_j + a_{5i} \cdot Elevation_j \end{aligned} \quad (1)$$

We modeled species-specific detection probabilities as a function of the 4 vegetation covariates only:

$$\begin{aligned} logit(\rho_{i,j,k}) &= \beta_0 + \beta_{1i} \cdot Conifer_j + \beta_{2i} \cdot Broadleaf_j \\ &+ \beta_{3i} \cdot Deciduous\ broadleaf_j + \beta_{4i} \cdot Hardwood_j \end{aligned} \quad (2)$$

Each covariate was centered and scaled prior to analysis. We included each of our primary variables of interest in both the occupancy and detection models. This approach is necessary to separate occupancy effects, which are of primary interest, from potential detection effects; otherwise, our estimates of richness could be biased.

Under the hierarchical community model, we assume that the species-specific effects for a given parameter are drawn from a common normal distribution, e.g., that$\alpha_{1,i} \sim N(\mu_1, \sigma_1^2)$for parameter $\alpha_1$of species i, where the mean and variance are population-level hyper-parameters. This population-level distribution provides a summary of community response, both in terms of the mean behavior as well as the variability in behavior. The extent to which information is shared across species depends on both the degree of uniformity across the population, as estimated by the population-level parameters, and the amount of information available for each species. For species for which we are less certain of the parameter estimates, e.g., those with low detection probabilities, estimates will tend to shrink toward the population mean value.

All computations were performed using WinBUGS [63] called from R [64] using package R2WinBUGS [65]. We ran 3 chains of length 100,000 each, with a burn-in of 50,000 and 1/50 thinning. We assessed convergence using the Gelman-Rubin statistic [22] and visual inspection of the chains, with both measures indicating a reasonable assumption of convergence. We provide WinBUGS code for this model in Text S1.

Species richness is not modeled directly in the Dorazio-Royle community occupancy model and we are not aware of any existing methods for quantifying the association between model covariates and species richness within their framework. Past approaches have focused on visual displays of estimated richness [45,56]. Such displays are useful, but inference is indirect and not quantitative. Here, we adopt the use of average predictive comparisons [66] to quantify directly the association (and uncertainty) between *predicted* species richness and each model covariate. Predictive comparisons evaluate the difference in expected response for a unit difference in an input covariate, using the fitted model, and averaging over the distribution of all other covariates. We extend this approach to species richness by summing over the species-specific predictions to obtain averaged expected differences in species count. For our dataset $(x,y)_j$, $j = 1,\ldots,n$, we denote our input of interest u, and all other inputs v, such that $x = (u,v)$, where n is the number of sites. We let $i = 1,\ldots,N$, be the index of species, where N is the total number of observed species. We estimated the average predictive comparison for species richness using Equation 3

$$\hat{\Delta}_u = \frac{\sum_{j=1}^{n} \sum_{k=1}^{n} \sum_{s=1}^{S} w_{jk} \sum_{i=1}^{N} \left(E(y|u_k, v_j, \theta^S) - E(y|u_j, v_j, \theta^S)\right) sign(u_k - u_j)}{\sum_{j=1}^{n} \sum_{k=1}^{n} \sum_{s=1}^{S} w_{jk} (u_k - u_j) sign(u_k - u_j)} \quad (3)$$

A set of $s = 1,\ldots,S$ simulations were sampled from the posterior distribution. We calculated predictive comparisons for all model inputs, treating each in turn as the input of interest. Standard errors for $\hat{\Delta}_u$ are estimated as described in Gelman and Pardoe [66], and account for the uncertainty in model parameter estimates, while treating all covariates as fixed. We note that unlike the richness estimator described in Dorazio and Royle [48], the approach described here is based on the model prediction, and does not directly consider the observed occupancy status of each site.

The predictive comparison approach may also be extended to post-hoc combinations of inputs, in addition to the individual model inputs. For a linear function f of q inputs of interest, we can estimate the average predictive comparison for species richness as in equation (4):

$$\hat{\Delta}_u = \frac{\sum_{j=1}^{n} \sum_{k=1}^{n} \sum_{s=1}^{S} w_{jk} \sum_{i=1}^{N} \left(E(y|u_{1k},\ldots,u_{qk}, v_j, \theta^S) - E(y|u_{1j},\ldots,u_{qj}, v_j, \theta^S)\right) sign\left(f(u_{1k},\ldots,u_{qk}) - f(u_{1j},\ldots,u_{qj})\right)}{\sum_{j=1}^{n} \sum_{k=1}^{n} \sum_{s=1}^{S} w_{jk} \left(f(u_{1k},\ldots,u_{qk}) - f(u_{1j},\ldots,u_{qj})\right) sign\left(f(u_{1k},\ldots,u_{qk}) - f(u_{1j},\ldots,u_{qj})\right)} \quad (4)$$

We used Equation (4) to examine the association between total cover of all 4 vegetation classes and both total species richness and leaf-gleaner richness. We provide all code used to estimate average predictive comparisons in Text S1.

Supporting Information

Figure S1 The correlation between conifer cover and hardwood cover by stand age class (split into two groups by the median stand of 8 years), for 212 forest stands, Oregon Coast Range, USA, 2008–2009.

Table S1 Common plant species with classification as conifer (*), broadleaf, deciduous broadleaf, or hardwood vegetation, Oregon Coast Range, USA, 2008–2009.

Table S2 Median species-specific estimates and 95% credibility intervals for occupancy and detection and parameter estimates and 95% credibility intervals for elevation and vegetation covariates on intensively managed forest stands, Oregon Coast Range, USA, 2008–2009. We included the total number of observations for each species and if species were classified as leaf-gleaners.

Text S1 WinBUGS code for hierarchical community model and average predictive comparisons of species richness.

Acknowledgments

We thank two anonymous reviewers for comments that improved the manuscript.

Author Contributions

Conceived and designed the experiments: JEJ AJK JG SDD TME MGB. Performed the experiments: JEJ AJK JG SDD TME MGB. Analyzed the data: JEJ. Contributed reagents/materials/analysis tools: JEJ AJK JG SDD TME MGB. Wrote the paper: JEJ AJK JG SDD TME MGB.

References

1. Beschta RL, Rhodes JJ, Kauffman JB, Gresswell RE, Minshall GW, et al. (2004) Postfire management on forested public lands of the western United States. Conservation Biology 18: 957–967.
2. Perrings C, Naeem S, Ahrestani F, Bunker DE, Burkill P, et al. (2010) Ecosystem services for 2020. Science 330: 323–324.
3. Kroll AJ, Arnett EB, Altman B (2010) Effects of salvage logging on avian nest survival in beetle-killed forests. Forest Ecology and Management 260: 1599–1606.
4. Bunnell FL, Kremsater LL, Wind E (1999) Managing to sustain vertebrate richness in forests of the Pacific Northwest: relationships within stands. Environmental Reviews 7: 97–146.
5. Arnett EB, Kroll AJ, Duke SD (2010) Avian foraging and nesting use of created snags in intensively-managed forests of western Oregon, USA. Forest Ecology and Management 260: 1773–1779.
6. Binkley CS, Aronow ME, Washburn CL, New D (2005) Global perspectives on intensively managed plantations: implications for the Pacific Northwest. Journal of Forestry 103: 61–64.
7. Berlik MM, Kittredge DB, Foster DR (2002) The illusion of preservation: a global environmental argument for the local production of natural resources. Journal of Biogeography 29: 1557–1568.
8. Hayes JP, Schoenholtz SH, Hartley MJ, Murphy G, Powers RF, et al. (2005) Environmental consequences of intensively managed forest plantations in the Pacific Northwest. Journal of Forestry 103: 83–87.
9. Harrington TB, Wagner RG, Radosevich SR, Walstad JD (1995) Interspecific competition and herbicide injury influence 10-year responses of coastal Douglas-fir and associated vegetation to release treatments. Forest Ecology and Management 76: 55–67.
10. Hagar JC (2007) Wildlife species associated with non-coniferous vegetation in Pacific Northwest conifer forests-a review. Forest Ecology and Management 246: 108–122.
11. Lindh B, Muir PS (2004) Understory vegetation in young Douglas-fir forests: does thinning help restore old-growth composition? Forest Ecology and Management 192: 285–296.
12. Ellis TM, Betts MG (2011) Bird abundance and diversity across a hardwood gradient within early seral plantation forest. Forest Ecology and Management 261: 1372–1381.
13. DeGraaf RM, Yamasaki M (2003) Options for managing early-successional forest and shrubland bird habitats in the northeastern United States. Forest Ecology and Management 185: 179–191.
14. Paquet J-Y, Vandevyvre X, Delahaye L, Rondeux J (2006) Bird assemblages in a mixed woodland–farmland landscape: the conservation value of silviculture-dependent open areas in plantation forest. Forest Ecology and Management 224: 59–70.
15. Enoksson B, Angelstam P, Larsson K (1995) Deciduous forest and resident birds: the problem of fragmentation within a coniferous forest landscape. Landscape Ecology 10: 267–275.
16. Kennedy RSH, Spies TA (2005) Dynamics of hardwood patches in a conifer matrix: 54 years of change in a forested landscape in Coastal Oregon, USA. Biological Conservation 122: 363–374.
17. Spies TA, Johnson KN, Burnett KM, Ohmann JL, McComb BC, et al. (2007) Cumulative ecological and socioeconomic effects of forest policies in Coastal Oregon. Ecological Applications 17: 5–17.
18. Betts MG, Hagar J, Rivers J, Alexander J, McGarigal K, et al. (2010) Thresholds in songbird occurrence in relation to early seral broadleaf forest. Ecological Applications 20: 2116–2130.
19. Sauer JR, Hines JE, Fallon J (2008) The North American Breeding Bird Survey, Results and Analysis 1966–2007. Version 5.15.2008. USGS Patuxent Wildlife Research Center. Laurel, MD.
20. Rich TD, Beardmore CJ, Berlanga H, Blancher PJ, Bradstreet MSW, et al. (2004) Partners in Flight North American Landbird Conservation Plan, Cornell Lab of Ornithology, Ithaca, NY, USA.
21. Morrison ML, Meslow EC (1984) Response of avian communities to herbicide-induced vegetation changes. Journal of Wildlife Management 48: 14–22.
22. Gelman A, Carlin JB, Stern HS, Rubin DB (2004) Bayesian data analysis, 2nd edition: Chapman & Hall/CRC, Boca Raton, FL, USA.
23. Swanson ME, Franklin JF, Beschta RL, Crisafulli CM, DellaSala DA, et al. (2011) The forgotten stage of forest succession: early-successional ecosystems on forest sites. Frontiers in Ecology and the Environment 9: 117–125.
24. Ellis TM, Kroll AJ, Betts MG (2012) Early seral hardwood vegetation increases adult and fledgling bird abundance in Douglas-fir plantations of the Oregon Coast Range, USA. Canadian Journal of Forest Research 42: 918–933.
25. Harrington TB, Madsen J (2005) Silvicultural technology and applications for forest plantation establishment west of the Cascade crest. In: Harrington CA, Schoenholtz SH, editors. Productivity of Western forests: a forest products focus: Gen. Tech. Rep. PNW-GTR-642, U.S. Department of Agriculture, Forest Service, Pacific Northwest Research Station, Portland, OR, USA: 176 p. 27–37.
26. Martin TE (1987) Food as a limit on breeding birds: a life-history perspective. Annual Review of Ecology and Systematics 18: 453–487.
27. Nagy LR, Holmes RT (2005) Food limits annual fecundity of a migratory songbird: an experimental study. Ecology 86: 675–681.
28. Marshall MR, DeCecco JA, Williams AB, Galeb GA, Cooper RJ (2003) Use of regenerating clearcuts by late-successional bird species and their young during the post-fledging period. Forest Ecology and Management 183: 127–135.
29. Martin TE (1993) Nest predation among vegetation layers and habitat types: revising the dogmas. The American Naturalist 141: 897–913.
30. Lima SL (2009) Predators and the breeding bird: behavioral and reproductive flexibility under the risk of predation. Biological Reviews of the Cambridge Philosophical Society 84: 485–513.
31. Carnus J-M, Parrotta J, Brockerhoff E, Arbez M, Jactel H, et al. (2006) Planted forests and biodiversity. Journal of Forestry 104: 65–77.
32. Martin TE (1992) Breeding productivity considerations: what are the appropriate habitat features for management? In: Hagan JM, Johnston DW, editors. Ecology and conservation of neotropical migrants. Washington, D.C.: Smithsonian Institution Press. 455–473.
33. Nájera A, Simonetti JA (2010) Enhancing avifauna in commercial plantations. Conservation Biology 24: 319–324.
34. Oliver CD, Larson BC (1996) Forest stand dynamics: John Wiley and Sons, New York, New York, USA.
35. Tappeiner JC, II, Zasada JC, Huffman DW, Ganio LM (2001) Salmonberry and salal annual aerial stem production: the maintenance of shrub cover in forest stands. Canadian Journal of Forestry Research 31: 1629–1638.
36. MacArthur RH, MacArthur JW (1961) On bird species diversity. Ecology 42: 594–598.
37. Willson MF (1974) Avian community organization and habitat structure. Ecology 55: 1017–1029.
38. Terborgh J, Weske JS (1975) The role of competition in the distribution of Andean birds. Ecology 56: 562–576.
39. Pearman PB (2002) The scale of community structure: habitat variation and avian guilds in tropical forest understory. Ecological Monographs 72: 19–39.
40. van Dorp D, Opdam PFM (1987) Effects of patch size, isolation and regional abundance on forest bird communities. Landscape Ecology 1: 59–73.
41. Lesak AA, Radeloff VC, Hawbaker TJ, Pidgeon AM, Gobakken T, et al. (2011) Modeling forest songbird species richness using LiDAR-derived measures of forest structure. Remote Sensing of Environment 115: 2823–2835.
42. Maas B, Putra DD, Waltert M, Clough Y, Tscharntke T, et al. (2009) Six years of habitat modification in a tropical rainforest margin of Indonesia do not affect bird diversity but endemic forest species. Biological Conservation 142: 2665–2671.
43. Bibby CJ, Buckland ST (1987) Bias of bird census results due to detectability varying with habitat. Acta Oecologica 8: 103–112.
44. Schieck J (1997) Biased detection of bird vocalizations affects comparisons of bird abundance among forested habitats. Condor 99: 179–190.
45. Zipkin EF, DeWan A, Royle JA (2009) Impacts of forest fragmentation on species richness: a hierarchical approach to community modelling. Journal of Applied Ecology 46: 815–822.
46. Linden DW, Roloff GJ (in review) Conserving avian richness through structure retention in managed forests of the Pacific Northwest. Forest Ecology and Management.
47. MacKenzie DI, Nichols JD, Lachman GB, Droege S, Royle JA, et al. (2002) Estimating site occupancy rates when detection probabilities are less than one. Ecology 83: 2248–2255.
48. Dorazio RM, Royle JA (2005) Estimating size and composition of biological communities by modeling the occurrence of species. Journal of the American Statistical Association 100: 389–398.
49. Etterson MA, Niemi GJ, Danz NP (2009) Estimating the effects of detection heterogeneity and overdispersion on trends estimated from avian point counts. Ecological Applications 19: 2049–2666.
50. Hutchinson GE (1959) Homage to Santa Rosalia or why are there so many kinds of animals? American Naturalist 93: 145–159.
51. Jetz W, Kreft H, Ceballos G, Mutke J (2009) Global associations between terrestrial producer and vertebrate consumer diversity. Proceedings of the Royal Society B: Biological Sciences 276: 269–278.
52. Kissling WD, Field R, Böhning-Gaese K (2008) Spatial patterns of woody plant and bird diversity: functional relationships or environmental effects? Global Ecology and Biogeography 17: 327–339.
53. Connor EF, Simberloff D (1979) The assembly of species communities: chance or competition? Ecology 60: 1132–1140.
54. Vitousek PM, Mooney HA, Lubchenco J, Melillo JM (1997) Human domination of Earth's ecosystems. Science 277: 494–499.
55. Flynn DFB, Gogol-Prokurat M, Nogeire T, Molinari N, Richers BT, et al. (2009) Loss of functional diversity under land use intensification across multiple taxa. Ecology Letters 12: 22–33.

56. Zipkin EF, Royle JA, Dawson DK, Bates S (2010) Multi-species occurrence models to evaluate the effects of conservation and management actions. Biological Conservation 143: 479–484.
57. Cam E, Nichols JD, Sauer JR, Hines JE, Flather CH (2000) Relative species richness and community completeness: birds and urbanization in the mid-Atlantic states. Ecological Applications 10: 1196–1210.
58. Harvey CA, Villalobos JAG (2007) Agroforestry systems conserve species-rich but modified assemblages of tropical birds and bats. Biodiversity and Conservation 16: 2257–2292.
59. Ralph CJ, Droege S, Sauer JR (1995) Monitoring bird populations by point counts. Albany: General Technical Report PSW-GTR-149, U.S. Dept of Agricultural Forest Service, Pacific Southwest Research Station. 187 p.
60. Russell RE, Royle JA, Saab VA, Lehmkuhl JF, Block WM, et al. (2009) Modeling the effects of environmental disturbance on wildlife communities: avian responses to prescribed fire. Ecological Applications 19: 1253–1263.
61. Terborgh J (1977) Bird species-diversity on an Andean elevational gradient. Ecology 58: 1007–1019.
62. Rahbek C (1997) The relationship among area, elevation, and regional species richness in neotropical birds. American Naturalist 149: 875–902.
63. Spiegelhalter DJ, Thomas A, Best NG, Lunn D (2003) WinBUGS Version 1.4 User Manual: MRC BiostatisticsUnit, Cambridge, UK.
64. R Development Core Team (2010) R: A language and environment for statistical computing: R Foundation for Statistical Computing, Vienna, Austria.
65. Sturtz S, Ligges U, Gelman A (2005) R2WinBUGS: a package for running WinBUGS from R. Journal of Statistical Software 12: 1–16.
66. Gelman A, Pardoe I (2007) Average predictive comparisons for models with nonlinearity, interactions, and variance components. Sociological Methodology 37: 23–51.

The Effect of Nitrogen and Glyphosate on Survival and Colonisation of Perennial Grass Species in an Agro-Ecosystem: Does the Relative Importance of Survival Decrease with Competitive Ability?

Christian Damgaard[1]*, Beate Strandberg[1], Solvejg K. Mathiassen[2], Per Kudsk[2]

1 Department of Bioscience, Aarhus University, Vejlsøvej 25, Silkeborg, Denmark, **2** Department of Agroecology, Aarhus University, Flakkebjerg, Slagelse, Denmark

Abstract

The ecological success of a plant species is typically described by the observed change in plant abundance or cover, but in order to more fully understand the fundamental plant ecological processes, it is necessary to inspect the underlying processes of survival and colonization and how they are affected by environmental conditions. A general ecological hypothesis on the effect of environmental gradients on demographic parameters is proposed and tested. The hypothesis is that decreasing fitness or competitive ability along an environmental gradient is associated with an increasing importance of survival for regulating the abundance of the species. The tested hypothesis is related to both the stress gradient hypothesis and whether the importance of competition increases along productivity gradients. The combined effect of nitrogen and glyphosate on the survival and colonization probability of two perennial grass species, *Festuca ovina* and *Agrostis capillaris*, which are known to differ in their responses to both glyphosate and nitrogen treatments, is calculated using pin-point cover data in permanent frames. We found that the relative importance of survival increased with the level of glyphosate for the glyphosate sensitive *A. capillaris* and decreased for the glyphosate tolerant *F. ovina*. Likewise, increasing levels of nitrogen increased the importance of survival for the relative nitrophobic *F. ovina*. Consequently, the proposed hypothesis was corroborated in this specific study. The proposed method will enable predictions of the effects of agricultural practices on community dynamics in a relatively simple setup eliminating the need to quantify all the interaction among the species in the plant community. The method will be immediately useful for the regulation of non-cultivated buffer strips between agricultural fields and semi-natural and natural biotopes such as hedgerows and waterways.

Editor: Kurt O. Reinhart, USDA-ARS, United States of America

Funding: The funders, The Danish Environmental Agency, had no role in study design, data collection and analysis, decision to publish, or preparation of the manuscript.

Competing Interests: The authors have declared that no competing interests exist.

* E-mail: cfd@dmu.dk

Introduction

Biodiversity within European agricultural areas is declining due to the intensification of the agricultural practices [1–5]. More specifically, the repeated application of fertilizers and pesticide usage are generally regarded to affect the botanical composition in small natural and semi-natural biotopes in the neighbourhood of agricultural fields (see references in [6]). This is unfortunate since these biotopes play an important role in maintaining biodiversity in the agro-ecosystem by providing key habitats for flora and fauna, dispersal corridors, and they offer an important source of ecosystem services such as pollination [7–10]. Consequently, in order to predict effects of agricultural practices on the ecosystem dynamics in the neighbouring natural and semi-natural biotopes, it is of interest to study how fertilizer and herbicides affect the ecological success of different plant species commonly found in the agro-ecosystem.

Substantive and increasing knowledge exists on the effects of applying either fertilizers or herbicides on the biodiversity of higher plants in the agro-ecosystem, but studies of the combined effect of fertilizers and herbicide drift on non-target vegetation are still scarce (see references in [6]), and as a response to this important knowledge gap a replicated long-time field experiment sown with a mixture 31 grassland species and treated with various rates of nitrogen and glyphosate was set-up in 2001 [6]. In the long-term field experiment, the vegetation has gradually changed over the years both in respect to species richness and species composition. Generally, application of nitrogen as well as glyphosate has affected species richness negatively. However, at the highest nitrogen level (100 kg N/ha), the application of low dosages of glyphosate to some extent counteracts the effect of nitrogen on species richness. Regardless of the treatment, grasses dominate the vegetation at the experimental field, and the three grasses, common bentgrass (*Agrostis capillaris*), sheep's fescue (*Festuca ovina*) and couch grass (*Elytrigia repens*) make up the main part of the vegetation, but the composition of the grass community depends on the treatment. Previously, it has been documented that both nitrogen and glyphosate affected the competitive interactions of *F.*

ovina and *A. capillaris* [6,11], and the competitive effect of *F. ovina* increased with glyphosate [12]. Generally, *A. capillaris* performed best at low and intermediate glyphosate and nitrogen concentrations, although it seemed to be sensitive to competition from both *F. ovina* and *E. repens*.

The ecological success of a plant species is typically described by the observed change in plant abundance or cover, but in order to more fully understand the fundamental plant ecological processes leading to a change in plant abundance and make ecological predictions of the effect of environmental drivers, it is necessary to investigate and quantify the underlying ecological processes. In principle, this means that all the different interactions between the species in the plant community need to be investigated [13], but this is typically an exceedingly demanding task in multi-species plant communities, and instead we investigate how the vital rates of selected key species are affected by the environmental gradients [14].

Here, we will examine the combined effect of nitrogen and glyphosate on the survival and colonization probability of the two perennial grass species *Festuca ovina* and *Agrostis capillaris*, which are known to differ in their responses to both glyphosate and nitrogen treatments as summarized above. The underlying ecological hypothesis tested in the study is that decreasing fitness or competitive ability of a specific species along an environmental gradient is associated with an increasing importance of survival for regulating the abundance of the species. More specifically, we expect that the importance of survival relative to colonization increases with glyphosate dose for the glyphosate sensitive *A. capillaris* and decreases with glyphosate dose for the glyphosate tolerant *F. ovina*. Furthermore, we expect that that the importance of colonization relative to survival increases with nitrogen for the relative nitrophilous *A. capillaris*, while the opposite will be the case for *F. ovina*. The tested hypothesis is motivated by the notion that in a stressed environment, the resources that may be allocated to reproduction and clonal growth are limited and, consequently, the change in abundance will mainly be regulated by differential survival.

The tested hypothesis is related to both i) the stress gradient hypothesis [15], which predicts that the frequency of facilitative and competitive interactions will vary inversely across abiotic stress gradients, with facilitation being more common in conditions of high abiotic stress relative to more benign abiotic conditions, as well as to ii) the long-standing discussion in plant ecology, known as the Grime-Tilman debate, whether the importance of competition increases along productivity gradients [16,17].

Typically, colonization and survival probabilities are estimated from demographic data of individual plants. However, in many natural plant communities dominated by perennial plants, e.g. grasslands, it is often difficult to distinguish individual plants due to their vegetative growth pattern and, consequently, to obtain reliable demographic data. Instead, it is possible to measure "colonization" and "survival" by considering the turnover of a species at a specific spatial point [18]. Considering such a specific spatial point, then if the species was present at time t but absent at time t+1, we may loosely speak of a mortality event, and if the species is absent from a specific pin-position at time t and present at time t+1, we may loosely speak of a colonization event [18]. We have chosen to use the term "colonization" rather than "recruitment", since the event is defined by a novel occurrence in space [19,20]. However, in the interpretation of the results it is important to remember that the concepts of colonization and mortality have a different meaning than usual in studies where individuals are considered. For perennial plant species that spread clonally by forming well-defined ramets, the concepts of colonization and mortality at a certain pin position make apparent biological sense, whereas for more plastic species with variable sizes, the concepts are inadequate descriptions of the underlying biological causes of a change in plant abundance. Bearing this issue of terminology in mind, we may, generally, assert that the cover of a plant species will increase with colonization and decrease with mortality.

Materials and Methods

Field experiment

The field experiment was designed to study the ecological processes, including establishment, survival and competitive interactions, of a semi-natural ecosystems affected by herbicide (glyphosate) and fertilizer (nitrogen) in a relatively realistic way.

The selected area was a former agricultural field on dry, nutrient poor sandy soil in Djursland, Denmark. The field laid fallow for a couple of years prior to the start of the experiment in 2001. The field is quadrangular and surrounded by forest on two sides (south and west) and separated from the neighbouring fields by 5 meter broad hedgerows on the other sides.

In 2001, the area was deep ploughed down to 60 cm to minimize establishment from the soil seed bank and prepared for the experiment by harrowing and rolling. Thirty-one selected grassland plant species covering different life form strategies (CRS strategies sensu [21]) were sown in spring 2001 [22]. Since 2001, the area has been undisturbed except for the application of the experimental treatments and removal of woody species (trees and bushes) every year prior to herbicide application.

No specific permits were required for the described field studies. The area is a privately own former agricultural field that has been rented on a long-time basis with the sole purpose of doing scientific experiments.

Treatments

The experimental manipulations were set up as a completely randomized block design with 10 replicates of twelve treatments. Each replicate plot was 7 m×7 m with a buffer zone of 1.5 m surrounding the plot. A buffer zone of 10 m separated the experiment from the surrounding vegetation. The buffer zones were also sown with the seed mixture.

The treatments included 4 glyphosate treatments (0; 14.4; 72 and 360 g a.i./ha RoundupBio®, Monsanto Crop Science, Denmark A/S, 360 g/L glyphosate as a isopropylamine salt) and 3 nitrogen treatments (0, 25 and 100 kg N/ha) applied in a full factorial design. The applied glyphosate doses were equal to 0, 1, 5 and 25%, respectively, of the dose recommended for pre- and post-harvest treatment against perennial weeds, which is the most common use of glyphosate in Denmark [23]. All 120 plots received phosphorus (53 kg/ha), potassium (141 kg/ha), sulphur (50 kg/ha) and copper (0.7 kg/ha) every year.

For the herbicide applications, experimental spraying equipment was used. The boom was fitted with Lurmark Low-drift LD 015 Green nozzles operated at a pressure of 2.0 bars delivering a spray volume of 300 L/ha. The wind speed on the days selected for spraying was very low (0–2 m/s). It did not rain during the days following the application. The amount of fertilizer was weighed individually for each plot and spread by hand. The plots were treated by glyphosate for the first time on 24 August 2001. Since then, glyphosate and fertilizer treatments were carried out once every year in mid-May [11].

Plant species

F. ovina and *A. capillaris* are perennial grasses. They both have a caespitose growth form, but the turfs formed by *F. ovina* generally are much denser than those formed by *A. capillaris*. In contrast to *A. capillaris* which is deciduous, *F. ovina* is winter green and productive through the winter as long as temperatures are above zero. Furthermore, *F. ovina* has narrow, curled and waxy leaves, whereas *A. capillaris* has broader, flat leaves without any wax cover.

F. ovina has previously been shown to be more tolerant to glyphosate than *A. capillaris* [11,24], whereas *A. capillaris* is a more nitrophilous species than *F. ovina* (the Ellenberg N values [25] of the two species are 4 and 1, respectively)

Sampling

In order to study the effect of nitrogen and glyphosate on *F. ovina* and *A. capillaris*, one permanent 0.5 m×0.5 m quadrate was placed within each of the 120 plots in 2007. The quadrate was not placed randomly, but in such a way that both *F. ovina* and *A. capillaris* were noticeably abundant in the quadrate. Local presence-absence data of the two species were determined by the pin-point method [26,27] using a pin-point frame with the same dimension as the quadrate. The frame had 25 pin-positions regularly placed at a distance of 10 cm. At each position, a sharply pointed pin with a diameter of 0.5 mm was passed vertically through the vegetation and presence-absence data of the two species were recorded for each pin.

The sampling was performed three times a year for three years in the period 2007–2009 [6], but in this analysis, the sampling that was made in the spring approximately two weeks after the herbicide application was used.

Model

The quantitative effect of survival and colonization processes on population growth of individuals is most effectively summarized using population models [28], and the calculation of elasticity (the relative contribution of different demographic parameters on population growth rates) has particularly been shown to be a powerful tool to investigate the importance of different demographic variables in determining population growth [18].

If absence-presence data of species *A* from two successive recordings from the same pin-position are considered, there are four possible transition events and corresponding probabilities (Table 1). These transition probabilities depend on i) the probability (p) that a plant of species *A* is present at time *t*, ii) survival; the probability (s) that a plant of species *A* is present at the pin-position at both time *t* and time *t*+1, and iii) colonization; the probability (c) that a plant of species *A* is present at the pin-position at time *t*+1 but was absent at time *t* [18]. The possibility of a combined mortality and colonization events is ignored.

The change in the probability that species *A* is present in year *t* to *t*+1 is denoted by π and may be defined, analogous to the population growth rate λ of individuals [28], as the ratio between the probability p' that species *A* is present at time *t*+1 and the probability p that species *A* is present at time t:

$$\pi = \frac{p'}{p} = \frac{p_{X_1} + p_{X_4}}{p_{X_1} + p_{X_3}} = \frac{c}{p} + s - c\,s, \tag{1}$$

where $p_{X\,i}$ are defined in Table 1.

From the above definition it is apparent that if $\pi < 1$, then the probability that species *A* is present decreases, and if $\pi > 1$, then the probability that species *A* is present increases. Furthermore, the change in the probability that species *A* is present is a function of

Table 1. The four possible events of two successive recordings of presence absence data and their corresponding probabilities, *p*: current probability, *c*: colonization probability, *s*: survival probability.

Event	Description	Probability ($p_{X\,i}$)
X_1: A_t, A_{t+1}	A plant of species A was present in year *t* and was also present in year *t*+1	$p(s+(1-s)c)$
X_2: nA_t, nA_{t+1}	A plant of species A was not present in year *t* and was also not present in year *t*+1	$(1-p)(1-c)$
X_3: A_t, nA_{t+1}	A plant of species A was present in year *t* but was not present in year *t*+1 (indicates a possibly event of mortality)	$p(1-s)(1-c)$
X_4: nA_t, A_{t+1}	A plant of species A was not present in year *t* but was present in year *t*+1 (indicates a possibly event of colonization)	$(1-p)c$

the current probability (p), the colonization probability and the survival probability probabilities (Fig. 1). Note also that the plant cover of species *A* at time *t* is estimated by taking the mean of p across space.

The change in the probability of being present, as defined above in equation (1), is always positive, which ensures that it is possible to calculate both the sensitivity and elasticity of the change [28]. The sensitivity of the change in the probability that species *A* is present is a function of the colonization and survival probabilities which are defined as:

$$\frac{\partial \pi}{\partial c} = \frac{1}{p} - s, \quad \frac{\partial \pi}{\partial s} = 1 - c. \tag{2}$$

Likewise, the elasticity of the change in the probability that species *A* is present is a function of the colonization and survival probabilities which are defined as:

$$\frac{c}{\pi}\frac{\partial \pi}{\partial c} = \frac{c - (p\,c\,s)}{c + (p\,s) - (p\,c\,s)}, \quad \frac{s}{\pi}\frac{\partial \pi}{\partial s} = 1 - \frac{c}{c + (p\,s)(1-c)}. \tag{3}$$

Estimation and statistical inferences

The colonization and survival probabilities may be estimated from data of the four possible events of two successive recordings of presence-absence data (X_1, X_2, X_3, X_4), which are defined in Table 1, by maximizing the likelihood function of the multinomial distribution $L(s,c)$ using the transition probabilities specified in Table 1 and assuming that the plants are distributed homogenously across the site of the investigation so that the probability that species *A* is present at time *t* ($p = p_{X_1} + p_{X_3}$) may be estimated as $\hat{p} = (X_1 + X_3)/n$:

$$L(s,c) = \frac{(X_1 + X_2 + X_3 + X_4)!}{X_1! X_2! X_3! X_4!} {p_{X_1}}^{X_1} {p_{X_2}}^{X_2} {p_{X_3}}^{X_3} {p_{X_4}}^{X_4}, \tag{4}$$

Different hypotheses on the treatment effect on the maximum likelihood estimates of elasticity of survival probabilities on cover change were tested using linear models where the residual error

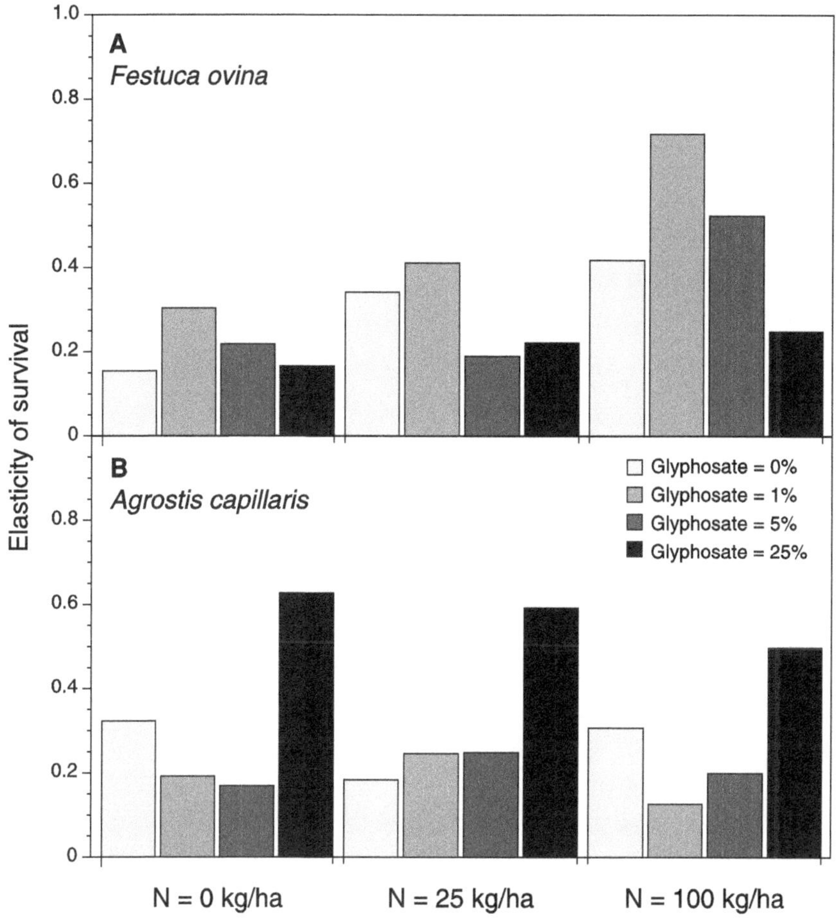

Figure 1. The elasticity of the survival probability of *Festuca ovina* (a) and *Agrostis capillaris* (b) at different treatment levels of nitrogen and glyphosate.

was assumed to be normally distributed (this assumption was checked using residual plots). Only the elasticity of the survival probability was analyzed, since the elasticity of colonization probability, due to the mathematical definition of elasticities (3), showed similar but opposite results than the elasticity of survival probability.

Results

The maximum likelihood estimates as well as the sensitivity and elasticity of the survival and colonization probability of the two grass species *F. ovina* and *A. capillaris* at different treatment levels of nitrogen and glyphosate are shown in Table 2 and 3 for both yearly changes and all years together. However, the main findings of the study are most easily communicated by the calculated elasticity of one of the complementing demographic parameters. Thus, the effect of nitrogen and glyphosate on the survival elasticity (the relative contribution of the survival probability of the change in regulating plant cover) is shown in Fig. 1 for both species for all years and analyzed using linear models (Table 4 and 5).

As expected, *F. ovina* and *A. capillaris* responded differently to the nitrogen and glyphosate treatments. When the level of glyphosate increased, survival became more important than colonization for regulating the cover of the glyphosate sensitive *A. capillaris* (Fig. 1a; Table 5: $P = 0.0002$), whereas survival tended to become less important for the glyphosate tolerant *F. ovina* (Fig. 1b; Table 4: $P = 0.047$). Increasing levels of nitrogen increased the importance of survival for regulating the cover of the relative nitrophobic *F. ovina* (Fig. 1a; Table 4: $P = 0.0052$), whereas there was no significant effect of nitrogen on the elasticity of survival for *A. capillaris* (Fig. 1b; Table 5: $P = 0.43$). There was no significant interaction effect of glyphosate and nitrogen on the elasticity of survival for the two investigated species.

Discussion

As mentioned previously, the calculation of elasticity of different birth and death processes on population growth rates has been shown to be a powerful tool for investigating the importance of different demographic variables in determining population growth. More specifically, in this study the elasticity of survival has proven useful for quantifying the relative importance of survival for regulating plant cover along an environmental gradient.

The general hypothesis tested in the study, i.e. that decreasing fitness or competitive ability of a specific species along an environmental gradient is associated with an increasing importance of survival for regulating the abundance of the species (as measured by the elasticity of survival), was corroborated in the study. More specifically, the relative importance of survival increased with the level of glyphosate for the glyphosate sensitive *A. capillaris* and decreased for the glyphosate tolerant *F. ovina*.

Table 2. The maximum likelihood estimates as well as the sensitivity and elasticity of the survival (s) and colonization (c) probability of *Festuca ovina* at different treatment levels of nitrogen and glyphosate.

Nitrogen	Glyphosate	Year	s	c	*p*	Sen. s	Sens. c	Elast. s	Elast. c
0	0	2007-2008	0.427	0.422	0.906	0.578	1.136	0.272	0.529
0	0	2008-2009	0.610	0.752	1.448	0.248	1.114	0.104	0.579
0	0	All	0.451	0.600	-	0.400	1.189	0.155	0.613
0	1	2007-2008	0.588	0.293	0.963	0.707	1.277	0.432	0.389
0	1	2008-2009	0.642	0.545	1.349	0.455	1.296	0.216	0.524
0	1	All	0.605	0.422	-	0.578	1.296	0.304	0.475
0	5	2007-2008	0.543	0.371	0.948	0.629	1.091	0.360	0.427
0	5	2008-2009	0.561	0.686	1.359	0.314	1.163	0.130	0.587
0	5	All	0.538	0.535	-	0.465	1.139	0.218	0.531
0	25	2007-2008	0.410	0.441	0.835	0.559	0.964	0.274	0.509
0	25	2008-2009	0.429	0.643	1.211	0.357	1.216	0.127	0.646
0	25	All	0.380	0.560	-	0.440	1.117	0.166	0.622
25	0	2007-2008	0.537	0.231	0.808	0.769	1.175	0.511	0.336
25	0	2008-2009	0.692	0.477	1.373	0.523	1.427	0.263	0.496
25	0	All	0.574	0.369	-	0.631	1.320	0.342	0.459
25	1	2007-2008	0.674	0.237	0.902	0.763	0.960	0.570	0.252
25	1	2008-2009	0.762	0.482	1.268	0.518	1.049	0.311	0.399
25	1	All	0.701	0.368	-	0.632	1.017	0.411	0.349
25	5	2007-2008	0.449	0.432	0.869	0.568	0.971	0.293	0.483
25	5	2008-2009	0.812	0.722	1.405	0.278	0.822	0.161	0.422
25	5	All	0.525	0.596	-	0.404	0.994	0.190	0.530
25	25	2007-2008	0.627	0.483	0.953	0.517	0.675	0.340	0.342
25	25	2008-2009	0.748	0.761	1.219	0.239	0.618	0.147	0.386
25	25	All	0.652	0.632	-	0.368	0.681	0.222	0.398
100	0	2007-2008	0.509	0.126	1.257	0.874	5.919	0.354	0.595
100	0	2008-2009	0.528	0.092	0.976	0.908	4.853	0.491	0.459
100	0	All	0.519	0.110	-	0.890	5.340	0.418	0.530
100	1	2007-2008	0.391	0.005	0.409	0.995	3.397	0.950	0.045
100	1	2008-2009	0.764	0.058	1.259	0.942	8.495	0.571	0.393
100	1	All	0.488	0.034	-	0.966	4.889	0.718	0.256
100	5	2007-2008	0.651	0.178	0.922	0.822	1.523	0.581	0.294
100	5	2008-2009	0.535	0.208	0.915	0.792	1.823	0.463	0.415
100	5	All	0.596	0.194	-	0.806	1.666	0.523	0.351
100	25	2007-2008	0.471	0.385	0.849	0.615	0.983	0.341	0.445
100	25	2008-2009	0.782	0.618	1.354	0.382	0.926	0.221	0.422
100	25	All	0.555	0.517	-	0.483	1.016	0.248	0.486

Likewise, increasing levels of nitrogen increased the importance of survival for the relative nitrophobic *F. ovina*. Naturally, the generality of the proposed hypothesis needs to be tested in other biological systems and along different environmental gradients.

As mentioned above, the tested hypothesis is related to both i) the stress gradient hypothesis [15], which predicts that the frequency of facilitative and competitive interactions will vary inversely across abiotic stress gradients, with facilitation being more common in conditions of high abiotic stress relative to more benign abiotic conditions, as well as to ii) the long-standing discussion in plant ecology, known as the Grime-Tilman debate, whether the importance of competition increases along productivity gradients [16,17]. However, assigning the frequency or importance to the competitive interactions is not trivial and e.g. depends on the investigated life stage [15] and is, typically, confounded with the underlying stress or productivity gradient [29]. Consequently, in the cases where a limited experimental design precludes the parameterization of a full population ecological model, it may be more robust to focus on the underlying demographic parameters, e.g. survival and colonization, in investigations of environmental gradients instead of calculating the importance of competitive interactions.

It is important to remember that the estimated colonization and survival probabilities as well as the connected elasticities are calculated from repeated pin-point cover data and only are meaningful for plant species with a constant or modular type of

Table 3. The maximum likelihood estimates as well as the sensitivity and elasticity of the survival (s) and colonization (c) probability of *Agrostis capillaris* at different treatment levels of nitrogen and glyphosate.

Nitrogen	Glyphosate	Year	s	c	*p*	Sen. s	Sens. c	Elast. s	Elast. c
0	0	2007–2008	0.179	0.072	0.595	0.928	5.774	0.279	0.699
0	0	2008–2009	0.519	0.084	1.320	0.916	9.481	0.360	0.607
0	0	All	0.304	0.079	-	0.921	7.159	0.323	0.649
0	1	2007–2008	0.087	0.144	0.727	0.856	4.458	0.103	0.880
0	1	2008–2009	0.319	0.119	1.025	0.881	5.931	0.274	0.689
0	1	All	0.189	0.131	-	0.869	5.075	0.192	0.779
0	5	2007–2008	0.199	0.108	0.661	0.892	4.265	0.269	0.699
0	5	2008–2009	0.071	0.127	0.919	0.873	6.685	0.068	0.922
0	5	All	0.147	0.118	-	0.882	5.230	0.169	0.808
0	25	2007–2008	0.057	0.005	0.082	0.995	5.046	0.689	0.308
0	25	2008–2009	0.247	0.004	0.500	0.996	62.253	0.492	0.506
0	25	All	0.071	0.004	-	0.996	9.363	0.627	0.370
25	0	2007–2008	0.101	0.082	0.350	0.918	3.024	0.265	0.712
25	0	2008–2009	0.067	0.081	0.786	0.919	8.861	0.079	0.914
25	0	All	0.093	0.082	-	0.918	4.537	0.184	0.800
25	1	2007–2008	0.242	0.129	0.524	0.871	2.185	0.402	0.539
25	1	2008–2009	0.227	0.209	1.148	0.791	4.402	0.157	0.802
25	1	All	0.220	0.175	-	0.825	2.964	0.246	0.702
25	5	2007–2008	0.157	0.093	0.485	0.907	3.520	0.292	0.678
25	5	2008–2009	0.015	0.046	0.364	0.954	7.561	0.040	0.958
25	5	All	0.119	0.068	-	0.932	4.832	0.248	0.734
25	25	2007–2008	0.102	0.010	0.148	0.990	4.528	0.681	0.312
25	25	2008–2009	0.367	0.012	0.750	0.988	30.883	0.483	0.510
25	25	All	0.135	0.011	-	0.989	7.929	0.592	0.401
100	0	2007–2008	0.125	0.037	0.342	0.963	5.796	0.352	0.634
100	0	2008–2009	0.148	0.033	0.588	0.967	13.146	0.243	0.749
100	0	All	0.133	0.035	-	0.965	8.060	0.307	0.682
100	1	2007–2008	0.041	0.014	0.135	0.986	6.716	0.296	0.700
100	1	2008–2009	0.000	0.020	1.000	0.980	50.000	0.000	1.000
100	1	All	0.031	0.017	-	0.983	11.874	0.127	0.871
100	5	2007–2008	0.029	0.047	0.175	0.953	3.096	0.160	0.832
100	5	2008–2009	0.051	0.021	0.429	0.979	17.806	0.117	0.880
100	5	All	0.044	0.032	-	0.968	5.275	0.200	0.794
100	25	2007–2008	0.092	0.000	0.092	1.000	3.197	1.000	0.000
100	25	2008–2009	0.000	0.020	0.714	0.980	35.143	0.000	1.000
100	25	All	0.073	0.012	-	0.988	5.943	0.498	0.496

growth. For plant species with individual size variation from year to year and with a variable spatial arrangement of tissues, the estimated variables will rather be a characterization of the growth form of the individual plant than an estimation of demographic parameters. Since grass species generally, including both *F. ovina* and *A. capillaris*, are relatively modular plant species, we think that the estimated parameters to a certain degree reflect the "population dynamic" of the ramets of the two grass species.

In many countries, legislation or incentives have been put in place to increase the area of non-cultivated buffer strips between agricultural fields and semi-natural and natural biotopes such as hedgerows and waterways. The empirical data available to determine e.g. the width of the buffer strips required to protect those biotopes is often inadequate, partly because it is resource demanding to conduct such experiments. Furthermore, it can be difficult to generalize on the basis of data originating from few sites. We propose an alternative approach making use of the model presented in this paper. If data were available on the susceptibility of the key species in the biotope in question and this information was combined with existing spray drift models (e.g. [30]), the impact of herbicides could be simulated for different widths of the buffer strips. If herbicide sensitivity data are not already available, they would be easy and relatively inexpensive to generate. Simulating effects of nitrogen would be more difficult because nitrogen transfer across a buffer strip is more complex and less understood than that of pesticides. Nonetheless, simple models

Table 4. Analysis of variance of the elasticity of the survival probability on the change in the cover of *Festuca ovina* for all years.

	DF	SS	MS	F-Statistic	P-Value
Nitrogen	1	0.1487	0.1487	14.4757	0.0052
Glyphosate	1	0.0566	0.0566	5.5079	0.0469
Glyphosate * Nitrogen	1	0.0237	0.0237	2.3059	0.1674
Error	8	0.0822	0.0103		
Total	11	0.3111			

Fitted model: 0.236+0.00342 Nitrogen−0.00248 Glyphosate−0.000103 Glyphosate * Nitrogen (adjusted r^2: 64%).

Table 5. Analysis of variance of the elasticity of the survival probability on the change in the cover of *Agrostis capillaris* for all years.

	DF	SS	MS	F-Statistic	P-Value
Nitrogen	1	0.0045	0.0045	0.6797	0.4336
Glyphosate	1	0.2574	0.2574	39.0347	0.0002
Glyphosate * Nitrogen	1	0.0045	0.0045	0.6825	0.4327
Error	8	0.0527	0.0066		
Total	11	0.3191			

Fitted model: 0.202−0.000106 Nitrogen+0.0163 Glyphosate−0.000045 Glyphosate * Nitrogen (adjusted r^2: 77%).

describing nitrogen gradients across buffer strips could be developed and incorporated in the model. Besides predicting the combined effect of herbicide and nitrogen exposure on the species composition, the model could also provide insight into the relative impact of herbicides and nitrogen.

Author Contributions

Conceived and designed the experiments: CD BS. Performed the experiments: BS. Analyzed the data: CD. Wrote the paper: CD BS SM PK.

References

1. Firbank LG, Petit S, Smart S, Blain A, Fuller RJ (2008) Assessing the impacts of agricultural intensification on biodiversity: a British perspective. Philosophical Transactions of the Royal Society B-Biological Sciences 363: 777–787.
2. Flohre A, Fischer C, Aavik T, Bengtsson J, Berendse F, et al. (2011) Agricultural intensification and biodiversity partitioning in European landscapes comparing plants, carabids, and birds. Ecological Applications 21: 1772–1781.
3. Schuepp C, Herrmann JD, Herzog F, Schmidt-Entling MH (2011) Differential effects of habitat isolation and landscape composition on wasps, bees, and their enemies. Oecologia 165: 713–721.
4. Stoate C, Baldi A, Beja P, Boatman ND, Herzon I, et al. (2009) Ecological impacts of early 21st century agricultural change in Europe - A review. Journal of Environmental Management 91: 22–46.
5. Storkey J, Meyer S, Still KS, Leuschner C (2011) The impact of agricultural intensification and land-use change on European arable flora. Proceedings of the Royal Society, Series B.
6. Damgaard C, Strandberg B, Mathiassen SK, Kudsk P (2011) The combined effect of nitrogen and glyphosate on the competitive growth, survival and establishment of *Festuca ovina* and *Agrostis capillaris*. Agriculture Ecosystems & Environment 142: 374–381.
7. Corbit M, Marks PL, Gardescu S (1999) Hedgerows as habitat corridors for forest herbs in central New York, USA. Journal of Ecology 87: 220–232.
8. Croxton PJ, Sparks TH (2002) A farm-scale evaluation of the influence of hedgerow cutting frequency on hawthorn (Crataegus monogyna) berry yields. Agriculture, Ecosystems & Environment 93: 437–439.
9. Hannon LE, Sisk TD (2009) Hedgerows in an agri-natural landscape: Potential habitat value for native bees. Biological Conservation 142: 2140–2154.
10. Roy Vr, de Blois S (2008) Evaluating hedgerow corridors for the conservation of native forest herb diversity. Biological Conservation 141: 298–307.
11. Strandberg B, Mathiassen SK, Bruus M, Kjær C, Damgaard C, et al. (2012) Effects of herbicides on non-target plants: How do effects in standard plant test relate to effects in natural habitats? : Miljøstyrelsen. 115 p.
12. Damgaard C, Strandberg B, Mathiassen SK, Kudsk P (submitted) The effect of glyphosate on the growth and competitive effect of *Festuca ovina* and *Agrostis capillaris* in semi-natural grasslands
13. Damgaard C (2003) Modelling plant competition along an environmental gradient. Ecological Modelling 170: 45–53.
14. Harper JL (1977) Population biology of plants. London: Academic Press.
15. Maestre FT, Callaway RM, Valladares F, Lortie CJ (2009) Refining the stress-gradient hypothesis for competition and facilitation in plant communities. Journal of Ecology 97: 199–205.
16. Grace JB (1991) A clarification of the debate between Grime and Tilman. Functional Ecology 5: 583–587.
17. Damgaard C, Fayolle A (2010) Measuring the importance of competition: a new formulation of the problem. Journal of Ecology 98: 1–6.
18. Damgaard C, Merlin A, Mesléard F, Bonis A (2011) The demography of space occupancy: measuring plant colonisation and survival probabilities using repeated pin-point measurements. Methods in Ecology and Evolution 2: 110–115.
19. Adler PB, HilleRisLambers J, Kyriakidis PC, Guan Q, Levine JM (2006) Climate variability has a stabilizing effect on the coexistence of prairie grasses. Proceedings of the National Academy of Sciences (USA) 103: 12793–12798.
20. Adler PB, HilleRisLambers J, Levine JM (2009) Weak effect of climate variability on coexistince in a sagebrush steppe community. Ecology 90: 3303–3312.
21. Grime P (2001) Plant Strategies, Vegetation Processes, and Ecosystem Properties: Wiley.
22. Bruus Pedersen M, Aude E, Tybirk K (2004) Adskillelse af effekter af herbicider og kvælstof på vegetation og leddyr i hegn og græslandsvegetation. 103 p.
23. Jensen PK, N JL, C NG, F NS, K P, et al. (2012) Vejledning i Planteværn 2012. Århus: Landbrugsforlaget.
24. Holst N, Axelsen JA, Bruus M, Damgaard C, Kudsk P, et al. (2008) Sprøjtepraksis i sædskifter med og uden glyphosattolerante afgrøder. Effekter på floraen i mark og hegn.
25. Ellenberg H, Weber HE, Düll R, Wirth V, Werner W, et al. (1992) Zeigerwerte von Pflanzen in Mitteleuropa. Scripta Geobotanica 18: 1–258.
26. Kent M, Coker P (1992) Vegetation description and analysis. London: Belhaven Press.
27. Levy EB, Madden EA (1933) The point method of pasture analyses. New Zealand Journal of Agriculture 46: 267–279.
28. Caswell H (2001) Matrix population models: Construction, analysis, and interpretation. Sunderland: Sinauer.
29. Rees M, Childs DZ, Freckleton RP (2012) Assessing the role of competition and stress: a critique of importance indices and the development of a new approach. Journal of Ecology 100: 577–585.
30. Rautmann D, Streloke M, Winkler R (2001) New basic drift values in the authorisation procedure for plant protection products. In: Forster R, Streloke M, editors; Biologischen Bundesanstalt für Land- und Fortwirtschaft, Berlin and Braunschweig, Germany.

10

Effect of Weed Management and Seed Rate on Crop Growth under Direct Dry Seeded Rice Systems in Bangladesh

Sharif Ahmed[1,2]*, Muhammad Salim[2], Bhagirath S. Chauhan[3]

1 Crop and Environmental Sciences Division, International Rice Research Institute, Metro Manila, Los Baños, Philippines, 2 Agronomy Division, Bangladesh Agricultural University, Mymensingh, Bangladesh, 3 Queensland Alliance for Agriculture and Food Innovation (QAAFI), The University of Queensland, Toowoomba, Queensland, Australia

Abstract

Weeds are a major constraint to the success of dry-seeded rice (DSR). The main means of managing these in a DSR system is through chemical weed control using herbicides. However, the use of herbicides alone may not be sustainable in the long term. Approaches that aim for high crop competitiveness therefore need to be exploited. One such approach is the use of high rice seeding rates. Experiments were conducted in the aman (wet) seasons of 2012 and 2013 in Bangladesh to evaluate the effect of weed infestation level (partially-weedy and weed-free) and rice seeding rate (20, 40, 60, 80, and 100 kg ha^{-1}) on weed and crop growth in DSR. Under weed-free conditions, higher crop yields (5.1 and 5.2 t ha^{-1} in the 2012 and 2013 seasons, respectively) were obtained at the seeding rate of 40 kg ha^{-1} and thereafter, yield decreased slightly beyond 40 kg seed ha^{-1}. Under partially-weedy conditions, yield increased by 30 to 33% (2.0–2.2 and 2.9–3.2 t ha^{-1} in the 2012 and 2013 seasons, respectively) with increase in seeding rate from 20 to 100 kg ha^{-1}. In the partially-weedy plots, weed biomass decreased by 41–60% and 54–56% at 35 days after sowing and at crop anthesis, respectively, when seeding rate increased from 20 to 100 kg ha^{-1}. Results from our study suggest that increasing seeding rates in DSR can suppress weed growth and reduce grain yield losses from weed competition.

Editor: Anil Shrestha, California State University, Fresno, CA, United States of America

Funding: International Fund for Agricultural Development (IFAD) and Cereal Systems Initiative for South Asia (CSISA, Bangladesh). The funders had no role in study design, data collection and analysis, decision to publish, or preparation of the manuscript.

Competing Interests: The authors have declared that no competing interests exist.

* Email: s.ahmed@irri.org

Introduction

As in many Asian countries, conventional puddled-transplanted rice (CPTR) is the major system of rice production in Bangladesh. This system is becoming less profitable and less sustainable because of its high labor, water, and energy requirements. Agricultural laborers are becoming involved in other non-farm jobs such as textile, garments, and other industries. As a result, it is often difficult to find labor during transplanting. In addition, the level of groundwater, the major source of irrigation for cultivation has been declining due to excessive withdrawal. [1]. It has been predicted that a significant amount of rice area in Asia may suffer from physical and economic water scarcity by 2025 [2]. To overcome water and labor problems, farmers in many Asian countries including Bangladesh are increasingly adopting dry-seeded rice (DSR) production systems.

DSR has several advantages over CPTR. DSR systems save labor by eliminating the need for managing a nursery bed, seedling uprooting, puddling, and transplanting. In DSR fields, there is no need to keep standing water at all the growth stages, which helps save a huge amount of water compared with CPTR fields [3]. DSR systems can also be harvested at least 7–10 days before CPTR systems [4], which facilitates the timely planting of subsequent crops [e.g., wheat (*Triticum aestivum* L.), mustard (*Brassica juncea* (L.) Czern), lentil (*Lens culinaris* Medic.), and potato (*Solanum tuberosum* L.)]. In DSR, however, weeds are a major constraint to achieving high yield [5]. The main reasons for the heightened weed problem in DSR are that weeds and rice seedlings emerge simultaneously, which reduces the competitive advantage of the crop, and that alternate events of wetting and drying enhances the growth of weeds [6]. If weeds are not controlled in DSR, yield losses could exceed 90%, depending on factors such as tillage system, type of cultivar, seeding rate, water management, types of weed flora, and field conditions [7]. Timely weed management is thus crucial to the success of DSR.

In Bangladesh, weeding is commonly done manually. The practice, however, is becoming less common due to labor scarcity. In recent years, chemical weed control has increased in Bangladesh, but reliance only on herbicides is not sustainable as the continuous use of the same mode of action over a long period may result in the evolution of herbicide-resistant weed biotypes [8], [9], [10]. In addition, it is not possible to control a range of weeds using a single herbicide. Weed management in DSR therefore needs an integrated approach [7], [11], [12].

Higher seeding rate is one approach that helps increase crop competitiveness against weeds [7], [13], [14]. High seeding rates facilitate quick canopy closure, which helps suppress weeds more effectively. At low seeding rates, crop plants take more time to close their canopy, which encourages weed growth [15]. High seeding rates improve the ability of crops to suppress weeds and can reduce yield loss under partially-weedy conditions. Similar results have been reported elsewhere for different crops, including barley (*Hordeum vulgare* L.) [16], wheat [17], and soybean [*Glycine max* (L.) Merr] [18]. In the Philippines, seeding rates of 100–300 viable seeds m^{-2} increased yield and decreased weed biomass in aerobic rice systems [19]. In lowland rice, higher seeding rates favored rice against weeds and, at the same time, increased yields under weedy conditions [20].

Studies in different countries suggest that crop seeding rate can affect weed growth and rice grain yield. Such information, however, is not available in Bangladesh. The objective of this study was, therefore, to evaluate the effect of weed management and seeding rate on weed density, weed biomass, crop yield, and yield components in DSR.

Materials and Methods

The authors confirm that no permission was needed to conduct the experiment, as well as that the field studies did not involve endangered or protected species.

Field experiments were conducted during the aman seasons (June to October) of 2012 and 2013 at the Regional Agricultural Research Station (RARS) of the Bangladesh Agricultural Research Institute (BARI) in Jessore (23°11′ N, 89°14′ E; 16 m above mean sea level). Historically, the area is known as the High Ganges River Floodplain and is predominantly highland. The climate of the area is subtropical, with an average annual rainfall of 1,590 mm, minimum temperatures of 6–9°C in January, and maximum temperatures of 36–44°C in April and May. Rainfall recorded at the site during the experimental periods is presented in Figure 1. Soil in the experimental fields at 0–15-cm depth was clay loam in texture with a bulk density of 1.60 Mg m^{-3}, pH of 7.8, organic carbon of 1%, sand of 31%, silt of 31%, and clay of 38%. The experimental area was dry-cultivated using a four-wheel tractor before crop sowing. The experiments in each year were arranged in a split-plot design with weed infestation level (partially-weedy and weed-free) in the main plots and seeding rate (20, 40, 60, 80, and 100 kg ha^{-1}) in the subplots. The area of each subplot was 13.5 m^2; three replications were made in each season. The rice cultivar BRRI dhan49 (135-d duration) was used in both seasons.

Dry rice seeds were sown by hand at different seeding rates with row spacing of 20 cm. In both years, the crop was sown on June 20 and harvested at the end of October. Fertilizers were applied at the rates of 120-15-48-12-2.2 kg ha^{-1} of N, P, K, S, and Zn in the form of urea, triple superphosphate (TSP), muriate of potash (MoP), gypsum, and zinc sulphate, respectively. Full doses of TSP, MoP, gypsum, and zinc sulphate were applied immediately before sowing. Urea was applied in four equal splits at 14 d after sowing (DAS), at the start of tillering (35 DAS), at maximum tillering (50 DAS), and at panicle initiation (70 DAS).

In weed-free plots, weeds were managed initially by the application of 80 g ai ha^{-1} of oxadiargyl (Topstar 400 SC, Bayer Crop Science Limited, Bangladesh) at 2 DAS and later with hand weeding (at 15, 25, 35, 50 and 70 DAS). The herbicide was applied using a knapsack-sprayer, attached with three flat-fan nozzles on a boom that delivered 450 L solution ha^{-1}. In the partially-weedy plots, emerged weeds were manually removed once at 36 DAS and then weeds were allowed to grow for the rest of the season. One hand weeding was done in these plots because allowing weeds to grow throughout the season could have resulted

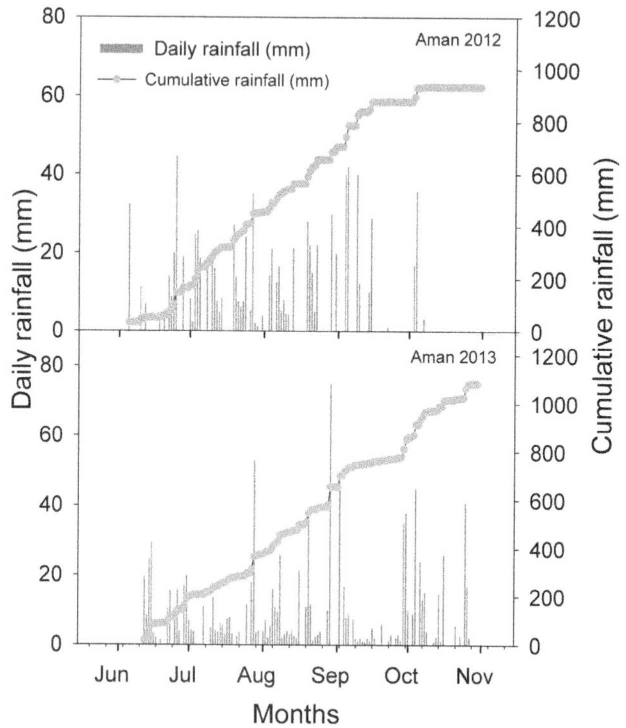

Figure 1. Daily and cumulative rainfall during the aman seasons of 2012 and 2013.

in more than 90% yield loss in DSR systems [7]. In addition, it is not common for farmers in irrigated areas to leave their rice fields infested with weeds throughout the season.

The field was irrigated lightly immediately after sowing, after which irrigation was made based on tensiometer readings using a threshold value of 15 kPa at 15 cm soil depth. At each irrigation, water was added until its depth on the soil surface reached 5 cm. At 50 DAS, 300 g ai ha^{-1} of the fungicide tebuconazole + trifloxystrobin 75 WP (Nativo 75 WP, Bayer Crop Science Limited, Bangladesh) was applied and, at 70 DAS, 300 g ai ha^{-1} of fipronil 3G (Regent 3GR, BASF Bangladesh Limited) was applied to control blast and stem borers, respectively.

Rice plant density was evaluated at 14 DAS from four randomly selected 1-m row lengths in each plot. Weed density and weed biomass were measured at 35 DAS and at anthesis. At each sampling time, two quadrats of 40 cm×40 cm were placed randomly in each plot and weeds were collected from each quadrat. Collected weeds were clustered by group (i.e., grass, broadleaf, and sedges) and counted. Weed biomass was measured group-wise after the samples were oven-dried at 70°C for 72 h. Weed density and biomass data were converted to density or biomass per m^2. The number of rice tillers and rice dry biomass were determined through collection of samples from the same quadrats used for weed sampling, and on the same date. At harvest, rice panicles were counted from four randomly placed 1-m row lengths in each plot. Grains per panicle (filled and unfilled) were counted by randomly sampling 20 panicles per plot. Rice grain yield was determined from the harvested area of 8.8 (4.0 m×2.2 m) m^2. Grain yield was converted to t ha^{-1} at 14% moisture content. Data were analyzed using ANOVA to evaluate differences between treatments and the means were separated using least significant differences (LSD) at 5% level of significance (Crop Stat 7.2; International Rice Research Institute, Philippines).

In a combined analysis of data, the interactions of years with the treatments were significant therefore, the data were presented separately. Weed density and biomass data were transformed using square-root transformation ($\sqrt{x+0.5}$) before analyses. Transformation, however, did not improve homogeneity. Therefore, original values were used for analysis and presentation. The relationship of seeding rate (kg ha^{-1}) with weed biomass (g m^{-2}) and rice biomass (g m^{-2}) were assessed using linear regression analysis (SigmaPlot 11, Systat Software Inc.). In the regression analysis, the values of all individual replications were included.

Results and Discussion

Rice plant density

Rice plant density was not influenced by weed infestation level but it was strongly influenced by seeding rate in both seasons (Table 1). Rice plant density ranged from 100 to 397 plants m^{-2} in 2012 and 100 to 415 plants m^{-2} in 2013 (data not shown). Plant density increased linearly with increase in seeding rates in both years (Figure 2).

Weed density and biomass

The common weed species found at the experimental site were tropic ageratum (*Ageratum conyzoides* L.), spiny amaranth (*Amaranthus spinosus* L.), scarlet pimpernel (*Anagalis arvensis* L.), fringed spiderflower (*Cleome rutidosperma* D.C.), celosia (*Celosia argentea* L), bermudagrass [*Cynodon dactylon* (L.) Pers.], purple nutsedge (*Cyperus rotundus* L.), crowfootgrass [*Dactyloctenium aegyptium* (L.) Willd.], southern crabgrass [*Digitaria ciliaris* (Retz.) Koel.], junglerice [*Echinochloa colona* (L.) Link], goosegrass [*Eleusine indica* (L) Gaertn], fringed quickweed [*Galinsoga ciliata* (Raf.) Blake], and niruri (*Phyllanthus niruri* L.). Results are presented for each weed group (i.e., grass, broadleaf, and sedges).

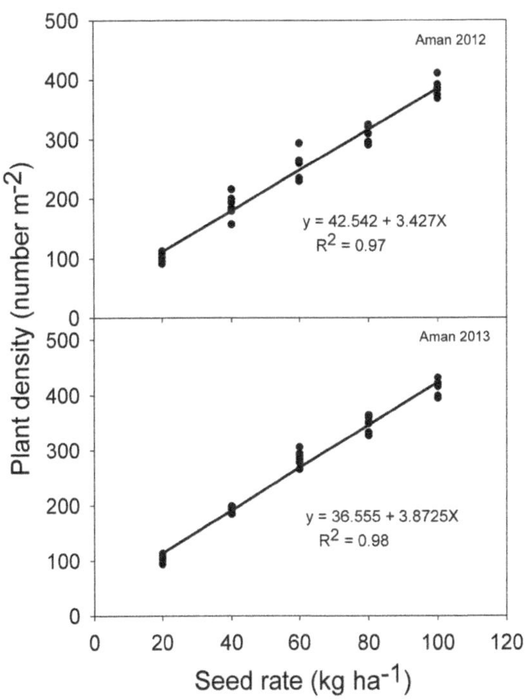

Figure 2. Relationship between rice seeding rate and rice plant density at 14 days after sowing in 2012 and 2013.

Table 1. Analysis-of-variance results for year, weed infestation level (weed) and seed rate and their interactions on plant density, rice biomass at 35 DAS and at anthesis, rice panicle m^{-2}, grains $panicle^{-1}$, and grain yield.

Source of variation	Degree of freedom	*P*-value					
		Plant density (no. m^{-2})	Rice biomass at 35 DAS	Rice biomass at anthesis	Panicle m^{-2}	Grains $panicle^{-1}$	Grain yield (t ha^{-1})
Year	4	0.01	0.01	0.02	0.01	0.25	0.04
Weed	1	<0.001	<0.001	<0.001	<0.001	<0.001	<0.001
Year×weed	1	0.05	0.12	0.01	0.19	0.09	0.02
Seed rate	4	<0.001	<0.001	<0.001	<0.001	0.30	<0.001
Weed×seed rate	4	0.82	<0.001	<0.001	0.59	<0.001	0.001
Year×seed rate	4	<0.001	0.02	0.86	0.53	0.95	0.74
Year×weed×seed rate	4	0.70	0.03	0.94	0.91	0.91	0.09

At 35 DAS (before hand weeding in the weedy plots), total weed density was influenced ($P<0.05$) by seeding rate (Table 2). Higher seeding rate resulted in lower weed density compared to the lowest seeding (20 kg ha^{-1}) (Table 3). With increase in seeding rate from 20 to 100 kg ha^{-1}, weed density decreased by 38% and 47% in the 2012 and 2013 seasons, respectively. Similar to weed density results at 35 DAS, weed biomass was also influenced by seeding rate (Table 4). Weed biomass decreased significantly (by 54% and 56% in 2012 and 2013, respectively) when seeding rate increased from 20 to 100 kg ha^{-1}. Grass weeds were found to be the dominant group in this study. The density and biomass of grass weeds decreased by 46–51% and 56–64% in 2012 and 2013, respectively, when seeding rate was increased to 100 kg ha^{-1} from 20 kg ha^{-1}.

The effect of seeding rate on weed density and biomass at anthesis are given in Tables 5 and 6. Similar to observations made at 35 DAS, total weed density and biomass was greatly influenced by seeding rate at anthesis (Table 7). However, the effects of seeding rate on density and biomass of broadleaf and sedges were not significant at this stage. In 2012, total weed density and biomass decreased by 40 and 41%, respectively, when seeding rate increased from 20 to 100 kg ha^{-1}; these values were 65 and 60%, respectively, in 2013.

The relationship between rice seeding rate and weed biomass are shown in Figures 3 and 4. In both seasons, weed biomass decreased linearly with increasing seeding rate. This explained 96% and 86% of variations at 35 DAS in 2012 and 2013, respectively. At anthesis, these values were 67% and 82%, respectively.

At low seeding rates, weeds may have had a favorable environment that provided them an opportunity for easier germination and faster growth and development. But at high seeding rates, the crop had a competitive advantage over weeds and closed canopy earlier, thus, reducing weed growth. [20], [21], [22]. A previous study explained that, due to increased crop density; the crop fraction of the total plant biomass (crop + weed) was increased, which resulted in higher weed suppression [23]. It has been reported that fresh weight of grass weeds was reduced by 18% when seeding rate of wheat was doubled from the recommended rate (80 kg ha^{-1}) [24]. Similarly, another study in an aerobic rice system found greater weed suppression when the seeding rate of rice was increased from 50 to 75 kg ha^{-1} [25].

Rice biomass

Rice biomass was strongly influenced by weed infestation level and seeding rate (Table 8). At 35 DAS, rice biomass increased significantly when seed rate increased from 20 to 100 kg ha^{-1}. At crop anthesis, however, rice biomass increased only up to the seeding rate of 40 kg ha^{-1}; there was no increase in biomass beyond this rate. Under partially-weedy conditions, however, rice biomass increased with each increment in the seeding rate past 20 kg ha^{-1}. This was probably because at low crop densities, there is less canopy cover early in the growing season, leaving more resources available for the weeds and thus enabling them to establish and grow quickly. At high crop density, the canopy closes quickly and shades out the weeds. In an earlier study on wheat, the highest crop biomass was obtained at a high crop density in the presence of weeds [23]. Under weed-free conditions, the plasticity in crop growth allows them to produce more tillers at low density. Previous studies on wheat [17], barley [16], and rice [26] also reported strong and consistent negative effects of increased crop density on weed biomass, as well as positive effects on crop biomass.

Table 2. Analysis-of-variance results for year and seed rate and their interactions on weed density and weed biomass at 35 DAS.

Source of variation	Degree of freedom	*P*-value							
		Density (no. m^{-2})				Biomass (g m^{-2})			
		Grass	Broadleaf	Sedges	Total	Grass	Broadleaf	Sedges	Total
Year	1	0.94	0.03	<0.001	<0.001	<0.001	0.86	0.01	0.02
Seed rate	4	4.02	0.37	0.11	<0.001	<0.001	<0.001	0.15	<0.001
Year×seed rate	4	0.06	<0.001	0.02	0.14	0.01	0.92	<0.001	0.56

Table 3. Effect of rice seeding rate on weed density (number m^{-2}) at 35 days after sowing in the aman seasons of 2012 and 2013.

Seed rate (kg ha^{-1})	Weed density							
	Grass		Broadleaf		Sedges		Total	
	2012	2013	2012	2013	2012	2013	2012	2013
	number m^{-2}							
20	631.3	575.0	47.9	98.9	45.8	328.1	725.0	1002.1
40	475.0	444.8	46.8	109.4	68.8	319.8	590.6	873.9
60	465.6	430.2	62.5	71.8	50.0	239.6	578.1	741.7
80	422.9	362.5	55.2	69.8	93.8	164.6	571.9	596.9
100	306.3	313.5	69.8	54.2	73.9	161.5	450.0	529.2
$LSD_{0.05}$	111.2	NS	NS	32.3	40.8	144.3	106.9	245.2

Abbreviations: $LSD_{0.05}$, least significant difference at 5% level of significance; NS, non-significant.

Table 4. Effect of rice seeding rate on weed biomass (g m^{-2}) at 35 days after sowing in the aman seasons of 2012 and 2013.

Seed rate (kg ha^{-1})	Weed biomass							
	Grass		Broadleaf		Sedges		Total	
	2012	2013	2012	2013	2012	2013	2012	2013
	g m^{-2}							
20	107.4	59.9	19.1	19.1	8.4	42.6	134.9	121.7
40	91.2	43.3	18.0	15.3	11.8	40.1	121.0	98.6
60	76.9	39.2	14.7	12.7	5.9	35.4	97.6	87.3
80	54.3	31.7	6.5	8.3	20.7	21.1	81.6	61.1
100	38.6	26.5	9.1	8.6	14.8	17.8	62.5	52.9
$LSD_{0.05}$	15.1	19.7	10.1	6.5	7.53	17	11.9	18.3

Abbreviations: $LSD_{0.05}$, least significant difference at 5% level of significance; NS, non-significant.

Table 5. Effect of rice seeding rate on weed density (number m^{-2}) at crop anthesis in the aman seasons of 2012 and 2013.

Seed rate ($kg\ ha^{-1}$)	Weed density							
	Grass		Broadleaf		Sedges		Total	
	2012	2013	2012	2013	2012	2013	2012	2013
	number m^{-2}							
20	233.3	256.3	82.3	78.1	3.1	20.8	318.8	355.2
40	221.9	182.3	52.1	75.0	2.1	10.4	276.0	267.7
60	158.3	196.9	43.8	47.9	0.0	0.0	202.1	244.8
80	178.1	131.3	39.6	42.7	0.0	0.0	217.7	173.9
100	146.9	104.2	41.7	15.6	0.0	3.1	188.5	122.9
$LSD_{0.05}$	NS	100.7	NS	NS	NS	NS	90.8	127.2

Abbreviations: $LSD_{0.05}$, least significant difference at 5% level of significance; NS, non-significant.

Table 6. Effect of rice seeding rate on weed biomass (g m^{-2}) at crop anthesis in the aman seasons of 2012 and 2013.

Seed rate ($kg\ ha^{-1}$)	Weed biomass							
	Grass		Broadleaf		Sedges		Total	
	2012	2013	2012	2013	2012	2013	2012	2013
	g m^{-2}							
20	218.8	156.9	33.8	50.2	1.4	3.2	254.0	210.4
40	182.8	126.9	25.6	41.3	0.6	3.1	208.9	171.2
60	143.4	108.5	31.6	38.2	0.0	0.0	175.0	146.7
80	149.2	83.5	14.3	32.5	0.0	0.0	163.5	115.9
100	133.7	70.8	16.6	13.1	0.0	0.7	150.3	84.6
$LSD_{0.05}$	58.3	41.2	NS	NS	NS	NS	57.1	34.8

Abbreviations: $LSD_{0.05}$, least significant difference at 5% level of significance; NS, non-significant.

Table 7. Analysis-of-variance results for year and seed rate and their interactions on weed density and weed biomass at anthesis.

Source of variation	Degree of freedom	*P*-value							
		Density (no. m-2)				Biomass (g m^{-2})			
		Grass	Broadleaf	Sedges	Total	Grass	Broadleaf	Sedges	Total
Year	1	0.05	1.00	0.26	0.03	<0.001	0.04	0.81	0.01
Seed rate	4	0.01	0.14	0.39	<0.001	<0.001	0.11	<0.001	<0.001
Year×seed rate	4	0.42	0.81	0.69	0.82	0.86	0.80	0.44	0.77

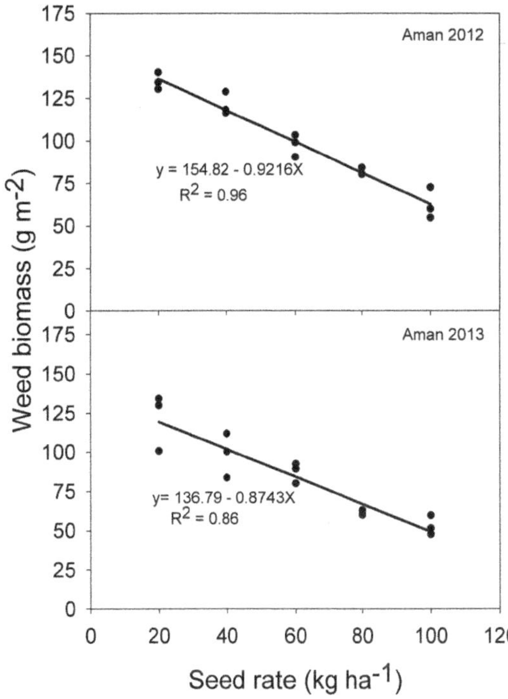

Figure 3. Relationship between rice seeding rate and weed biomass at 35 days after sowing in 2012 and 2013.

Rice Panicle

The interaction between weed infestation level and seeding rate had a significant effect on the number of rice panicles in both seasons (Figure 5). In both partially-weedy and weed-free

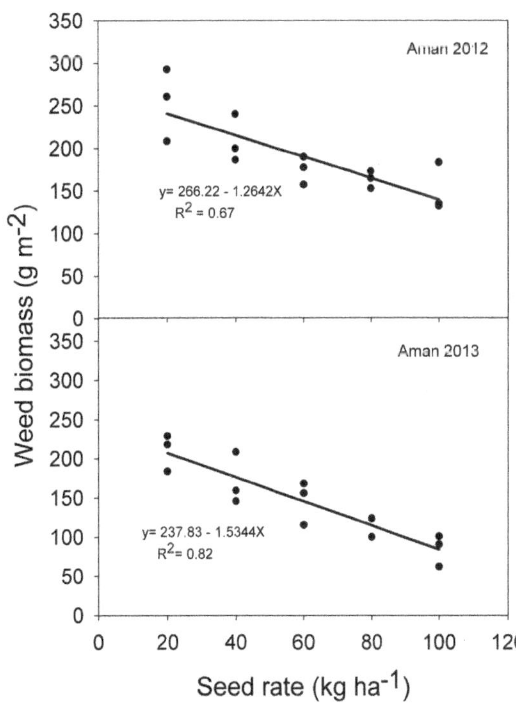

Figure 4. Relationship between rice seeding rate and weed biomass at crop anthesis in 2012 and 2013.

Table 8. Effect of rice seeding rate and weed infestation level on rice biomass (g m^{-2}) in the aman seasons of 2012 and 2013.

Seed rate (kg ha^{-1})	Rice biomass							
	At 35 days after sowing				At crop anthesis			
	2012		2013		2012		2013	
	Weed-free	Partially-weedy	Weed-free	Partially-weedy	Weed-free	Partially-weedy	Weed free	Partially-weedy
	g m^{-2}							
20	22.3	15.2	21.2	11.6	796.9	408.4	937.4	320.1
40	48.5	22.2	34.1	20.3	912.7	472.7	1053.2	435.2
60	52.1	25.4	43.6	28.7	917.4	585.8	1061.5	572.7
80	58.6	33.5	47.4	34.4	897.3	619.9	1037.3	616.3
100	63.6	45.1	50.5	38.4	901.6	655.2	1055.4	632.2
$LSD^a_{0.05}$	17.4		1.2		123.5		68.9	
$LSD^b_{0.05}$	4.0		3.7		91.1		64.2	

[a]LSD = compared within the same seeding rate.
[b]LSD = compared within the same level of weed infestation.
Abbreviations: $LSD_{0.05}$, least significant difference at 5% level of significance; NS, non-significant.

treatments, panicle numbers increased as seeding rates were increased from 20 kg ha^{-1} to 100 kg ha^{-1}. However, increase in panicle number tended to be greater in the partially-weedy plots than in the weed-free plots. Under weed-free conditions, panicle number at 100 kg seed ha^{-1} increased by 15 and 18% in 2012 and 2013, respectively, compared with panicle number at 20 kg seed ha^{-1}. Under partially-weedy conditions, these values were 26 and 47%, respectively. In the partially-weedy plots, panicle number decreased by 30 to 36% compared with the weed-free plots. Similar results were observed in an earlier study conducted in the Philippines in which the weed-free treatment posted 50–67% higher panicle numbers than the partially-weedy treatment at 15 kg ha^{-1} seeding rate, while at 100 kg ha^{-1} these values were only 16–25% [27].

Rice Grains Panicle^{-1}

The number of grains panicle^{-1} was influenced by the interaction between weed infestation level and rice seeding rate. Under weed-free conditions, the number of grains panicle^{-1} decreased with increase in seeding rate; however, under partially-weedy conditions, this result was reversed (Figure 6). The greater number of grains panicle^{-1} (97–102) was recorded under weed-free conditions at seeding rate of 20 kg ha^{-1} whereas, under partially-weedy conditions, the greatest number of grains panicle^{-1} (81–83) was recorded at seeding rate of 80 kg ha^{-1}. Regardless of the seeding rate, the number of grains panicle^{-1} decreased by 9–11% under weed-free conditions when seed rate increased from 20 to 100 kg ha^{-1}. Under partially-weedy conditions, the number of grains panicle^{-1} increased by 7–9%. In a previous study, it was reported that increase in seeding rate from 200 to 400 seeds m^{-2} decreased the number of grains panicle^{-1} by 17% [22].

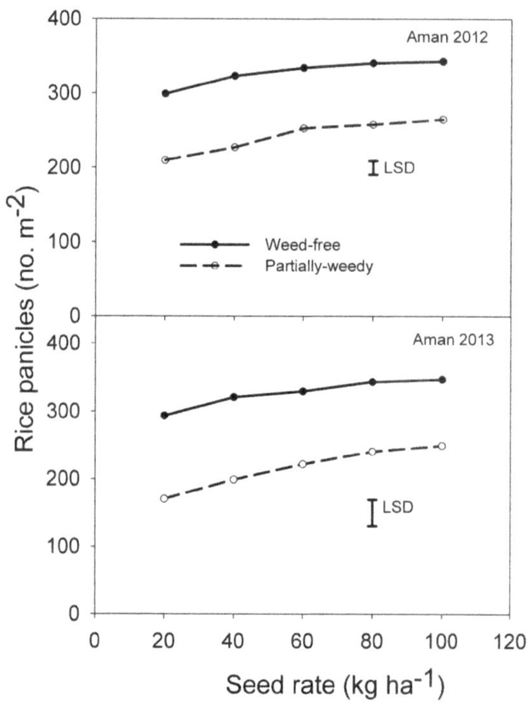

Figure 5. Effect of rice seeding rate and weed infestation level on rice panicle (number m^{-2}) at crop harvest in 2012 and 2013.

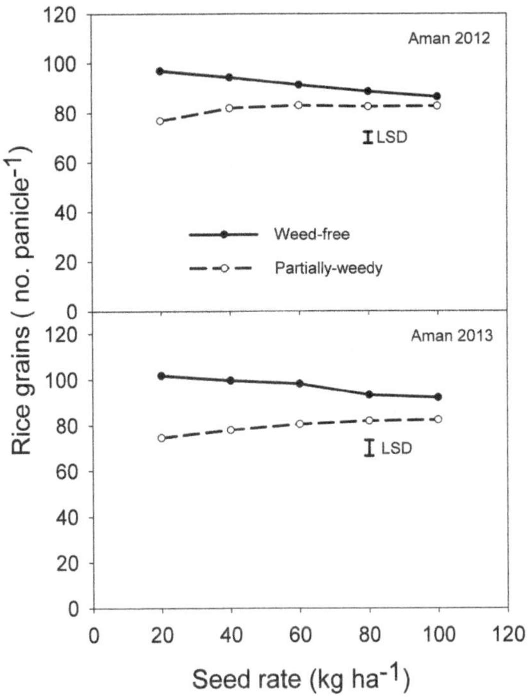

Figure 6. Effect of rice seeding rate and weed infestation level on the number of rice grains panicle^{-1} at crop harvest in 2012 and 2013.

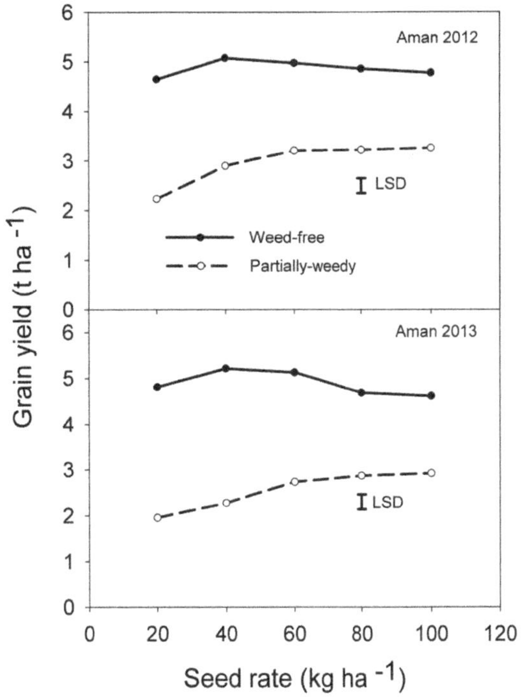

Figure 7. Effect of rice seeding rate and weed infestation level on rice grain yield (t ha^{-1}) in 2012 and 2013.

Rice Grain Yield

Grain yield was significantly influenced by the interaction between weed infestation level and seed rate (Figure 7). Regardless of the seeding rate, yield was always higher in the weed-free plots (4.6–5.2 t ha^{-1}) than in the partially-weedy plots (2.0–3.2 t ha^{-1}). In the weed-free plots, the highest yield (5.1–5.2 t ha^{-1}) was recorded at a seeding rate of 40 kg ha^{-1} in both seasons. Under these conditions, grain yield increased by 7–8% with increase in seeding rate from 20 to 40 kg ha^{-1}; beyond this rate, grain yield declined slightly. Under partially-weedy conditions, however, yield continued to increase across the entire range of seeding rates, whereas in the presence of weeds, yield increased by 30–33% when seeding rate increased from 20 to 100 kg ha^{-1}.

Under weedy or partially-weedy conditions, a higher seeding rate suppresses weed growth and increases grain yield. Such results, however, depend on environmental conditions, crop genotypes, and the growing season. In a previous study on wheat, under weed-free environments grain yield ranged from 2.2 to 2.3 t ha^{-1}, when seeding rate ranged from 100 to 200 plant m^{-2} and, beyond this, yield declined slightly (by about 0.1 t ha^{-1}). However, in the presence of weeds, yield continued to increase (17 and 23% increase when seeding rate increased to 200 and 300 wheat plant m^{-2}, respectively, compared with 100 plant m^{-2}) across the entire range of densities [21]. In another study, no differences were found in wheat yield at different crop densities in herbicide-treated plots whereas, in weedy plots, grain yield increased by 22% from low to medium and high crop density [28]. Our results also support the findings of another study on aerobic rice in the Philippines, in which the yield of an inbred variety under partially-weedy conditions increased from 0.8 to 2.4 t ha^{-1} (70% yield increase) with increase in seeding rate from 15 to 125 kg ha^{-1} [28].

In DSR systems in Bangladesh, farmers use seeding rates ranging from 50 to 100 kg ha^{-1}, depending on season, variety, seed quality, and field conditions. However, there is no specific recommendation available in literature for Bangladesh conditions. Seeding rate is an important factor for optimum yield in DSR. High seeding rate helps to suppress weeds; however, yield does not always increase with high seeding rate and there are some risks associated with this such as crop lodging, diseases, and insect infestation. On the other hand, low seeding rates reduces seed cost. However, there are also some risks associated with the use of low seeding rate such as losses due to weed competition and poor rice seedling establishment [29]. Therefore, seed price, seed quality, field situations, and chances of seed loss by insects, and diseases need to be considered before recommending a seeding rate [28].

The present study showed that seeding rate of 40 to 60 kg ha^{-1} in the weed-free and 80 to 100 kg ha^{-1} in the partially-weedy condition resulted in better crop. Under weed-free conditions, yield increased up to the seeding rate of 40 kg ha^{-1} and decreased slightly beyond this rate. In the partially-weedy plots, grain yield continued to increase up to the highest tested seeding rate (i.e., 100 kg ha^{-1}), mainly because increasing crop density suppressed weed growth and reduced losses in grain yield. Our results are also supported by those of an earlier study, in which increasing rice seeding rates suppressed weed growth and reduced losses due to weeds [19].

Weeds are the major constraint to the success of DSR. In our study, the average yield loss under partially-weedy conditions ranged from 40 to 48% compared with that under weed-free conditions, which confirms that weeds are a crucial yield-limiting factor in DSR and that weed management should be properly addressed to make DSR cultivation more profitable.

Due to labor scarcity and its high cost, herbicides are considered the best option for managing weeds in DSR. But sole dependence on herbicides is not sustainable because of its harmful effect on the environment and toxicity risk to crops if not applied at optimum dose. In addition, excess use of herbicides can result in the evolution of resistant biotypes. Weed management in DSR therefore needs to be considered and integrated with all possible agronomic practices. High seeding rates play an important role in increasing the competitiveness of the crop against weeds, which minimizes weed pressure and reduces dependence on herbicide use in DSR. The results of our study indicate that, in Bangladesh, farmers can use a seeding rate of 40 kg ha^{-1} where weed management is not a problem. However, in the absence of effective weed control, farmers should use relatively higher seeding rates (80 to 100 kg ha^{-1}) to reduce yield loss due to weed infestation.

Acknowledgments

We would like to thank Ms. Maria Leah Cruz for providing comments on the manuscript.

Author Contributions

Conceived and designed the experiments: BSC SA. Performed the experiments: SA. Analyzed the data: SA BSC. Contributed reagents/materials/analysis tools: BSC SA. Contributed to the writing of the manuscript: SA BSC MS.

References

1. Shamsudduha M, Chandler RE, Taylor RG, Ahmed KM (2009) Recent trends in groundwater levels in a highly seasonal hydrological system: the Ganges-Brahmaputra-Meghna delta. Hydrol Earth Syst Sci 13: 2373–2385.
2. Tuong TP, Bouman BAM (2003) Rice production in water-scarce environments. In: Kijne JW, Barker R, Molden D (editors). Water Productivity in Agriculture: Limits and Opportunities for Improvements. CABI Publishing, UK, 53–67.
3. Choudhury BU, Bouman BAM, Singh AK (2009) Yield and water productivity of rice–wheat on raised beds at New Delhi, India. Field Crops Res 100: 229–239.
4. Singh S, Ladha JK, Gupta RK, Bhushan L, Rao AN (2008) Weed management in aerobic rice systems under varying establishment methods. Crop Prot 27: 660–669.
5. Chauhan BS (2012) Weed ecology and weed management strategies for dry-seeded rice in Asia. Weed Technol 26: 1–13.
6. Rao AN, Johnson DE, Sivaprasad B, Ladha JK, Mortimer AM (2007) Weed management in direct-seeded rice. Adv Agron 93: 153–255.
7. Chauhan BS, Johnson DE (2011) Row spacing and weed control timing affect yield of aerobic rice. Field Crops Res 121: 226–231.
8. Fischer AJ, Ateh CM, Bayer DE, Hill JE (2000) Herbicide-resistant *Echinochloa oryzoides and E. phyllopogon* in California *Oryza sativa* fields. Weed Sci 48: 225–230.
9. Buhler DD, Liebman M, Obrycki JJ (2002) Review: theoretical and practical challenges to an IPM approach to weed management. Weed Sci 48: 274–280.
10. Singh Y, Singh VP, Chauhan BS, Orr A, Mortimer AM, et al. (2008) Direct seeding of rice and weed management in the irrigated rice-wheat cropping system of the Indo-Gangetic Plains. Los Baños (Philippines): International Rice Research Institute, and Pantnagar (India): Directorate of Experiment Station, G.B. Pant University of Agriculture and Technology. 272 p.
11. Azmi M, Chin DV, Vongsaroj P, Johnson DE (2005) Emerging issues in weed management of direct-seeded rice in Malaysia, Vietnam, and Thailand. In: Toriyama K, Heong KL, Hardy, B. (editors). Rice is Life: Scientific Perspectives for the 21st Century. Los Baños, Philippines/Japan, International Rice Research Institute and Tsukuba/Japan International Research Center for Agricultural Sciences. 196–198.
12. Chauhan BS, Johnson DE (2010) Implications of narrow crop row spacing and delayed *Echinochloa colona* and *Echinochloa crus-galli* emergence for weed growth and crop yield loss in aerobic rice. Field Crops Res 117: 177–182.
13. Mortensen DA, Dieleman JA, Johnson GA (1998) Weed spatial variation and weed management. In: Hatfield JL, Buhler DD, Stewart BA (editors). Integrated weed and soil management. Ann Arbor Press, Chelsea, Michigan, USA, 293–310.
14. Gibson KD, Fischer AJ, Foin TC, Hill JE (2002) Implications of delayed *Echinochloa spp.* germination and duration of competition for integrated weed management in water-seeded rice. Weed Res 42: 351–358.
15. Guillermo DA, Pedersen P, Hartzler RG (2009) Soybean seeding rate effects on weed management. Weed Technol 23: 17–22.
16. O'Donovan JT, Harker KN, Newman JC, Hall LM (2001) Barley seeding rate influences the effects of variable herbicide rates on wild oat. Weed Sci 49: 746–754.
17. Olsen JM, Kristensen L, Weiner J, Griepentrog HW (2005) Increased density and spatial uniformity increases weed suppression by spring wheat *(Triticum aestivum* L.). Weed Res 45: 316–321.
18. George TP, Samuel CRH, Jim ED, Adam NS (2009) Seeding rate effects on weed control and yield for organic soybean production. Weed Technol 23: 497–502.
19. Zhao DL, Bastiaans L, Atlin GN, Spiertz JHJ (2007) Interaction of genotype x management on vegetative growth and weed suppression of aerobic rice. Field Crops Res 100: 327–340.
20. Phuhong LT, Denich M, Vlek PLG, Balasubramanian V (2005) Suppressing weeds in direct seeded lowland rice: effects of methods and rates of seeding. J Agron Crop Sci 191: 185–194.
21. Lemerle D, Cousens RD, Gill GS, Peltzer SJ, Moerkerk M, et al. (2004) Reliability of higher seeding rates of wheat for increased competitiveness with weeds in low rainfall environments. J Agric Sci 142: 395–409.
22. Anwar MP, Juraimi AS, Puteh A, Selamat A, Man A, et al. (2011) Seeding method and rate influence on weed suppression in aerobic rice. Afr J Biotechnol 10(68): 15259–15271.
23. Weiner J, Griepentrog H, Kristensen L (2001) Suppression of weed by spring wheat increases with crop density and spatial uniformity. J Appl Ecol 38: 784–790.
24. Kirkland KJ, Holm FA, Stevenson FC (2000) Appropriate crop seeding rate when herbicide rate is reduced. Weed Technol 14(4): 692–698.
25. Khaliq A, Mahmood S, Matloob A, Khan MB, Awan IU (2012) Optimizing seeding density and tank mixture of herbicides help reduce yield losses in dry seeded fine rice. Pak J Weed Sci Res 18(2): 167–181.
26. Ni HW, Moody K, Robles RP (2004) Analysis of competition between wet-seeded rice and barnyardgrass (*Echinochloa crus-galli* L.) using a response-surface model. Weed Sci 52: 142–146.
27. Chauhan BS, Singh VP, Kumar A, Johnson DE (2011) Relations of rice seeding rates to crop and weed growth in aerobic rice. Field Crops Res 121: 105–115.
28. Kristensen L, Olsen J, Weiner J (2008) Crop density, sowing pattern, and nitrogen fertilization effects on weed suppression and yield in spring wheat. Weed Sci 56: 97–102.
29. Chauhan BS, Singh RG, Mahajan G (2012). Ecology and management of weeds under conservation agriculture: a review. Crop Prot 38: 57–65.

11

Physiological and Morphological Responses of *Ischaemum rugosum* Salisb. (Wrinkled Grass) to Different Nitrogen Rates and Rice Seeding Rates

Tahir Hussain Awan[1]*, Bhagirath Singh Chauhan[1], Pompe C. Sta. Cruz[2]

1 Weed Science, Crop and Environmental Sciences Division, International Rice Research Institute (IRRI), Los Baños, Philippines, **2** Crop Science Cluster, Department of Agronomy, College of Agriculture, University of Philippines, Los Baños, Philippines

Abstract

Ischaemum rugosum is a competitive weed in direct-seeded rice systems. Developing integrated weed management strategies that promote the suppression of weeds by crop density, cultivar selection, and nutrition requires better understanding of the extent to which rice interferes with the growth of this weed and how it responds to resource limitation due to rice interference. The growth of *I. rugosum* was studied when grown with four rice seeding rates (0, 25, 50, and 100 kg ha^{-1}) and four nitrogen (N) rates (0, 50, 100, and 150 kg ha^{-1}). Compared to the weed plants grown alone, weed tiller number was reduced by 63–80%, leaf number by 68–77%, leaf area by 69–77%, leaf biomass by 72–84%, and inflorescence biomass by 81–93% at the rice seeding rates of 25–100 kg ha^{-1}. All these parameters increased with increasing rates of N from 0 to 150 kg ha^{-1}. At weed maturity, *I. rugosum* plants were 100% taller than rice at 0 kg N ha^{-1}, whereas, with added N, the weeds were only 50% taller than rice. Weed biomass increased by 82–160%, whereas rice biomass increased by 92–229%, with the application of 50–150 kg N ha^{-1}. Added N favored rice biomass production more than it did the weed. Rice interference reduced the height and biomass of *I. rugosum*, but did not suppress its growth completely. *I. rugosum* showed the ability to reduce the effects of rice interference by increasing leaf area, leaf weight ratio, and specific leaf area, and by decreasing the root-shoot weight ratio in comparison to the weed plants grown alone. The results suggest that rice crop interference alone may reduce *I. rugosum* growth but may not provide complete control of this weed. The need for integrated weed management practices to effectively control this weed species is highlighted.

Editor: Manuel Reigosa, University of Vigo, Spain

Funding: The authors have no support or funding to report.

Competing Interests: The authors have declared that no competing interests exist.

* E-mail: t.awan@irri.org

Introduction

Majority of the people living in Asia consume rice as their staple food and 90% of the world's rice is produced and consumed in Asia. Here, the major rice establishment method is manual transplanting of seedlings. However, this method is more laborious and requires more water [1]. Shortage of labour and water compels researchers and growers to develop alternative methods of crop establishment, such as dry-seeded rice (DSR). Acceptance of DSR is increasing among farmers in South and Southeast Asian countries. DSR is a resource-conserving technology compared with puddled transplanted rice, except that it is prone to heavy weed infestation [1]. Weeds are principal biotic constraints to rice production in DSR [2] With DSR, rice and weeds emerge approximately at the same time and, therefore, greater effort is needed to control weeds. Manual hand weeding is very expensive, time-consuming, and, sometimes, it is not feasible. Therefore, farmers have to rely mainly on herbicides. Non-judicious use of herbicides, however, is linked with the evolution of herbicide resistance in weeds and concerns over soil and environmental pollution [3,4].

Efficient weed control in DSR is still a main concern, and strategies are needed to reduce the weed problem. If there is an interest in reducing reliance on herbicides, additional weed management tools (i.e., use of competitive cultivars, high seeding rates, and narrow row spacing) are needed to achieve sustainable weed control [2,5]. Comprehension of weed biology and ecology is important to develop cultural weed management strategies.

Ischaemum rugosum Salisb. is a noxious weed in at least 26 countries of the world [6] and to many crops, including rice[7]. This weed is adapted to a wide variety of habitats. However, there is a scarcity of information in literature on the biology and ecology of this weed. It can emerge even when buried at a 10-cm depth [8] because of its long coleoptile length. At harvest time, it contaminates rice seeds because it has a similar size and shape [9]. *I. rugosum* has a high level of seed dormancy because of the presence of glumes, which delay seed germination after shedding. Therefore, its control by herbicides is difficult in rice. *I. rugosum* causes considerable yield losses in rice—by 50% [7] to 60% [9]. Because of the continuous use of herbicides, this weed has developed multiple resistance to herbicides belonging to different modes of action, that is, ACCase inhibitors, ALS inhibitors, urease, and amides [10]. Thus, nowadays *I.rugosum* has become a serious weed of rice.

Several researchers have projected the use of high rice seeding rates in DSR to suppress weeds and to achieve high rice yield

[4,11]. In an earlier study, weeds severely reduced yield (71%) at a low seeding rate (40 kg ha^{-1}), whereas high seeding rates of 80–160 kg ha^{-1} produced high rice yield and minimized losses caused by the weeds [12]. At high seeding rates, the crop may need more nutrients to produce high yield. The effect of N fertilizer on the *I. rugosum*-rice competitive interaction has not yet been studied in Asia. Crop-weed interference can be affected by fertilizer management and seeding rates, and N is one of the crucial components for crop-weed competitive interactions [13]. Some weeds consume high quantities of N, reduce crop N uptake, and suppress growth, biomass, and yield of rice [14]. Other researchers claimed that high doses of N fertilizer enhanced crop growth and yield compared with weeds, and that weed response to added N decreased when they are under shaded conditions [15,16]. Some scientists found that increasing N rates had little effect on crop-weed competition [17]. Therefore, the effect of N may be species-specific [18,19]. The effect of crop seeding rate on weed suppression could be affected by N rates. Effects of high crop seeding rate could be more prominent at low N levels because weeds grow slowly in that condition [19].

Weed control practices that enhance the competitive ability of crops over weeds should be a fundamental part of an integrated weed management (IWM) strategy. Before IWM strategies that rely on crop competitiveness can be improved, there is a need to better understand how rice interferes with weed growth, and how weeds compete with rice for resource use. Therefore, a study was conducted to evaluate the physiological and morphological responses of *I. rugosum* to different N and rice seeding rates.

Materials and Methods

Seeds of *I. rugusom* were collected in 2011 from upland rice fields around Los Baños, Philippines. The first two authors work at this nistitue, and thus no permission was needed to collect the weed seeds. They also confirmed that the study did not involve endangered or protectd species.

Test rice variety is NSICRc222 (IR154). Experiments were conducted in a screenhouse at the International Rice Research Institute (IRRI), Los Baños, Philippines. The screenhouse is made up of a large iron steel frame covered with a 2-mm steel mesh from all sides to maintain environmental conditions similar to field conditions.

The experiments were conducted by growing the weed and rice plants in plastic pots [20,21] with holes at the bottom filled with sieved soil (8.3 kg pot^{-1}). The pots were 25 cm in diameter and 30 cm in height. The soil was sieved through a 3-mm aperture and it was analyzed before use in the experiments. The soil had 22% sand, 38% silt, 40% clay, a pH of 6.0, 0.99% organic carbon, 0.12% Kjeldahl N, 43 mg kg^{-1} soil of available P_2O_5, and 1.26 meq 100^{-1} g soil of available K.

In this study, there were 16 treatment combinations of two factors: four N rates—0, 50, 100, and 150 kg ha^{-1}—and four seeding rates—0 (0 plants pot^{-1}), 25 (5 plants pot^{-1}), 50 (10 plants pot^{-1}), and 100 kg ha^{-1} (20 plants pot^{-1}). Phosphorus and potash fertilizers were applied at sowing at 40 kg P_2O_5 ha^{-1} and 40 kg K_2O ha^{-1}, respectively. N in the form of urea was applied in two equal splits at 20 and 40 days after sowing (DAS). Two to three weed seeds were planted at the centre of each pot and covered with a thin layer (2 mm) of soil. In the case of five rice plants per pot (i.e., 25 kg seed ha^{-1}), rice seeds were planted in a single circle around the weed plant at a distance of 5 cm. In the case of 10 and 20 rice plants per pot, the plants were planted in two circles. The placement of the first circle was the same as described above and the distance of the second circle was 2.5 cm away from the first circle. In each circle, rice seeds were planted at an equal distance from each other. At 7 DAS, thinning was carried out to maintain the required density of rice and weed plants per pot. Only one plant of *I. rugosum* was maintained at the centre of each pot. A randomized complete block design with three replications was used to arrange the pots. Weeds other than *I. rugosum* were removed manually, as and when required. The pots were placed at a distance of 30 cm from each other to avoid the effect of shading and, fortnightly, pots were rotated to new positions to reduce experimental errors. Pots were irrigated 2–3 times a day with a sprinkler system.

The study was conducted two times with a gap of two and a half months in-between. The first and second experiments began on 4 May and 13 July 2012, respectively. These experiments were harvested on 30 July and 5 October, respectively. Plant height, number of leaves per plant, number of tillers per plant, and SPAD values were measured at 14, 28, 42, 56, 70, and 84 DAS. Plant

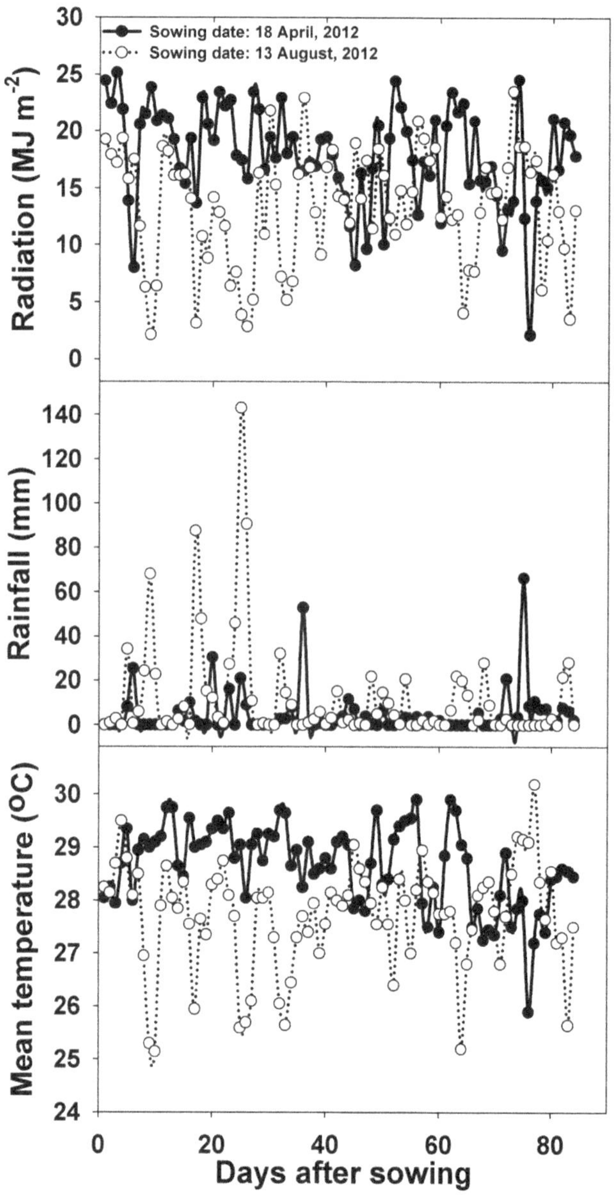

Figure 1. (a) Radiation (MJ m-2), (b) rainfall (mm), and (c) average temperature during the experimental period.

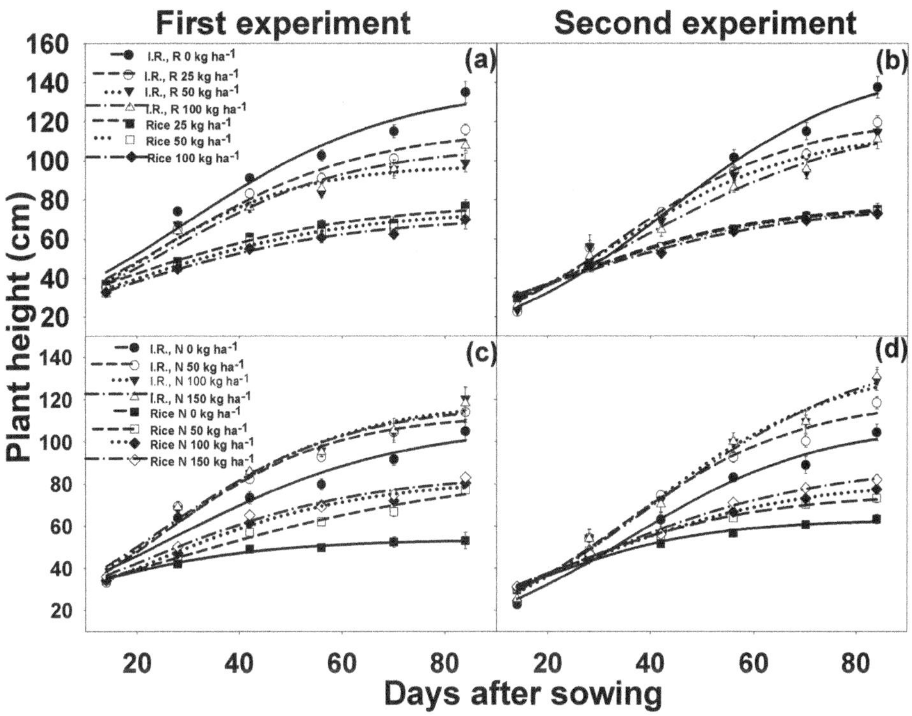

Figure 2. (a, b) Plant height of *Ischaemum rugosum* (I.R.) when grown alone (I.R., R0) or in competition with rice at different seeding rates (R), i.e., 25 (I.R., R25), 50 (I.R., R50), and 100 kg ha^{-1} (I.R., R100); Rice plant height at different seeding rates: 25 kg ha^{-1}, 50 kg ha^{-1}, and 100 kg ha^{-1}; and (c, d) *Ischaemum rugosum* (I.R.) and rice plant height under different nitrogen levels, i.e., 0 (I.R., N0, and rice N0), 50 (I.R., N50, and rice N50), 100 (I.R., N100, and rice N100), and 150 kg ha^{-1} (I.R., N150, and rice N150). The vertical bars represent standard error of means. The lines represent a three-parameter sigmoid model ($y = a/\{1+e^{[-(x-x50)/b]}\}$) fitted to the data. Parameter estimates are shown in Table 1.

height of the weed and rice plants were measured from the ground to the tip of the longest leaf.

Plants were harvested at weed maturity from the ground level at 84 DAS, with the criterion of weed seed maturity. After detaching leaves for leaf area measurements, shoots were separated into stems and inflorescence. Leaf area was measured using a leaf area meter (LI-COR, model LI-3100, USA). After measuring the leaf area, stem, leaves, and inflorescence were placed separately in paper bags and dried in an oven at 70°C for 72 hours to achieve dry biomass. From each pot, roots were removed with the soil intact, then, were washed through a steel strainer. Weed and rice roots were separated and placed in paper bags for oven-drying and biomass measurements. The following ratios were calculated: root to shoot weight ratio (RSWR), specific leaf area (SLA), leaf weight ratio (LWR), and leaf area ratio (LAR). LAR is the ratio of leaf area to total plant (above- and belowground) biomass and it indicates how photosynthates are being partitioned within the plant parts.

$$LAR = {}^{Leaf.area}/_{Total.plant.weight} \quad (1)$$

$$LWR = {}^{Leaf.weight}/_{Total.plant.weight} \quad (2)$$

$$SLA = {}^{Leaf.area}/_{Total.leaf.weight} \quad (3)$$

$$RSWR = {}^{Root.weight}/_{Shoot.weight} \quad (4)$$

Radiation (MJ m^{-2}), rainfall (mm), and mean temperature (°C) during the study period for both experiments are shown in Figure 1.

Statistical analyses

Data from both experiments were subjected to analysis of variance (ANOVA) (GenStat 8.0, 2005). Data variance was visually inspected by plotting residuals to confirm homogeneity of variance before statistical analysis. The ANOVA results indicated that there were significant interactions between treatments and experimental 'runs'; therefore, the data were separately analyzed for both experimental runs. Treatment means for SPAD, LAR, LWR, RSWR, and SLA were separated using standard error of difference (SED) at a 5% level of significance. Data were analyzed using regression analysis for plant height, and leaf and tiller numbers per plant of *I. rugosum* and a three-parameter sigmoid model was used:

$$y = a/\{1 + \exp[-(x - d50)/b]\} \quad (5)$$

Where: y is the estimated plant height, leaf number, or tiller number per plant of *I. rugosum* as a function of rice seeding rate or DAS at time x; a is the maximum of the parameter; $d50$ is the time to reach 50% of the final plant height, leaf number, or tiller number per plant; and b is the slope. The data of leaf area and leaf, stem, inflorescence, and total aboveground biomass (g plant^{-1})

Table 1. Parameter estimates of a three-parameter sigmoid model ($y = a/\{1{+}e^{[-(x-x50)/b]}\}$, where *a* is the maximum height, leaf number, or tiller per plant; *d50* is the time (days after planting) to reach 50% of final height, leaf number, or tiller; and *b* is the slope at time *x*) fitted to the plant height (rice and weed), leaf number, or tiller per plant of *Ischaemum rugosum* when grown alone (I.R., R0) or in competition with a varying density of rice plants (R25, R50, or R100 kg ha^{-1}).

	Plant height							
	First experiment				**Second experiment**			
Treatments	*a*	*b*	*d50*	R^2	*a*	*b*	*d50*	R^2
I.R., R 0 kg ha^{-1}	137.83(16.18)	20.26 (6.56)	29.97 (6.32)	0.96	149.51 (9.47)	18.43 (2.29)	43.41 (3.23)	0.99
I.R., R 25 kg ha^{-1}	116.46 (9.50)	19.21 (5.01)	26.60 (4.29)	0.97	121.80 (9.21)	16.94 (3.52)	34.44 (3.86)	0.98
I.R., R 50 kg ha^{-1}	97.82 (6.12)	15.20 (4.41)	20.81 (3.34)	0.96	115.69 (12.82)	17.82 (5.38)	33.43 (5.76)	0.96
I.R., R 100 kg ha^{-1}	107.44 (9.10)	18.25 (5.28)	25.77 (4.42)	0.97	121.70 (10.80)	21.39 (3.91)	38.23 (4.94)	0.99
Rice 25 kg ha^{-1}	78.21 (4.61)	22.50 (5.23)	16.19 (3.05)	0.98	77.61(2.95)	19.60 (2.76)	21.74 (1.98)	0.99
Rice 50 kg ha^{-1}	74.81 (4.10)	21.75 (4.67)	17.32 (2.84)	0.98	77.29(0.87)	20.59 (0.82)	21.84 (0.59)	0.99
Rice 100 kg ha^{-1}	71.19 (3.67)	22.13 (4.50)	16.65 (2.67)	0.98	77.62 (3.90)	22.47 (3.69)	22.08 (2.67)	0.99
I.R., N 0 kg ha^{-1}	108.22(15.21)	22.25 (8.90)	26.67 (7.69)	0.95	108.59 (7.62)	18.36 (3.19)	35.78 (3.66)	0.99
I.R., N 50 kg ha^{-1}	113.03 (8.04)	16.60 (4.43)	25.05 (3.69)	0.97	120.19 (9.62)	17.52 (3.79)	34.10 (4.13)	0.98
I.R., N 100 kg ha^{-1}	119.43 (9.64)	17.99 (4.89)	26.57 (4.21)	0.97	137.05 (11.84)	18.57 (3.58)	39.02 (4.50)	0.99
I.R., N 150 kg ha^{-1}	117.81 (8.23)	17.97 (4.41)	25.27 (3.64)	0.97	143.99 (15.02)	20.39 (4.09)	41.61 (5.64)	0.99
Rice N 0 kg ha^{-1}	53.69 (1.16)	18.72 (3.26)	2.48 (2.52)	0.99	63.20 (2.05)	16.95 (2.96)	14.85 (1.89)	0.99
Rice N 50 kg ha^{-1}	85.74 (10.50)	30.49 (8.79)	24.70 (7.84)	0.98	75.44 (3.30)	19.62 (3.36)	20.14 (2.28)	0.99
Rice N 100 kg ha^{-1}	82.23 (3.75)	20.65 (3.42)	21.05 (2.38)	0.99	81.98 (0.62)	21.28 (0.50)	24.54 (0.40)	0.99
Rice N 150 kg ha^{-1}	84.84 (4.98)	21.68 (4.59)	19.91 (3.06)	0.98	89.18(4.23)	22.43 (2.95)	27.36 (2.62)	0.99
Number of tillers plant^{-1}								
I.R., R 0 kg ha^{-1}	20.35 (2.04)	12.19 (3.80)	39.05(4.56)	0.96	20.83(0.34)	8.05 (0.56)	40.73 (0.65)	0.99
I.R., R 25 kg ha^{-1}	6.78 (0.93)	16.50 (6.46)	33.74 (6.97)	0.94	7.92 (0.42)	14.45 (2.06)	39.35 (2.52)	0.99
I.R., R 50 kg ha^{-1}	8.63 (3.55)	26.67(13.17)	52.26 (25.70)	0.94	6.42 (0.73)	21.39 (3.89)	46.58 (6.22)	0.99
I.R., R 100 kg ha^{-1}	8.63 (3.55)	26.67(13.17)	52.26 (25.70)	0.94	6.91 (3.11)	32.07 (10.25)	69.93 (30.18)	0.98
I.R., N 0 kg ha^{-1}	5.95 (2.01)	26.81(13.83)	43.59 (21.57)	0.93	6.82 (0.35)	16.17 (2.07)	38.81 (2.53)	0.99
I.R., N 50 kg ha^{-1}	10.14 (0.74)	14.77 (2.73)	40.85 (3.45)	0.99	7.79(0.26)	9.74 (1.29)	35.56 (1.50)	0.99
I.R., N 100 kg ha^{-1}	11.98 (0.91)	14.32 (2.90)	40.07 (3.59)	0.98	10.80(0.26)	11.23 (0.86)	41.38 (1.05)	0.99
I.R. N 150 kg ha^{-1}	11.93 (0.94)	14.25(3.28)	36.97 (3.82)	0.98	13.64 (0.44)	11.13 (1.05)	44.92 (1.34)	0.99
Number of leaves plant^{-1}								
I.R., R 0 kg ha^{-1}	102.70 (2.18)	10.88 (0.70)	44.08(0.88)	0.99	95.85 (1.75)	9.33 (0.56)	46.39 (0.70)	0.99
I.R., R 25 kg ha^{-1}	42.36 (12.25)	25.86 (8.13)	56.17 (17.36)	0.98	48.46 (1.27)	18.82 (0.66)	55.63 (1.26)	0.99
I.R., R 50 kg ha^{-1}	39.95 (14.63)	27.02 (9.61)	59.52 (22.37)	0.97	112.80 (53.84)	29.83 (2.90)	112.52 (22.13)	0.99
I.R., R 100 kg ha^{-1}					95.61 (247.58)	34.28 (14.9)	124.77 (131.8)	0.98
I.R., R N 0 kg ha^{-1}	89.28 (66.67)	39.48 (6.93)	118.27(47.13)	0.99	68.79 (24.22)	29.25 (4.57)	86.43 (19.39)	0.99
I.R., N 50 kg ha^{-1}	44.84 (3.15)	14.59 (2.37)	44.44 (3.22)	0.99	46.10 (1.35)	16.23 (0.80)	52.09 (1.32)	0.99

Table 1. Cont.

	Plant height							
	First experiment				Second experiment			
Treatments	*a*	*b*	*d50*	R^2	*a*	*b*	*d50*	R^2
I.R., N 100 kg ha^{-1}	61.15 (4.95)	15.41 (2.51)	47.65 (3.69)	0.99	52.45 (2.80)	11.67 (1.62)	47.86 (2.16)	0.99
I.R., N 150 kg ha^{-1}	62.62 (2.97)	14.07 (1.57)	45.01(2.13)	0.99	66.46 (2.86)	11.13 (1.21)	50.58 (1.65)	0.99

were fitted to a three-parameter exponential model:

$$y = y0 + [a * \exp(-b.x)] \quad (6)$$

Where: y is the leaf area and leaf, stem, inflorescence, and total aboveground biomass (g plant^{-1}) at seeding rate x; $y0$ and a are the constants; and b is the slope. Root biomass (g plant^{-1}) was fitted to an exponential model:

$$y = a * \exp(-b.x) \quad (7)$$

Where: y is the estimated root biomass (g plant^{-1}) at seeding rate (kg ha^{-1}) x; a is the constant; and b is the slope. R^2 values were used to determine goodness-of-fit for the selected model. Parameter estimates were compared using their standard error.

Results

Analysis showed that there were interactions ($p<0.05$) among experimental run, seed rate, and N rate. Interactions were also significant for experimental run by N rate and experimental run by seed rate for most parameters. Therefore, both experiments were examined separately for N and effects of seed rate.

Plant height

Ischaemum rugosum and rice were almost of the same height at 14 DAS (Figures 2a–d). After that, the height of *I. rugosum* was significantly affected by increasing the seed rates from 25 to 100 kg ha^{-1}, and plants grown alone were taller than plants grown with rice (Figures 2a and b). *I. rugosum* was 117–122, 98–116, and 107–122 cm tall when grown with rice at seeding rates of 25, 50, and 100 kg ha^{-1}, respectively; whereas, its height was 138–150 cm when grown alone (Table 1).

Nitrogen levels did not affect weed height within 14–28 DAS, although an apparent effect of N was observed within 42 to 54 DAS. Compared to the height at 14 DAS, the rates of increase in height at 84 DAS were 218, 244, 248, and 223% for *I. rugosum* when fertilized with 0, 50, 100, and 150 kg N ha^{-1}, respectively. These values for rice were 51, 126, 135, and 130%, respectively. Plant height increased with increasing N rates of up to 100 kg ha^{-1}. Beyond this rate, N had no positive effect on the height of *I. rugosum*, whereas it had a positive effect on the height of rice. In the first experiment, the weeds were 49, 31, and 51% taller than rice at seeding rates of 25, 50, and 100 kg ha^{-1}, respectively; whereas at 0, 50, 100, and 150 kg N ha^{-1}, the weeds were 101, 32, 45, and 39% taller than rice, respectively. A similar trend was observed for the second experiment.

Tiller density

The number of weed tillers per plant was not affected by varying rice densities until 14 DAS, although, it declined significantly ($p<0.001$) beyond 14 DAS when grown with rice (Figures 3a and b). At 84 DAS, the maximum number of tillers per plant (parameter *a*) were 20.4 to 20.8, when weed was grown alone. As the rice seeding rates increased from 0 to 100 kg ha^{-1}, the number of weed tillers decreased (Table 1). In the first experiment (Figure 3a), rice seeding rates of 25 and 50 kg ha^{-1} had a similar effect and suppressed the tiller number by 67% (6.8 tillers plant^{-1}); whereas 150 kg ha^{-1} of seeding rates suppressed tiller density by 79% (4.7 tillers plant^{-1}). In the second experiment (Figure 3b), however, all seeding rates suppressed the tiller density significantly beyond 14 DAS. In terms of percent reduction, the

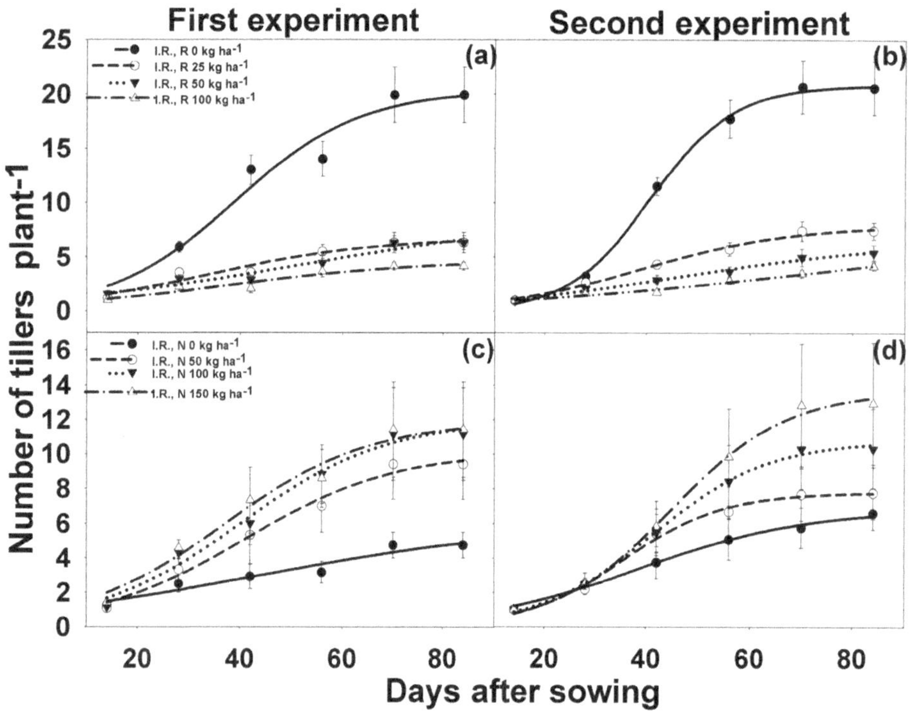

Figure 3. (a, b) Number of tillers per plant of *Ischaemum rugosum* (I.R.), rice seeding rates (R) when grown alone without rice (I.R., R0) or in competition with rice at different seeding rates (R), i.e., 25 (I.R., R25), 50 (I.R., R50), and 100 kg ha^{-1} (I.R., R100); and (c, d) number of tillers per plant under different nitrogen levels, i.e., 0 (I.R., N0), 50 (I.R., N50), 100 (I.R., N100), and 150 kg ha^{-1} (I.R., N150). The vertical bars represent standard error of means. The lines represent a three-parameter sigmoid model ($y = a/\{1+e[-(x-x50)/b]\}$) fitted to the data. Parameter estimates are shown in Table 1.

tiller number plant^{-1} was reduced by 62%, 74%, and 80% at seeding rates of 25, 50 and 100 kg ha^{-1}, respectively (Table 1).

The increased levels of N significantly increased the number of tillers of *I. rugosum* after 14 DAS (Figures 3c and d). In the first experiment, N application at 0 to 100 kg ha^{-1} significantly increased the number of tillers per plant, whereas no significant differences were observed between 100 and 150 kg N ha^{-1}. However, in the second experiment, all N levels had a significant effect on tiller density compared with tiller density (6.0 and 6.8 tillers plant^{-1} in the first and second experiments, respectively) without added N. Nitrogen at 50, 100, and 150 kg ha^{-1} increased tiller number by 70% (10.1 tillers plant^{-1}), 101% (12 tillers plant^{-1}), and 101% (12 tillers plant^{-1}), respectively, for the first experiment; and 14% (7.8), 58% (10.8), and 100% (13.6), respectively, for the second experiment (Table 1).

Leaf number

The number of *I. rugosum* leaves per plant declined significantly with the increase in rice seeding rates from 25 to 100 kg ha^{-1} (Figures 4a and b). In the first experiment, rice seeding rates of 25, 50, and 100 kg ha^{-1} reduced the leaf number of *I. rugosum* by 68%, 72%, and 77%, respectively, compared with the leaf number in weeds grown alone (Table 1). A similar reduction (57–76%) was also noted in the second experiment. The number of leaves significantly increased with increasing N levels after 14 DAS (Figures 4c and d Table 1). On average over seeding densities and compared with 0 kg N ha^{-1}, the increasing level of N (50, 100, and 150 kg N ha^{-1}) increased leaf number by 59%, 111%, and 122%, respectively, in the first experiment; and 22%, 48%, and 90%, respectively, in the second experiment.

Leaf area

Interaction between different rice seeding rate and N levels for leaf area plant^{-1} was significant for both the experiments. With increase in rice seeding rate there was a significant reduction in leaf area of *I. rugosum* (Figure 5a Table 2). Compared to the leaf area of weed plants grown without rice competition, rice seeding rates of 25, 50, and 100 kg ha^{-1} reduced leaf area by 69%, 73%, and 77%, respectively (Figure 5a). Leaf area increased with rising levels of N (Figure 5a Table 2). At each rice seeding rate, leaf area (cm^2 plant^{-1}) was always lower in the absence of added N than where N was applied, whereas 150 kg N ha^{-1} produced maximum leaf area (2,442 cm^2 plant^{-1}). In the first experiment, there was an increase of 105% (1,073+501 cm^2 plant^{-1}), 167% (1,622+431 cm^2 plant^{-1}), and 218% (2058+385 cm^2 plant^{-1}) leaf area with added N of 50, 100, and 150 kg ha^{-1}, respectively, compared to 0 kg N ha^{-1} (430+339 cm^2 plant^{-1}) (Figure 5a Table 2). A similar trend (55–232% increase) was observed in the second experiment (Figure 5b Table 2).

Leaf biomass

At each rice seeding rate, the leaf biomass of *I. rugosum* was always lower in the absence of added N in comparison to the treatments where N was applied (Figure 6 Table 3). Interaction between seeding rate and N levels for leaf biomass area plant^{-1} was significant for both the experiments The application of 150 kg N ha^{-1} produced maximum leaf biomass (20.6 g plant^{-1}). In the first experiment, there was an increase of 119%, 180%, and 284% leaf biomass with N rates of 50, 100, and 150 kg ha^{-1}, respectively, compared to 0 kg N ha^{-1}(Figures 6a and b Table 3). A similar response was observed for the second

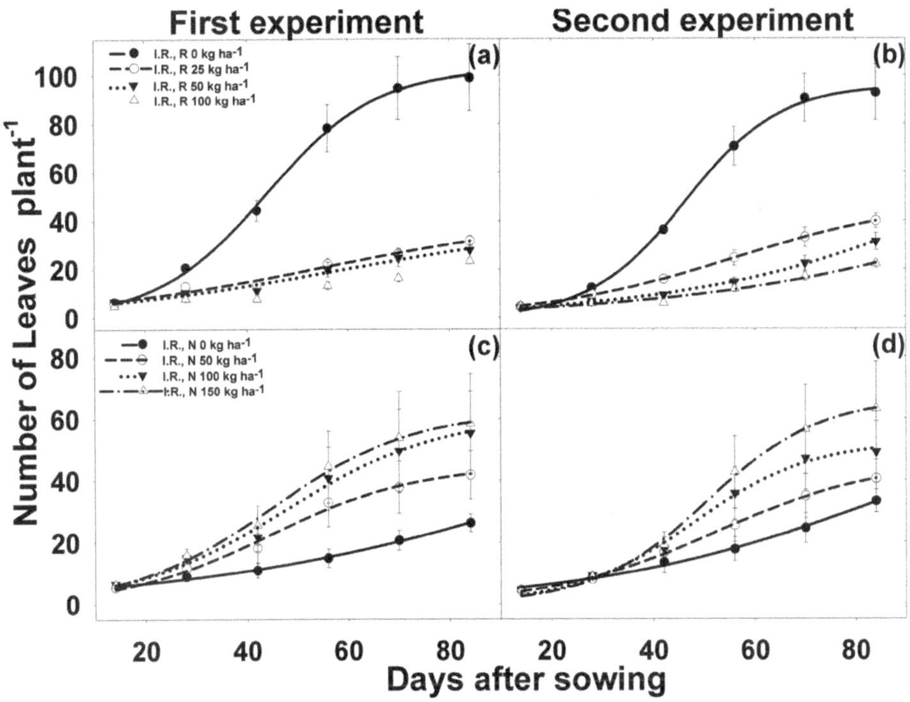

Figure 4. (a, b) Number of leaves per plant of *Ischaemum rugosum* (I.R.), rice seeding rates (R), when grown alone without rice (I.R., R0) or in competition with rice at different seeding rates (R), i.e., 25 (I.R., R25), 50 (I.R., R50), and 100 kg ha^{-1} (I.R., R100); and (c, d) number of leaves per plant when grown under different nitrogen levels, i.e., 0 (I.R., N0), 50 (I.R., N50), 100 (I.R., N100), and 150 kg ha^{-1} (I.R., N150). The vertical bars represent standard error of means. The lines represent a three-parameter sigmoid model ($y = a/\{1+e^{[-(x-x50)/b]}\}$) fitted to the data. Parameter estimates are shown in Table 1.

experiment, in which leaf biomass decreased with the increase in rice seeding rate. Compared with the weed plants grown without rice competition, leaf biomass was reduced by 72%, 77%, and 84% with rice seeding densities of 25, 50, and 100 kg ha^{-1}, respectively.

Stem biomass

For stem biomass different rice seeding rate and N levels have significant interaction for both experiments. Regardless of N rate, increasing the seeding rate decreased the stem biomass of *I. rugosum* (Figure 6c). Compared with the plants grown alone, stem biomass decreased by 80%, 85%, and 89% when grown in competition, with rice seeding densities of 25, 50, and 100 kg ha^{-1}, respectively. Increasing the level of N increased stem biomass significantly at all seeding rates. In the first experiment, stem biomass of *I. rugosum* increased 2.8, 4.2, and 5.8 times with the application of 50, 100, and 150 kg N ha^{-1}, respectively, compared with no added N (Figure 6c Table 3). In the second experiment, the corresponding values were 1.8, 2.9, and 4.1, respectively (Figure 6d Table 3).

Inflorescence biomass

Different rice seeding rate and N levels have significant interaction for Inflorescence biomass plant^{-1} for both experiments. With the increase in seeding rate, there was a significant decline in the inflorescence biomass of *I. rugosum*, whereas, with the

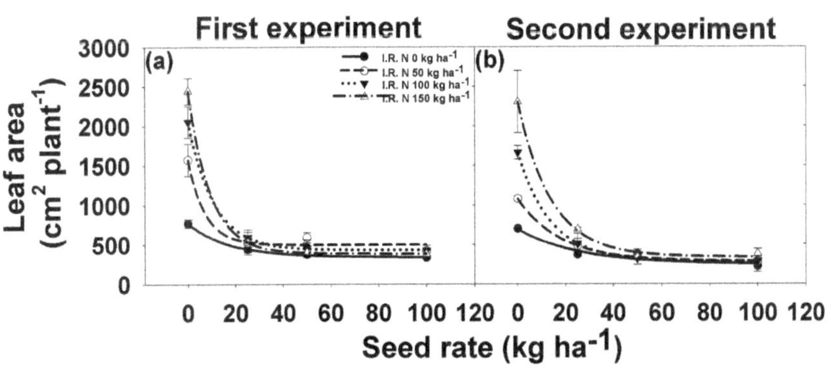

Figure 5. (a, b) Leaf area of *Ischaemum rugosum* under different seeding rates and nitrogen levels. The vertical bars represent standard error of means. The lines represent an exponential model fitted to the leaf area data. Parameter estimates are shown in Table 2.

Table 2. The effect of different seeding rates and nitrogen levels on the leaf area of *Ischaemum rugosum*.

N (kg ha^{-1})	First experiment				Second experiment			
	y0	*A*	*B*	R^2	*y0*	*a*	*B*	R^2
0	338.7(10.5)	429.8(15.1)	0.06 (0.01)	0.99	244.9 (107.1)	449.9 (134.2)	0.04 (0.03)	0.99
50	501.1(89.0)	1072.8 (150.7)	0.14 (0.18)	0.99	282.4 (4.9)	796.4 (7.0)	0.06 (0.00)	0.99
100	431.0(10.3)	1622.0 (16.6)	0.09 (0.00)	0.99	282.1 (0.3)	1380.9 (0.4)	0.07 (0.01)	0.99
150	384.5 (6.6)	2058.4 (10.8)	0.11 (0.00)	0.99	338.9 (60.4)	1968.6 (92.5)	0.07 (0.01)	0.99

Leaf area data were fitted to an exponential model: $l = y0 + a \times e^{-bx}$, where l is the leaf area (cm^2 plant^{-1}) at different seed rates (kg ha^{-1}) x, $y0$ is the minimum leaf area at seed rate 100 kg ha^{-1}, $y0+a$ is the maximum leaf area at seed rate 0 kg ha^{-1}, and b is the slope. Parameter estimates are followed by standard error in parentheses.

increase in N rate, inflorescence biomass increased significantly (Figures 6e and f Table 3). Compared with the rice seeding rate of 0 kg ha^{-1}, the average reduction in inflorescence biomass was 81%, 83%, and 93% for the first experiment (Figure 6e) and 59%, 71%, and 87% for the second experiment (Figure 6f), with rice densities of 25%, 50%, and 100 kg ha^{-1}, respectively. Compared to 0 kg N ha^{-1}, there was an increase of 173%, 584%, and 759% in the inflorescence biomass with N levels of 50, 100, and 150 kg ha^{-1}, respectively, in the first experiment (Figure 6e). In the second experiment, the corresponding values were 82%, 130%, and 294%, respectively (Figure 6f Table 3).

Total shoot and root biomass

Interaction between different rice seeding rate and N levels for Total shoot and root biomass plant^{-1} was significant for both the experiments. Total shoot and root biomass of *I. rugosum* was affected by rice seeding rates and N levels (Figure 7). With increased seeding rate, there was a decline in shoot and root biomass of *I. rugosum*. Shoot and root biomass decreased by 78% to 99% with rice seeding rates of 25 to 100 kg ha^{-1} compared with the biomass of plants grown alone. The rising level of N increased the shoot and root biomass of both rice and weeds significantly (Figures 7a–d Tables 4 and 5). Results revealed that the effect of N

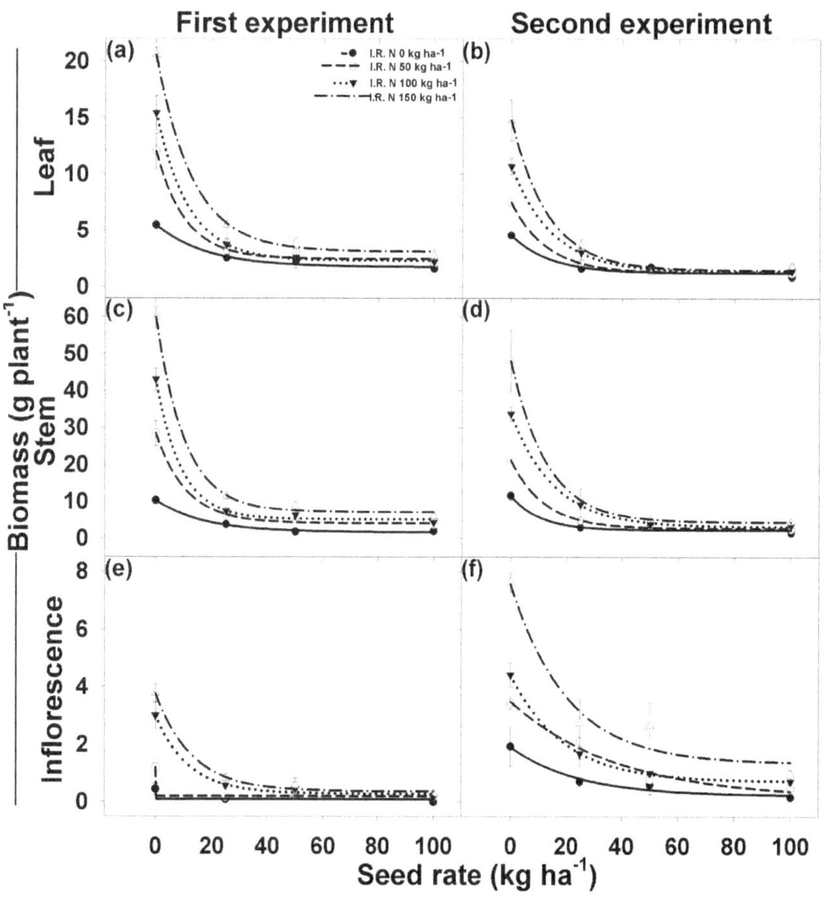

Figure 6. (a, b) Leaf biomass, (c, d) stem biomass, and (e, f) inflorescence biomass of *Ischaemum rugosum* under different seeding rates and nitrogen levels. The vertical bars represent standard error of means. (g) Stem biomass. The lines represent an exponential model fitted to the seedling biomass data.

Table 3. The effect of different seeding rates and nitrogen levels on leaf, stem, inflorescence, and root biomass (g plant^{-1}) of *Ischaemum rugosum*.

N (kg ha^{-1})		*Ischaemum rugosum* biomass (g plant^{-1})											
		Leaf				Stem				Inflorescence			
		y0	*a*	*b*	R^2	*y0*	*a*	*B*	R^2	*y0*	*a*	*B*	R^2
First experiment	0	1.77 (0.32)	3.72 (0.47)	0.06 (0.02)	0.99	1.67 (0.40)	8.70 (0.59)	0.06 (0.01)	0.99	0.09 (0.00)	0.35 (0.00)	9964.89 (+inf)	0.84
	50	2.48 (0.51)	9.55 (0.83)	0.10 (0.04)	0.99	4.07 (0.55)	24.44 (0.88)	0.09 (0.02)	0.99	0.19 (0.00)	1.01 (0.00)	808.97 (+inf)	0.99
	100	2.32 (0.13)	13.07 (0.20)	0.09 (0.01)	0.99	5.18 (0.97)	37.84 (1.61)	0.11 (0.03)	0.99	0.28 (0.14)	2.73 (0.23)	0.09 (0.03)	0.99
	150	3.12 (0.53)	17.44 (0.83	0.08 (0.01)	0.99	7.14 (1.39)	52.85 (2.26)	0.11 (0.02)	0.99	0.37 (0.12)	3.41 (0.19)	0.08 (0.02)	0.99
Second experiment	0	1.24 (0.48)	3.33 (0.74)	0.07 (0.06)	0.96	2.30 (0.99)	9.32 (1.61)	0.10 (0.08)	0.96	0.23 (0.25)	1.69 (0.32)	0.04 (0.02)	0.97
	50	1.26 (0.07)	6.25 (0.11)	0.08 (0.01)	0.99	2.69 (0.26)	18.55 (0.42)	0.08 (0.01)	0.99	0.14 (0.64)	3.35 (0.67)	0.03 (0.01)	0.98
	100	1.32 (0.16)	9.38 (0.24)	0.07 (0.01)	0.99	3.11 (0.53)	30.62 (0.79)	0.07 (0.01)	0.99	0.72 (0.04)	3.69 (0.06)	0.05 (0.00)	0.99
	150	1.47 (0.52)	13.30 (0.81)	0.08 (0.02)	0.99	4.40 (0.10)	43.75 (0.16)	0.08 (0.00)	0.99	1.34 (0.99)	6.23 (1.36)	0.05 (0.03)	0.96

Data were fitted to an exponential decay model: $Y = y0 + a \times e^{-bx}$ where *Y* is the leaf, stem, and inflorescence biomass (g plant^{-1}) at different seed rates (kg ha^{-1}) *x*; *y0* is the minimum leaf, stem, and inflorescence biomass at seed rate 100 kg ha^{-1}; *y0+a* is the maximum leaf, stem, and inflorescence biomass at seed rate 0 kg ha^{-1}; and *b* is the slope. Parameter estimates are followed by standard error in parentheses.

levels on shoot and root biomass of *I. rugosum* was more prominent when weed plants were grown alone than they were grown with rice interference. Increasing levels of N increased the biomass of rice more than it did the weeds (Table 4).

Relationship between rice and weed biomass

Total plant biomass of rice increased with the increase in rice densities and N levels (Table 5), which suppressed the biomass of *I. rugosum* (Figure 8). The increased seeding rate of rice from 0 to 100 kg ha^{-1} reduced ($p<0.001$) the total weed biomass, whereas the increasing rate of N increased ($p<0.001$) the biomass of *I. rugosum*. Compared with the plants grown alone, the biomass of *I. rugosum* was reduced by 85%, 89%, and 92% for the first experiment, and 77%, 89%, and 91% for the second experiment with seeding rates of 25, 50, and 100 kg ha^{-1}, respectively. There was an increase in *I. rugosum* biomass by 106%, 196%, and 293% for the first experiment and 59%, 125%, and 206% for the second experiment at N rates of 50, 100, and 150 kg ha^{-1}, respectively, compared with the no N-added treatment.

Over the average of two experiments, compared with no added N, increase in weed biomass was 82%, 160%, and 150% and increase in rice biomass was 92%, 155%, and 229% with the addition of 50, 100, and 150 kg N ha^{-1}.

Correlation and regression analysis between the biomass of rice and weeds showed that every unit increase in rice biomass decreased weed biomass by 1.25 times in the first experiment and 0.93 times in the second experiment (Figure 8).

Leaf area ratio

With the increase in seeding rate, there was an increasing ($p< 0.001$) trend in leaf area ratio (LAR) irrespective of N rates (Figure 9a). At each rice seeding density, LAR decreased ($p=0.006$ and 0.003 for the first and second experiments, respectively) with the increase in N for both experiments. For the first experiment, the increase in LAR was 234%, 258%, and 283% at rice densities of 25, 50, and 100 kg ha^{-1}, respectively, compared to 0 kg ha^{-1}. In contrast, compared with 0 kg N ha^{-1}, there was a decrease in LAR by 9%, 25%, and 34% at N rates of 50, 100, and 150 kg ha^{-1}, respectively. Similarly, LAR increased with increasing seeding rates and decreased with increasing N rates in the second experiment (Figure 9b).

Root to shoot weight ratio

The interaction between seeding and N rates was not significant for root to shoot weight ratio (RSWR). Seeding rate had a significant effect ($p<0.001$), whereas N rate had a non-significant effect on RSWR in both experiments (Figure 9c). RSWR was highest (0.733 and 0.312 in the first and second experiments, respectively) when weed plants were grown alone and the ratio reduced to 0.09 and 0.21 for the first and second experiments, respectively, when weed plants were grown with rice plants at 100 kg seed ha^{-1} (Figure 9d).

Leaf weight ratio

Increased seeding rate from 0 to 100 kg ha^{-1} increased ($p< 0.001$) leaf weight ratio (LWR) of *I. rugosum* (Figure 9e). Compared with the plants grown alone, LWR increased by 195%, 181%, and 199% for the first experiment and 129%, 135%, and 169% for the second experiment at rice seeding rates of 25, 50, and 100 kg ha^{-1}, respectively (Figure 9e). Compared to 0 kg N ha^{-1}, the increased levels of N (50–150 kg ha^{-1}) reduced LWR by 11–21% for the first experiment and 18–28% for the second experiment (Figure 9f).

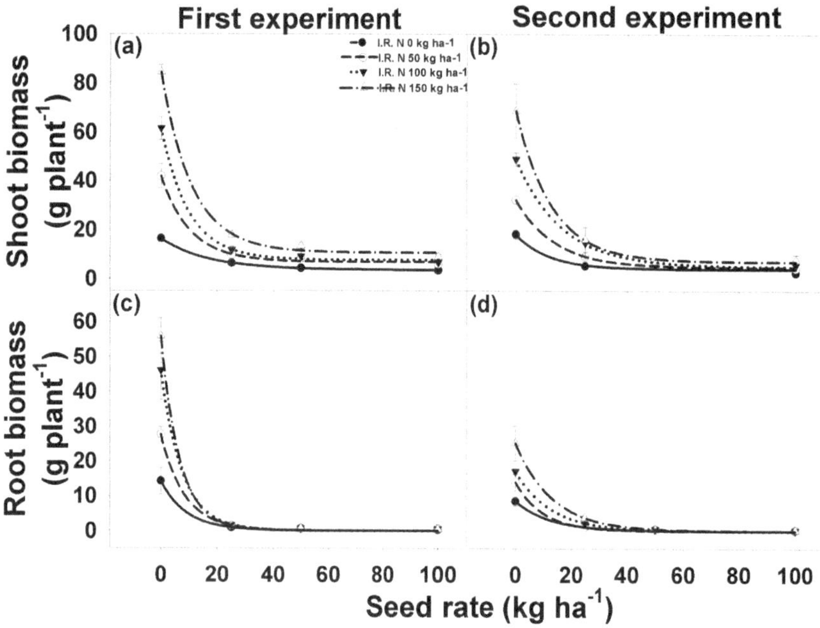

Figure 7. (a, b) Shoot biomass, and (c, d) root biomass of *Ischaemum rugosum* under different seeding rates and nitrogen levels. The vertical bars represent standard error of means. The lines represent an exponential model fitted to the seedling biomass data.

Specific leaf area

In both experiments, increase in the rice seeding rate increased ($p<0.001$) the specific leaf area (SLA) of *I. rugosum*, whereas N rates did not affect the SLA of the weeds. Compared with the plants grown alone, increase in SLA was 122%, 144%, and 143% for the first experiment, and 147%, 146%, and 161% for the second experiment at rice seeding rates of 25, 50 and 100 kg ha^{-1}, respectively (Figure 9g). Compared with the treatment without added N, the decrease in SLA was 3%, 7%, and 23% for the first experiment and 4%, 13%, and 17% for the second experiment at N rates of 50,100, and 150 kg ha^{-1}, respectively (Figure 9h).

Discussion

The height of *I. rugosum* was affected by high rice seeding rates and N rates. N application had a more positive effect on the rice plant height than it had on the weed. N effect was more pronounced on the height of *I. rugosum* plants grown alone compared to the plants grown in competition with rice. Similar findings were reported in an early study, in which the growth of *Equistem arvense* L. increased at high rates of N supplied [22]. With the application of N, *I. rugosum* growth can be suppressed by rice interference. *Ischaemum rugosum* was always taller than rice owing to the reason that the weed has a pronounced shade-avoiding syndrome. With increasing seeding rates (0–100 kg ha^{-1}), weed

Table 4. The effect of different seeding rates and nitrogen levels on aboveground shoot and root biomass (g plant^{-1}) of *Ischaemum rugosum*.

Experiment	N (kg ha^{-1})	*Ischaemum rugosum* biomass (g plant^{-1})						
		Shoot				**Root**		
		y0	*a*	*b*	R^2	*A*	*b*	R^2
First	0	3.59 (0.17)	12.70 (0.25)	0.06 (0.00)	0.99	14.20 (0.51)	0.11 (0.02)	0.99
	50	6.82 (1.16)	34.92 (1.89)	0.10 (0.03)	0.99	27.55 (0.56)	0.12 (0.02)	0.99
	100	7.66 (1.11)	53.74 (1.82)	0.10 (0.02)	0.99	45.97 (0.60)	0.13 (0.01)	0.99
	150	10.64 (2.02)	73.68 (3.26)	0.09 (0.02)	0.99	55.85 (0.56)	0.15 (0.02)	0.99
Second	0	3.95 (1.85)	14.16 (2.93)	0.08 (0.07)	0.96	8.55 (0.29)	0.08 (0.01)	0.99
	50	4.59 (0.49)	27.59 (0.75)	0.07 (0.01)	0.99	13.99 (0.31)	0.10 (0.01)	0.99
	100	4.86 (1.24)	44.02 (1.83)	0.06 (0.01)	0.99	17.09 (0.34)	0.08 (0.01)	0.99
	150	7.03 (1.27)	61.69 (1.98)	0.08 (0.01)	0.99	25.17 (0.18)	0.08 (0.00)	0.99

Shoot data were fitted to an exponential decay model: $Y = y0 + a \times e^{-bx}$, where Y is the shoot biomass (g plant^{-1}) at different seed rates (kg ha^{-1}) x, $y0$ is the minimum shoot biomass at seed rate 100 kg ha^{-1}, $y0+a$ is the maximum shoot biomass at seed rate 0 kg ha^{-1}, and b is the slope. Root biomass data were fitted to an exponential decay model: $Y = ae^{-bx}$, where Y is the root biomass (g plant^{-1}) at seed rate x kg ha^{-1}, a is the maximum root biomass (g plant^{-1}) at seed rate 0 kg ha^{-1}, and b is the slope. Parameter estimates are followed by standard error in parentheses.

Table 5. Data on rice tillers per unit area (25 cm^2); biomass of rice shoot, root, and total plant and root weight ratio (RWR) of the rice plant under different seeding rates, i.e., 0, 25, 50, or 100 kg ha^{-1}; and under different nitrogen levels, i.e., 0, 50, 100, and 150 kg ha^{-1} for both experiments.

Seed rate (kg ha^{-1})	First experiment					Second experiment				
	Biomass (g pot^{-1})			Tillers $plant^{-1}$	RWR (g g^{-1})	Biomass (g pot^{-1})			Tillers $plant^{-1}$	RWR (g g^{-1})
	Shoot	Root	Total plant			Shoot	Root	Total plant		
25	39.9	13.8	53.7	19.3	0.381	34.16	11.78	45.94	16.42	0.3529
50	49.5	15.79	65.3	28.6	0.371	41.24	12.28	53.52	23.58	0.3088
100	46.7	15.54	62.3	30.9	0.362	46.56	13.06	59.12	29	0.2775
S.E.	4.47	1.20	5.18	2.92	0.032	1.24	0.96	1.98	1.034	0.03
P	0.111	0.216	0.09	0.002	0.843	<0.001	0.716	<0.001	<0.001	0.011
N rate (kg ha^{-1})										
0	15.3	7.67	23.0	15.8	0.512	23.45	7.36	30.14	18.11	0.3201
50	37.2	12.46	49.6	22.1	0.340	38.06	12.73	50.79	22	0.3486
100	56.2	17.03	73.2	31.3	0.313	45.74	12.28	58.01	24.89	0.2707
150	72.8	23.02	95.8	35.9	0.320	55.37	17.13	72.50	27.00	0.3130
S.E.	5.16	1.38	5.98	3.37	0.037	1.65	1.11	2.29	1.20	0.03
P	<0.001	<0.001	<0.001	<0.001	<0.001	<0.001	<0.001	<0.001	<0.001	0.052

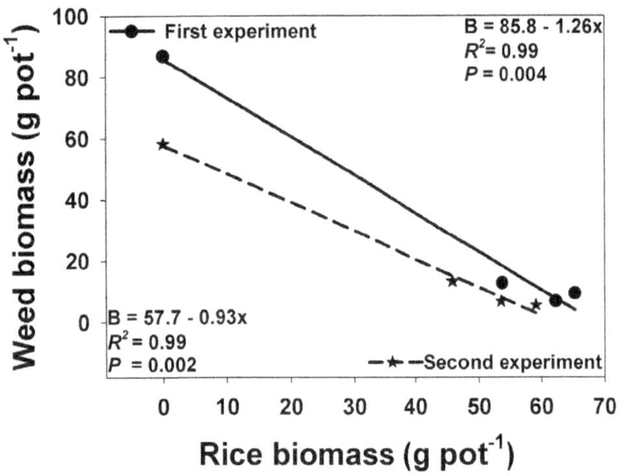

Figure 8. Relationship between rice and *Ischaemum rugosum* biomass (g pot^{-1}).

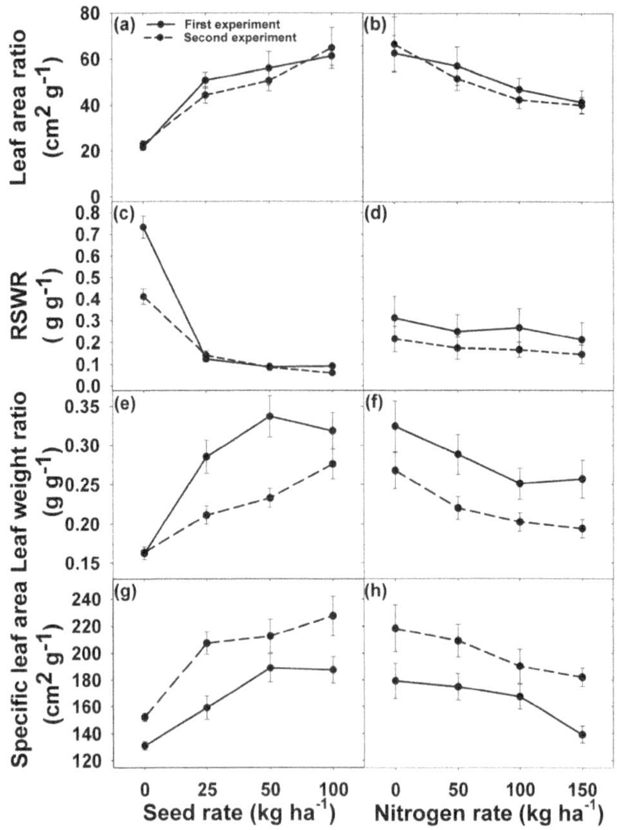

Figure 9. (a, b) Leaf area ratio (LAR); (c, d) root-to-shoot weight ratio (RSWR); (e, f) leaf-weight ratio (LWR); and (g, h) specific leaf area (SLA) of *Ischaemum rugosum* under different seeding rates and nitrogen levels. The vertical bars represent standard error of means.

height increased, irrespective of N levels, to overcome the effects of shade for capturing sunlight for photosynthesis [23]. Similar results were reported for *Echinochloa phyllopogon* (Stapf) Koss., whose height increased with the increase in shade [21].

With increasing rice seeding rate, there was a decline in the biomass of aboveground (stem, leaves, and inflorescence) and belowground (root) plant parts of *I. rugosum*. Results are in line with an earlier study on *E. crus-galli*, in which aboveground biomass decreased (38–48%) significantly with increasing rice density and increased with the increase in N rates [24]. Earlier, researchers reported that *E. phyllopogon* biomass decreased under shaded conditions and increased with the increase in N rates from 0 to 224 kg ha^{-1} [21]. In our study, as the shading by high rice density increased, root biomass decreased. This proved that under the sunniest environments, the roots of plants preferred sinks for photosynthates, whereas under the shadiest circumstances, the shoots were the most important sinks during almost the entire life of the plant [25].

Increasing N rates did not increase *I. rugosum* biomass at the same rate as of rice biomass. In earlier studies in corn [16] and soybean [26] the adverse effects of weed competition was severe on crops at low N rates than high N rates. Soybean was more responsive to N than weeds, which indicates that high N rate is more beneficial to crops than weeds. At initial crop growth stages, low N rate make crops vulnerable to weed competition. Higher N rates can enhance rapid crop growth and canopy closure, required to suppress the growth of weeds [26]. Thus, rice-weed interactions can be strongly affected by N availability [14,21].

With the increase in seeding rate, there was a reduction in leaf area and leaf biomass of *I. rugosum* irrespective of N rates; whereas, with the increase in N rate, there was an increase in leaf area and leaf biomass. This increase can be attributed to changes in total biomass. Total plant or shoot biomass increased with the increase in N and this is highly correlated with leaf area. Similar results were also reported for *E. phyllopogon*, in which leaf area increased as N increased from 0 to 224 kg ha^{-1} because more biomass was available for allocation to leaves [21]. In *C. selloana*, leaf area per plant was five times higher where N was applied as compared to the control [27].

LAR and RSWR indicate how photosynthates were being apportioned among the plant parts. LAR and SLA also demonstrate the utilization of photosynthates available for the growth of leaves (leaf expansion vs. leaf thickness). Results revealed that higher seeding rates increased LAR, LWR, and SLA and reduced RSWR, whereas higher N rates reduced LAR, LWR, SLA, and RSWR. *Ischaemum rugosum* plants grown with rice at 100 kg seed ha^{-1} (shade) had a lower RSWR and higher LAR, LWR, and SLA than plants grown without rice interference (no shade), indicating that the weed has a phenotypic plasticity to allocate more photosynthates to aboveground plant parts than those belowground when grown in competition. It seems that partitioning of biomass in *I. rugosum* was affected more by limitations in light than by N availability. Weed RSWR decreased more than rice RSWR with increase in rice seeding rates (shade) and N rates. Increased partitioning to roots is a common compensatory response of plants to low N availability [21]. Reynolds and Antonio made 77 studies representing 129 species. They found that RSWR decreased with increase in N availability [28]. Bonifas *et al.* reported that RSWR decreased for *Abutilon theophrasti* Medik. and corn with increase in N application [29].

The production of higher SLA by the weed under high rice seeding rates demonstrates that *I. rugosum* has shade-avoiding syndrome characteristics and produced thinner leaves to avoid the shading effects of rice leaves [30]. This weed produced thinner leaves to increase leaf area per allocated photosynthates to intercept light and maintain photosynthesis, which are crucial for the life of plants. Thinner leaves enhanced the amount of radiation absorbed per chloroplast in the leaf to increase the rate of photosynthesis [31]. *Ischaemum rugosum* was able to change leaf thickness to adapt to reduction in sunlight by producing thin leaves under shady conditions and thicker leaves under full light conditions. This phenotypic plasticity made the weed extremely competitive with rice.

Conclusions and Implication of the study

Ischaemum rugosum showed plasticity in SLA, LAR, LWR, and RSWR in response to competition with rice plants. Increase in rice seeding rates increased SLA, LAR, and LWR and reduced RSWR, whereas all these ratios declined with increase in N rate. *Ischaemum rugosum* plants are phenotypically plastic and can alter their morphology and physiology to improve the acquisition of a resource, when lack of that resource limits the growth of the plant [21]. Increased partitioning of biomass to shoot vs. roots (low RSWR) is a common response of shade-avoiding plants to shade and, in addition to physiological and morphological changes in leaf (increased LAR, LWR, and SLA), these plants can reduce the effects of shade. The weed had a higher capacity for N uptake compared to rice plants. But when the weed was grown with rice interference, rice plants caused reduced N uptake of the weed (data not shown).

Our study suggests that management strategies that depend on shade due to crop interference may not provide complete suppression of *I. rugosum*. This was highlighted by observations on its plant height, LAR, LWR, SLA, RSWR, and inflorescence biomass when grown alone or in competition with a varying density of rice plants. Although *I. rugosum* cannot be completely suppressed by crop interference, high seeding rates can reduce the biomass of *I. rugosum* appreciably and make it weak and vulnerable and thus can be easily controlled. This study was conducted in controlled conditions and the results can suggest possile approaches for weed management that are worthy of corroboration by further field study. However, the results of this study can be used to aid the conceptualization of weed management strategies for *Ischaemum rugosum*

Acknowledgments

The author would like to thank the IRRI Climate Unit for providing climate data.

Author Contributions

Conceived and designed the experiments: THA BSC PCSC. Performed the experiments: THA. Analyzed the data: THA BSC. Contributed reagents/materials/analysis tools: THA BSC. Wrote the paper: THA BSC PCSC.

References

1. Chauhan BS (2012) Weed ecology and weed management strategies for dry-seeded rice in Asia. Weed Tech 26: 1–13.
2. Chauhan BS, Mahajan G, Sardana V, Timsina J, Jat ML (2012) Productivity and sustainability of the rice-wheat cropping system in the Indo-Gangetic Plains of the Indian subcontinent: problems, opportunities, and strategies. Advan Agron 117: 315–369.
3. Buhler DD, Liebman M, Obrycki JJ (2002) Review: theoretical and practical challenges to an IPM approach to weed management. Weed Sci 48: 274–280.
4. Chauhan BS, Johnson DE (2011) Ecological studies on *Echinochloa crus-galli* and the implications for weed management in direct-seeded rice. Crop Protn 30: 1385–1391.
5. Mahajan G, Chauhan BS, Gill MS (2013) Dry-seeded rice culture in Punjab state of India: lessons learned from farmers. Field Crops Res 144: 89–99.
6. Holm LG, Plucknett DL, Pancho JV, Herberger JP (1977) The World's Worst Weeds: Distribution and Biology University of Hawaii Press, Honolulu, Hawaii (USA). 609 p.
7. Itoh K. (Ed.) (1991) Life cycles of rice field weeds and their management in Malaysia. Peang, Malaysia, Tropical Agriculture Research Center, p. 55–59.
8. Bakar BH, Nabi LNA (2003) Seed germination, seedling establishment and growth patterns of wrinklegrass (*Ischaemum rugosum* Salisb.). Weed Bio Mnt 3: 8–14.
9. Marenco RA, Santos RVCD (1999) *Ischaemum rugosum* and Rice Intra and Interspecific Competition1revista Brasileira De Fisiologia Vegetal, 11: 107–111,.
10. Valverde BE (2007) Status and management of grass-weed herbicide resistance in Latin America. Weed Tech 21: 310–323.
11. Zhao D (2006) Weed competitiveness and yielding ability of aerobic rice genotypes, PhD thesis. Wageningen University, The Netherlands.
12. Phoung LT, Denich M, Vlek PLG, Balasubramanian V (2005) Suppressing weeds in direct-seeded lowland rice: effects of methods and rates of seeding. J Agron Crop Sci 191: 185–194.
13. Liebman M, RJ . Janke (1990) Sustainable weed management practices. Pp. 111–143 *in* Francis, C. A., C. B. . Flora, and L. D. . King, editors. Sustainable Agriculture in Temperate Zones. New York John Wiley and Sons.
14. Ampong-Nyarko K, S K . DE Datta (1993) Effects of nitrogen application on growth, nitrogen use efficiency and rice-weed interaction. Weed Res 33: 269–276.
15. Mishra JS, Kurchania SP (2001) Nutrient concentration in mustard and associated weeds as influenced by nitrogen levels, planting geometry and weed control methods. Indian J Plant Physiol 6: 386–389.
16. Evans SP, Knezevic SZ, Lindquist JL, Shapiro CA, Blankenship EE (2003) Nitrogen application influences the critical period for weed control in corn. Weed Sci 51: 408–417.
17. Ponce RG (1998) Competition between barley and *Lolium rigidum* for nitrate. Weed Res 38: 453–460.
18. Jørnsgard BK, Rasmussen JH, Christiansen JL (1996) Influence of nitrogen on competition between cereals, their natural weed populations. Weed Res 36: 461–470.
19. Blackshaw RE, Brandt RN, Janzen HH, Grant CA, Derksen DA (2003) Differential response of weed species to added nitrogen. Weed Sci 51: 532–539.
20. Gibson KD, Fischer AJ, Foin TC (2001) Shading and the growth and photosynthetic responses of Ammannia coccinnea. Weed Res 41: 59–67.
21. Gibson KD, Fischer AJ, Foin TC (2004) Compensatory responses of late watergrass (*Echinochloa phyllopogon*) and rice to resource limitations. Weed Sci 52: 271–280.
22. Andersson TN, Lundegardh B (1999) Field horsetail (*Equisetum arvense*)-effects of potassium under different light and nitrogen conditions. Weed Sci 47: 47–54.
23. Lubigan RT, Moody K (1989) Effects of herbicides on *Ischaemum rugosum*. International Rice Research Newsletter 14, 49–53.
24. Chauhan BS, Abugho SB (2013) Effects of water regime, nitrogen fertilization, and rice plant density on growth and reproduction of lowland weed *Echinochloa crus-galli*. Crop Protn 54: 142–147.
25. Marenco RA, Reis ACS (1998) Shading as an environmental factor affecting the growth of *Ischaemum rugosum*. Revista Brasileira de Fisiologia Vegetal 10: 107–112.
26. Kolvanagh JS, Salmasi SZ, Javanshir A, Moghaddam M, Nasab ADM (2008) Effects of nitrogen and duration of weed interference on grain yield and SPAD (chlorophyll) value of soybean (*Glycine max* (L.) Merrill.). J Food, Agri Envir 6: 368–373.
27. Vourlitis GL, Kroon JL (2013) Growth and resource use of the invasive grass, pampagrass (*Cortaderia selloana*), in response to nitrogen and water availability. Weed Sci 61: 117–125.
28. Reynolds HL, Antonio CD (1996) The ecological significance of plasticity in root weight ratio in response to nitrogen: opinion. Plant Soil 185: 75–97.
29. Bonifas KD, Walters DT, Cassman KG, Lindquist JL. (2005) Nitrogen supply affects root:shoot ratio in corn and velvetleaf (*Abutilon theophrasti*) Weed Sci 53: 670–675.
30. Knops JMH, Reinhart K (2000) Specific leaf area along a nitrogen fertilization gradient. American Mid Natur 144, 265–272.
31. Terashima I, Evans JR (1988) Effects of light and nitrogen nutrition on the organization of the photosynthetic apparatus in spinach. Plant Cell Physiol 29: 143–155.

12

A Powerful Molecular Engineering Tool Provided Efficient *Chlamydomonas* Mutants as Bio-Sensing Elements for Herbicides Detection

Maya D. Lambreva[1]*, Maria Teresa Giardi[1], Irene Rambaldi[1], Amina Antonacci[1], Sandro Pastorelli[1¤], Ivo Bertalan[2], Ivan Husu[1], Udo Johanningmeier[2], Giuseppina Rea[1]

1 Institute of Crystallography, National Research Council, Monterotondo Scalo, Rome, Italy, **2** Martin-Luther-University, Plant Physiology Institute, Halle (Saale), Germany

Abstract

This study was prompted by increasing concerns about ecological damage and human health threats derived by persistent contamination of water and soil with herbicides, and emerging of bio-sensing technology as powerful, fast and efficient tool for the identification of such hazards. This work is aimed at overcoming principal limitations negatively affecting the whole-cell-based biosensors performance due to inadequate stability and sensitivity of the bio-recognition element. The novel bio-sensing elements for the detection of herbicides were generated exploiting the power of molecular engineering in order to improve the performance of photosynthetic complexes. The new phenotypes were produced by an *in vitro* directed evolution strategy targeted at the photosystem II (PSII) D1 protein of *Chlamydomonas reinhardtii*, using exposures to radical-generating ionizing radiation as selection pressure. These tools proved successful to identify D1 mutations conferring enhanced stability, tolerance to free-radical-associated stress and competence for herbicide perception. Long-term stability tests of PSII performance revealed the mutants capability to deal with oxidative stress-related conditions. Furthermore, dose-response experiments indicated the strains having increased sensitivity or resistance to triazine and urea type herbicides with I_{50} values ranging from 6×10^{-8} M to 2×10^{-6} M. Besides stressing the relevance of several amino acids for PSII photochemistry and herbicide sensing, the possibility to improve the specificity of whole-cell-based biosensors, via coupling herbicide-sensitive with herbicide-resistant strains, was verified.

Editor: Kostas Bourtzis, International Atomic Energy Agency, Austria

Funding: This research was performed within the FP7-SME-2008-1 SENSBIOSYN project (ID: 232082, http://www.sensbiosyn.com), FILAS Microbiosis project (POR FESR-LAZIO 2007.2013) and the CMST COST TD1102 Action PHOTOTECH (http://www.phototech.eu). The funders had no role in study design, data collection and analysis, decision to publish, or preparation of the manuscript.

Competing Interests: The authors have declared that no competing interests exist.

* E-mail: maya.lambreva@ic.cnr.it

¤ Current address: Neotron S.p.a., Santa Maria di Mugnano, Modena, Italy

Introduction

Recent expansion of biosensor technology has been the direct answer to the increase in worldwide demand for cost-efficient, fast and reliable methods for monitoring chemical species, additives and xenobiotics in clinical chemistry, environmental sciences and food related processes. R&D efforts are mainly focused on the improvement of specificity, sensitivity and stability of the biosensor sensing element, which is involved in the analyte recognition process, providing easily detectable changes in its physico-chemical properties. The nature and the intrinsic properties of the bio-recognition element determine the reliability and application fields of the biodevice.

Photosynthetic microalgae are among the most preferred microorganisms for environmental monitoring and screening of food and agricultural products for hazards compounds [1,2]. The unique features and structural constituents of the photosynthetic systems make them a suitable sensing element in the emerging biosensoristic field. This is largely due to their ability to conduct charge separation and electron transfer sensitive to the presence of different classes of pesticides and heavy metals, as well as some drugs and explosive compounds [3–6]. In contrast to the conventional analytical techniques, the photosynthetic biosensors offer several advantages, such as suitability for both laboratory and field applications, easy use without need of a skilled operator, minimum needs for maintenance and sample pre-treatments [7]. In addition, photosynthetic whole-cell-based biosensors allow for easy propagation and handling of the bio-recognition element, fast adaptation to changes in the working medium, possibility for regeneration and no need for extensive sample preparation [2,5]. Nonetheless, due to laboratory manipulation, interfacing with transducer elements, or repeated cycles of operational activity, the photosynthetic microorganisms can undergo a reduction of the specific performance and/or half-life. As a consequence, the storage- and operational-stability of the bio-sensing element is a crucial point that should be addressed. Different approaches have been adopted to increase the stability of the biological elements and preserve their functionality. In the case of whole-cell-based biosensors, entrapment and immobilisation of the microorganisms in inert organic matrices are among the most widely used methods [2,8].

Although biosensor technology is a winning alternative to the classical quantitative methods for fast and field-based pre-screenings, these biodevices are still challenged by the identification of a single analyte into the complex matrix of a real sample. The research for biosensor specificity improvement concerning microalgae is focused on screening for sensitive microalgae species [9,10], selection of microalgal resistant genotypes [5], or generation of genetically modified species [7,11]. In this respect, the use of biodevices exploiting an array of several bio-recognition elements specifically sensitive or resistant to a certain toxic compound represents a promising strategy.

Many biosensors, using photosynthetic recognition elements, specifically exploit the inhibition of PSII photochemical reaction in presence of different toxins. The PSII is a multiple subunit protein-chlorophyll complex responsible for the light-driven water oxidation; its core is assembled by the D1 and D2 proteins, which host all cofactors involved in the photochemical electron transport reaction [12]. A typical example of contaminants detected by PSII-based biosensor devices are the photosynthetic herbicides, which are widely used in the agricultural practice. Due to their extensive exploitation, high persistence and mobility, some of these compounds are widespread water and food contaminants, bringing about considerable ecological damage, as well as posing a threat for human health [6]. These toxic substances can be straightforwardly detected by PSII, since they interact with the D1 core complex protein coupling to the binding site of the secondary electron acceptor, the plastoquinone (Q_B) [13]. The herbicides successfully compete with the native electron acceptor due to their high binding affinity, blocking the PSII electron transport rate in a concentration dependent manner. The presence of PSII herbicides is easily detectable by the rapid decline of the primary plastoquinone (Q_A) reoxidation rate, reflected by changes in the chlorophyll *a* fluorescence, reduction of oxygen production and the photosynthetic electron transport rate [7,8,14].

In the present study, we have used the unicellular green alga *Chlamydomonas reinhardtii*, which has become one of the most-commonly exploited model microorganisms in various fundamental and applied studies. It offers the benefits of high growth rate and easy cultivation of the microorganisms, and the ability to perform post-transcriptional and post-translational modifications distinctive for the higher plants. All these advantages and the complete annotation of its genome [15] turned *C. reinhardtii* into a robust platform for synthesis of bioactive compounds, recombinant protein expression [16,17], biomass and biofuels production [18,19], and a main protagonist in bioremediation and biological sensing research fields [8,11,20,21]. In particular, the rational design and *ad hoc* production of *C. reinhardtii* mutants for biosensor purposes has been truly facilitated by the advances in its chloroplast transformation and resolution of PSII crystal structure [7,11,20,22].

This research was focused on the identification and characterisation of novel bio-recognition elements suitable for herbicide sensing by whole-cell-based biosensors. In particular, it aimed at overcoming core limitations affecting biosensor performance due to inadequate stability and sensitivity of the biological component. Coupling random mutagenesis targeted to the PSII D1 protein with exposure to ionizing radiation as selection pressure, we aimed to isolate several radical-stress tolerant mutants, which also had modified sensitivity to different classes of herbicides. The final objective was to unravel the possibility to develop biosensor arrays constituted by herbicide-sensitive and herbicide-resistant strains. The results discussed in this study demonstrated the suitability of the adopted directed evolution approach for identification of mutants with enhanced stability under free radocal-assosiated stress conditions, and their use for improving the selectivity of biosensor devices.

Results and Discussion

The *in vitro* directed evolution strategy is a unique molecular tool used to generate genetic diversity in a specific *locus*, mimicking and speeding-up the natural evolution process. This approach is particularly useful to evidence crucial sequences in genes characterised by low genetic variability, and when coupled to a functional selection allows for the easy discovery of novel traits. Exploiting directed evolution, we produced a random mutant library of *C. reinhardtii* with improved stability to oxidative stress-related conditions. The approach was targeted to the highly evolutionary conserved D1 PSII core protein, hosting a binding niche for the secondary electron acceptor Q_B and most of its photosynthetic herbicides competitive inhibitors. The newly isolated D1 random mutants were characterised under physiological and stressful conditions and their relevance for biosensor purposes was demonstrated by testing the strain's resistance/sensitivity toward triazines and urea type herbicides.

Generation and selection of radical tolerant D1 random mutants of *C. reinhardtii*

Random mutagenesis was performed by ep-PCR into a 400 bps nucleotide sequence of the *psb*A gene. This sequence encodes a D1 functionally important region (Ala153-Ala294) including the tyrosine residue at position 161, the primary electron donor for P_{680} and the transmembrane helices IV and V building up the Q_B binding niche [12]. Following biolistic chloroplast transformation, positive D1-transformants were selected for their competence for photoautotrophic growth. This method brought the advantage of obtaining marker-free transgenic strains, avoiding introduction of any foreign or antibiotic-resistant genes in the genome. Further on, the library of photosynthetically competent D1 mutants were utilized to isolate strains with improved tolerance to free-radical-associated stress conditions by exposures to additional selective pressure, such as neutron and proton bombardments. One of the main impact that these high-energy particles could cause to any biological material is derived from the secondary generated radical species and the subsequent oxidative chain reactions [23]. Table 1 lists mutants, which survived to the ionizing radiation exposures and were further analysed in the present study. The power of this approach to isolate strains with an improved tolerance to radical associated stress conditions has been demonstrated in a previously published work [24]. The common trend observed in the D1 aminoacidic substitutions was the replacement of less polar by more polar amino acids. In other words, the applied selection pressure forced replacement of residues, more sensitive to oxidative damage, with less sensitive ones. The relevance of this finding in the cellular context was demonstrated by *de novo* produced D1 site-directed mutants of *C. reinhardtii* carrying some of the aminoacidic substitutions selected by the ionizing radiation experiments. In comparison with the referent IL, the new strains displayed reduced photosynthetic performance under growth conditions and increased electron transport efficiency and oxygen evolution capacity in stressful high-light conditions. These findings enriched the abundant evidence present in the literature for the central role of the aminoacidic composition and transcriptional and post-translational regulation of the D1 protein in ruling the whole photosynthetic process under variety of environmental conditions [25].

In the present work the tolerance of the D1 random mutants of *Chlamydomonas* toward radical-associated conditions was also

Table 1. Amino acid substitutions and their localization in the D1 protein of the analysed *C. reinhardtii* random mutants survived after neutrons and protons exposures.

Strain	Amino acid substitution		Localization of the
	wild type→	**mutated**	**mutation in D1 protein**
L159M	leucine	methionine	near to Tyr_{161}
L159I/I184V	leucine	isoleucine	near to Tyr_{161}
	isoleucine	valine	near to EOC
P162S/F211S	proline	serine	near to Tyr_{161}
	phenylalanine	serine	helix IV of D1
I163T/I224F	isoleucine	threonine	near to Tyr_{161}
	isoleucine	phenylalanine	helix IV of D1
M172L	methionine	leucine	near to EOC
M172T	methionine	threonine	near to EOC
S177P	serine	proline	near to EOC
F197C/F285C	phenylalanine	leucine	helix IV of D1
	phenylalanine	leucine	helix V of D1
L200I	leucine	isoleucine	helix IV of D1
G207S	glycine	serine	helix IV of D1
F274Y	phenylalanine	tyrosine	helix V of D1
I281T	isoleucine	threonine	helix V of D1

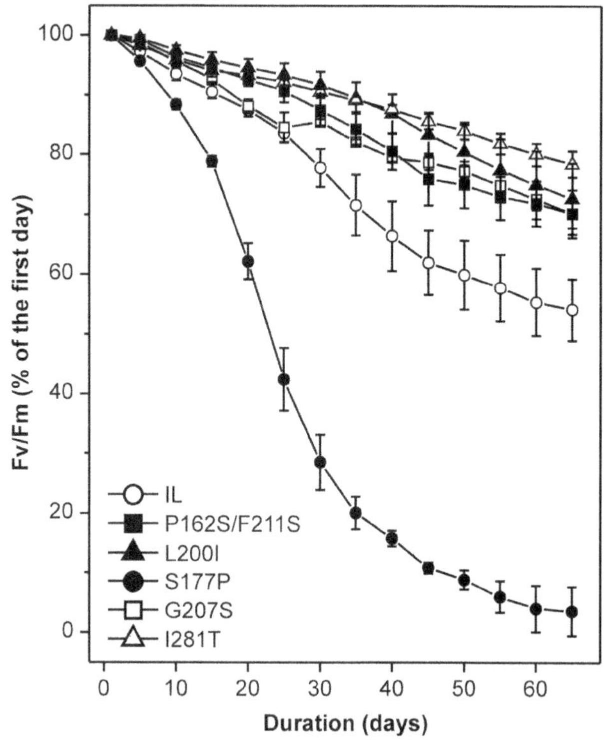

Figure 1. Long-term stability test of *C. reinhardtii* strains under oxidative stress conditions. The resistance of the strains is expressed as a reduction of their maximum quantum yield of the photochemical reaction (F_v/F_m) over time. Fluorescence measurements were performed hourly during the 7 h light/17 h dark photoperiod; herein are reported F_v/F_m values measured at the end of the light phase. Average values from two independent experiments are presented, ±SE, n = 4.

questioned by the following long-term stability test. Hence, the algal cultures were exposed to low temperature (13°C) and low light intensity (20 µmol m^{-2} s^{-1}) and their PSII performance was examined hourly for approximately two months (Figure 1). The low temperature stress slows down all catabolic reactions, decreasing the demand of absorbed energy and impeding the photosynthetic electron transport. These phenomena lead to accumulation of long-lasting reduced forms of the PSII primary quinone (Q_A) [26]. As a result, the probability for oxygen radical species generation and oxidative damage increases considerably [27]. The optimal growth temperature for most *Chlamydomonas* species varies in the range of 20–25°C and few of them could tolerate temperatures as low as 15°C [28]. Moreover, it has been shown that short (hours) or long (days) treatment with low temperature stress could reduce the overall fitness and viability of the *Chlamydomonas* cells, the chlorophyll stability and membrane integrity [29,30]. After 65 days exposure, the D1 mutants P162S/F211S, L200I, G207S and I281T preserved more that 70% of their initial photosynthetic activity, as determined by the F_v/F_m ratio, representing the maximum quantum yield of PSII photochemical reaction (Figure 1). In contrast, under these non-lethal stress conditions, the reference strain IL maintained only 54±5% of its initial PSII performance. Similarly to IL, the M172L, F197L/F285L and L159M mutants retained 66±2%, 57±3% and 55±3% of their initial F_v/F_m values, respectively (Table S1). On the contrary, the mutants F274Y, S177P and M172T were very sensitive to the applied stress conditions and at the end of the treatment they were able to keep only 25±1%, 4±2% and 0% of their F_v/F_m, respectively (Figure 1 and Table S1). Indeed, the mutant M172T did not resist the experimental conditions more than 30 days. I281T stood out against all strains as the most resistant to radical-related stress conditions, preserving 76±2% of its F_v/F_m value even after 110 days treatment (IL retained only 46±4%) (Table S1). When maintained at physiological temperature the average reduction of PSII performance in the same strains did not exceed 20±5%.

In addition, the tolerance to radical-associated stress conditions of some of these strains was verified by free radical scavenging capacity assay, based on the cell extract potential to reduce the stable DPPH radicals present in a reaction mixture. The cell extracts of the tolerant strains P162S/F211S, L200I, G207S and I281T showed higher antioxidant power in comparison with the reference IL. This trait was estimated by the extract concentration (mg dry weight mL^{-1}) necessary to reduce 50% of the DPPH free radicals (Table S2). Moreover, the radical scavenging capacity assay on the strain F274Y, characterized by low performance in the long-term stability test (Table S1), showed lower ability to reduce the DPPH free radicals compared to IL. These results clearly demonstrate the improved ability of the selected tolerant strains to deal with radical associated stress conditions. Hereafter in this study, in-depth photosynthesis analyses were performed only on mutants showing high photosynthetic stability in stressful conditions (Table 2); in particular, mutants demonstrating reduction of F_v/F_m value lesser than 50% over a 2 months period, were characterised in physiological conditions and in the presence of herbicides.

Physiological characterization of selected mutants

The growth of selected *Chlamydomonas* strains was characterized measuring culture optical density and chlorophyll content for a period of 88 h (Figure S1). During all growth phases, the OD_{750}

Table 2. Physiological parameters of the reference strain IL and D1 random mutants of *C. reinhardtii* selected for their PSII long-term stability under oxidative stress-related conditions.

Strains	Total Chl	F_v/F_m	$1\text{-}V_J$	O_2 evolution
	μg mL^{-1}			μmol O_2 mg Chl^{-1} h^{-1}
IL	7.7±0.9	0.773±0.002	0.548±0.003	92±3
L159I/I184V	3.5±0.6	0.691±0.011	0.499±0.008	112±4
P162S/F211S	3.7±0.0	0.664±0.022	0.518±0.017	120±3
M172L	2.5±0.0	0.731±0.006	0.511±0.012	112±3
F197L/F285L	3.3±0.8	0.665±0.011	0.472±0.009	141±3
L200I	3.9±0.4	0.686±0.009	0.537±0.006*	110±6
G207S	3.7±0.2	0.731±0.003	0.556±0.009*	129±4
I281T	3.9±0.3	0.707±0.005	0.545±0.004*	115±3

The total chlorophyll content, the maximum quantum yield of PSII photochemical reaction (F_v/F_m), the efficiency of the electron transport through PSII primary and secondary quinones ($1\text{-}V_J$) and the oxygen evolution rate (measured at 350 μmol m^{-2} s^{-1} light intensity) were determined in cell cultures in exponential growth phase (OD_{750} = 0.45±0.02). Average values are presented, ±SE, 4<*n*<12; the asterisks indicate values that are not significantly different from the IL strain at p≤0.05 (Mann-Whitney U Test).

was similar among the mutants and a little bit higher in comparison with the reference strain IL during the late growth phase. On the contrary, the D1 mutants accumulated much lower amount of chlorophyll per mL than IL. In the exponential growth phase, the mutant total chlorophyll content was half compared to IL (Table 2). This deviation gradually increased with the time resulting in very low chlorophyll/OD_{750} ratios for the mutants in the late growth phase (Figure 2).

Amino acid replacements in the D1 selected region influenced the PSII function, lowering the mutants maximum photochemical quantum yield as reported in Table 2. On the other hand, the efficiency of the mutants electron transport between the primary (Q_A) and secondary (Q_B) quinones in PSII ($1\text{-}V_J$) was lower, but still similar to IL. This finding suggested that even though the electron transfer through the PSII was altered by the amino acid substitutions, no severe interruption of the PSII functioning occurred in the selected D1 mutants. In support of this statement is the higher capacity of the mutants to evolve oxygen under saturating light intensity (350 μmol m^{-2} s^{-1}) in comparison with the reference strain (Table 2). Previous experiments showed that a light intensity of 200 μmol m^{-2} s^{-1} saturates the photosynthetic reactions in IL and a further increase in radiation intensity was not correlated to an increased in the amount of the oxygen released [24]. In contrast to IL, under saturating light the D1 mutants were able to produce from 20% (L200I) up to 54% (F197L/F285L) more oxygen per mg chlorophyll than the reference strain.

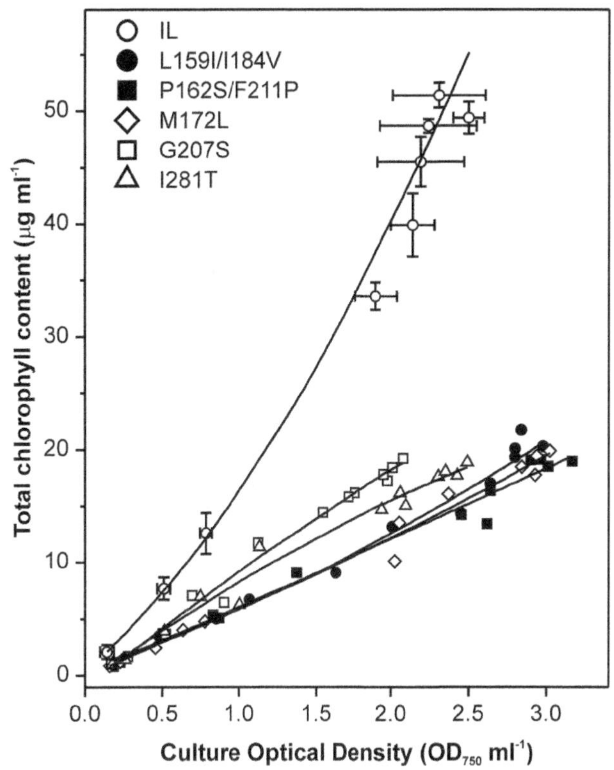

Figure 2. Comparison of the *C. reinhardtii* strains culture chlorophyll and optical density during growth. The development of the reference strain, IL, and D1 random mutants is presented as a ratio of the total chlorophyll mL^{-1} and the corresponding OD_{750} mL^{-1}. The cell cultures were grown for a period of 88 h in TAP medium under continues illumination of 50 μmol m^{-2} s^{-1} at 24°C and 150 rpm agitation. Average values from four experiments are presented, ±SE, n = 4. For the sake of clarity the standard error bars of the mutants values are omitted.

Relevance of selected mutants as bio-recognition elements

The response of the selected strains to triazines and urea type herbicides was studied by chlorophyll fluorescence induction curves (*O-J-I-P*) [31], normalised by the maximum variable fluorescence ($F_v = F_m - F_0$), as reported in Materials and Methods. This normalisation allows for a direct comparison between the samples, excluding any absorbance or heat dissipation differences (Figure 3). The relative variable fluorescence at the characteristic point *J* (V_J) gives a measure of relative amount of the reduced Q_A. Therefore, $1\text{-}V_J$ represents the fraction of oxidised Q_A or the efficiency by which the electron is transferred from Q_A to Q_B [32]. Using this approach, we evaluated the sensitivity of the newly generated strains to two different classes of PSII herbicides. In the absence of inhibitor, the V_J registered in the mutant P162S/F211S is higher than in the IL strain, thus reflecting the lower efficiency of the electron transport through PSII ($1\text{-}V_J$) under physiological conditions. In the presence of PSII herbicides, the Q_B binding site is permanently occupied by the inhibitor and electron transport does not extend beyond Q_A. The accumulation of reduced forms of Q_A is proportional to the herbicide concentration and can be directly evaluated by the increase of the variable fluorescence level at point *J* (Figure 3), making the parameter $1\text{-}V_J$ one of the most sensitive for herbicide toxicity detection [10,14,20].

The herbicide sensitivity or resistance of the newly characterized strains, was assessed by measuring the herbicide-induced reduction of the parameter $1\text{-}V_J$ (Figure 4), and evaluated by the I_{50} and corresponding R/S values (Table 3). R/S>1.0 points out herbicide resistance, while R/S<1.0 refers to herbicide "super-sensitivity" [33]. In this study, only 0.9≤R/S≥1.1 values were considered to indicate significant changes in the mutants tolerance to photosynthetic herbicide. The mutants herbicide sensitivity was evaluated also by the LOD parameter, which represented the minimal analyte concentration inducing significant reduction of the algal PSII electron transport efficiency (Table 4). The L200I

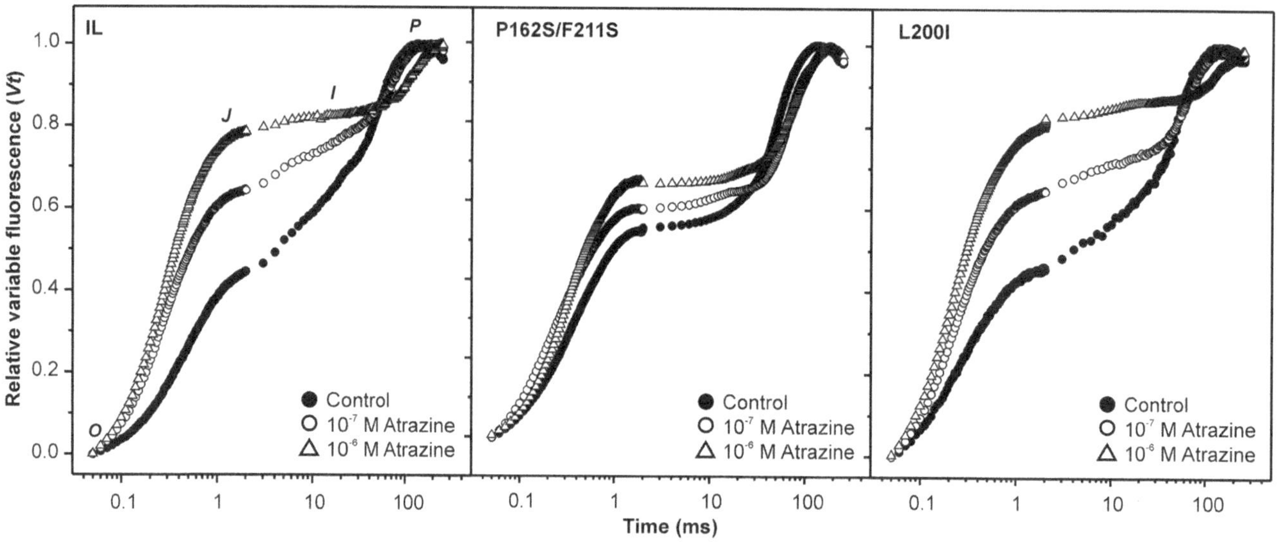

Figure 3. Polyphasic chlorophyll fluorescence transient under different atrazine concentrations. The fluorescence transients of reference strain IL and P162S/F211S and L200I mutants are presented as curves of the relative variable fluorescence, $V_t = (F_t - F_0)/(F_m - F_0)$. Typical V_t curves for each strain and herbicide treatment are presented; the characteristic points *O-J-I-P* are also indicated.

mutation conferred supersensitivity to both triazine and urea type herbicides. In this mutant, the R/S values were 0.7 for atrazine, 0.6 for terbuthylazine and 0.8 for linuron. The L200I LOD values for terbuthylazine and linuron were 3-fold and 4-fold lower than IL, respectively, confirming the supersensitivity of this mutant. The substitution of methionine 172 with leucine did not cause significant changes in I_{50} and R/S values for the tested herbicides, but led to approximately 3 times lower LOD values for atrazine and linuron compared to the reference strain. This result could be ascribed to the M172L supersensitivity to very low herbicide concentrations (within nano-molar range), which might be advantageous for its use as biosensor bio-recognition element. In contrast, the double mutant P162S/F211S showed reduced affinity to all tested inhibitors; its herbicide resistance decreased in the order atrazine (R/S = 8.2)> terbuthylazine (R/S = 6.2)> linuron (R/S = 1.9). As evident from the LOD values, the P162S/F211S minimal detectable concentration for the tested herbicides declined in the same order (Table 4). While the D1 aminoacidic substitutions in L200I and P162S/F211S strains resulted in an increase in their sensitivity or resistance to the tested inhibitors, the replacement of isoleucine 281 with threonine (I281T strain) determined opposed responses to triazines and ureas. The I281T mutant was supersensitive to atrazine and terbuthylazine (R/S = 0.8) and resistant to linuron (R/S = 1.2), even though both triazines and urea type herbicides target the same D1 amino acid residue (S264) in the Q_B binding niche [13]. In contrast, the G207S mutant, hosting a serine instead of a glycine, showed reduced affinity to terbuthylazine and linuron (R/S values 4.4 and 1.2 respectively), while its I_{50} for atrazine was very similar to the inhibition constant of the reference strain (R/S = 0.9). It has been previously demonstrated that the same point mutations in D1 aminoacidic composition, which cause resistance towards one class of herbicides, can cause a supersensitivity to other classes of photosynthetic inhibitors [22,33]. This finding was explained by the possibility for different interactions between the herbicide molecules and the amino acids lining the cavity of the Q_B binding pocket, although the inhibitors interface common regions of the binding pocket.

Extensive literature proved the relevance of aminoacidic substitutions in the Q_B binding niche of D1 protein in the modulation of the herbicide response in higher plants, algae and cyanobacteria. Most of these mutations were localised in the D1 primary structure encompassing amino acids from phenylalanine 211 to leucine 275 [12,22,33,34]. In *C. reinhardtii*, it has been demonstrated that aminoacidic mutations outside this region can also modify the herbicide sensitivity, leading to hypothesise long-range effects in the arrangement of the D1 Q_B hosting cavity [22]. In the present study, excluding the Phe 211, all the selected mutations were localised outside the Q_B binding niche (M172L, L200I, G207S and I281T), and were not previously described. Comparative analyses of triazine and urea type herbicides binding indicated I_{50}values in the same order of magnitude of those previously reported by Wilski et al. for similar D1 mutations in *C. reinhardtii* [22].

Molecular dynamics simulations evidenced an active role of Phe 211 in the atrazine binding to the Q_B pocket [6], while an atomic structure resolution of cyanobacterial PSII provided evidence on terbutryn binding to the same residue [13]. In *C. reinhardtii*, aminoacidic replacements of Phe 211 with alanine, glycine, threonine or isoleucine were usually associated to additional mutations, and were responsible for an increased resistance to different classes of herbicides. The literature reported R/S values were similar or somewhat lower compared to those determined in our study for the P162S/F211S strain [33]. Similar herbicide responses and R/S values were also reported in *Synechococcus* PCC7002 and *Synechocystis* PCC6714 AzI and AzV mutants [33].

The assembly of herbicide resistant and sensitive genotypes on multiarray transducers or biosensor platforms could be a very useful strategy to increase the specificity of these analytical devices. For instance, when P162S/F211S, G207S and L200I strains are exposed to terbuthylazine, only L200I will demonstrate fast and high reduction in its photosynthetic activity, while in the presence of atrazine only the double mutant P162S/F211S will maintain high PSII electron transport efficiency. In a recent review, Hernández-Allica et al. [5] described the successful development of a fibre optic biosensor based on the comparative response of herbicides sensitive and resistant genotypes. This data reiterates

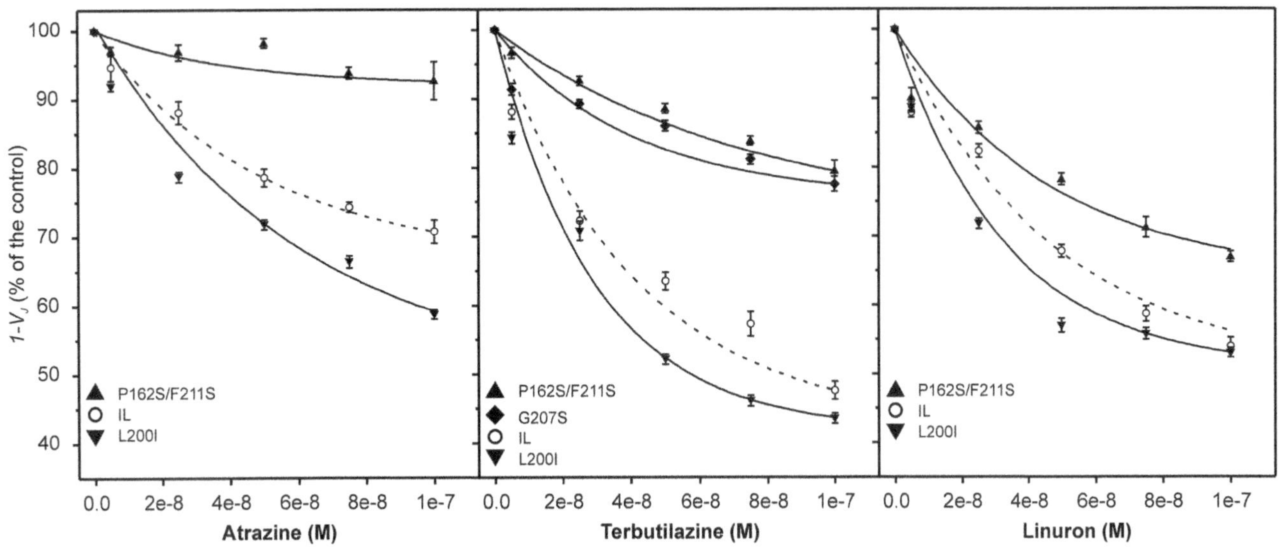

Figure 4. Herbicide dose-response curves of PSII electron transport efficiency. Fluorescence transients of the reference strain IL and D1 random mutants of *C. reinhardtii* were registered after 10 min of incubation with increasing concentrations of atrazine, terbutilazine or linuron. The parameter *1-V_J* was calculated as *1-V_J* = 1-(F_{2ms}-F_0)/(F_m-F_0) according to Strasser et al. [32]. Average values from three independent experiments are presented, ±SE, n = 6–9.

the benefits of using resistant strains to improve the specificity of whole-cell based biosensors. In addition, we proposed a molecular bio-engineering strategy for fast and efficient generation of resistant genotypes.

Using a conventional instrument for prompt fluorescence registration, very short incubation time and selected *Chlamydomonas* mutants, it was possible to distinguish nano-molar concentrations of triazines and urea type herbicides that are very close to the limits imposed by the European legislation [6]. A significant increase of the proposed biomediators sensitivity can be expected when the newly generated bio-recognition elements are combined with last generation biosensor devices [35], or microfluidic chip with organic light emitting diode and photodetector [8]. In addition, the biosensors exploiting multiarray platforms [1,5,7,9] offer the possibility to test the same sample by several sensing elements working in series. Coupled with *ad hoc* designed or selected sensitive and resistant strains this approach can significantly contribute to increasing whole-cell-based biosensor's specificity.

Conclusions

Random mutagenesis targeted at the PSII D1 protein combined with an *in vitro* directed evolution strategy succeeded in the selection of *Chlamydomonas* strains with reduced susceptibility to free-radical related stressful conditions and altered herbicide sensitivity. Among the twelve D1 random mutants which survived to protons/neutrons exposures, seven demonstrated high ability to maintain functional photosynthetic apparatus for more than 65 days. Out of the tolerance strains, five showed an increased sensitivity/resistance to herbicides, and revealed to be promising bio-recognition elements. Furthermore, the significance of new D1 aminoacidic residues for the PSII electron transfer and herbicide

Table 3. Herbicides toxicity effects on the PSII electron transport efficiency (*1-V_J*) of IL and selected D1 random mutants, and corresponding R/S values.

Strains	Atrazine		Terbuthylazine		Linuron	
	I_{50}×10^{-7} M (±SE)	*R/S*	I_{50}×10^{-7} M (±SE)	*R/S*	I_{50}×10^{-7} M (±SE)	*R/S*
IL	2.4±0.11	/	1.0±0.08	/	1.1±0.05	/
P162S/F211S	19.7±1.57	8.2	5.9±0.26	6.2	2.0±0.17	1.9
M172L	2.2±0.03*	0.9	1.1±0.02*	1.1	1.1±0.06*	1.0
L200I	1.8±0.06	0.7	0.6±0.04	0.6	0.9±0.06	0.8
G207S	2.1±0.09*	0.9	4.3±0.18	4.4	1.3±0.09	1.2
I281T	2.0±0.07	0.8	0.7±0.02	0.8	1.3±0.05	1.2

The I_{50} is the molar herbicide concentration, which induces 50% inhibition of the parameter *1-V_J*. The R/S were calculated as ratios of mutant I_{50} and I_{50} of the reference strain; the asterisks indicate values that do not differ significantly from the IL strain at p≤0.05 (Mann-Whitney U Test). We considered that R/S<0.9 values define increased herbicide sensitivity, R/S>1.1 – increased herbicide resistance and 0.9≤R/S≤1.1 – no significant alterations in the mutants herbicide sensitivity relative to the reference strain.

Table 4. Limits of detection of IL and selected D1 random mutants to three different herbicides, representing the lower herbicide molar concentration inducing detectable inhibition of strains PSII electron transport efficiency.

Strains	Limit of detection ($\times 10^{-9}$ M)		
	Atrazine	Terbuthylazine	Linuron
IL	10	4	6
P162S/F211S	75	24	8
M172L	3	3	3
L200I	8	1	2
G207S	9	23	5
I281T	5	8	8

sensing was demonstrated and the perspectives for improving whole-cells based biosensors specificity and sensitivity have been discussed.

Materials and Methods

Generation and selection of the *Chlamydomonas* strains

The libraries of PSII D1 protein mutants of *C. reinhardtii* were produced by random mutagenesis of the *psbA* gene, coupled with biolistic transformation of the Del1 deletion mutant as previously described [24]. Briefly, a pool of randomly mutated *psbA* gene-fragments was produced by error-prone (ep) PCR using a pSH5 plasmid as a template, which hosts an intronless *psbA* gene [36]. Positive transformants were selected for their photoautotrophic growth on high salt (HS) medium. In order to isolate algal strains tolerant to free-radical-induced oxidative stress, a coctail of photoauthotrophycally competent D1 random mutants were subjected to an additional selection pressure applied by bombardments with protons and neutrons [24]. The IL strain of *C. reinhardtii* with intronless *psbA* gene [37] was utilized as a reference strain in all the experiments evaluating the functionality and herbicide sensitivity of the newly produced D1 random mutants.

Growth conditions and media

The *C. reinhardtii* strains were grown in Tris-acetate-phosphate (TAP) or HS media [28] under continuous illumination of 50 µmol m^{-2} s^{-1}, 24±1°C and agitation (150 rpm). When necessary, the media were solidified with 1.5% agar. In all analyses algae cultures in exponential growth phase (optical density, OD_{750} = 0.45±0.02) were used.

Physiological characterisation of the selected bio-recognition elements

The growth rate of the strains was tracked by measuring the culture OD_{750} and total chlorophyll content during an 88 hour period as described previously [24].

PSII performance was evaluated by registering the chlorophyll *a* fluorescence induction curves at 24°C in TAP liquid cell cultures by the Plant Efficiency Analyser (PEA, Hansatech Instr. Ltd, Norfolk, UK). The fluorescence transient was induced by 3 s saturated pulse (600 W m^{-2}) with peak at 650 nm. Before each measurement the samples were kept for 10 min in dark in order to register PSII performance in dark adapted steady state. To allow for direct comparison between the different samples the OJIP fluorescence transients were normalised to the maximum variable fluorescence ($F_v = F_m - F_0$), providing the curves of the relative variable fluorescence $V_t = (F_t - F_0)/(F_m - F_0)$. The maximum quantum yield of PSII photochemistry was calculated as $F_v/F_m = (F_m - F_0)/F_m$; the efficiency of the electron transport between PSII primary (Q_A) and secondary (Q_B) quinone electron acceptors was evaluated by the parameter $1\text{-}V_J$, where $1\text{-}V_J = 1-(F_{2ms} - F_0)/(F_m - F_0)$ [32]. F_0, F_m and F_{2ms} are the fluorescence level at 50 µs, the maximum fluorescence and the fluorescence level at 2 ms after the onset of the illumination, respectively. The measurements were performed on at least three separate cultures, with an average of four repetitions for each culture.

The oxygen evolution capacity was measured at 24°C by Clark-type oxygen electrode (Chlorolab 2, Hansatech, Instr. Ltd, Norfolk, UK). The oxygen evolution rate was determined under saturating illumination (350 µmol m^{-2} s^{-1}), continuous stirring and in presence of 10 mM $NaHCO_3$, as additional carbon source [38]. To compare the photosynthetic capacity between the different samples, the same amount of chlorophyll (15 µg mL^{-1}) was loaded into oxygen electrode chamber. Each strain was assayed in three separate cultures, with three repetitions for each culture.

Long-term stability test

The strains long-term resistance test to stressful conditions was performed using the Photo II fluorometer [39]. For each strain calculated amount of cell culture was harvested by weak centrifugation to reach a final OD_{750} = 9 and layered on TAP agar medium solidified in the instrument containers. The algal cells were exposed to 20 µmol m^{-2} s^{-1} light intensity with 7 h light/17 h dark photoperiod at 13±1°C. Photosynthetic performance of the cultures was followed for approximately 2 months by hourly registration of chlorophyll *a* fluorescence induction curves. Each fluorescence measurement was performed on 15 min dark-adapted sample by a 6 s pulse of red light.

Herbicides treatment of *Chlamydomonas* cell cultures

The triazine herbicides atrazine (2-chloro-4-(ethylamine)-6-(isopropylamine)-s-triazine) and therbuthylazine (6-chloro-*N*-(1,1-dimethylethyl)-*N'*-ethyl-1,3,5-triazine-2,4-diamine) and the phenylurea herbicide linuron (1-methoxy-1-methyl-3-(3,4-dichlorophenyl)urea) were purchased from Sigma Aldrich (USA). 50 mM and 5 mM herbicide stock solutions were prepared in 100% methanol and water-diluted to prepare the working solutions. The experimental data set for each tested herbicide included control and at least nine different concentrations (10^{-9} M, 5×10^{-9} M, 2.5×10^{-8} M, 5×10^{-8} M, 7.5×10^{-8} M, 10^{-7} M, 2.5×10^{-7} M, 5×10^{-7} M, 10^{-6} M). In order to obtain at least 90% PSII electron transport inhibition in some of the mutants the herbicide dose-response curves were registered up to 2×10^{-5} M herbicide concentration. For the treatment, algal cultures (15 µg chlorophyll mL^{-1}) were incubated with herbicide for 10 min under 50 µmol m^{-2} s^{-1} irradiation, 24°C and constant stirring. The herbicides toxicity was evaluated following the inhibition level of the parameter $1\text{-}V_J$, measured and calculated as described above. The maximum final concentration of the methanol in algal samples was 0.02% and by itself did not alter the algal performance (data not shown). The effect of each herbicide was tested by performing the dose-response curve on three independent biological cultures, with at least two repetitions for each culture.

Evaluation of the bio-recognition elements herbicide response

The inhibition constant I_{50} (herbicide concentration required for 50% inhibition of the parameter $1\text{-}V_J$) and the limit of detection (LOD) were determined for all herbicides and examined *Chlamydomonas* strains. The inhibition of PSII electron transport efficiency at increasing herbicide concentration was calculated as $100\text{-}(1\text{-}V_J)_{herbicide}/(1\text{-}V_J)_{control}$. The data points followed a typical Langmuir isothermal ligand binding curve that can be fitted by the equation (1): $y=(I_{MAX}\times x)/(I_{50}+x)$, where x and y stand for the molar herbicide concentration and the percentage of inhibition of the parameter $1\text{-}V_J$ respectively; I_{50} is the herbicide concentration required for 50% inhibition of $1\text{-}V_J$ and I_{MAX} is the maximum registered inhibition of $1\text{-}V_J$. I_{50} and I_{MAX} are free parameters attained by the fitting of the experimental data. Since the experimentally obtained reduction of the parameter $1\text{-}V_J$ did not exceed 90%, the I_{50} achieved from the fit of the dose-response curve referred to an I_{MAX} value significantly lower than 100% inhibition. Therefore, the actual value of I_{50} was calculated by solving the equation (1) for y = 50% using the I_{50} and I_{MAX} obtained from the fit. The herbicide sensitivity or resistance was assessed also by R/S values calculated as a ratio of mutant I_{50} and I_{50} of IL strain. As a statistical parameter, the LOD represents an estimation of the lowest concentration of herbicide analyte that may be experimentally detected with a certain confidence level. It was calculated using the actual value of I_{50} according the equations reported by Maly et al. [40]: $LOD=(2.6\times\sigma\times I50)/(100-2.6\times\sigma)$, where σ is the mean standard error of the measurements, while $2.6\times\sigma$ indicates the confidence interval in which there is a 99% probability of finding measurement results following the assumption of a normal distribution.

Statistical analyses

Presented data are means of at least three independent experiments with at least two repetitions each. The differences between the reference strain and each of the D1 mutants were assessed by a non-parametric Mann-Whitney U test for comparison of two independent samples. The statistical significance of the differences was evaluated at $p\leq0.05$ values.

Supporting Information

Figure S1 Growth rate curves of the reference strain IL and D1 random mutants of *C. reinhardtii* selected for their PSII long-term stability under oxidative stress-related conditions. The time courses of cell culture growth were followed for a period of 88 h under growth conditions by measuring: **A**) culture optical density (OD_{750}) and **B**) total chlorophyll content (µg/ml). Average values from four different cultures are presented, ±SE, n = 4.

Table S1 Fv/Fm values demonstrating the long-term stability of the *C. reinhardtii* strains under oxidative stress-related conditions. Average values from two independent experiments are presented, ±SE, n = 4; n.s. – not survive; n.d. – not detected.

Table S2 Free radical scavenging capacity of some *C. reinhardtii* strains extracts determined by DPPH assay*. The value of I_{50}, the concentration of algal extract (mg dry weight mL^{-1}) that reduces 50% of the DPPH free radical, was estimated for each strain. TROLOX-Equivalent Antioxidant Capacity (TEAC) for each extract was calculated as ratio of algal I_{50} and I_{50} of the well-known antioxidant TROLOX (0.0032 ± 0.0005 mg mL^{-1}). Average values of 3 different cultures are presented, ±SE, n = 6. The values are significantly different from the reference strain at $p\leq0.05$ (Mann-Whitney U Test).

Author Contributions

Conceived and designed the experiments: GR MDL MTG UJ. Performed the experiments: MDL IR AA SP IB. Analyzed the data: MDL GR IB IH IR. Contributed reagents/materials/analysis tools: GR MTG UJ. Wrote the paper: MDL GR.

References

1. Del Carlo M, Compagnone D (2010) Recent strategies for the biological sensing of pesticides: from the design to the application in real samples. Bional Rev 1: 159–176.
2. Brayner R, Couté A, Livage J, Perrette C, Sicard C (2011) Micro-algal biosensors. Anal Bioanal Chem 401: 581–597.
3. Rouillon R, Piletsky SA, Breton F, Piletska EV, Carpentier R (2006) Photosystem II biosensors for heavy metals monitoring. In: Giardi MT, Piletska EV editors, Biotechnological Applications of Photosynthetic Proteins: Biochips, Biosensors and Biodevices. Springer, New York, pp. 166–174.
4. Bhalla V, Zhao X, Zazubovich V (2011) Detection of explosive compounds using Photosystem II-based biosensor. J Electroanal Chem 657: 84–90.
5. Hernández-Allica J, Carrera-Martínez D, López-Rodas V, Flores-Moya A, Costas E (2011) Highly specific biosensors to herbicides, based on sensitive- and resistant-mutants of microalgae. In: Kortekamp A. Editors. Herbicides and Environment, INTECH Publisher, Vienna, pp. 433–442.
6. Rea G, Polticelli F, Antonacci A, Lambreva M, Pastorelli S, et al. (2011) Computational biology, protein engineering, and biosensor technology: a close cooperation for herbicides monitoring, in: Soloneski S, Larramendy M editors. Herbicides, Theory and Applications. INTECH Publisher, Vienna, pp. 93–120.
7. Buonasera K, Pezzotti G, Scognamiglio V, Tibuzzi A, Giardi MT (2010) New platform of biosensors for prescreening of pesticide residues to support laboratory analyses. J Agric Food Chem 58: 5982–5990.
8. Lefèvre F, Chalifour A, Yu L, Chodavarapu V, Juneau P, et al. (2012) Algal fluorescence sensor integrated into a microfluidic chip for water pollutant detection. Lab Chip 12: 787–793.
9. Podola B, Melkonian M (2005) Selective real-time herbicide monitoring by an array chip biosensor employing diverse microalgae. J Appl Phycol 17: 261–271.
10. Chalifour A, Spearb PA, Boilyb MH, DeBloisc C, Girouxd I, et al. (2009) Assessment of toxic effects of pesticide extracts on different green algal species by using chlorophyll *a* fluorescence. Toxicol Environ Chem 91: 1315–1329.
11. Rea G, Polticelli F, Antonacci A, Scognamiglio V, Katiyar P, et al. (2009) Structure-based design of novel *Chlamydomonas reinhardtii* D1-D2 photosynthetic proteins for herbicide monitoring. Prot Sci 18: 2139–2151.
12. Kern J, Renger G (2007) Photosystem II: Structure and mechanism of the water: plastoquinone oxidoreductase. Photosynth Res 94: 183–202.
13. Broser M, Glöckner C, Gabdulkhakov A, Guskov A, Buchta J, et al. (2011) Structural basis of cyanobacterial photosystem II inhibition by the herbicide terbutryn. J Biol Chem 286: 15964–15972.
14. Dewez D, Didur O, Heroux JV, Popovic R (2008) Validation of photosynthetic-fluorescence parameters as biomarkers for isoproturon toxic effect on alga *Scenedesmus obliquus*. Environ Pollut 151: 93–100.
15. Merchant SS, Prochnik SE, Vallon O, Harris EH, Karpowicz SJ, et al. (2007) The Chlamydomonas genome reveals the evolution of key animal and plant functions. Science 318: 245–250.
16. Mayfield SP, Manuell AL, Chen S, Wu J, Tran M, et al. (2007) *Chlamydomonas reinhardtii* chloroplasts as protein factories. Curr Opin Biotechnol 18: 1–8.
17. Rasala BA, Muto M, Lee PA, Jager M, Cardoso RMF, et al. (2010) Production of therapeutic proteins in algae, analysis of expression of seven human proteins in the chloroplast of *Chlamydomonas reinhardtii*. Plant Biotechnol J 8: 719–733.
18. Singh A, Olsen SI (2011) A critical review of biochemical conversion, sustainability and life cycle assessment of algal biofuels. Appl Energy 88: 3548–3555.
19. Hoshino T, Johnson DJ, Cuello JL (2012) Design of new strategy for green algal photo-hydrogen production: spectral-selective photosystem I activation and photosystem II deactivation. Bioresour Technol 120: 233–240.
20. Giardi MT, Scognamiglio V, Rea G, Rodio G, Antonacci A, et al. (2009) Optical biosensors for environmental monitoring based on computational and biotechnological tools for engineering the photosynthetic D1 protein of *Chlamydomonas reinhardtii*. Biosens Bioelectron 25: 294–300.

21. Yang WW, Miao AJ, Yang LY (2012) Cd^{2+} Toxicity to a green alga *Chlamydomonas reinhardtii* as influenced by its adsorption on TiO_2 engineered nanoparticles. PLoSONE 7, e32300.
22. Wilski S, Johanningmeier U, Hertel S, Oettmeier W (2006) Herbicide binding in various mutants of the photosystem II D1 protein of *Chlamydomonas reinhardtii*. Pest Biochem Physiol 84: 157–164.
23. Sanche L (2002) Nanoscopic aspects of radiobiological damage: fragmentation induced by secondary low-energy electrons. Mass Spectrom Rev 21: 349–369.
24. Rea G, Lambreva M, Polticelli F, Bertalan I, Antonacci A, et al. (2011) Directed evolution and *in silico* analysis of reaction centre proteins reveal molecular signatures of photosynthesis adaptation to radiation pressure. PLoSONE 6(1): e16216.
25. Mulo P, Sakurai I, Aro EM (2012) Strategies for *psb*A gene expression in cyanobacteria, green algae and higher plants: from transcription to PSII repair. Biochim Biophys Acta 1878: 247–257.
26. Huner NPA, Öquist G, Sarhan F (1998) Energy balance and acclimation to light and cold. Trends Plant Sci 3: 224–230.
27. Allen DJ, Ort DR (2001) Impacts of chilling temperatures on photosynthesis in warm-climate plants. Trends Plant Sci 6: 36–42.
28. Harris EH (1989) The *Chlamydomonas* sourcebook. A Comprehensive Guide to Biology and Laboratory Use. Academic Press, San Antonio.
29. Goho S, Bell G (2000) Mild environmental stress elicits mutations affecting fitness in *Chlamydomonas*. Proc R Soc Lond B 267: 123–129.
30. Hema R, Senthil-Kumar M, Shivakumar S, Chandrasekhara Reddy P, Udayakumar M (2007) *Chlamydomonas reinhardtii*, a model system for functional validation of abiotic stress responsive genes. Planta 226: 655–670.
31. Lazár D (2006) The polyphasic chlorophyll a fluorescence rise measured under high intensity of exciting light. Funct Plant Biol 33: 9–30.
32. Strasser RJ, Srivastava M, Tsimilli M (2000) The fluorescence transient as a tool to characterize and screen photosynthetic samples. In: Mohanty P, Yunus Pathre U, editors, Probing photosynthesis: Mechanism, regulation and adaptation. Taylor and Francis, London, pp. 445–483.
33. Oettmeier W (1999) Herbicide resistance and supersensitivity in photosystem II. Cell Mol Life Sci 55: 1255–1277.
34. Johanningmeier U, Sopp G, Brauner M, Altenfeld U, Orawski G, et al. (2000) Herbicide resistance and supersensitivity in Ala250 or Ala251 mutants of the D1 protein in *Chlamydomonas reinhardtii*. Pest Biochem Physiol 66: 9–19.
35. Shitanda I, Takanatsu S, Watanabe K, Itagaki M (2009) Amperometric screen-printed algal biosensor with flow injection analysis system for detection of environmental toxic compounds. Electrochim Acta 54: 4933–4936.
36. Preiss S, Schrader S, Johanningmeier U (2001) Rapid, ATP-degradation of a truncated D1 protein in the chloroplast. Eur J Biochem 268: 4562–4569.
37. Johanningmeier U, Heiss S (1993) Construction of a *Chlamydomonas reinhardtii* mutant with an intronless *psb*A gene. Plant Mol Biol 22: 91–99.
38. Melis A, Neidhardt J, Benemann JR (1999) *Dunaliella salina* (Chlorophyta) with small chlorophyll antenna sizes exhibit higher photosynthetic productivities and photon use efficiencies than normally pigmented cells. J Appl Phycol 10: 515–525.
39. Cano J, Giannini D, Pezzotti G, Rea G, Giardi MT (2011) Space impact and technological transfer of a biosensor facility to Earth application. Recent Patents Space Technol 1: 18–25.
40. Maly J, Masojidek J, Masci A, Ilie M, Cianci E, et al. (2005) Direct mediatorless electron transport between the monolayer of photosystem II and poly(mercapto-p-benzoquinone) modified gold electrode—new design of biosensor for herbicide detection. Biosens Bioelectron 21: 923–932.

13

Broad Resistance to ACCase Inhibiting Herbicides in a Ryegrass Population Is Due Only to a Cysteine to Arginine Mutation in the Target Enzyme

Shiv Shankhar Kaundun*, Sarah-Jane Hutchings, Richard Paul Dale, Eddie McIndoe

Syngenta, Jealott's Hill International Research Centre, Biological Sciences, Bracknell, United Kingdom

Abstract

Background: The design of sustainable weed management strategies requires a good understanding of the mechanisms by which weeds evolve resistance to herbicides. Here we have conducted a study on the mechanism of resistance to ACCase inhibiting herbicides in a *Lolium multiflorum* population (RG3) from the UK.

Methodology/Principal Findings: Analysis of plant phenotypes and genotypes showed that all the RG3 plants (72%) that contained the cysteine to arginine mutation at ACCase codon position 2088 were resistant to ACCase inhibiting herbicides. Whole plant dose response tests on predetermined wild and mutant 2088 genotypes from RG3 and a standard sensitive population indicated that the C2088R mutation is the only factor conferring resistance to all ten ACCase herbicides tested. The associated resistance indices ranged from 13 for clethodim to over 358 for diclofop-methyl. Clethodim, the most potent herbicide was significantly affected even when applied on small mutant plants at the peri-emergence and one leaf stages.

Conclusion/Significance: This study establishes the clear and unambiguous importance of the C2088R target site mutation in conferring broad resistance to ten commonly used ACCase inhibiting herbicides. It also demonstrates that low levels "creeping", multigenic, non target site resistance, is not always selected before single gene target site resistance appears in grass weed populations subjected to herbicide selection pressure.

Editor: Pär K. Ingvarsson, University of Umeå, Sweden

Funding: These authors have no support or funding to report.

Competing Interests: Please note that all authors of the manuscript are employees of Syngenta. This does not alter the authors' adherence to all the PLoS ONE policies on sharing data and materials.

* E-mail: deepak.kaundun@syngenta.com

Introduction

Modern herbicides are very effective tools for controlling weeds in agricultural production. Their extensive use over time, however, has resulted in evolution of weed resistance to key herbicide modes of action [1]. This is often the case when herbicides are employed as the sole method of weed control combined with little to no diversity in agronomic practices [2]. Weeds prone to evolve resistance are generally highly prolific, genetically diverse, out crossing species [3]. This is exemplified by ryegrasses which have evolved resistance to practically all herbicide modes of action that are active against sensitive populations of this species [4,5]. Particularly affected are highly effective single site herbicides such as inhibitors of acetolactate synthase (ALS) [6,7,8] and acetyl CoA carboxylase (ACCase) [9,10,11].

The first case of ryegrass resistance to an ACCase herbicide was reported in 1982 in a wheat field in Australia [12]. Over the last 30 years many more ryegrass populations have evolved resistance to ACCase herbicides across five continents, including France, UK, Germany, Spain, Italy, Canada, USA, Chile, Brazil South Africa, Tunisia, Greece, Israel, Iran, Saudi Arabia and Australia [5]. The situation is particularly alarming in Australia with confirmed resistance in over 70% of ryegrass populations sampled randomly in the wheat belt. A majority of these populations are also resistant to other herbicide modes of action thus complicating weed management strategies [13].

Herbicides targeting acetyl CoA carboxylase inhibit the synthesis of *de novo* fatty acids which are essential components in cell membranes and secondary plant metabolites [14]. Three classes of ACCase herbicides have been developed, namely cyclohexanediones (DIMs), aryloxyphenoxypropionate (FOPs) and phenylpyraxoline (DEN) [15]. All three herbicide classes are active on the chloroplastic form of most grass ACCase with little to no activity on broadleaved species [16]. The difference in herbicide activity results from the dissimilar forms of chloroplastic ACCase in grass and broadleaves weeds. In dicotyledonous plants, plastidic ACCase is multisubunit consisting of biotin carboxylase (BC), biotin-carboxyl carrier protein (BCCP) and carboxyl-transferase (CT). In contrast chloroplastic ACCase is multidomain with a large functional polypeptide comprising of BC, BCCP and CT in most grass weeds [17].

ACCase inhibiting herbicides bind to the target enzyme in a near competitive manner with respect to the substrate acetyl-CoA. Early double inhibition studies showed that the FOPs and DIMs are mutually exclusive indicating that they share a common binding site [18]. Recently, crystal structures of the CT domain in complex with haloxyfop and tepraloxydim revealed that the two herbicides are bound in the active site of the CT domain, particularly at the interface of the dimer [19,20]. The DIM and FOP herbicides probed distinct regions of the dimer interface sharing only two main anchoring points on the ACCase enzyme. Pinoxaden and tepraloxydim were found to bind in a very similar location on the ACCase target in spite of their very different chemical structures [21].

Resistance to ACCase herbicides can be due to enhanced metabolic degradation of the toxophore or insensitivity of the target enzyme. Metabolism is complex and involves several genes that are gradually selected and combined over several generations [22,23,24,25]. Target site resistance results from a single amino acid change in the ACCase enzyme. Mutations at seven ACCase codons have been reported to date and include positions 1781, 1999, 2027, 2041, 2078, 2088 and 2096 (*Alopecurus myosuroides* equivalent) [26,27]. Resistance conferred by target site mutations can be broad or specific and strong or weak, in part, reflecting the different binding modes of the three classes of ACCase herbicides [26]. The cross resistance patterns between current commercial ACCase herbicides have been established for some but not all these resistance mutations and weed species. In many cases, the contribution of additional underlying non target site resistance mechanisms have not been taken into consideration. Investigating the mechanisms of resistance and exploiting the differences in activities between ACCase inhibiting herbicides is of importance given the limited options for controlling grass weeds in cereal and broad leaved crops [28]. Here we have conducted a thorough biological and molecular study on the mechanism of ACCase resistance in a *Lolium multiflorum* population (RG3) from the UK. Ten most commonly used ACCase herbicides were investigated for their efficacies against the suspected resistant ryegrass population. Additionally the influence of plant growth stages was investigated on wild and mutant RG3 genotypes for the most potent ACCase herbicide, and a DNA based marker method was developed for large scale identification of the target site resistant mutation identified in this study.

Results

Resistance confirmation test

The standard sensitive population STD1 was killed with field rates of clodinafop-propargyl, cycloxydim and pinoxaden. In contrast the three FOP, DIM and DEN herbicides provided only partial and similar levels of control for population RG3. On average the percentage biomass reduction was around 30% relative to the untreated control for all three herbicides. There were either dead or very healthy plants in the treated pots indicating that RG3 is segregating for a strong resistance mechanism(s) to all three ACCase herbicides.

Selection of RG3 plants sensitive and resistant to ACCase herbicides

Of the 128 RG3 plants tested, 92 survived the ACCase herbicide treatments. The two tiller replicates for each plant and herbicide treatments behaved in a very similar way demonstrating the robustness of the tiller method for investigating the comparative efficacies of ACCase herbicides on ryegrass plants. The performance of the three FOP, DIM and DEN herbicides was highly correlated on a single plant basis indicating that a similar genetic factor may be involved in conferring resistance to all three herbicides. As cycloxydim is not known to be significantly affected by non target site resistance in ryegrass species, a target site mutation was suspected to be associated with resistance in RG3. As expected all 20 standard sensitive plants tillered and treated in the same manner as RG3 were killed with all three herbicides.

ACCase gene analysis on sensitive and resistant RG3 and STD1 plants

PCR amplification and sequencing of eight plants each from the two sensitive and resistant RG3 subpopulations and the standard sensitive population STD1 generated an expected gene fragment of 2274 bp, showing an overall 96% homology with published ACCase ryegrass sequences. Twenty three nucleotide changes were observed among the 24 plants sequenced. Nine non-synonymous changes were detected, resulting in E1773Q, R1870P, E1874A, H1878N, E1946N, W1948R, L2012M, T2054I and C2088R mutations. Of these nine mutations, eight were present among both sensitive and resistant plants. On the other hand the cysteine to arginine mutation at codon position 2088 was present in the eight resistant RG3 plants and absent in the eight sensitive plants each from RG3 and STD1. The cysteine to arginine change results from a thymine to cytosine transition at nucleotide position 6262 of the *ACCase* gene (*Alopecurus myosuroides* equivalent). Among the eight resistant RG3 plants, five were heterozygous CR2088 and three homozygous RR2088. The combined whole plant herbicide test and sequencing results suggested that the C2088R ACCase mutation conferred dominant and strong resistance to practical field rates of clodinafop-propargyl, cycloxydim and pinoxaden in RG3.

CAPS assay for large scale population genotyping and for confirming the association between resistance and the C2088R ACCase mutation

A DNA based CAPS assay was developed for genotyping a large number of RG3 plants and ascertaining the association of the C2088R mutation with clodinafop-propargyl, cycloxydim and pinoxaden resistance in RG3. PCR generated a 161 bp fragment among all the 128 RG3 and 20 plants from STD1. Upon restriction with the enzyme *Hha*I, mutant RR2088 plants generated a 126 bp fragment while wild type plants showed the undigested 161 bp fragment (Figure 1). Heterozygous plants displayed a copy each of the 161 bp and 126 bp DNA fragments. The results obtained with the CAPS and sequencing assays were completely correlated for the 24 plants sequenced earlier, thus demonstrating the accuracy of the 2088 DNA assay developed here. Genotyping of the remaining 112 RG3 plants from 128 individuals that were tillered and treated with clodinafop-propargyl, cycloxydim and pinoxaden revealed overall genotypic frequencies of 27.6% 46.9% and 25.5% for homozygous wild type CC2088, heterozygous CR2088 and homozygous mutant RR2088 plants respectively. Importantly genotype and phenotype comparison confirmed the complete association between one or two copies of the R2088 mutant allele and resistance to all three herbicides in population RG3.

The CAPS primer used here amplified successfully a corresponding ACCase gene fragment from eight plants each of the three *A. myosuroides* and *A. fatua* populations. As expected the DNA fragments were not digested upon treatment with the enzyme *Hha*I for the wild type plants from *A. myosuroides* and *A. fatua* and a clearly restricted 126 bp restricted fragment was visible on agarose gel for the resistant *A. fatua* characterised by the C2088R mutation.

Production of pure CC2088 and RR2088 seeds and whole dose response assays with ten commonly used ACCase herbicides

Of the 387 untreated RG3 plants genotyped with the CAPS assay prior to seed bulking, 106 and 100 plants were homozygous

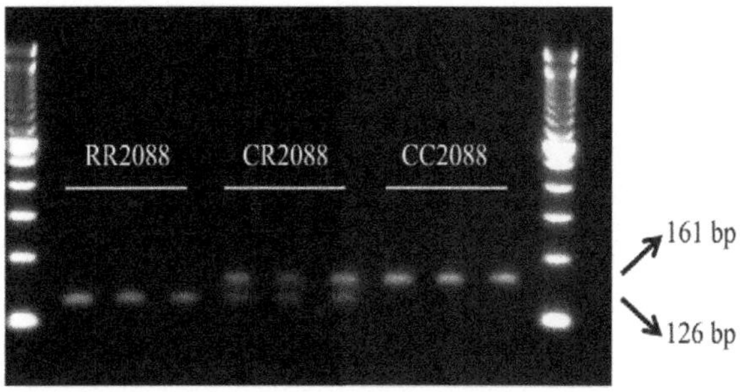

Figure 1. CAPS procedure for the detection of a thymine to cysteine change (C2088R mutation) at nucleotide position 6262 in *Lolium* spp. The *Hha* I digested fragment (126 bp) correspond to the resistant R2088 allele and the *Hha* 1 undigested fragment (161 bp) correspond to the C2088 allele. Heterozygous plants display both the 126 bp and 161 resistant and sensitive alleles respectively. Lanes 1 and 10: DNA ladder, lanes 2, 3 and 4: homozygous mutant RR2088, lanes 4, 5, 6: heterozygous CR2088 plants, lanes 7, 8, 9: homozygous wild CC2088 plants.

CC2088 and homozygous RR2088 respectively. The frequencies of the CC2088, CR2088 and RR2088 were consistent with the earlier characterisation of the 128 plants treated with three ACCase herbicides. Separate crosses of 100 each of CC2088 and RR2088 plants produced similar amounts of wild and mutant seeds under optimal glasshouse conditions. Analysis of 32 individual germinated plants from each of the two lots confirmed purity of the CC2088 and RR2088 seed batches.

Using split rates of herbicides for wild and mutant 2088 plants (Table 1), GR50 values could be estimated for all but mutant genotype RR2088 treated with diclofop-methyl. At the highest rate of 32000 g ai ha^{-1} diclofop-methyl, the percentage biomass reduction was 23% only. Overall, the ACCase herbicides were very active on the standard sensitive CC2088-STD1 population and the wild type CC2088-RG3 subpopulation with GR50 values ranging from less than 2 g ai ha for pinoxaden to 105 g ai ha^{-1} for diclofop-methyl (Table 2). With the exception of fluazifop (requiring the full recommended rate of herbicide), the wild type plants were controlled with half or even below, the recommended field rates of herbicides. In contrast none of the ten herbicides fully controlled the mutant RR2088-RG3 subpopulation at the common use rates and beyond for most herbicides. Corresponding resistance indices ranged from 13 and 15 for clethodim and tepraloxydim to 317 and over 358 for clodinafop-propargyl and diclofop-methyl respectively (Table 3). Interestingly, the estimated resistance factors between wild type genotypes CC2088-STD1 and CC2088-RG3 were close to 1 for all compounds, thus implying that RG3 is not characterised by additional ACCase resistance mechanisms even at very low rates of herbicides. Examples of whole plant dose responses obtained for a FOP, DIM and DEN herbicides are shown on figures 2, 3 and 4 respectively. It is noteworthy that for the most potent herbicide clethodim, the percentage biomass reduction of mutant RR2088 plants at the two-leaf stage was 55%, 85% and 90% for the commonly use rates of 60, 120 and 240 g ai ha^{-1}. Though significant biomass reduction was achieved with clethodim at 120 and 240 g ai ha^{-1}, several plants survived the treatments and would produce seeds.

Efficacy of clethodim on wild and mutant plants at the peri-emergence and one leaf stage

In an attempt to achieve full control of population RG3, clethodim the most potent herbicide was applied on smaller plants

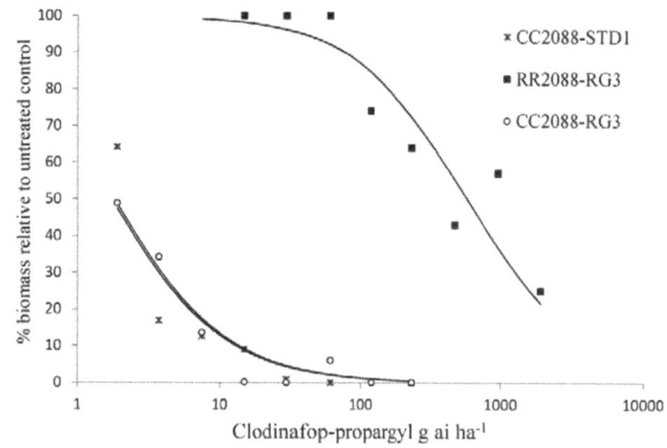

Figure 2. Clodinafop-propargyl whole plant dose response assay on wild and mutant 2088 genotypes.

at the peri-emergence and one leaf stage. Lower levels of

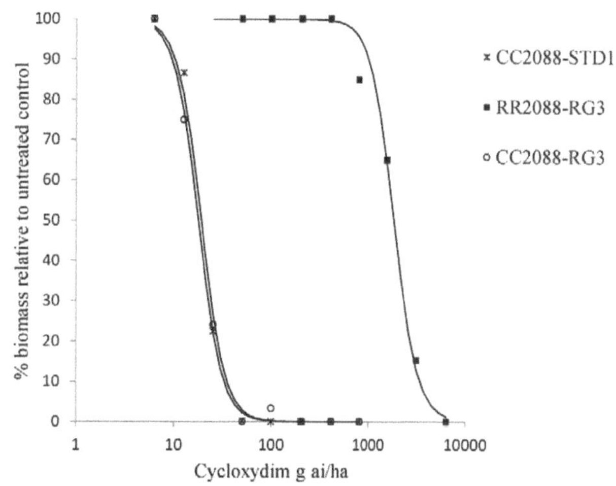

Figure 3. Cycloxydim whole plant dose response assay on wild and mutant 2088 genotypes.

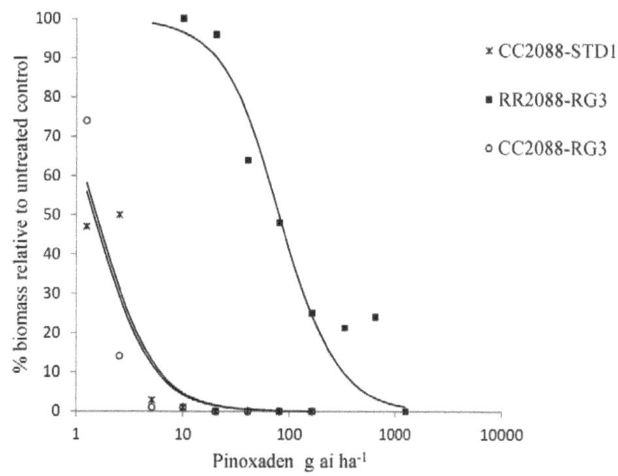

Figure 4. Pinoxaden whole plant dose response assay on wild and mutant 2088 genotypes.

Table 1. Herbicide rates used in whole plant dose response assays.

Herbicide	Rate in g ai ha^{-1} in dose response assays	
	CC2088-STD1; CC2088-RG3	RR2088-RG3
Clodinafop-propargyl	1.88, 3.75, 7.5, 15, 30, 60, 120, 240	15, 30, 60, 120, 240, 480, 960, 1920
Diclofop-methyl	31.25, 62.5, 125, 250, 500, 1000, 2000, 4000	250, 500, 1000, 2000, 4000, 8000, 16000, 32000
Fluazifop	2.11, 4.22, 8.44, 16.88, 33.75, 67.5, 125, 250, 500	8.44, 16.88, 33.75, 67.5, 125, 250, 500, 1000, 2000
Haloxyfop	0.58, 1.15, 2.34, 4.69, 9.38, 18.75, 37.5, 75	9.38, 18.75, 37.5, 75, 150, 300, 600, 1200
Quizalofop	1.17, 2.34, 4.68, 9.37, 18.75, 37.5, 75, 150, 300	4.68, 9.37, 18.75, 37.5, 75, 150, 300, 600, 1200
Cycloxydim	6.25, 12.5, 25, 50, 100, 200, 400, 800	50, 100, 200, 400, 800, 1600, 3200, 6400
Sethoxydim	9.38, 18.75, 37.5, 75, 150, 300, 600, 1200	75, 150, 300, 600, 1200, 2400, 4800, 9600
Clethodim	1.88, 3.75, 7.5, 15, 30, 60, 120, 240	15, 30, 60, 120, 240, 480, 960, 1920
Tepraloxydim	0.19, 0.39, 0.78, 1.56, 3.13, 6.25, 12.5, 25, 50	1.56, 3.13, 6.25, 12.5, 25, 50, 100, 200, 400
Pinoxaden	1.25, 2.5, 5, 10, 20, 40, 80, 160	10, 20, 40, 80, 160, 320, 640, 1280

CC2088-STD1: wild type homozygous 2088 genotype from standard sensitive population.
CC2088-RG3: wild type homozygous 2088 genotype from population RG3.
RR2088-RG3: mutant homozygous 2088 genotype from population RG3.

clethodim control were observed on the standard sensitive population at the peri-emergence stage compared to plants at the one leaf stage for all three herbicide rates tested (Table 4). Similarly partial weed control was achieved on the wild type CC2088-RG3 genotype at the peri-emergence stage for the lower clethodim rate of 60 g ai ha^{-1}. This indicates that the efficacy of clethodim was suboptimal on smaller plants, especially at the peri-emergence stage, given that clethodim rates as low as 30 g ai ha^{-1} provided full control of the two wild type genotypes CC2088-STD1 and CC2088-RG3 at the two leaf stage in the previous test. In any case clethodim provided less than 40% biomass reduction on mutant plants at the peri-emergence and one leaf stages even at the highest recommended field rate of 240 g ai/ha. Pairwise comparisons using the non parametric Wilcoxon rank sum test indicated a significant difference in herbicide responses between the wild type and mutant genotypes for all but one (due to sub-optimal conditions) herbicide rate and plant growth stages, thus confirming the association between the R2088 ACCase allele and resistance to clethodim in RG3 (Table 4).

Discussion

The evolution of resistance to ACCase herbicides is seriously limiting grass weed management options in small grain cereal and dicot crops. The main objectives of this study were to conduct an integrated biological and molecular study on the mechanism of resistance to ACCase inhibiting herbicides in a *Lolium multiflorum* population and to develop a DNA based marker method for the quick identification of the target site mutation identified here.

Mechanism of resistance to ACCase herbicides in population RG3

The study revealed a cysteine to arginine mutation in the target enzyme associated with resistance to all herbicides tested. Increasing levels of resistance were detected from clethodim to diclofop-methyl through fluazifop and pinoxaden. This is the first report of the C2088R mutation in a *Lolium multiflorum* population from the UK. The C2088R mutation has thus far been detected in five other ryegrass populations from Australia [11] and one ryegrass population from Italy [10]. The mutation was also detected in an *A. fatua* population from the USA [29]. However, in the previous studies the potential for non target site mechanisms to

Table 2. Estimated GR_{50} values for 10 ACCase inhibiting herbicides and different 2088 ACCase genotypes.

Herbicide	CC2088-STD1	CC2088-RG3	RR2088-RG3
Clodinafop-propargyl	2.28 (1.67–3.11)	1.96 (1.08–3.53)	620.69 (358.93–1073.36)
Diclofop-methyl	105.39 (83.57–132.90)	89.34 (63.91–124.91)	>32000
Fluazifop	27.14 (25.05–29.41)	23.48 (21.09–26.13)	505.30 (430.30–592.61)
Haloxyfop	3.92 (3.50–4.40)	3.64 (3.21–4.11)	75.89 (65.65–87.72)
Quizalofop	5.07 (4.67–5.50)	5.70 (5.32–6.10)	228.17 (230.30–256.08)
Cycloxydim	18.98 (16.93–21.27)	17.57(15.58–19.81)	1873.19 (1650.73–2125.63)
Sethoxydim	23.02 (21.31–24.86)	19.76 (17.83–21.89)	2046.16 (1845.68–2268.41)
Clethodim	4.29 (4.10–4.48)	4.33 (4.18–4.49)	57.93 (55.62–60.34)
Tepraloxydim	3.43 (3.09–3.82)	3.45 (3.05–3.89)	52.00 (44.75–60.43)
Pinoxaden	1.47 (1.10–1.96)	1.61 (1.40–1.85)	83.55 (65.66–106.31)

Table 3. Estimated resistance factors for 10 ACCase herbicides and different 2088 ACCase genotypes.

Herbicide	RG3-CC2088 vs STD1-CC2088	RG3-RR2088 vs RG3-CC2088	RG3-RR2088 vs STD1-CC2088
Clodinafop-proparlgyl	0.86 (0.44–1.67)	317.27 (141.84–709.66)	271.98 (145.00–510.18)
Diclofop-methyl	0.85 (0.56–1.27)	>358	>304
Fluazifop	0.86 (0.76–0.99)	21.52 (17.77–26.08)	18.62 (15.58–22.26)
Haloxyfop	0.93 (0.78–1.10)	20.87 (17.26–25.25)	19.34 (16.07–23.26)
Quizalofop	1.12 (1.01–1.25)	40.06 (35.04–45.80)	45.01 (39.07–51.85)
Cycloxydim	0.93 (0.78–1.10)	106.64 (89.58–126.95)	98.71 (83.24–117.05)
Sethoxydim	0.86 (0.75–0.98)	103.57 (89.54-–119.79)	88.89 (78.15–101.11)
Clethodim	1.01 (0.96–1.07)	13.37 (12.66–14.12)	13.52 (12.74–14.35)
Tepraloxydim	1.00 (0.85–1.18)	15.09 (12.45–18.30)	15.14 (12.60–18.21)
Pinoxaden	1.10 (0.79–1.51)	51.87 (39.26–68.54)	56.84 (39.00–82.86)

contribute to resistance were overlooked as the resistant populations were compared to sensitive populations from different genetic backgrounds. Here, comparison of wild and mutant RG3 subpopulations has clearly and unambiguously established the importance of the C2088R in conferring resistance to all the ten herbicides tested at the whole plant level. In addition, contrasting wild type CC2088-RG3 and CC2088-STD1 genotypes indicated that RG3 is not characterised by non target site resistance, even at very low doses of ACCase herbicides. This is a rather unique case since when analysed in detail target site resistant populations have always been found to contain underlying additional non target site resistance [9,10,30,31]. For instance, in the Australian *Lolium rigidum* population SLR31, an I1781L target site resistance mutation was identified in addition to metabolic resistance [31,32]. Subsequent genetics studies have shown that P450 mediated metabolism in this population is governed by at least two genes [22]. Similarly the *Lolium multiflorum* population UK24 was found to contain very high levels of non target site resistance to clodinafop-propargyl and diclofop-methyl in addition to the D2078G ACCase target site mutation [9]. Additional non target site resistance in populations for which a target site mutation have been uncovered could also be inferred from large number of *Lolium* spp. samples from Greece [33], USA [34], UK [35] and Italy [10] on the basis of differential responses to metabolisable and non-metabolisable ACCase herbicides. Weed populations containing additional non target site resistance mechanisms are not restricted to grass weeds as these have recently been identified in broad leaved weed populations characterised by known ALS mutations [36].

Dynamics of target site and non target site resistance evolution under herbicide selection pressure

The presence of non target site resistance to ACCase herbicides in grass weed populations is often explained by multiple genes that individually confer low levels of resistance to herbicides, but when accumulated results in resistance to practical field rates of herbicides [23,25]. This was demonstrated by recurrent selection at low doses of diclofop-methyl of increasingly tolerant *Lolium rigidum* plants from an initially sensitive population. Starting from a genetic pool of a few hundred plants in all, resistance to practical field rates of diclofop-methyl was acquired after three generations only. [23]. A similar scenario is believed to operate under field conditions given that not all plants receive effective rates of herbicides due to staggered germination and shading under high

Table 4. Efficacy of clethodim on different 2088 ACCase genotypes and plant growth stages. Genotypes compared using Wilcoxon rank sum test.

Clethodim rate g ai ha^{-1}	Plant growth Stage	Median herbicide response per genotype (% of visual control)			P-values		
		CC2088-STD1	CC2088-RG3	RR2088-RG3	CC2088-RG3 vs CC2088-STD1	RR2088 RG3 vs CC2088-STD1	RR2088-RG3 vs CC2088-RG3
60	peri-em	30	0	0	0.028	0.002	0.455
	1_leaf	45	65	35	0.054	0.34	0.009
120	peri-em	55	100	5	0.251	0.017	0.015
	1_leaf	70	100	40	0.123	0.162	0.013
240	peri-em	90	100	30	1	0.002	0.028
	1_leaf	100	100	25	0.848	0.058	0.032

plant densities. Under field conditions the accumulation of minor non target site resistance genes ultimately results in significant levels of resistance to ACCase herbicides [25]. In contrast target site resistance mutations are often viewed as a rare event and present at different frequencies depending on the herbicide target [37]. Here we demonstrate that target site resistance can be selected before any non target resistance gene appears in a *Lolium multiflorum* population following years of selective ACCase herbicide selection pressure. Our observation therefore questions the idea of a large number of genes acting additively to confer non target site resistance in weeds or that ACCase target site mutations exist a low frequencies in *Lolium* spp. populations.

Insights into herbicide binding at the ACCase target site

Crystallographic studies have recently showed that the binding mode of pinoxaden and tepraloxydim were very similar in spite of their very different chemical structures [20,21]. Haloxyfop and diclofop on the other hand were found to share a similar binding point at a rather different region on the ACCase target [19]. Our study implies that subtle differences in binding modes of the different herbicides can have a profound effect on their efficacies in the presence of specific mutations. This can be inferred from the large differences in resistance indices between herbicides belonging to the same chemical or binding groups. Though belonging to the FOP family and found to bind very closely on ACCase, the resistance factors associated with the C2088R mutation were 20 and over 350 fold for haloxyfop and diclofop-methyl respectively. Similarly pinoxaden and tepraloxydim, identified as sharing a very close binding mode on ACCase [21], were characterised with resistance indices of 15 and 51 respectively. The differences observed between herbicide efficacies were not due to their relative potencies on *Lolium multiflorum* but due to their binding modes. This is because the resistance indices were derived based on comparison between sensitive and resistant sub-populations sharing similar genetic backgrounds except for the 2088 ACCase mutation.

Importance of cysteine for herbicide binding and catalytic activity of ACCase

At ACCase position 2088 three different amino acids have thus far been identified. Two are sensitive to ACCase herbicides, including cysteine in *Lolium* spp. and most other grass weed species and a phenylalanine residue that is naturally present in barley, wheat and several *Bromus* species [38]. Therefore cysteine *per se* does not appear to be critical for the catalytic activity of ACCase and the effective binding of ACCase inhibiting herbicides. While this study has clearly established the detrimental effect of the C2088R on ACCase herbicide efficacies, the impact of the arginine mutation on ACCase catalytic activity and plant fitness remains to be established. In this respect, the specific activity of the wild type ACCase enzyme was higher than for the homozygous 2088 mutant enzyme [11]. It is to be noted though that the sensitive and resistant enzymes originated from different populations and that the differences in activities could be due to population to population variations. At the whole plant level and when grown under optimal glasshouse conditions, we found no visual differences in growth characteristics between the wild and mutant CC2088-RG3 and RR2088-RG3 sub-populations. The wild and mutant genotypes established here will be very useful for studying whether the C2088R ACCase mutation is affected by a fitness penalty under field conditions.

Efficacy of clethodim on the C2088R mutant plants

Clethodim was by far the most effective herbicide on the homozygous mutant RR2088 genotypes investigated here. At the highest recommended rate of 240 g ai ha^{-1} most plants were killed. However the few individuals that survived the clethodim treatment would re-grow and produce seeds. This is in agreement with a previous study on Australian samples for which clethodim provided insufficient levels of control on C2088R resistant plants at the commonly used rate of 60 g ai ha^{-1}[11]. On the other hand clethodim at 216 g ai ha^{-1} killed all heterozygous and homozygous 2088 mutant plants from an Italian *Lolium multiflorum* population [10]. These results should nevertheless be interpreted with caution given that herbicides tend to be very active on soft (i.e. non-weathered) plants at low densities under optimal glasshouse conditions. In this respect 50% biomass reduction was achieved on sensitive CC2088-RG3 and CC2088-STD1 genotypes with less than 5 g ai ha^{-1}clethodim. This level of activity is never attained under field conditions. Also it is often recommended that herbicides be applied on weeds when they are still at an early growth stage given that metabolic resistance tends to be accentuated in larger weeds. In this study however, the activity of clethodim was lower on plants at the one leaf and especially peri-emergence stage as compared to the two leaf stages. This can be explained by the fact that clethodim is a post-emergence herbicide that works optimally on plants with adequate foliage ensuring good herbicide absorption. Therefore clethodim will very unlikely control *Lolium* spp. populations containing the 2088 mutations irrespective of the plant growth stage.

Detection of the C2088R mutation

A DNA based CAPS assay was developed for detecting the mutant R2088 allele in *Lolium* spp. The assay proved to be very robust as sequencing and CAPS results were totally correlated. The methodology used relies on the conservation of the third guanine on the 2087 ACCase codon for the identification of the R2088 allele. The 2087 codon is a glutamic acid and is particularly conserved among plant species. Glutamic acid is coded by either the GAG or GAA nucleotide triplets. Alignment of 200 ACCase sequences available in the GenBank showed that glutamate is very conserved at the 2087 position and that the codon triplet is GAG for all but one Oryza clone (CZ119516). Thus it is very unlikely that the CAPS 2088 assay developed here will be affected by false negatives for the detection of mutant R2088 alleles in *Lolium* spp. The CAPS assay complements two other dCAPS methods aimed at detecting the wild type cysteine 2088 allele and based on the positive identification of the thymine on the first base of the codon [10,39]. The method developed by Scarabel et al [10] also relies on the conservation of the last base of the 2087 codon for unambiguous detection of the wild type cysteine allele while the reverse primer from the Delye et al [39] assay encompass all the bases of the recognition site of the restriction enzyme used except for the variable first base of the 2088 codon. However, the latter assay uses a nested PCR strategy for successful amplification of the ACCase gene fragment, thus adding an additional and costly step in the detection procedure. The interest of the current CAPS 2088 assay is that it is highly transferable to other grass weed species namely *A. fatua* and *A. myosuroides*, prone to evolve resistance to ACCase herbicides. It employs a very efficient restriction enzyme which is two and four times cheaper than the ones used in the assays proposed by Scarabel et al [10] and Delye et al [39] respectively. It is thus particularly suited for large scale genotyping of ryegrass populations as is the case in this study. Additionally, because it is a CAPS assay, it is less likely to be affected by priming stability compared to dCAPS assays with one [10] and two [39] forced mutations in the reverse primers.

In conclusion, we identified a C2088R target site mutation conferring high levels of resistance to a broad range of ACCase herbicides in the *Lolium multiflorum* population from the UK. The population is rather unique as it does not appear to contain any additional underlying non target site resistance mechanisms affecting the efficacies of all ACCase herbicides studied here. The C2008R mutation can be detected using a robust CAPS assay that is readily transferable to two other grass weed species. Additionally the pure CC2088 and RR2088 sub-populations created as part of this study will be helpful in determining whether the R2088 allele confers a fitness penalty in competition under field conditions.

Materials and Methods

Plant materials

Seeds from the suspected resistant *Lolium multiflorum* population (RG3) were collected in 2003 from a wheat field in Widnes, Cheshire, UK. The seeds were randomly sampled from a large number of surviving plants from across the entire field. A standard sensitive *Lolium multiflorum* population (STD1) commercially available from Herbiseed (Twyford, Berkshire, UK) was used for comparison in all glasshouse and DNA tests. Additionally, a sensitive *Alopecurus myosuroides* population and a sensitive and resistant *Avena fatua* populations were used for testing the transferability of the DNA based Cleaved Amplified Polymorphic Sequence (CAPS) marker developed in this study to other species.

"No specific permissions were required for the location where the ryegrass seeds were collected. This study did not involve any endangered or protected species".

Initial resistance confirmation test

Around 50 seeds for each of STD1 and RG3 populations were sown in individual pots containing a 1:1 ratio of compost and peat and irrigated as necessary. The germinated plants were maintained in a controlled greenhouse set at 24°C/16 hr day, 18°C night, 65% relative humidity conditions, and a photon flux density of approximately 250 μmol quanta m^{-2} s^{-1}. The plants at the two leaf stage were treated with a single rate of commercial formulations of clodinafop-propargyl (60 g ha^{-1}), cycloxydim (100 g ha^{-1}) and pinoxaden (45 g ha^{-1}). Three replicate pots were used and the level of control was visually assessed 21 days after treatment.

Segregation of the RG3 population into plants sensitive and resistant to ACCase herbicides

Seeds were sown as described earlier. One hundred and twenty eight seedlings were transplanted in individual pots and allowed to tiller by trimming the plants regularly. Two months later the 128 mother plants were separated into seven genetically identical tillers. The roots and leaves were trimmed and the tillers from each mother plant were potted separately. After one week, two separate tillers each from the 128 mother plants were treated with clodinafop-propargyl (60 g ha^{-1}), cycloxydim (100 g ha^{-1}) and pinoxaden (45 g ha^{-1}), whilst the 7th tiller was used as untreated control. Twenty plants from the standard sensitive *Lolium multiflorum* population STD1 were also tillered and sprayed in the same way as the resistant population. Twenty one days after treatment the 128 mother plants were classified as either dead and sensitive or alive and resistant to the ACCase herbicides.

Sequencing of the carboxyl transferase (CT) domain of the ACCase gene

A gene fragment comprising the entire carboxyl transferase domain of *ACCase* was sequenced using an RT-PCR procedure. Eight plants each of sensitive and resistant RG3 subpopulations and eight other plants from the standard sensitive population STD1 were sequenced. For each of the 24 plants, 1 g of frozen leaf tissue was ground using a pestle and a mortar in the presence of liquid nitrogen and extracted using the Trizol procedure (Invitrogen, CA, USA). One μg of RNA from each plant was cleared from contaminating DNA using a DNase treatment (Invitrogen, CA, USA). First strand DNA syntheses were carried out using purified RNA and Oligo-dT and reverse transcriptase superscript II (Invitrogen, CA, USA). Gene specific primers were used for the second DNA strand synthesis, and were based on available Genbank *Lolium* spp. ACCase sequences (AF359513, AF359514, AF359515 and AF359515) for the forward primer (5′TGGTCCTTTGCATGGTGTCGC3′) and consensus reverse primer (5′GCTCTCTTAGAGGGATCCATCTTATC3′) deduced following alignment of *Setaria italica* (AF294805), *Alopecurus myosuroides* (AJ310767), and *Zea mays* (ZMU19183) ACCase sequences. The final PCRs for second strand DNA synthesis were carried out using puReTaq Ready-To-Go PCR beads (Amersham Biosciences, Bucks, UK) in a total volume of 25 μL containing 0.8 μM of each primer and 1 μL product from 1st strand DNA synthesis. PCRs were performed on an Eppendorf Master Cycle Gradient Thermocycler Model 96 programmed for an initial denaturation step of 94°C of 2 min followed by 40 cycles of 30 s at 94°C, 30 s at 64°C and 2 min at 72°C. A final extension step for 10 min at 72°C was also included. ACCase sequence alignment and comparison was carried out with the Seqman software (Lasergene, USA).

Development of a Cleaved Amplified Polymorphic Sequence (CAPS) method for detecting the C2088R ACCase mutation

A DNA based Cleaved Amplified Polymorphic Sequence (CAPS) procedure was developed for large scale genotyping of the 128 individuals treated with clodinafop-propargyl, cycloxydim and pinoxaden and for confirming the association of the C2088R mutation with resistance to ACCase herbicides. The wild and mutant amino acid residues at ACCase position 2088 are coded by the triplets TGC and CGC respectively. The discriminant enzyme employed in the CAPS assay was *Hha*I. This enzyme recognises the quartet GCGC (arginine) from mutant 2088 genotypes but not the corresponding GTGC (cysteine) sequence from the wild type plants. As cysteine can be potentially coded by the triplets TGC or TGT and the second base of the 2088 codon is always a G in *Lolium* spp., the 2nd and 3rd bases of the 2088 codon was enforced with the last two bases at the 3′ end of the reverse CAPS primer. For each plant, a 2 cm leaf segment was ground on a Spex Certiprep (Metuchen, NJ, USA) 2000 model Genogrinder. DNA from the ground material was subsequently extracted on a Biomek FX robot (Beckman Coulter, High Wycombe, Buckinghamshire, UK) using a Wizard Magnetic 96 DNA Plant System kit (Promega, Madison, WI, USA).

PCR amplification was carried out with the forward 5′TCAACAATTGTTGAGAACCTTAGG3′ and reverse 5′AGAACATTCCCTTTTGCAGTTGTCTCAGCATAGC3′ primers to generate a 161 bp PCR fragment. PCRs were carried out using puReTaq Ready-To-Go PCR beads (Amersham Biosciences, Bucks, UK) in a total volume of 25 μL containing 0.8 μM of each primer and from 10–50 ng genomic DNA. PCRs were performed on an Eppendorf Master Cycle Gradient Thermocycler Model 96 programmed for an initial denaturation step of 94°C of 2 min followed by 40 cycles of 30 s at 94°C, 30 s at 64°C and 1 min at 72°C. A final extension step for 10 min at 72°C was also included.

Two µl aliquots of neat PCR products were digested with 2 units of *Hha*I (New England Biolabs, Hertfordshire, UK) in a total volume of 20 µL according to the manufacturer's recommendations and analysed on 2% agarose/1 X TBE buffer gels containing 0.5 µg mL^{-1} ethidium bromide. Upon restriction with *Hha*I the mutated allele is digested into a 126 bp and 35 bp (invisible on the 2% agarose gel) fragments.

To test the transferability of the CAPS method to two other major grass weed species affected by ACCase herbicide resistance, eight plants each from a sensitive *A. myosuroides*, a sensitive *A. fatua* and a resistant *A. fatua* populations characterised by the C2088R mutation were extracted and analysed as described for the ryegrass populations.

Production of pure homozygous CC2088 and RR2088 seeds and whole plant dose response tests with 10 different ACCase herbicides

Three hundred and eighty seven RG3 plants were grown and genotyped with the DNA based CAPS method as described above. One hundred plants each of the CC2088 and RR2088 genotypes were separated and allowed to cross freely to produce pure wild type and mutant 2088 seeds. Subsequently 32 seeds from each of the two batches were germinated and tested for purity using the 2088 CAPS marker. These pure wild and mutant seed lots were subsequently used in whole plant dose response assays for assessing the precise resistance indices (RI) associated with the C2088R mutation. The standard sensitive population STD1 was tested alongside to determine whether other additional underlying mechanisms contribute to resistance to the ACCase herbicides. The ACCase herbicides and the rates used for each genotype in dose response assays are given in table 1. For each herbicide rate, four replicates pots containing six plants at the two leaf stage each were tested. Visual biomass relative to untreated controls was assessed 21 days after treatment.

The resulting percentages were analysed using a non-linear least squares regression model with the objective of estimating the resistance factors between the different genotypes for each herbicide. Separate regression lines were fitted to the data for each genotype but with the slope of each line constrained to be identical for all genotypes. For each genotype, the fitted model is sigmoidal in shape and described by the equation:

$$P = \frac{100}{1 + e^{-\beta(x-\mu_{ij})}}$$

where x denotes $\log_{10}$(Rate), μ_{ij} denotes the logGR50 for genotype i and β denotes the common slope fitted to all regression lines. The estimated resistance index between a pair of genotypes is calculated as the ratio of their estimated GR50 values. Confidence intervals around the estimated resistance indices were calculated using the error term derived from a separate analysis of variance having excluded from the analysis those treatments that did not contribute meaningfully to the estimated residual sum of squares. All statistical analyses were carried out using SAS software, version 9.2.

Influence of plant growth stage on clethodim efficacy

To investigate whether effective weed control could be achieved on smaller RG3 plants, wild and mutant 2088 genotypes were treated at the peri-emergence (around the time of emergence) and one leaf stage with clethodim, the most potent herbicide on 2088 mutant plants as identified with the whole plant dose response assays. Six replicates pots of ten plants each of the standard sensitive CC2088-STD1, wild type CC2088-RG3 and mutant type RR2088-RG3 genotypes were grown at the appropriate growth stages and tested at 60, 120 and 240 g ha^{-1} of clethodim. The pots were arranged according to a randomised complete block (RCB) design. Visual biomass relative to untreated controls was assessed 21 days after treatment. The data for each treatment and genotype were analysed using the nonparametric Wilcoxon rank sum test. Pairwise comparisons were carried out between two genotypes for each herbicide rate and plant growth stage. A p-value of less than 0.05 indicates a statistically significant result at the 5% probability level and therefore provides evidence that the median levels of control produced by the two genotypes in question are genuinely different.

Acknowledgments

The authors are grateful to Peter Glanfield for harvesting the ryegrass samples, the Syngenta plant production and herbicide screening teams for their help in carrying out the glasshouse experiments in this study and present and past technical and research managers for their encouragements for conducting this study.

Author Contributions

Conceived and designed the experiments: SSK. Performed the experiments: SSK SJH RD. Analyzed the data: EM. Contributed reagents/materials/analysis tools: SSK SJH RD. Wrote the paper: SSK.

References

1. Wrubel RP, Gressel J (1994) Are herbicide mixtures useful for delaying the rapid evolution of resistance? A case study. Weed Technol 8: 635–648.
2. Beckie HJ, Hall LM, Meers C, Laslo JJ, Stevenson FC (2004) Management practices influencing herbicide resistance in wild oat. Weed Technol 18: 853–859.
3. Maxwell BD, Mortimer AM (1994) Selection for herbicide resistance. In: Powles SB, Holtum JAM, editors. In herbicide resistance in plants. Boca Raton, FL: Lewis Publishers. 1–26.
4. Burnet MWM, Hart Q, Holtum JAM, Powles SB (1994) Resistance to Nine Herbicide Classes in a Population of Rigid Ryegrass. Weed Sci 42: 369–377.
5. Heap IM (2012) International Survey of Herbicide-Resistant Weeds. Available: http://www.weedscience.com. Accessed 3 May 2012.
6. Delye C, Boucansaud K, Pernin F, Le Corre V (2009) Variation in the gene encoding acetolactate-synthase in *Lolium* species and proactive detection of mutant, herbicide-resistant alleles. Weed Res 49: 326–336.
7. Yu Q, Han H, Powles SB (2008) Mutations of the ALS gene endowing resistance to ALS-inhibiting herbicidesin *Lolium rigidum* populations. Pest Manag Sci 64: 1229–1236.
8. Tan MK, Preston C, Wang GX (2007) Molecular basis of multiple resistance to ACCase-inhibiting and ALS-inhibiting herbicides in *Lolium rigidum*. Weed Res 47: 534–541.
9. Kaundun SS (2010) An aspartate to glycine change in the carboxyl transferase domain of acetyl CoA carboxylase is in part associated with resistance to ACCase inhibitor herbicides in a Lolium multiflorum population. Pest Manag Sci 66: 1249–1256.
10. Scarabel L, Panozzo S, Varotto S, Sattin M (2011) Allelic variation of the ACCase gene and response to ACCase-inhibiting herbicides in pinoxaden-resistant Lolium spp. 67: 932–941.
11. Yu Q, Collavo A, Zheng M-Q, Owen M, Sattin M, et al. (2007) Diversity of Acetyl-Coenzyme A Carboxylase Mutations in Resistant Lolium Populations: Evaluation Using Clethodim. Plant Physiol 145: 547–558.
12. Heap I, Knight R (1982) A population of ryegrass tolerabnt to the herbicide diclofop-methyl. Journal of the Australian Institute of Agricultural Science 48: 156–157.

13. Owen MJ, Walsh MJ, Llewellyn RS, Powles SB (2007) Widespread occurrence of multiple herbicide resistance in Western Australian annual ryegrass (*Lolium rigidum*) populations. Australian Journal of Agricultural Research 58: 711–718.
14. Secor J, Cséke C (1988) Inhibition of acetyl-CoA carboxylase activity by haloxyfop and tralkoxydim. Plant Physiol 86: 10–12.
15. Hofer U, Muehlebach M, Hole S, Zoschke A (2006) Pinoxaden – for broad spectrum grass weed management in cereal crops. J Plant Dis Prot 113: 989–995.
16. Konishi T, Sasaki Y (1994) Compartmentalization of two forms of acetyl-CoA carboxylase in plants and the origin of their tolerance toward herbicides. Proc Nati Acad Sci USA 91: 3598–3601.
17. Sasaki Y, Nagano Y (2004) Plant acetyl-CoA carboxylase: structure, biosynthesis, regulation, and gene manipulation for plant breeding. Biosci Biotechnol Biochem 68: 1175–1184.
18. Rendina AR, Craig-Kennard AC, Beaudoin JD, Breen MK (1990) Inhibition of acetyl-coenzyme A carboxylase by two classes of grass-selective herbicides. J Sci Agric Food Chem 38: 1282–1287.
19. Zhang H, Tweel B, Tong L (2004) Molecular basis for the inhibition of the carboxyltransferase domain of acetyl-coenzyme-A carboxylase by haloxyfop and diclofop. Proc Natl Acad Sci USA 101: 5910–5915.
20. Xiang S, Callaghan MM, Watson KG, Tong L (2009) A different mechanism for the inhibition of the carboxyltransferase domain of acetyl-coenzyme A carboxylase by tepraloxydim. Proc Natl Acad Sci USA 106: 20723–20727.
21. Yu LPC, Kim YS, Tong L (2010) Mechanism for the inhibition of the carboxyltransferase domain of acetyl-coenzyme A carboxylase by pinoxaden. Proc Natl Acad Sci USA 107: 22072–22077.
22. Busi R, Vila-Aiub MM, Powles SB (2011) Genetic control of a cytochrome P450 metabolism-based herbicide resistance mechanism in Lolium rigidum. Heredity 106: 817–824.
23. Neve P, Powles SB (2005) Recurrent selection with reduced herbicide rates results in the rapid evolution of herbicide resistance in *Lolium rigidum*. Theor Appl Genet 110: 1154–1166.
24. Petit C, Duhieu B, Boucansaud K, Délye D (2010) Complex genetic control of non-target-site-based resistance to herbicides inhibiting acetyl-coenzyme A carboxylase and acetolactate-synthase in Alopecurus myosuroides Huds. Plant Science 178: 501–509.
25. Manalil S, Busi R, Renton M, Powles SB (2011) Rapid evolution of herbicide resistance by low herbicide dosages. Weed Sci 59: 210–217.
26. Powles S, Yu Q (2010) Evolution in action: plants reistant to herbicides. Annu Rev Plant Biol 61: 317–347.
27. Délye C (2005) Weed resistance to acetyl coenzyme A carboxylase inhibitors: an update. Weed Sci 53: 728–746.
28. Rueegg WT, Quadranti M, Zoschke A (2007) Herbicide research and development: challenges and opportunities. Weeds res 47: 271–275.
29. Christoffers MJ, Pederson SN, Kandikonda AV. Herbicide dose response of wild-oat with altered acetyl-CoA carboxylase genes; 2005. 60: 35.
30. de Prado JL, Osuna MD, Heredia A, de Prado R (2005) *Lolium rigidum*, a pool of resistance mechanisms to ACCase inhibitor herbicides. J Agric Food Chem 53: 2185–2191.
31. Tardiff FJ, Powles SB (1994) herbicide multiple-resistance in a Lolium rigidum biotype is endowed by multiple mechanisms; isolation f a subset with resistant acetyl-CoA carboxylase. Physiol Plant 91: 488–494.
32. Zhang XQ, Powles SB (2006) The molecular bases for resistance to acetyl co-enzyme A carboxylase (ACCase) inhibiting herbicides in two target-based resistant biotypes of annual ryegrass (*Lolium rigidum*). Planta 223: 550–557.
33. Kotoula-Syka E, Tal A, Rubin B (2000) Diclofop-resistant *Lolium rigidum* from northern Greece with cross-resistance to ACCase inhibitors and multiple resistance to chlorsulfuron. Pest Manag Sci 56: 1054–1058.
34. Kuk YI, Burgos NR, Scott RC (2008) Resistance profile of diclofop-resistant Italian ryegrass (Lolium multiflorum) to ACCase- and ALS-inhibiting herbicides in Arkansas, USA. Weed Sci 56: 614–623.
35. Cocker KM, Northcroft DS, Coleman JO, Moss SR (2001) Resistance to ACCase-inhibiting herbicides and isoproturon in UK populations of *Lolium multiflorum*: mechanisms of resistance and implications for control. Pest Manag Sci 57: 587–597.
36. Délye C, Pernin F, Scarabel L (2011) Evolution and diversity of the mechanisms endowing resistance to herbicides inhibiting acetolactate-synthase (ALS) in corn poppy (Papaver rhoeas L.). Plant Science 180: 333–342.
37. Sammons R, Heering D, Danicola N, Glick H, Elmore G (2007) Sustainability and stewardship of glyphosate and glyphosate-resistant crops. Weed Technol 21: 347–354.
38. Délye C, Michel S (2005) 'Universal' primers for PCR-sequencing of grass chloroplastic acetyl-CoA carboxylase domains involved in resistance to herbicides. Weed Res 45: 323–330.
39. Delye C, Pernin F, Michel S (2011) Universal' PCR assays detecting mutations in acetyl-coenzyme A carboxylase or acetolactate synthase that endow herbicide resistance in grass weeds. Weed Res 51: 353–362.

14

Increasing Minimum Daily Temperatures Are Associated with Enhanced Pesticide Use in Cultivated Soybean along a Latitudinal Gradient in the Mid-Western United States

Lewis H. Ziska*

Crop Systems and Global Change Laboratory, United States Department of Agriculture, Agricultural Research Service, Beltsville, Maryland, United States of America

Abstract

Assessments of climate change and food security often do not consider changes to crop production as a function of altered pest pressures. Evaluation of potential changes may be difficult, in part, because management practices are routinely utilized *in situ* to minimize pest injury. If so, then such practices, should, in theory, also change with climate, although this has never been quantified. Chemical (pesticide) applications remain the primary means of managing pests in industrialized countries. While a wide range of climate variables can influence chemical use, minimum daily temperature (lowest 24 h recorded temperature in a given year) can be associated with the distribution and thermal survival of many agricultural pests in temperate regions. The current study quantifies average pesticide applications since 1999 for commercial soybean grown over a 2100 km North-South latitudinal transect for seven states that varied in minimum daily temperature (1999–2013) from −28.6°C (Minnesota) to −5.1°C (Louisiana). Although soybean yields (per hectare) did not vary by state, total pesticide applications (kg of active ingredient, ai, per hectare) increased from 4.3 to 6.5 over this temperature range. Significant correlations were observed between minimum daily temperatures and kg of ai for all pesticide classes. This suggested that minimum daily temperature could serve as a proxy for pesticide application. Longer term temperature data (1977–2013) indicated greater relative increases in minimum daily temperatures for northern relative to southern states. Using these longer-term trends to determine short-term projections of pesticide use (to 2023) showed a greater comparative increase in herbicide use for soybean in northern; but a greater increase in insecticide and fungicide use for southern states in a warmer climate. Overall, these data suggest that increases in pesticide application rates may be a means to maintain soybean production in response to rising minimum daily temperatures and potential increases in pest pressures.

Editor: Matthew Germino, US Geological Survey, United States of America

Funding: This author has no support or funding to report.

Competing Interests: The authors have declared that no competing interests exist.

* E-mail: l.ziska@ars.usda.gov

Introduction

Considerable research effort has focused on determining the impact of anthropogenic climate change on global agriculture [1–4]. Of merited interest in this regard are the physical aspects of climate (e.g. carbon dioxide, temperature, precipitation, extreme weather events) that directly alter crop biology (e.g. growth, phenology, sterility, yields) and the resulting consequences for food security [5–9].

However, research efforts related to assessing the agricultural impacts of rising CO_2 and climate change do not always consider trophic interactions. Overall, changes to the biology and competitive abilities of agricultural pests (insects, pathogens, weeds) relative to potential crop yield losses has not been well quantified [10–11]. This is an important omission as the role of pests on constraining crop production is significant and well recognized. For example, weed competition can result in potential crop losses of ~34% globally, with insect pests and pathogens resulting in additional losses of ~18 and 16%, respectively [12].

Such omissions may reflect the complex challenges in relating atmospheric CO_2 and climate variables to potential reductions in crop production related to increased pest pressures. For example, weed growth and fecundity can be directly affected by increasing atmospheric CO_2 as well as rising temperature; insects and pathogens can also be directly affected by temperature, but indirectly by CO_2 and/or climate induced changes to their weed hosts [12–13]. Overall, while a number of pest studies have been conducted, empirical evidence has been eclectic, although it has been suggested that pest pressures will probably increase with climate change (e.g., [14]).

Yet, even if pest pressures regarding crop production were unequivocal and well-characterized in regard to climate change, it would still be difficult to quantify yield reductions *in situ*. This is because there are strong economic incentives at the field level to manage agro-ecosystems to prevent or minimize pest damage. While management methods vary greatly and may include cultural, mechanical, chemical and biological options, among developed countries, such as Australia and the United States, application of chemicals, usually as pesticides, represents the most widely used method for pest control.

But could quantification of pesticide usage in turn, provide an alternative means to gauge changes in pest pressures associated with changes in climate? Among climate variables, it is generally recognized that in temperate regions, the distribution and survival of agricultural pests is often limited by low winter temperatures; i.e., minimum thermal thresholds) [15]. Rising minimum temperatures associated with anthropogenic climate change could extend the potential geographic range of pest species and/or alter their demographics, although long-term changes in species diversity are unclear [16–17]. Climate change assessments have also emphasized that the current and projected increases in global warming are not uniform, and enhanced land-surface temperatures (relative to the global average) are more probable for minimum (Winter) than maximum (Summer) temperatures [18–19].

To ascertain if minimum temperatures reflect pesticide usage, regression analysis of the interrelationship between insecticide, fungicide and herbicide application rates and the minimum daily (24 h) observed temperature was conducted on a commercial crop, soybean, grown over a latitudinal transect of seven mid-western states within a humid region. This analysis was performed using a multi-year (1999–2013), multi-state (Minnesota, Wisconsin, Iowa, Missouri, Arkansas, Mississippi and Louisiana) data series obtained from USDA-NASS pesticide usage surveys in conjunction with state based minimum temperature. If significant correlations were observed between minimum temperature and usage for a given class of pesticide, longer trends (1977–2013) in minimum temperature were analyzed and then used to project near-term (decadal, to 2023) changes in pesticide rates by state.

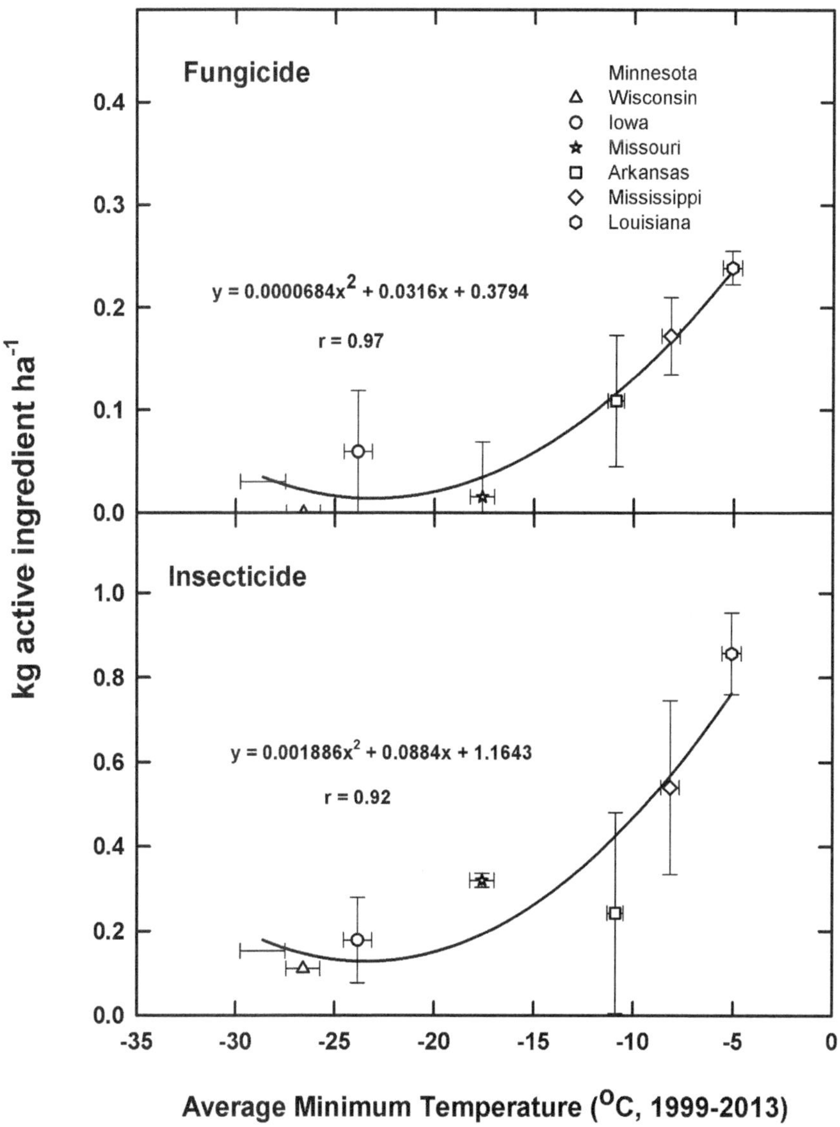

Figure 1. Minimum temperature by state averaged for the period 1999–2013 as a function of insecticide and fungicide usage (kg active ingredient, ai, per hectare) as determined from the National Agricultural Statistical Service survey for the years 1999, 2000, 2001, 2002, 2004, 2006, and 2012) for a north-south transect of seven Midwestern states where commercial soybean is grown. Minimum temperatures were determined as the lowest recorded temperature for a 24 h period during a given calendar year for four locations for a soybean growing area within a state. Lines represent the "best fit" second order polynomial. Bars are ±SE. See Methods for additional details.

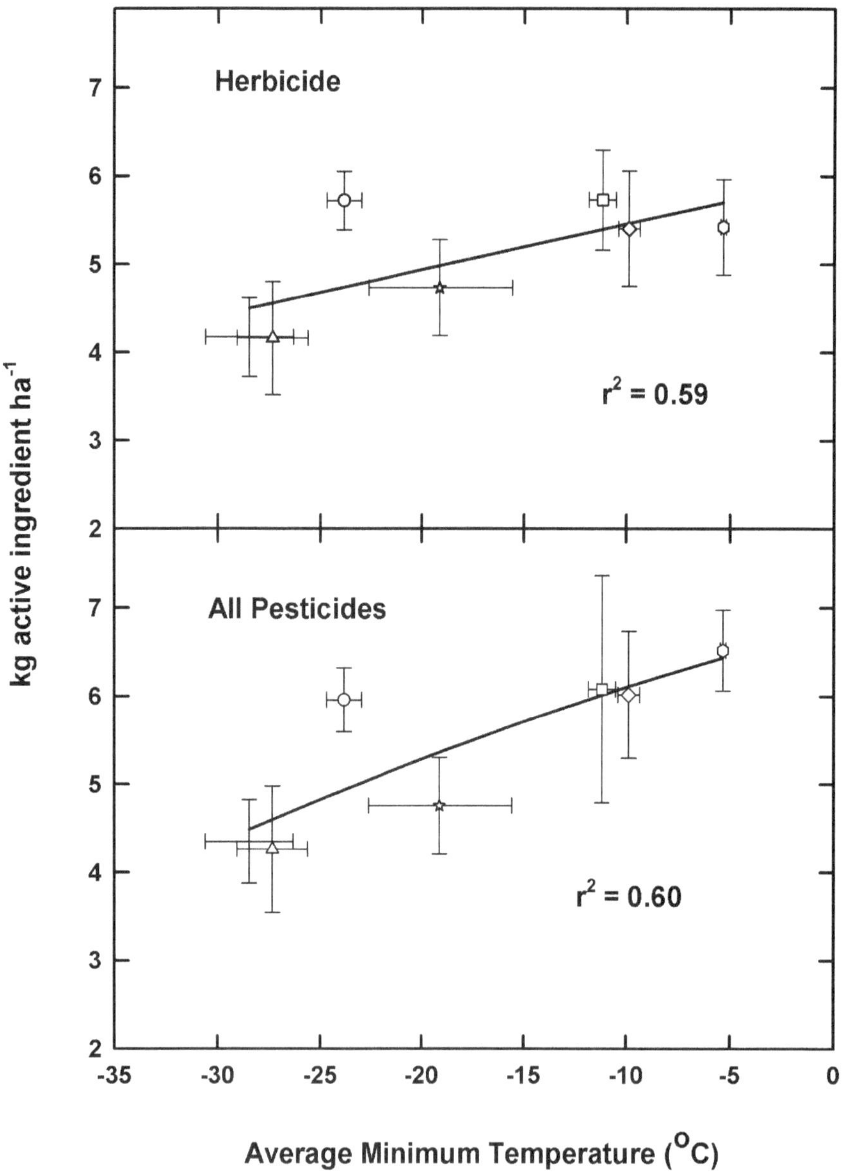

Figure 2. Symbols and regression determined as for figure 1, but for herbicide use and total pesticide (insecticide, fungicide and herbicide) usage for soybean.

Results

For two classes of pesticides, fungicides and insecticides, a second order quadratic function provided the 'best-fit' for minimum temperature and amounts of active ingredient applied for the period 1999–2012. There was a significant correlation (r values of 0.97 and 0.92, **,P<0.01) for fungicide and insecticide soybean applications, respectively over the north-south transect (**Figure 1**). A linear function between minimum daily temperature and herbicide application rates was also observed to be positively correlated and significant (r value of 0.84: *,P<0.05) for herbicide use (**Figure 2**). As most pesticide applications are herbicides, a similar relationship (r = 0.84) was observed for total pesticide use (**Figure 2**).

As the correlation between minimum daily temperature and pesticide usage was significant in all cases, states within the transect were examined to quantify longer term temperature changes (1977 through 2013) in order to assess the increase in minimum winter temperature per decade. Overall, these data indicated that the rise in minimum temperatures was a function of latitude, with states such as Minnesota showing a more rapid increase in minimum temperatures than southern states (e.g., Louisiana) (**Table 1, Figure 3**). However, average soybean yields (2009–2013, MT Ha^{-1}) did not significantly vary as a function of the north-south transect (**Table 1**).

The average decadenal increase in minimum daily temperature was used to estimate potential near term (to 2023) changes in pesticides by category. Because a first order regression was the best-fit for herbicide use, increases in minimum temperature result in a relatively greater increase in herbicide applied for northern compared to southern states. In contrast, because the relationship between insecticide/fungicides and minimum temperatures is best described by a second degree polynomial, temperature changes above a critical threshold, (ca −20°C, **Figure 2**) resulted in

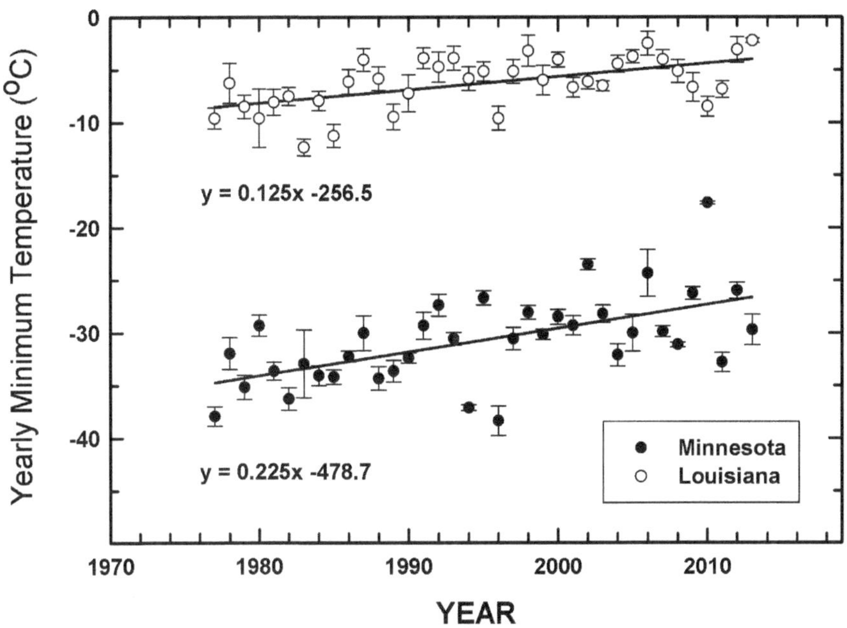

Figure 3. Change in average (±SE) minimum temperature (lowest recorded temperature for a 24 h period during a given calendar year for four locations) for the northernmost and southernmost states (Minnesota, closed circles) and Louisiana (open circles) used in the Midwestern transect. Slope of the regression line indicates the average increase in minimum temperature (°C) per year for each location.

proportionally larger changes in insecticide/fungicide usage. Hence, for those states with minimum temperatures above this threshold, further increases in temperature result in disproportionally larger projected changes in insecticide and fungicide usage (**Table 1**).

Discussion

The need to assess the relative impact of pests on agricultural production in a changing climate has long been recognized [20]; however, few quantitative *in situ* assessments are available. An earlier economic analysis [21] indicated increased pesticide costs as a function of increased rainfall and hotter weather during the 1990s. Application of this analysis to projected climate scenarios to 2090 also indicated general increases in average pesticide cost for a range of crop species [21]. However, this study, while useful, did not distinguish between classes of pesticide, or quantify changes in application amounts as a function of a specific climate variable.

Although agriculture is widespread, there are, in fact, only a handful of crops that are grown over a wide range of climates. For the U.S., soybean, corn and wheat fall into this category; however wheat is grown temporally over a latitudinal gradient as spring (e.g. North Dakota) or winter wheat (e.g. Kansas) with significant differences in pest management. In addition, wheat is not grown widely in the southern states. In contrast, corn is often grown in conjunction with soybean, but data on pesticide use for corn in the southern states (e.g. Mississippi, Arkansas, and Louisiana) is currently not available through NASS. Alternatively, the distribution of soybean in this study encompassed a broad range of minimum daily temperatures (i.e., a ~23°C difference from Minnesota to Louisiana). Such a range would include projections of future surface temperatures associated with anthropogenic climate change, e.g. if minimum daily temperatures in Minnesota became like those of Louisiana [22]; hence soybean was selected for analysis.

There is, as shown here, a significant positive correlation between the range of minimum daily temperatures and pesticide usage for soybean, particularly in regard to fungicides and insecticides. The second order functions for soybean are consistent with insect and fungal biology; i.e., once temperature has reached a critical thermal threshold, it is a significant driver of shifts in insect and pathogen demography [13,23]. For example, in the UK increasing winter temperatures above a biological minimum have been associated with increased northward migration of aphids in Scotland and increased genotype variation among aphid populations [24]. Similar correlations have also been observed between minimum winter temperatures and the southern pine beetle [25].

Patterson et al. [14] provide an extensive list of temperature thresholds in this regard including associated phenological responses and potential shifts in the expansion of insect ranges. These data indicate that for temperate regions, warming would result in increased winter survival as well as an increase in growth and insect fecundity. Similarly, it has been recognized that pathogen development is affected by warm conditions, with mild winters and warmer weather associated with increased outbreaks of powdery mildew; leaf spot disease, leaf rust and rizomania disease [14]. Treharne [26] in turn, has also suggested that increases in minimum daily temperatures would shift the occurrence of plant disease into cooler regions. The relationship between insecticide and fungicide use observed here for soybean in response to rising minimal temperatures is, overall, consistent with the biology of thermal limits and expected shifts in insect and fungal populations in a warmer climate (e.g., [14,23,27]).

For the current study, herbicide applications were a linear function of rising minimum daily temperatures in soybean. In contrast to insect and fungal populations, thermal limits for the occurrence of weeds in soybean were not evident. However, if climate is suitable for crop production, it will also be, de facto, suitable for weed growth.

Although herbicide tolerance in agriculture is increasing [28], for the current study the observed increase in herbicide

Table 1. States included in the latitudinal transect.

State	Soybean (Ha *1000)	Soybean (MT Ha^{-1})	T_{min}/decade	r value	Insecticide (2023)	Fungicide (2023)	Herbicide (2023)
Minnesota	2,513	2.81	+2.25	0.58,**	0	0	+160.2
Wisconsin	628	2.91	+1.99	0.57,***	0	0	+140.0
Iowa	3,741	3.20	+1.58	0.51,**	+20.3	0	+85.1
Missouri	2,247	2.52	+1.56	0.51,**	+35.6	+12.0	+73.5
Arkansas	1,308	2.62	+1.31	0.54,***	+110.0	+36.5	+70.0
Mississippi	806	2.77	+1.31	0.52,***	+117.7	+39.3	+70.3
Louisiana	449	2.81	+1.25	0.55,***	+170.0	+41.1	+58.6

Soybean area is given as hectares (*1000), average yield was determined from 2009–2013 as metric tons (MT) per hectare, the T_{min} is the increase in minimal daily (24 h) temperature (°C) per decade (determined from 1977–2013). The r value is the correlation coefficient of T_{min} over time, with $P<0.05$, *; $P<0.01$, ** and $P<.001$, ***. The values for insecticide, fungicide and herbicide are determined for 2023 (10 year projection) using the functions for figures 1–3 and are shown as changes in grams of active ingredient (ai) per hectare, with a value of zero indicating no change in application rates over 2013 levels.

applications with minimum temperatures did not appear to be related to a greater number of herbicide resistant weeds along the north-south transect (**Table 2**). Nor was there any apparent relationship with photosynthetic pathway (i.e. the number of C_4 weeds is greater in Minnesota than Louisiana). This suggests no difference in recent CO_2 increases on latitudinal weed selection between C_3 and C_4 weedy species *per se.* However the number of perennial or facultative perennial weed species was significantly greater for the southern soybean locations (e.g. Louisiana, **Table 2**). In weed management, it is generally recognized that perennial weeds can be more competitive and more difficult to control. More competitive because of their ability to quickly regenerate from below-ground structures in the spring and more difficult to control chemically because the elimination of perennial weeds requires the killing of all plant parts (including belowground stems, rhizomes, tubers, etc.) that are capable of producing new shoots [29]. Potential increases in perennial weed establishment with increasing minimum temperatures could, in turn, result in concomitant increases in herbicide usage.

The quantitative differences in decadenal increases in minimum temperature reported here (e.g., 2.25 vs. 1.25°C per decade for Minnesota and Louisiana, respectively since 1977) are consistent with the Intergovernmental Panel on Climate Change (IPCC) projections regarding enhanced warming as a function of latitude [19]. Because of the variation in pesticide category by minimal temperature for soybean, near-term projections (2023) suggest a greater relative increase in herbicide use for the upper Midwest, whereas greater insecticide and fungicide use is projected for the southern states for this same period. This may be due, in part, to the southern states having already surpassed the minimum winter temperatures needed to support insect and pathogen populations for soybean in the southern region. However, if minimum temperatures continue to rise, then similar increases in use for all pesticide categories could occur for the upper Midwest.

At present, soybean in the mid-western United States is grown over a ~2100 km north-south transect that provides a wide range of yearly temperatures. Although such a range of temperatures should, ostensibly, increase pest pressures and subsequent crop loss, no production losses are evident in average yield per hectare (e.g., average yields in Minnesota and Wisconsin do not differ from those of Mississippi and Louisiana). However, insecticide, fungicide and herbicide use increase significantly in conjunction with minimum daily temperatures along the latitudinal transect. Assuming that prophylactic use is not widespread in any one region, this increase should reflect increased pest pressures *per se.*

It should be emphasized however, that the biological basis for rising minimum temperatures and changes in pesticide usage are likely to be complex. While there are empirical correlations between population demographics and temperature, temperature by itself does not reflect the complexity of pest population causations in agriculture. For example, pest-pest interactions, insects as disease vectors, temporal asynchrony between insects and host plants, soil management and herbicide efficacy (e.g. precipitation, windspeed), *inter alia* [10,23,30] will also need to be considered. Rising levels of carbon dioxide are also likely to directly alter secondary plant chemistry and alter plant-arthropod interactions depending on plant species and insect group [31]. In addition, the current analysis assumes future monotonic poleward shifts in minimum temperatures influencing pesticide usage; whereas an increase in extreme climatic events may be the norm [32].

However, the ability to include realistic impacts of agricultural pests in future assessments of climate change and food security is imperative. While there are numerous challenges remaining before

Table 2. Top ten lists of most troublesome weed species in soybean for three states along the North-South transect.

State	Common Name	Scientific Name	Photosynthetic Pathway	Growth Habit
Louisiana	Morning glory spp.	*Ipomea spp.*	C_3	Annual or Perennial
	Pigweed spp.	*Amaranthus spp.*	C_4	Annual or Perennial*
	Browntop millet	*Urochloa ramose*	C_4	Perennial
	Nutsedge spp.	*Cyperus spp.*	C_4	Perennial
	Redvine	*Brunnichia ovate*	C_3	Perennial
	Henbit	*Lamium amplexicaule*	C_3	Annual*
	Dayflower	*Commelina spp.*	C_3	Annual or Perennial
	Johnson grass	*Sorghum halapense*	C_4	Perennial
	Prickly sida	*Sida spinosa*	C_3	Annual
	Red rice	*Oryza sativa*	C_3	Annual
Missouri	Waterhemp	*Amaranthus rudis*	C_4	Annual*
	Morning glory	*Ipomea spp.*	C_3	Annual or Perennial
	Palmer amaranth	*Amaranthus palmeri*	C_4	Annual*
	Giant ragweed	*Ambrosia trifida*	C_3	Annual*
	Johnson grass	*Sorghum halapense*	C_4	Perennial
	Asiatic dayflower	*Commelina communis*	C_3	Annual
	Horseweed	*Conyza canadensis*	C_3	Biennial*
	Prickly sida	*Sida spinosa*	C_3	Annual
	Common ragweed	*Ambrosia artemisiifolia*	C_3	Annual*
	Eastern black nightshade	*Solanum ptycanthum*	C_3	Annual or Perennial
Minnesota	Lambsquarters	*Chenopodium album*	C_3	Annual
	Giant foxtail	*Setaria faberii*	C_4	Annual
	Waterhemp	*Amaranthus rudis*	C_4	Annual*
	Wooly cupgrass	*Eriochloa villosa*	C_4	Annual
	Giant ragweed	*Ambrosia trifida*	C_3	Annual*
	Yellow foxtail	*Setaria viridis*	C_4	Annual
	Green foxtail	*Setaria lutescens*	C_4	Annual
	Quackgrass	*Elytrigia repens*	C_3	Perennial
	Common ragweed	*Ambrosia artemisiifolia*	C_3	Annual*
	Wild proso millet	*Panicum miliaceum*	C_4	Annual

Lists are generated by farmer surveys as reported by the Southern Weed Science Society (SWSS) for Louisiana (LA) and Missouri (MO) [37] and the University of Minnesota extension service for Minnesota (MN) (Frank Forcella, USDA-ARS Personal Communication). C_3/C_4 refers to photosynthetic pathway.
*Indicates herbicide resistant populations within that state.

the impacts of anthropogenic climate change on pest pressures and crop yields can be completely quantified, the current study suggests that farm management, through the increased use of chemical applications, may negate pest pressures on crop production that are incurred as a result of changing climate variables such as temperature. As such, pest-induced reductions in agricultural yield of soybean associated with climate change may be difficult to quantify where pesticides are widely used. At present, while the current analysis suggests that minimum daily temperatures could be used as a proxy to understand climate change impacts on pesticide use in soybean; critical information regarding the environmental and economic consequences of such impacts will require further study.

Materials and Methods

Soybean

For all seven states examined along the north-south transect, soybean is considered a major crop (>300,000 ha planted) [34]. Genetically modified (gmo) soybean (e.g., Round-Up Ready^{tm}) was introduced in 1995–1996, mainly as a means for blanket control of post-emergent weeds. Following its introduction it was quickly adopted by growers, with gmo soybean acreage >50% by 1999 [33] (**Table 3**). There is no evidence of different rates of adoption of gmo soybean over the period 1999–2012. For example, by 2001, the amount of soybean area plant to gmo was 63% for both Minnesota and Mississippi (**Table 3**). Since 2005, approximately 90% of all soybeans grown in the U.S. have been herbicide resistant gmo (**Table 3**).

Table 3. Genetically engineered soybean varieties for herbicide tolerance for states included in the north-south transect.

STATE	2000	2001	2002	2003	2004	2005	2006	2007	2008	2009	2010	2011	2012
Minnesota	46	63	71	79	82	83	88	92	91	92	93	95	91
Wisconsin	51	63	78	84	82	84	85	88	90	85	88	91	92
Iowa	59	73	75	84	89	91	91	94	95	94	96	97	93
Missouri	62	69	72	83	87	89	93	91	92	89	94	91	91
Arkansas	43	60	68	84	92	92	92	92	94	94	96	95	94
Mississippi	48	63	80	89	93	96	96	96	97	94	98	98	95
U.S.	54	68	75	81	85	87	89	91	92	91	93	94	93

Values are the percent of all soybeans planted. Data are available at:
www.usda.mannlib.cornell.edu/usda/nass/Acre/2010s/2011/Acre-06-30-2011.pdf#page = 27.

Pesticide Use Rates

State level pesticide usage rate are available for 1999, 2000, 2001, 2002, 2004, 2006 and 2012 from the USDA National Agricultural Statistical Service Survey (NASS) data online (www.quickstats.nass.usda.gov/#5072A3CD-64E7-384A-8FF0-DDC931164C3B) [34] These data include chemical usage by compound for insecticides, fungicides and herbicides for soybean for the states included in the north-south transect. Although data are also available from 1991 through 1998 from this same source, the rapid rates of gmo adoption for herbicide resistant soybean, and the shift in herbicide usage during this time period was thought to, potentially, obfuscate any potential changes related to climate variables such as temperature. This is because application rates often vary by herbicide mode of action; hence, differential herbicide use would result in different amounts of active ingredient (ai) being applied. However, the large-scale shift to gmo soybean provided greater uniformity in herbicide type over a large regional area (**Table 3**). This is an important consideration as herbicides represent the largest component of total pesticide use (e.g. **Figure 2**). However, the temporal range of yearly averages (1999–2013) was considered insufficiently long to evaluate any potential increase in minimum daily temperatures associated with anthropogenic surface warming and pesticide usage (relative to year to year variability) for any one state.

Temperature Data

A software program developed by Texas A&M University [35] was used to identify at least four weather stations located within soybean growing regions of a given state used in the transect. Information on the lowest recorded (24 h) minimum temperature for a given year was determined and averaged for all weather stations within a state for the 1999–2013 period. These same stations and data sets were used to generate minimum daily temperature over a longer time period, from 1977 (i.e. the start of the rapid increase in global land-ocean temperature index) through 2013 [36].

Statistical Analysis

In examining pesticide use it is important to determine whether there may be fixed or random regional effects within the data. For example, are there state specific factors beyond temperature that would result in systematic shifts in pesticide usage? (e.g. one state using irrigation while another state does not, change in rotation rates, etc.) This analysis has been previously done by Chen and McCarl [21] who found with 99% confidence that a random state effect existed for pesticide use in corn, potatoes, soybeans and wheat [21].

A step-wise regression program (ver. 10.0 Statview, Cary NC, USA) was used to determine the best fit regression line for average minimum daily temperature and pesticide application rate (insecticides, fungicides and herbicides) for all states within the latitudinal transect. If the correlation coefficient was determined to be significant for a given pesticide category (e.g. insecticides); then a longer term minimum daily temperature data set (1977–2013) was generated for each state to quantify increases in minimum temperature by decade. The increase in minimum daily temperature by decade was then used to project potential short-term (2014–2023) pesticide use (by category and state) using the quadratic or linear functions provided in figures 1 and 2.

Author Contributions

Conceived and designed the experiments: LHZ. Performed the experiments: LHZ. Analyzed the data: LHZ. Contributed reagents/materials/analysis tools: LHZ. Wrote the paper: LHZ.

References

1. Adams RM, Rosenzweig C, Peart RM, Ritchie JT, McCarl BA, et al. (1990) Global climate change and U.S. agriculture. Nature 345: 219–224.
2. Reilly J, Tubiello F, McCarl B, Abler D, Darwin R, et al. (2003) U.S. Agriculture and climate change. Clim Change 57: 43–69.
3. Howden SM, Soussana JF, Tubiello FN, Cherti N, Dunlop M, et al. (2007) Adapting agriculture to climate change. Proc Nat Acad Sci (USA) 104: 19691–19696.
4. Lobell DB, Burke MB, Tebaldi C, Mastrandrea MD, Falcon WP, et al. (2008) Prioritizing climate change adaptation needs for food security in 2030. Science 319: 607–610.
5. Tao F, Yokozawa M, Xu Y, Hayashi Y, Zhang Z (2006) Climate changes and trends in phenology and yields of field crops in China, 1981–2000. Ag Forest Met 138: 82–92.
6. Schmidhuber J, Tubiello FN (2007) Global food security under climate change. Proc Nat Acad Sci 104: 19703–19708.
7. Schlenker W, Roberts MJ (2009) Nonlinear temperature effects indicate severe damages to US crop yields under climate change. Proc Nat Acad Sci 106: 15594–15598.
8. Ziska LH, Bunce JA, Shimono H, Gealy DR, Baker JT, et al. (2012) Food security and climate change: On the potential to adapt global crop production by active selection to rising atmospheric carbon dioxide. Proc Royal Soc (B) 279: 4097–4105.
9. Lobell DB, Gourdji SM (2012) The influence of climate change on global crop productivity. Plant Physiol 160: 1686–1697.
10. Scherm H (2004) Climate change: can we predict the impacts on plant pathology and pest management? Can J Plant Pathol 26: 267–273.
11. Gregory PJ, Johnson SN, Newton AC, Ingram JSI (2009) Integrating pests and pathogens into the climate change/food security debate. J of Exp Bot 60: 2827–2838.
12. Oerke E-C (2006) Crop losses to pests. J Ag Sci 144: 31–43.
13. Ziska LH, Runion GB (2007) Future Weed, Pest and Disease Problems for Plants. In: Newton PCD, Carran A, Edwards GR, Niklaus PA, Editors. Agroecosystems in a Changing Climate. CRC Press, Boston, MA, 262–279.
14. Patterson DT, Westbrook JK, Joyce RJV, Lingren PD, Rogasik J (1999) Weeds, insects, and diseases. Climatic Change 43: 711–727.
15. Cammel ME, Knight JD (1992) Effects of climatic change on the population dynamics of crop pests. Adv Ecol Res 22: 117–162.
16. Bradley BA, Blumenthal DM, Wilcove DS, Ziska LH (2010) Predicting plant invasions in an era of global change. TREE 25: 310–318.
17. McDonald AJ, Riha S, Ditommaso A, DeGaetano A (2009) Climate change and the geography of Weed damage: Analysis of US maize systems suggests the potential for significant range transformations. Ag Ecosys Environ 130: 131–140.
18. Vose RS, Easterling DR, Gleason B (2005) Maximum and minimum temperature trends for the globe: An update through 2004. Geophys Res Let 32: 1–5.
19. Intergovernmental Panel on Climate Change (2007) Summary for policymakers. In: Climate Change 2007: The Physical Science Basis. Contribution of Working Group I to the Fourth Assessment Report of the Intergovernmental Panel on Climate Change (Solomon, S. et al., eds), Cambridge University Press.
20. Lough JM, Wigley TML, Palutikof JP (1983) Climate and climate impact scenarios for Europe in a warmer world. J. Clim. Appl. Meteorol. 22: 1673–1684.
21. Chen C-C, McCarl BA (2001) An investigation of the relationship between pesticide usage and Climate change. Climatic Change 50: 475–487.
22. Intergovernmental Panel on Climate Change (2012) Managing the risks of extreme events and disasters to advance climate change adaptation. A special report of Working Groups I and II of the Intergovernmental Panel on Climate Change. Cambridge University Press, Cambridge.
23. Fuhrer J (2003) Agroecosystem responses to combinations of elevated CO_2, ozone, and global climate change. Ag Ecosys Environ 97: 1–20.
24. Malloch G, Highet F, Kasprowicz L, Pickup J, Beilson R, et al. (2006) Microsatellite marker analysis of peach-potato aphids (*Myzus persicae*, Homoptera: Aphididae) from Scottish suction traps. Bull Entomol Res 96: 573–582.
25. Tran JK, Yliqja T, Billings RF, Regniere J, Ayres MP (2007) Impact of minimum winter temperatures on the population dynamics of *Dendroctonus frontalis*. Ecol Appl 17: 882–889.
26. Treharne K (1989) The implications of the 'greenhouse effect' for fertilizers and agrochemicals. In: Bennet RC editor, Ministry of Agriculture, Fisheries and Food, London, UK 67–78.
27. Porter JH, Parry ML, Carter TR (1991) The potential effects of climatic change on agricultural insect pests. Ag For Meteor 57: 221–240.
28. Delye CD, Jasieniuk M, Le Core V (2013) Deciphering the evolution of herbicide resistance in weeds. Trends in Gen 29: 649–655.
29. Zimdahl RL (2007) Weed-crop competition: a review. John Wiley & Sons, New York, NY, 220 pages.
30. Ziska LH (2010) Global Climate Change and Carbon Dioxide: Assessing Weed Biology and Management. In: Rosenzweig C, Hillel D, Ediotrs. Handbook of Climate Change and Agro-ecosystems: Impacts, adaptation and mitigation. World Scientific Publishing, Hackensack, NJ, 191–208.
31. Robinson EA, Ryan GD, Newman JA (2012) A meta-analytical review of the effects of elevated CO_2 on plant-arthropod interactions highlights the importance of interacting environmental and biological variables. New Phytologist 194: 321–336.
32. Hansen J, Sato M, Ruedy R (2012). Perception of climate change. Proc Nat Acad Sci 109: E2415–E2423.
33. Carpenter J, Gianessi L (1999) Herbicide tolerant soybeans: Why growers are adopting Round-up Ready varieties. AgBioForum 2: 65–72.
34. National Agricultural Statistical Service, (NASS) (www.quickstats.nass.usda.gov/#5072A3CD-64E7-384A-8FF0-DDC931164C3B) (Accessed, January, 2014).
35. Yang Y, Wilson LT, Wang J (2010) Development of an Automated Climatic Data Scraping, Filtering and Display System. Comp Elect Ag 71: 77–87.
36. Hansen J, Ruedy R, Glascoe J, Sato M (1999). GISS analysis of surface temperature change. J Geophysical Res: Atmos (1984–2012), 104: 30997–31022.
37. Southern Weed Science Society (SWSS) (2013) SWSS Proceedings. 66: 275–287.

15

Anilofos Tolerance and Its Mineralization by the Cyanobacterium *Synechocystis* sp. Strain PUPCCC 64

D. P. Singh*, J. I. S. Khattar, Mandeep Kaur, Gurdeep Kaur, Meenu Gupta, Yadvinder Singh

Department of Botany, Punjabi University, Patiala, Punjab, India

Abstract

This study deals with anilofos tolerance and its mineralization by the common rice field cyanobacterium *Synechocystis* sp. strain PUPCCC 64. The organism tolerated anilofos up to 25 mg L^{-1}. The herbicide caused inhibitory effects on photosynthetic pigments of the test organism in a dose-dependent manner. The organism exhibited 60, 89, 96, 85 and 79% decrease in chlorophyll a, carotenoids, phycocyanin, allophycocyanin and phycoerythrin, respectively, in 20 mg L^{-1} anilofos on day six. Activities of superoxide dismutase, catalase and peroxidase increased by 1.04 to 1.80 times over control cultures in presence of 20 mg L^{-1} anilofos. Glutathione content decreased by 26% while proline content was unaffected by 20 mg L^{-1} anilofos. The test organism showed intracellular uptake and metabolized the herbicide. Uptake of herbicide by test organism was fast during initial six hours followed by slow uptake until 120 hours. The organism exhibited maximum anilofos removal at 100 mg protein L^{-1}, pH 8.0 and 30°C. Its growth in phosphate deficient basal medium in the presence of anilofos (2.5 mg L^{-1}) indicated that herbicide was used by the strain PUPCCC 64 as a source of phosphate.

Editor: Paul Jaak Janssen, Belgian Nuclear Research Centre SCK/CEN, Belgium

Funding: This work was supported by Council of Scientific and Industrial Research (CSIR), New Delhi, India (scheme no. (38)1152/07/EMR-II). The funders had no role in designing the experiment, collection of data, decision to publish and preparation of this manuscript.

Competing Interests: The authors have declared that no competing interests exist.

* E-mail: dp.khokhar@rediffmail.com

Introduction

Weeds are a nuisance in field crops as they compete for soil and water resources, reduce crop quality, interfere with irrigation and harvesting operations, and reduce land value. The sole price of agricultural products gets lowered when weed seeds get mixed with harvested crops, which require extra cleaning operations. Besides this, weeds also serve as alternate host to many pathogens, insects and nematodes, resulting in increased costs of crop protection operations [1]. It is estimated that weeds account for 45% of the total annual loss of agricultural production from various pests in India [2].

Large scale use of herbicides that selectively kill weeds is an integral component of modern agriculture since these chemicals reduce labour and machine requirements. Discovery of 2,4-dichlorophenoxy acetic acid after second world war revolutionized agriculture in developed countries [1]. Since then, the use of herbicides has increased enormously across the world. In India, 240 pesticides are registered for use, out of which 51 are herbicides [3].

The herbicide use in India increased from 15 to 11000 tonnes during 1970 to 2002 [1]. The average consumption of herbicide in India is 60 g per hectare which accounts for 13% of the total global consumption (46% of pesticides). Punjab and Haryana are the leading states in India in terms of herbicide consumption [1]. Although herbicides helped the farmer community by increasing the crop yields, their use also resulted in a number of problems i.e. contamination of soil and water [4,5,6,7,8], negative effects on non-target organisms, such invertebrates [9,10,11,12], fish [13,14], amphibians [15] and microorganisms, including cyanobacteria [16,17,18,19].

Cyanobacteria are an integral component of paddy field ecosystems where they play an important role in building of and maintaining soil fertility [20,21,22]. In the soil, they are reported to improve organic content, water holding capacity, nitrogen status, and release vitamins, plant stimulating hormones, extra cellular polysaccharides and also solubilize phosphates [23,24]. More than half of the total nitrogen used in paddy crop derives from the native soil nitrogen pool which is maintained through biological nitrogen fixation by both hetero- and autotrophs in soil [25]. Thus, use of cyanobacterial biofertilizer is considered to be a good management of paddy fields since their use not only increases fertility of the soil but is also eco-friendly. The utilization of cyanobacterial biofertilizer in paddy fields requires that strains be tolerant or able to degrade a variety of routinely used agrochemicals, including herbicides.

The earlier work on the influence of herbicide on cyanobacteria has been extensively reviewed [26,27]. Herbicides cause deleterious effects on cyanobacteria by influencing their growth, photosynthesis, nitrogen fixation, biochemical composition and metabolic activities [28–30]. Studies on the cyanobacterial tolerance to herbicides and the degradation of these herbicides are available in literature [31–33].

Anilofos (S-[2-[(4-chlorophenyl) (1-methylethyl) amino]-2-oxoethyl] O O-dimethyl phosphorodithioate) is the active chemical of the commercial grade herbicide ANILOGUARD (Gharda Chemical Limited, Ratangiri, Maharashtra, India). It belongs to the organophosphate group and is widely used as a pre-emergence and early post-emergence herbicide for the control of annual grasses, sedges and some broad-leaved weeds in transplanted and direct seeded rice crops. The herbicide is mixed with dry sand and is spread in the standing water after paddy plantation at the rate of 300 to 450 g active ingredient per hectare. Anilofos has 5 to 25 days (d) half-life in water flooded soil [34].

We have previously shown that *Synechocystis* sp. strain PUPCCC 64 is able to degrade chlorpyrifos [19]. In the present investigation we aim to understand the mechanism of anilofos tolerance and its degradation potential by this cyanobacterium strain so that it might be used in Indian biofertilizer technology programmes.

Materials and Methods

Microorganism and Culture Conditions

The cyanobacterium *Synechocystis* sp. strain PUPCCC 64, was isolated from rice fields of village Derabassi (30° 58′ 72″ N; 76° 8′ 28″ E) of district Mohali, Punjab [19]. It was grown in slightly modified Chu-10 medium [35] supplemented with micronutrients [36]. One liter nutrient medium contained 0.232 g $CaCl_2.2H_2O$, 0.025 g $MgSO_4.7H_2O$, 0.02 g Na_2CO_3, 0.044 g $Na_2SiO_3.5H_2O$, 0.01 g K_2HPO_4, 0.0035 g of ferric citrate and citric acid each and 1 g, KNO_3. Cultures were maintained in a room at 28±2°C and illuminated 14 h daily with a light intensity of 44.5 μE (μmoles of photons m^{-2} s^{-1}). Exponentially growing cultures (8 d old) were used throughout our study and each experiment was repeated three times.

Growth of *Synechocystis* sp. Strain PUPCCC 64 in Anilofos

The tolerance limit of *Synechocystis* sp. against herbicide was determined by growing it in graded concentrations of anilofos (5–30 mg L^{-1}). These concentrations were prepared in 250 mL Erlenmeyer flasks containing 100 mL Chu-10 medium from the stock solution of commercial grade ANILOGUARD (30% Emulsifiable Concentrate). After two washings with sterilized double distilled water, exponentially growing stock cultures were inoculated in anilofos containing media to get the initial absorbance of 0.1 at 680 nm. At regular intervals of 2 d, extending up to 12 d, 10 mL samples were withdrawn and growth was measured spectrophotometrically as the increase in absorbance of the cultures with a Thermo Spectronic model 20D+ (Thermo Spectronic, Rochester, NY).

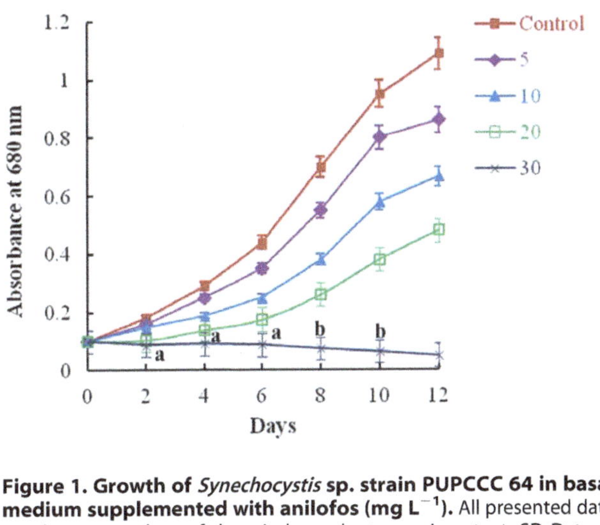

Figure 1. Growth of *Synechocystis* sp. strain PUPCCC 64 in basal medium supplemented with anilofos (mg L^{-1}). All presented data are the mean values of three independent experiments ± SD Data at different time intervals with same lower case letters are not significantly different from each other at the 95% confidence level ($p<0.05$).

Biochemical Analysis

Photosynthetic pigments. Cyanobacterial suspensions (10 mL) were centrifuged at 5000 rpm, washed thrice with double distilled water and a same volume of acetone (80%) was added to the cell pellet. The mixture was shaken vigorously and kept at room temperature for 12 h to extract acetone soluble photosynthetic pigments. The extracts were centrifuged at 5000 rpm, attained the same volume of supernatant by adding sufficient amount of acetone and absorbance was checked at 660, 645 and 450 nm. The amount of chlorophyll a (Chl a) was calculated according to Holm [37] whereas total carotenoids (Car) were quantified as per Myers and Kratz [38].

Phycobiliproteins (PBS) were extracted in phosphate buffer (50 mM, pH 7.0) by freeze-thawing as described elsewhere [39]. Pellet of cyanobacterial cells obtained after centrifugation at 5000 rpm were washed with double distilled water and suspended in the same buffer containing lysozyme (1 g L^{-1}). The mixture was incubated at 37°C in a water bath for 1 h with continuous stirring. Obtained spheroplasts were gently washed by centrifugation at 3000 rpm and resuspended in phosphate buffer. The resultant solution was then subjected to 10–12 freeze-thawing cycles until all the pigments were released from the cells. After centrifugation at 5000 rpm, the absorbance of the supernatant was recorded at 565, 615 and 652 nm.

Antioxidant enzymes. Superoxide dismutase (SOD) activity was measured as inhibition of reduction of nitroblue tetrazolium chloride (NBT) photochemically [40]. Thick cell suspension, obtained by centrifugation at 5000 rpm, was washed twice with double distilled water, resuspended in a small volume of 100 mM phosphate buffer (pH 7.8) and disintegrated with a sonicator (Soniprep 150, Sanyo, USA), by giving 5 pulses (5 micron amplitude) each of 1 min, with 30 s intervals. Supernatant obtained after centrifugation (15,000 rpm for 20 min, at 4°C) was used as an enzyme extract. The assay mixture for the estimation of SOD was prepared by mixing together 27 mL of 100 mM sodium phosphate buffer (pH 7.8), 1.5 mL of methionine (3%), 1 mL of NBT (0.014%), 1 mL of Na_2CO_3 (23.4%) and 1.5 mL of EDTA (2 mM). To 2.7 mL of reaction mixture, 0.1 mL enzyme extract and 0.2 mL of riboflavin (0.015%) were added. After mixing, test tubes were illuminated for 5 min under fluorescent lights (44.5 μE). The reaction was terminated by transferring the reaction mixture to the dark. Tubes containing reaction mixture with enzyme extract served as blank and were kept in the dark. Control tubes without enzyme extract served as controls and were kept in the light. The absorbance of solution was measured at 560 nm. One unit of SOD activity is the amount of enzyme required to inhibit the reduction of NBT in light by 50%.

Peroxidase (POD) activity was measured according to the method of Gahagen et al. [41]. The reaction mixture contained 2.0 mL double distilled water, 0.35 mL phosphate buffer (100 mM, pH 6.8), 0.2 mL of 0.5% H_2O_2, 0.3 mL of 5% pyrogallol and 0.15 mL of enzyme extract. The increase in absorbance at 420 nm was measured up to 5 min after mixing the contents by inversion.

Catalase (CAT) activity was determined by measuring O_2 release from the dissociation of H_2O_2 in darkness [42]. Three mL of 50 mM phosphate buffer (pH 7.0) containing 50 mM H_2O_2 were directly added to 1 mL cell suspension in the reaction vessel kept in dark. Oxygen released due to enzymatic dissociation of H_2O_2 was measured for 5 min using oxygen analyzer (YSI, USA). One unit of catalase is the amount of enzyme required to produce one nmol O_2 min^{-1}.

Glutathione and proline contents. Glutathione content (GSH) was determined according to the modified Ellman's method

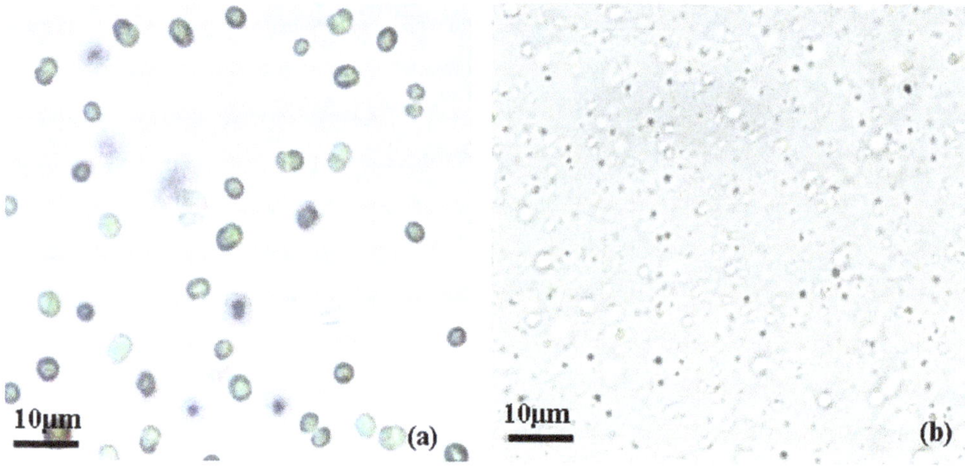

Figure 2. Photomicrograph of *Synechocystis* sp. strain PUPCCC 64 culture on day 4. (a): control; (b): anilofos (30 mg L^{-1}).

[43]. Ten mL of cell suspension was centrifuged (5000 rpm for 10 min at 28°C), washed twice with double distilled water and sonicated in 5 mL of 0.2 M phosphate buffer (pH 8.0) for 30 min. The supernatant obtained after centrifugation at 20,000 rpm for 20 min and 4°C was used for estimation of GSH. The reaction mixture was prepared by adding 0.2 mL of enzyme extract, 3 mL of phosphate buffer (pH 8.0) and 0.5 mL of 5–5′-dithiobis-2-nitrobenzoic acid. The development of colour intensity was measured at 412 nm. GSH was expressed in μg mg^{-1} protein. A standard curve was prepared by using GSH.

Proline content was measured according to the method of Bates et al. [44]. Twenty mL of culture was centrifuged at 5000 rpm, washed with double distilled water, resuspended in 10 mL of 3% aqueous sulphosalicylic acid, sonicated for 3–5 min and centrifuged at 20,000 rpm for 20 min at 4°C. Two mL of extract were taken in a test tube and 2 mL each of glacial acetic acid and acidic ninhydrin were added. The mixture was heated in a water bath at 100°C for 1 h and the reaction was stopped by shifting test tubes to an ice-bath. Then, 4 mL toluene was added to the reaction mixture. After stirring for 20–30 s, the toluene layer was separated and warmed to room temperature. Absorbance was measured at 520 nm. A standard curve was prepared using proline.

Anilofos Removal and Mineralization by *Synechocystis* sp. PUPCCC 64

Uptake experiments were conducted in 250 mL Erlenmeyer flasks containing 100 mL medium supplemented with 10 mg L^{-1} anilofos (concentration inhibiting 20% growth). Exponential growing cultures were harvested, washed twice with double distilled water and inoculated to obtain an initial absorbance of 0.5 at 680 nm. At regular intervals, 10 mL of cyanobacterial cells was withdrawn and centrifuged. For the estimation of intracellular herbicide and its degradation product(s) cells treated with anilofos for 4 days were harvested and sonicated in Chu-10 medium as described above in section antioxidant enzymes. Anilofos and its product(s) were extracted with hexane dichloromethane partitioned thrice [19]. Anilofos removal was optimized by studying its removal by varying cultural conditions i.e. pH (5.0–9.0) of medium, biomass load (25–200 mg L^{-1} protein), and temperature (20–40°C). At a time, one parameter was varied keeping others (optimized) constant.

Quantification of anilofos. Anilofos was quantified by HPLC (Waters, USA) injecting 20 μL of extracted sample of herbicide in acetonitrile. HPLC operating parameters were: column: C-18; detector: Ultra Violet-Visible; running phase: 70% acetonitrile; flow rate 0.5 mL min^{-1}; running time 7 min.

Anilofos as a Source of Phosphate

Three 250 ml Erlenmeyer flasks, each containing 100 mL Chu-10 medium with phosphate (0.01 g L^{-1}), minus phosphate, and minus phosphate plus 2.5 mg L^{-1} anilofos were inoculated with washed cyanobacterial suspension to attain the initial absorbance of 0.5 at 680 nm. At regular interval of 2 d extending up to 12 d, 10 mL of cultures were withdrawn and absorbance was measured at 680 nm.

Protein content was determined according to Lowry et al. [45].

Chemicals

All chemicals used in media preparation and enzymatic assays were obtained from Merck, India. Commercial grade anilofos (ANILOGUARD 30% w/v) was from Gharda Chemicals Limited, Ratangiri, Maharashtra, India.

Statistical Analysis

All the data are the average of three independent experiments ± Standard Deviation (SD). Data were statistically analyzed by applying one way analysis of variance and Tukey's honest significance difference test. All statistical analyses were tested against the probability value at 95% confidence level ($p<0.05$) using GraphPad Prism 6.0 version 6.0 (www.graphpad.com).

Results and Discussion

Growth of *Synechocystis* sp. Strain PUPCCC 64 in the Presence of Anilofos

The growth of cyanobacterial cells in graded concentrations (5–30 mg L^{-1}) of anilofos is shown in Fig. 1. The growth of control cultures increased from 0.1 on zero day to 1.09 on 12 day. In the presence of anilofos, a concentration-dependent decreased growth was observed. The cyanobacterium survived in anilofos up to 20 mg L^{-1} and did not grow in 30 mg L^{-1} anilofos. In this concentration of anilofos, the cells lysed, pigments were released in the medium and cultures became colourless (Fig. 2). On day 12,

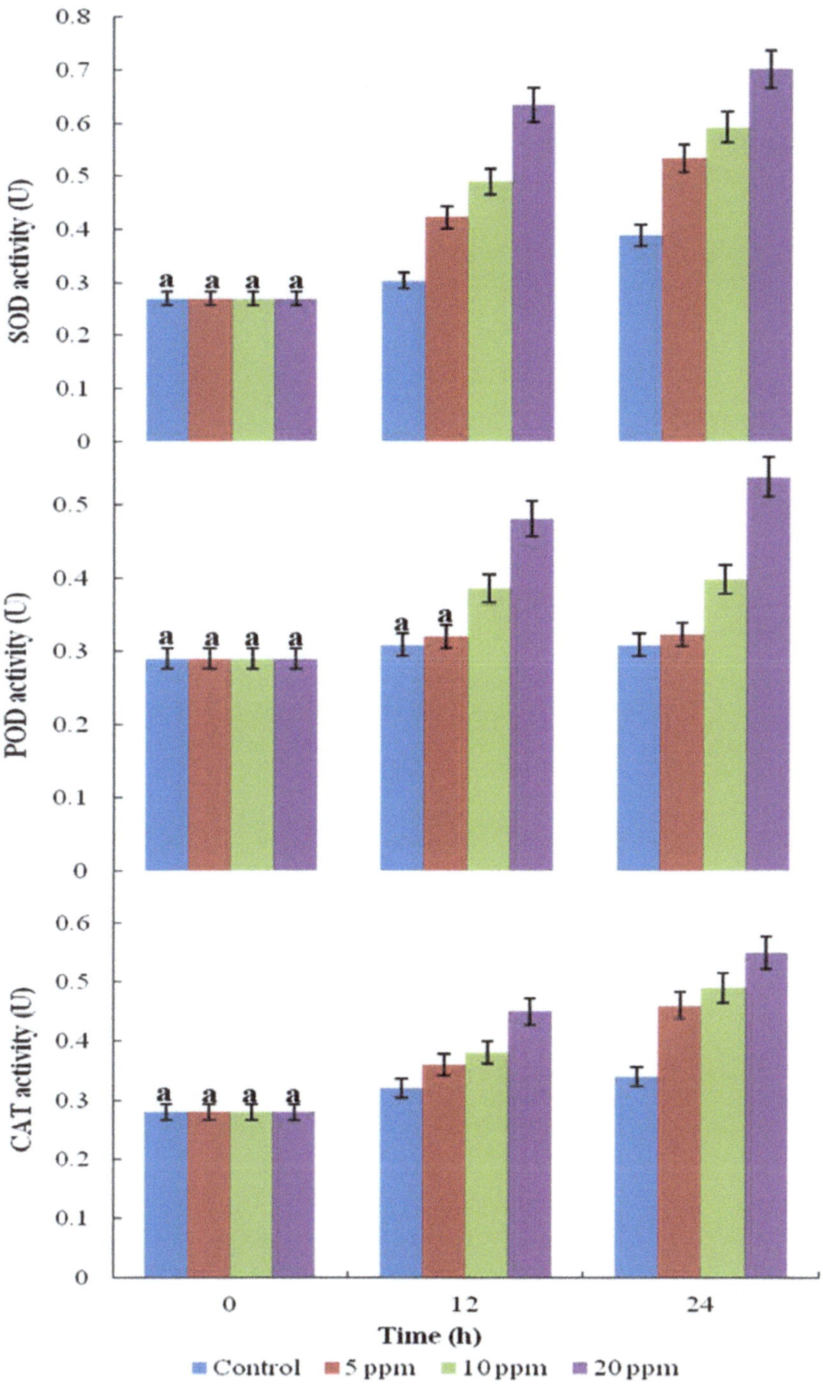

Figure 3. Activity of SOD, POD and CAT of *Synechocystis* sp. strain PUPCCC 64 in presence of anilofos (mg L^{-1}). All presented data are the mean values of three independent experiments ± SD Data at different time intervals with same lower case letters are not significantly different from each other at the 95% confidence level ($p<0.05$).

the microorganism exhibited 61, 44 and 25% growth as compared to control cultures in 5, 10 and 20 mg L^{-1} of anilofos, respectively. Growth inhibition by anilofos in *Anabaena torulosa* [46], butachlor, bensulfuron-methyl in *Nostoc* sp. [47,48], glyphosate in *Anabaena* sp., *Leptolyngbya boryana*, *Nostoc punctiforme*, and *Microcystis aeruginosa* [31,49], and, molinate and bentazon in *Anabaena cylindrica* and *Nostoc muscorum* [29,50,51] has recently been reported. These herbicides affect photosynthesis (bentazon), cell division and inhibition of synthesis of long chain fatty acids (anilofos, butachlor and molinate), the shikimate pathway (glyphosate), and acetolactate synthase (bensulfuron-methyl) of target organisms [52].

Photosynthetic Pigments

The presence of anilofos caused an inhibitory effect on the growth of *Synechocystis* sp. strain PUPCCC 64 in a concentration-dependent manner. On day six, the microorganism exhibited 21,

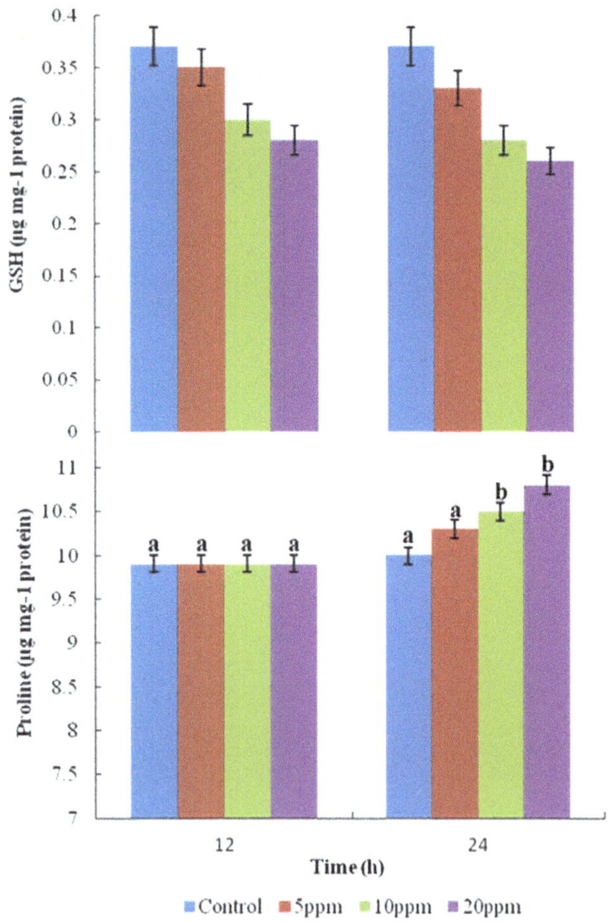

Figure 4. Effect of anilofos (mg L^{-1}) on GSH and Proline content of *Synechocystis* sp. strain PUPCCC 64. All presented data are the mean values of three independent experiments ± SD Data at different time intervals with same lower case letters are not significantly different from each other at the 95% confidence level ($p<0.05$).

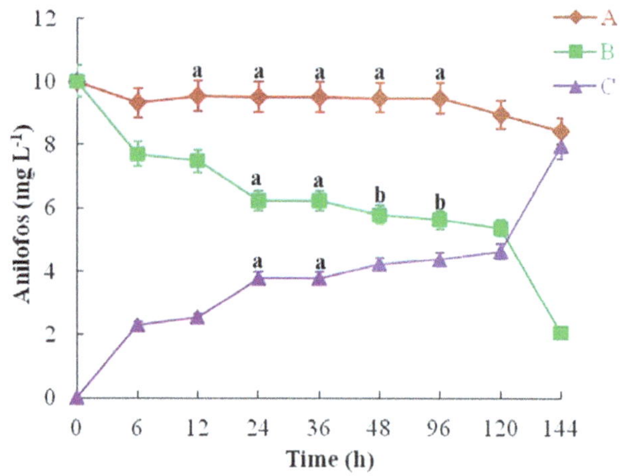

Figure 5. Uptake of anilofos by *Synechocystis* sp. strain PUPCCC 64. (A) control (without PUPCCC 64 cells); (B) anilofos in medium; (C) anilofos uptake All presented data are the mean values of three independent experiments ± SD Data at different time intervals with same lower case letters are not significantly different from each other at the 95% confidence level ($p<0.05$).

41 and 60% decrease in chlorophyll a (Chl a) content over control cultures with 5, 10 and 20 mg L^{-1} anilofos, respectively, whereas carotenoid (Car) content decreased by 32, 64 and 89%, respectively (Table 1). Phycocyanin (PC) content decreased by 55, 85 and 96% compared to control cultures with 5, 10 and 20 mg L^{-1} of anilofos, respectively, on 6 d. Allophycocyanin (APC) content decreased by 25, 75 and 85% while phycoerythrin (PE) contents were decreased by 47, 50 and 79% with 5, 10 and 20 mg L^{-1} of anilofos, respectively, on day six. Decreasing order of production of photosynthetic pigments by anilofos on day six was: PC>Car>APC>PE>Chl a. It is interesting to note that with 5 mg L^{-1} anilofos, which is 10 times the field concentration, levels of photosynthetic pigments decreased in the range of 21–55% (Table 1). Growth inhibition of PUPCCC 64 strain with increase of anilofos concentration can be explained on the basis of damaging effects of herbicide on photosynthetic pigments. Butachlor, bensulfuron-methyl and dimethoate inhibited photosynthetic pigments of *Nostoc* [48]. Treatment with bentazone (2 mM) for 72 h decreased total phycobiliproteins content of *Anabaena cylindrica* by 58% whereas molinate at highest tested concentration (2 mM) completely suppressed phycobiliproteins after 24 h treatment [29]. Anilofos inhibited all photosynthetic pigments of *Anabaena torulosa* [46] and *Oscillatoria simplicissima* [33]. Herbicides such as 3-(3,4-dichlorophenyl)-1,1-dimethyl urea (DCMU), 2-tert-butylamino-4-ethylamino-6-methylthio-1,3,5-triazine (terbutryn) and 4-hydroxy-3,5-diiodobenzonitrile (ioxynil) have been shown to inhibit PSII of photosynthetic organisms [53–56]. It has been shown that herbicides bind to the Q_B site of D1 subunit of PSII and block electron transfer from Q_A to Q_B [53,55–56]. This has been evidenced by studying the crystal structure of PSII of *Thermosynechococcus elongatus* [56] and by causing mutations in the proteins binding to PSII subunits of cyanobacteria [54–55]. Mutational studies at the acceptor side of PSII have identified the role of D1-Ser264 in herbicide binding especially triazines [57]. Herbicides are mainly responsible for inhibiting PS II but Anilofos has also been reported to cause inhibition in cell division and synthesis of long chain fatty acids [52]. The present study suggests that anilofos caused inhibition of synthesis/degradation of photosynthetic pigments. To elucidate the exact mechanism of anilofos inhibition in *Synechocystis* sp. strain PUPCCC 64, additional studies are needed.

Antioxidant Enzymes

Toxicity of herbicides may lead to the generation of free radicals and cyanobacteria may respond to this stress by inducing antioxidant enzymatic defense mechanism. Superoxide dismutase (SOD) is involved in the neutralization of highly reactive oxygen species (ROS) such as superoxide radicals and singlet oxygen resulting in the generation of the lesser toxic hydrogen peroxide (H_2O_2) that is still harmful to cells [58] requiring removal by catalase (CAT) and/or peroxidase (POD) enzymes. Thus, the activity of these antioxidant enzymes in the PUPCCC 64 strain grown in anilofos for 24 h was studied.

SOD, POD and CAT activities were stimulated significantly with higher anilofos concentration (Fig. 3). The cyanobacterium showed 137, 152 and 180% activities of SOD in 5, 10 and 20 mg L^{-1} of anilofos, respectively, at 24 h compared to control without herbicide (Fig 3a). POD activity was increased by 104, 128 and 174% (Fig. 3b) whereas activity of CAT increased by 109, 117 and 131% (Fig 3c) in 5, 10 and 20 mg L^{-1} anilofos compared to controls, respectively, at 24 h. The results showed that these enzymes played a key role in overcoming the oxidative stress caused by anilofos. The increase or decrease in level of antioxidant

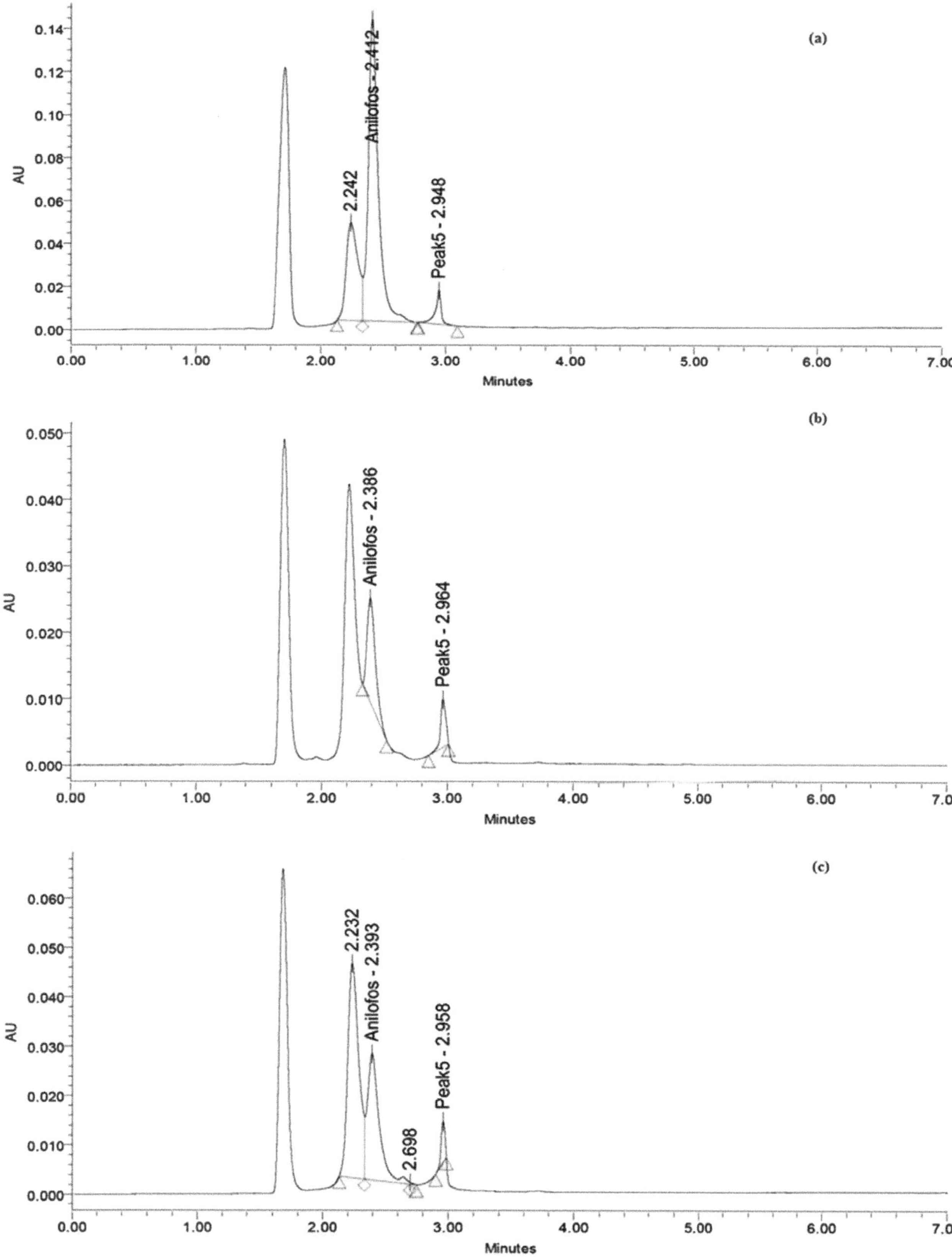

Figure 6. HPLC chromatogram showing the peaks of (a) standard anilofos, (b) cell extract and (c) supernatant. First large peak in the HPLC chromatogram represents solvent used to dissolve extracted anilofos.

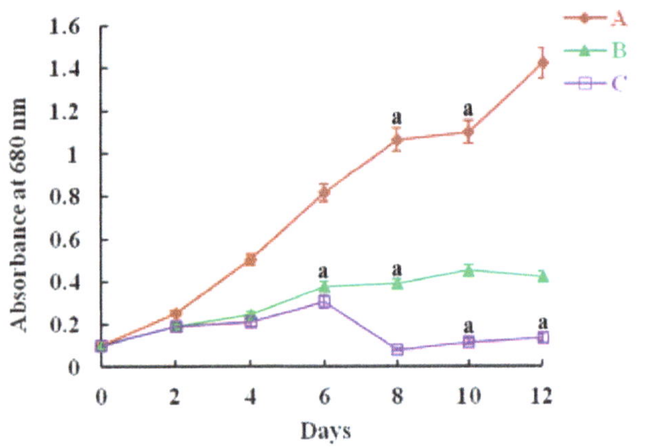

Figure 7. Growth of *Synechocystis* sp. strain PUPCCC 64 in basal medium (A) with 0.01 g K_2HPO_4 L^{-1}, (B) with 2.5 mg anilofos L^{-1} without phosphate and (C) without phosphate. All presented data are the mean values of three independent experiments ± SD Data at different time intervals with same lower case letters are not significantly different from each other at the 95% confidence level ($p<0.05$).

Table 1. Amount of photosynthetic pigments (mg L^{-1}) of *Synechocystis* sp. strain PUPCCC 64 in anilofos-grown cultures at day 6.

Pigment	Anilofos (mg L^{-1})			
	Control	5.0	10	20
Chl a	1.2±0.05	0.94±0.01	0.71±0.01	0.48±0.01
Car	5.6±0.25	3.8±0.04	2.0±0.02	0.6±0.01
PC	20±0.2	9.0±0.15	3.0±0.01	0.8±0.01
APC	4.0±0.05	3.0±0.01	1.0±0.01	0.6±0.01
PE	16±0.1	8.5±0.05	8.0±0.02	3.3±0.02

Data in rows are significantly different from each other at the 95% confidence level ($p<0.05$).
Data are means of three observations ± SD.
Chl a: Chlorophyll a; Car: Carotenoid; PC: Phycocyanin; APC: Allophycocyanin; PE: Phycoerythrin.

enzyme in cyanobacteria depends upon the type of organism or the nature of herbicides. The increase of antioxidant enzymes by anilofos (1.5–20 mg L^{-1}) stress has been reported in cyanobacteria *A. torulosa* [46] and *O. simplicissima* [33]. Other herbicides such as glufosinate [59] and paraquat [60], at 0.5 and 0.75 μM, respectively, increased SOD and CAT activities by 3–4 times over control cultures in the unicellular green alga *Chlorella vulgaris*. Molinate decreased antioxidant enzymes (SOD, CAT and POD) and increased levels of lipid peroxidation leading to electrolyte leakage in the diazotrophic cyanobacterium *Nostoc muscorum* [51]. The increased activities of SOD, POD and CAT in the PUPCCC 64 strain, on the other hand, indicate that anilofos stress may have stimulated the generation of reactive oxygen species which were scavenged by the elevated levels of these enzymes and helped the organism to tolerate herbicide stress.

Glutathione and Proline Contents

Low molecular weight compounds like GSH and proline also play a key role to counter abiotic stress in microorganisms and plants. Cellular GSH content of the PUPCCC 64 strain significantly decreased by 19 and 24% with 10 and 20 mg L^{-1} of anilofos, respectively. (Fig. 4a). A similar observation was reported in a rice field cyanobacterium *Nostoc muscorum* for which the GSH content decreased with an increase of 0.75 to 2.0 mM of molinate [51]. Glutathione is known to help in lowering the levels of H_2O_2 through the action of glutathione peroxidase [58]. Lowering of glutathione in strain PUPCCC 64 in presence of anilofos indicates that it may be helping in reducing the levels of H_2O_2.

Proline accumulation under stress conditions is either due to its increased synthesis or decrease in degradation [61,62]. Proline content did not significantly increase with 5 mg L^{-1} anilofos and only increased by 12% with 20 mg L^{-1} anilofos. (Fig. 4b). Thus, proline levels in strain PUPCCC 64 appeared to be unrelated to the organism's herbicide tolerance. Similar observations were made for other cyanobacteria [63,64] and it is unlikely that, of the two non-enzymatic defence systems, proline levels are related to herbicide-induced stress.

Anilofos Removal

Anilofos removal by the PUPCCC 64 strain was studied to test whether the microorganism was able to take up and degrade the herbicide intracellularly. The organism showed continuous uptake of herbicide intracellularly up to six days (Fig. 5). The results revealed that the removal of anilofos was rapid during the initial hours (2.3 mg L^{-1} in 6 h) and afterwards removal slowed down up to 4 d. The cyanobacterium removed 4.2 and 7.9 mg L^{-1} anilofos in 4 and 6 d, respectively. It appears that anilofos is initially adsorbed on the surface of biomass followed by slow intracellular uptake. Orús and Marco [65] reported removal of trichlorfon from culture media supplemented with 300 mg L^{-1} of insecticide by *Gloeothece* PCC 6501, *Plectonema calothricoides*, *Anabaena* PCC 7119, *Nostoc* UAM 2005 and *Chlorogloeopsis* PCC 6912. Rapid uptake of the herbicides such as atrazine and terbutryn has been reported for the cyanobacterium *Synechococcus elongatus* and green alga *Chlorella vulgaris* [66]. Fast removal of chlorpyrifos in the initial 12 hours by *Synechocystis* sp. has also been reported [19]. The only slight decrease in anilofos concentration observed on day six in the control flask without *Synechocystis* sp. PUPCCC 64 cells (Fig. 5) indicated that anilofos is relatively stable and that its decline is mainly due to cyanobacterial uptake of anilofos and intracellular degradation.

Optimal Conditions for Anilofos Removal

A number of factors such as pH, biomass and temperature have been shown to influence microbial degradation of chlorpyrifos [19,67–69]. Thus, the effect of pH, biomass and temperature on anilofos removal by *Synechocystis* sp. strain PUPCCC 64 was studied. Maximum uptake of anilofos by PUPCCC 64 strain was observed at pH 8.0 (Table 2). Isolated and purified alkaline phosphatase from the cyanobacterium *Spirulina platensis* degraded chlorpyrifos at pH 10.0 [70]. *Anabaena inaequalis* degraded isoproturon at a fast rate at pH 5.5 [30]. Sorption of the herbicide paraquat by *Oscillatoria* sp. dominated cyanobacterial mat was not significant at pH 2.0–7.0 [71]. Thus different microorganisms have different pH optima for herbicides/pesticides degradation. Temperature is also reported to influence the rate of pesticide degradation [66,67]. Thus, the effect of temperature on herbicide uptake by the test organism was studied. *Synechocystis* sp. strain PUPCCC 64 exhibited maximum uptake of herbicide at 30°C (Table 2). Our results are in agreements with the studies of Xu et al. [69] who reported maximum degradation of chlorpyrifos at

30°C by mixed cultures of bacterium *Serratia* sp. and fungus *Trichosporon*. To optimize biomass for maximum uptake of anilofos by the test organism, five biomass concentrations (25, 50, 100, 150, and 200 mg protein L^{-1}) were selected (Table 2). *Synechocystis* sp. strain PUPCCC 64 exhibited increased anilofos uptake at a concentration of up to 100 mg L^{-1}. Beyond this concentration anilofos uptake stagnated. This may be due to the fact that, at higher biomass loads, not enough herbicide molecules were available to saturate all the absorptive sites on the cyanobacterial cells. We earlier reported on the ability of *Synechocystis* sp. strain PUPCCC 64 to degrade chlorpyrifos at 100 mg protein L^{-1} biomass load [19]. With the current study we show for this same strain a similar capability for anilofos degradation. Such a display of dual tolerance against both a pesticide (chlorpyrifos) and an herbicide (anilofos) is at this time rather uncommon in cyanobacteria.

Anilofos Mineralization and its Use as a Nutrient

The overall pattern of the HPLC chromatograms for biomass extract and supernatant were similar to pure anilofos used as standard (Fig. 6). No additional peaks were observed in HPLC chromatograms neither for biomass nor for supernatant (Fig. 6b and 6c). These results indicated that *Synechocystis* sp. strain PUPCCC 64 did not form any degradation products nor it released into the medium. Of the total 1000 µg anilofos (peak area 7.9×10^5 AU (absorbance units), Fig. 6a) present in culture medium at the start of the experiment, only 69.7 (peak area 0.55×10^5 AU, Fig. 3b) and 91.6 µg (peak area 0.73×10^5 AU, Fig. 3c) herbicide was recovered from biomass and supernatant, respectively, after 4 days. From these results, it follows that nearly 839 µg anilofos seems to have been mineralized by the organism into CO_2 and H_2O. It is possible that *Synechocystis* sp. strain PUPCCC 64 uses the released phosphate component of anilofos as a nutrient. To test this hypothesis, the PUPCCC 64 strain was grown for 12 day in anilofos (2.5 mg L^{-1}) phosphate deficient basal medium. Whereas strain PUPCCC 64 did not grow in phosphate deficient medium after day 4 it survived at a slow growth rate in presence of anilofos until day 12 (Fig. 7). This confirms that strain PUPCCC 64, in the absence of phosphate in the medium, used the phosphate component of anilofos as a nutrient. Various cyanobacteria and algae are able to degrade pesticides or herbicides and utilize the break-down products to sustain growth. Organophosphate pesticides malathion, monochrotophos and quinalphos metabolized by the cyanobacterium *Anabaena fertilissima* could be used as phosphorus source [72]. Phosphonate herbicide glyphosate was mineralized by cyanobacterium *Spirulina* spp. and used as phosphorus and nitrogen source [32]. A number of filamentous cyanobacteria and unicellular green algae (Chlorophyceae) were reported to degrade fenamifos pesticide into fenamifos sulfoxide as a primary product which was further transformed into fenamifos sulfoxide phenol [73]. And four cyanobacterial strains, *Anabaena* sp. PCC 7120, *Leptolyngbya boryana* PCC6306, *Microcystis aeruginosa* PCC7941 and *Nostoc punctiforme* PCC73102 were able to utilise the herbicide glyphosate as the sole phosphate source [31]. Limited data are available on the environmental fate of anilofos. Five reactions of anilofos degradation, namely hydrolytic cleavage of P-S-aryl linkage, N-dealkylation, O-demethylation, aryl hydroxylation and conjugation, were observed in the rat [74]. No reports indicating route(s) of anilofos degradation in microorganisms are available. Since *Synechocystis* sp. strain PUPCCC 64 grew in phosphate deficient anilofos supplemented medium, it is possible that this strain causes hydrolytic cleavage of P-S aryl linkage in anilofos releasing phosphate to be utilized as a nutrient. From the above discussion, it is clear that many environmental cyanobacteria and microalgae have developed a tolerance to otherwise toxic pesticides including herbicides and even are able to utilize the degradation products to sustain growth.

Table 2. Effect of pH, temperature, and biomass on anilofos uptake (mg L^{-1}) by *Synechocystis* sp. strain PUPCCC 64.

pH				
5	6	7	8	9
2.5±0.01	3.3±0.02	3.7±0.02	4.2±0.03	3.2±0.01
Temperature (°C)				
20	25	30	35	40
2.5±0.01	2.9±0.01	3.8±0.02	3.4±0.01	1.8±0.01
Biomass (mg L^{-1})				
25	50	100	150	200
2.0±0.01	2.5±0.01	3.8^{a} ±0.03	$3.7^{a,b}$ ±0.02	3.6^{b} ±0.01

Data in rows with same lowercase letter are not significantly different from each other at the 95% confidence level (p<0.05).
One parameter was changed at a time keeping other parameters (optimum value) constant.
Data are means of three observations ± SD.
Anilofos used during experiments: 10 mg L^{-1}.

Conclusions

We have demonstrated in this study that *Synechocystis* sp. strain PUPCCC 64 tolerated anilofos herbicide stress by increasing the level of antioxidant enzymes. Additionally, the strain PUPCCC 64 mineralized anilofos intracellularly and utilized released phosphate. This is the first report on anilofos degradation by a cyanobacterium. Since this strain has a potential to degrade both anilofos as well as chlorpyrifos, thus, it might be employed for bioremediation of these harmful chemicals from Indian paddy field ecosystems through biofertilizer programmes.

Acknowledgments

The authors thank Head and Coordinator, DRS SAP-II, Department of Botany, Punjabi University, Patiala for infrastructure and laboratory facilities.

Author Contributions

Finalized the manuscript: DPS JISK. Conceived and designed the experiments: DPS JISK. Performed the experiments: MK GK MG YS. Analyzed the data: JISK DPS MK GK MG YS. Contributed reagents/materials/analysis tools: DPS JISK. Wrote the paper: MK GK MG YS.

References

1. Yaduraju NT, Prasad Babu MBB, Chandla P (2006) Herbicide Use. In: Chadha KL, Swaminathan MS (Editors). Environment and Agriculture. Malhotra Publishing House, New Delhi, 192–210.
2. Rao VS (2000) Principles of Weed Science. Science Publishers Inc., New Hampshire, USA. 555p.
3. Central Insecticides Board & Registration Committee (2012) Available: http://cibrc.nic.in.Accessed 2012 Sep 25.
4. Funari E, Donati L, Sandroni D, Vighi M (1995) Pesticide levels in ground water: Value and limitations of monitoring. In: Vighi M, Funari E (Editors). Pesticides Risk in Groundwater. CRC, Lewis, Boca Raton, Florida. 3–44.
5. Janardhan G, Muniyappa TV, Ramachandra C, Bhaskar S, Krishnamurthy N (1999) Residual effect of pre-emergence herbicides applied to transplanted rice on succeeding crops. Curr Res 28: 12–14.

6. Kulshreshta G, Singh SB, Lal SP, Yaduraju NT (2000) Effect of long term field application of pendimethalin: enhanced degradation in soil. Pest Manag Sci 56: 202–206.
7. Marchesan E, Zanella R, de Avila LA, Camargo ER, Machado SLdO, et al. (2007) Rice herbicide monitoring in two Brazilian rivers during the rice growing season. Scientia Agricol 64: 131–137.
8. Phong TK, Nhung DTT, Motobayashi T, Watanabe H (2009) Behavior of simetryn and thiobencarb in the plough zone of rice fields. Bull Environ Contam Toxicol 83: 794–798.
9. Faria MS, Nogueira AJA, Soares AMVM (2007) The use of *Chironomus riparius* larvae to assess effects of pesticides from rice fields in adjacent freshwater ecosystems. Ecotoxicol Environ Saf 67: 218–226.
10. Leitão S, Pinto P, Pereira T, Brito MF (2007) Spatial and temporal variability of macroinvertebrate communities in two farmed Mediterranean rice fields. Aquatic Ecol 41: 373–386.
11. Pereira T, Cerejeira MJ, Espírito-Santo J (2000) Use of microbiotests to compare the toxicity of water samples fortified with active ingredients and formulated pesticides. Environ Toxicol 15: 401–405.
12. Stoate C, Baldi A, Beja P, Boatman ND, Herzon I, et al. (2009) Ecological impacts of early 21st century agricultural change in Europe - A review. J Environ Manag 91: 22–46.
13. Clay JW (2004) World Agriculture and the Environment: A Commodity -by-Commodity Guide to Impacts and Practices, Island Press, Washington D.C. 555p.
14. Moraes BS, Loro VL, Pretto A, da Fonseca MB, Menezes C, et al. (2009) Toxicological and metabolic parameters of the teleost fish (*Leporinus obtusidens*) in response to commercial herbicides containing clomazone and propanil. Pestic Biochem Physiol 95: 57–62.
15. Kang HS, Park CJ, Gye M (2009) Effect of molinate on survival and development of *Bombina orientalis* (Boulenger) embryos. Bull Environ Contam Toxicol 82: 305–309.
16. Chen WC, Yen JH, Cheng CS, Wang YS (2009) Effect of herbicide butachlor on soil microorganisms and on nitrogen-fixing abilities in paddy soil. Ecotoxicol Environ Saf 72: 120–127.
17. Lü Z, Sang L, Li Z, Min H (2009) Catalase and superoxide dismutase activities in a *Stenotrophomonas maltophilia* WZ2 resistant to herbicide pollution. Ecotoxicol Environ Saf 72: 136–143.
18. Min H, Ye YF, Chen ZY, Wu WX, Yufeng D (2001) Effects of butachlor on microbial populations and enzyme activities in paddy soil. J Environ Sci Health 36: 581–595.
19. Singh DP, Khattar JIS, Nadda J, Singh Y, Garg A, et al. (2011) Chlorpyrifos degradation by the cyanobacterium *Synechocystis* sp. strain PUPCCC 64. Environ Sci Pollut Res 18: 1351–1359.
20. Fernandez-Valiente E, Ucha A, Quesda A, Leganes F, Carreres R (2000) Contribution of N_2 fixing cyanobacteria to rice production: availability of nitrogen from ^{15}N-labelled cyanobacteria and ammonium sulphate to rice. Plant Soil 221: 107–112.
21. Singh S, Datta P (2006) Screening and selection of most potent diazotrophic cyanobacterial isolate exhibiting natural tolerance to rice field herbicides for exploitation as biofertilizer. J Basic Microbiol 46: 219–225.
22. Singh S, Datta P (2007) Outdoor evaluation of herbicide resistant strains of *Anabaena variabilis* as biofertilizer for rice plants. Plant Soil 296: 95–102.
23. Tiwari DN, Kumar A, Mishra AK (1991) Use of cyanobacterial diazotrophic technology in rice agriculture. Appl Biochem Biotechnol 28/29: 387–396.
24. Whitton BA, Potts M (2000) Introduction to cyanobacteria. In: Whitton BA, Potts M (Editors). The Ecology of Cyanobacteria: Their Diversity in Time and Space. Kluwer Academic, Dordretch, Netherlands. 1–11.
25. Roger PA, Ladha JK (1992) Biological N_2 fixation in wetland rice fields: estimation and contribution to nitrogen balance. Plant Soil 141: 41–55.
26. Padhy RN (1985) Cyanobacteria and pesticides. Residue Rev 95: 1–44.
27. Pingali PL, Roger PA (1995) Impact of Pesticides on Farmer Health and the Rice Environment. Kluwer Academic Publishers, Boston, Massachusetts, USA. 659 p.
28. Eladel HM, Henley WJ, Kobbia IA (1999) Effect of thiobencarb on growth and photosynthesis of the soil alga *Protosiphon botryoides* (chlorophyta). J Appl Phycol 10: 547–554.
29. Galhano V, Peixoto F, Gomes-Laranjo J, Fernandez-Valiente E (2010) Differential effects of bentazon and molinate on *Anabaena cylindrica*, an autochthonous cyanobacterium of Portuguese rice field agro-ecosystem. Water Air Soil Pollut 197: 211–222.
30. Mostaffa FI, Helling CS (2002) Isoproturon degradation as affected by the growth of two algal species at different concentrations and pH values. J Environ Sci Health 36: 709–727.
31. Forlani G, Pavan M, Gramaek M, Kafarski P, Lipok J (2008) Biochemical bases for a wide spread tolerance of cyanobacteria to the phosphonate herbicide glyphosate. Plant Cell Physiol 49: 443–456.
32. Lipok J, Owsiak T, Mlynarz P, Forlani G, Kafarski P (2007) Phosphorus NMR as a tool to study mineralization of organophosphonates-the ability of *Spirulina* spp. to degrade glyphosate. Enz Microbiol Technol 41: 286–291.
33. Singh DP, Sandhu BS (2010) Effect of anilofos on growth, photosynthetic pigments and stress enzymes of cyanobacterium *Oscillatoria simplicissima*. Res J Biotechnol 5: 27–32.
34. Rai AK, Chhonkar PK, Agnihotri NP (2000) Persistence and degradation of pendimethalin and anilofos in flooded versus non-flooded soils. J Ind Soc Soil Sci 48: 57–62.
35. Safferman RS, Morris ME (1964) Growth characteristics of the blue-green algal virus. LPP-I. J Bacteriol 88: 771–775.
36. Allen MB, Arnon DI (1955) Studies on nitrogen-fixing blue-green algae. I. Growth and nitrogen fixation by *Anabaena cylindrica* Lemm. Plant Physiol 30: 366–372.
37. Holm G (1954). Chlorophyll mutations in barley. *Acta Agric Scand* 4: 457–471.
38. Myers J, Kratz WA (1955) Relation between pigment content and photosynthetic characteristics in a blue-green alga. J Gen Physiol 39: 11–22.
39. Bennett A, Bogorad L (1973) Complementary chromatic adaptation in a filamentous blue-green alga. J Cell Biol 58: 419–435.
40. Beauchamp C, Fridovich I (1971) Superoxide dismutase improved assay and an assay applicable to acrylamide gels. Analyt Biochem 44: 276–287.
41. Gahagen HE, Holm RE, Abeles FB (1968) Effect of ethylene on peroxidase activity. Physiol Plantarum 21: 1270–1279.
42. Egashira T, Takahama U, Nakamura K (1989) A reduced activity of catalase as a basis for light dependent methionine sensitivity of a *Chlamydomonas reinhardtii* mutant. Plant Cell Physiol 30: 1171–1175.
43. Ellman GL (1959) Tissue sulfhydryl groups. Arch Biochem and Biophys 82: 70–77.
44. Bates B, Waldern RP, Teare ID (1973) Rapid determination of free proline for water stress studies. Plant and Soil 39: 205–207.
45. Lowry OH, Rosebrough NJ, Farr AL, Randall RJ (1951) Protein measurement with the folin phenol reagent. J Biol Chem 193: 265–275.
46. Singh DP, Khattar JIS, Kaur K, Sandhu BS, Singh Y (2012) Toxicological impact of anilofos on some physiological processes of a rice field cyanobacterium *Anabaena torulosa*. Toxicol Environ Chem 94: 1304–1318.
47. Dowidar SMA, Osman MEH, El-Naggar AH, Khalefa AE (2010) Effect of butachlor and thiobencarb herbicides on protein content and profile and some enzyme activities of *Nostoc muscorum*. *J* Genet *Engin Biotechnol* 8: 89–95.
48. Chen Z, Juneau P, Qiu B (2007) Effects of three pesticides on the growth, photosynthesis and photoinhibition of the edible cyanobacterium Ge-Xian-Mi (*Nostoc*). Aquatic Toxicol 18: 256–265.
49. Lopas-Rodas V, Flores-Moya A, Manerio E, Perdigones N, Marva F, et al. (2006) Resistance to glyphosate in the cyanobacterium *Microcystis aeruginosa* as result of pre selective mutations. Evol Ecol 21: 535–547.
50. Galhano V, Peixoto F, Gomes-Laranjo J, Fernández-Valiente E (2010) Comparative toxicity of bentazon and molinate on growth, photosynthetic pigments, photosynthesis and respiration of the Portuguese rice field cyanobacterium *Nostoc muscorum*. Environ Toxicol 25: 147–156.
51. Galhano V, Gomes-Laranjo J, Peixoto F (2011) Exposure of the cyanobacterium *Nostoc muscorum* from Portuguese rice fields to molinate (Ordram®): Effects on the antioxidant system and fatty acid profile. Aquatic Toxicol 101: 367–376.
52. Mallory-Smith CA, Retzinger Jr EJ (2003) Revised classification of herbicides by site of action for weed resistance management strategies. Weed Technol 17: 605–619.
53. Narusaka Y, Narusaka M, Kobayashi H, Satoh K (1998) The herbicide-resistant species of the cyanobacterial DI protein obtained by thorough and random in vitro mutagenesis. Plant Cell Physiol 39: 620–626.
54. Bagchi SN, Pistorius EK, Michel K-P (2003) A *Synechococcus* sp. PCC7942 mutant with a higher tolerance towards bentazone. Photosynth Res 75: 171–182.
55. Zimmermann K, Heck M, Frank J, Kern J, Vass I, et al. (2006) Herbicide binding and thermal stability of photosystem II isolated from *Thermosynechococcus elongatus*. Biochim Biophysic Acta 1757: 106–114.
56. Broser M, Glöckner C, Gabdulkhakov A, Guskov A, Buchta J, et al. (2011) Structural basis of cyanobacterial photosystem II inhibition by the herbicide terbutryn. J Biol Chem 286: 15964–15972.
57. Oettmeier W (1999) Herbicide resistance and supersensitivity in photosystem II. Cell Mol Life Sci 55: 1255–1277.
58. Lesser MP (2006) Oxidative stress in marine environment: Biochemistry and physiological ecology. Annu Rev Physiol 68: 253–278.
59. Qian H, Chen W, Sheng GD, Xu X, Liu W, et al. (2008) Effects of glufosinate on antioxidant enzymes, subcellular structure and gene expression in unicellular green alga *Chlorella vulgaris*. Aquatic Toxicol 88: 301–307.
60. Qian H, Chen W, Sun L, Jin Y, Liu W, et al. (2009) Inhibitory effect of paraquat on photosynthesis and the response to oxidative stress in *Chlorella vulgaris*. Ecotoxicol 18: 537–543.
61. Bohnert HJ, Jensen RG (1996) Strategies for engineering water stress tolerance in plants. Trends Biotechnol 14: 89–97.
62. Kavi Kishor PB, Sangam S, Amrutha RN, Sri Lakshmi P, Naidu KR, et al. (2005) Regulation of proline biosynthesis, degradation, uptake and transport in higher plants: Its implications in plant growth and abiotic stress tolerance. Curr Sci 88: 424–38.
63. Fatma T, Khan MA, Choudhary M (2007). Impact of environmental pollution on cyanobacterial proline content. J Appl Phycol 19: 625–629.
64. Kumar S, Habib K, Fatma T (2008) Endosulfan induced biochemical changes in nitrogen-fixing cyanobacteria. Sci Total Environ 403: 130–138.
65. Orús MI, Marco E (1991) Disappearance of trichlorfon from cultures with different cyanobacteria. Bull Environ Contam Toxicol 47: 392–397.

66. Gonzalez-Barreiro O, Rioboo C, Herrero C, Cid A (2006) Removal of triazine herbicide from freshwater systems using photosynthetic microorganisms. Environ Pollut 144: 266–271.
67. Liu X, You MS, Wei YZ, Liao JY, Ye LL, et al. (2003). Isolation of chlorpyrifos-degrading *Aspergillus* sp. Y and measurements of degradation efficiency. Chinese J Appl Environ Biol 9: 78–80.
68. Wang X, Chu XQ, Yu YL, Feng H, Chen J, et al. (2006) Characteristics and function of *Bacillus laterosporus* DSP in degrading chlorpyrifos. Acta Pedologic Sinic 43: 648–654.
69. Xu G, Li Y, Zheng W, Peng X, Li W, et al. (2007) Mineralization of chlorpyrifos by co-culture of *Serratia* and *Trichosporon* spp. Biotechnol Lett 29: 1469–1473.
70. Thengodkar RRM, Sivakami S (2010) Degradation of chlorpyrifos by an alkaline phosphatase from the cyanobacterium *Spirulina platensis*. Biodegradation 21: 637–644.
71. Kumar D, Prakash B, Pandey LK, Gaur JP (2010) Sorption of paraquat and 2, 4-D by an *Oscillatoria* sp. dominated cyanobacterial mat. Appl Biochem Biotechnol 160: 2475–2485.
72. Subramanian G, Sekar S, Sampoornam S (1994) Biodegradation and utilization of organophosphorus pesticides by cyanobacteria. Internat Biodeterior Biodegrad 33: 129–143.
73. Caceres TP, Megharaj M, Naidu R (2008) Biodegradation of the pesticide fenamiphos by ten different species of green algae and cyanobacteria. Curr Microbiol 57: 643–646.
74. Roberts TK (1998) Metabolic Pathways of Agrochemicals: Part I. Herbicides and Plant Growth Regulators. Royal Society of Chemistry, Cambridge, UK. 847p.

16

Mechanisms of Tolerance and High Degradation Capacity of the Herbicide Mesotrione by *Escherichia coli* Strain DH5-α

Luiz R. Olchanheski[1], Manuella N. Dourado[2], Flávio L. Beltrame[3], Acácio A. F. Zielinski[4], Ivo M. Demiate[5], Sônia A. V. Pileggi[1], Ricardo A. Azevedo[2], Michael J. Sadowsky[6], Marcos Pileggi[1]*

1 Universidade Estadual de Ponta Grossa, UEPG, Departamento de Biologia Estrutural, Molecular e Genética, Ponta Grossa, PR, Brazil, **2** Escola Superior de Agricultura Luiz de Queiroz, ESALQ, Universidade de São Paulo, USP, Piracicaba, SP, Brazil, **3** Universidade Estadual de Ponta Grossa, UEPG, Departamento de Ciências Farmacêuticas, Ponta Grossa, PR, Brazil, **4** Programa de Pós-Graduação em Engenharia de Alimentos, Universidade Federal do Paraná, Curitiba, PR, Brazil, **5** Universidade Estadual de Ponta Grossa, UEPG, Departamento de Engenharia de Alimentos, Ponta Grossa, PR, Brazil, **6** Department of Soil, Water, and Climate, and BioTechnology Institute, University of Minnesota, St. Paul, Minnesota, United States of America

Abstract

The intensive use of agrochemicals has played an important role in increasing agricultural production. One of the impacts of agrochemical use has been changes in population structure of soil microbiota. The aim of this work was to analyze the adaptive strategies that bacteria use to overcome oxidative stress caused by mesotrione, which inhibits 4-hydroxyphenylpyruvate dioxygenase. We also examined antioxidative stress systems, saturation changes of lipid membranes, and the capacity of bacteria to degrade mesotrione. *Escherichia coli* DH5-á was chosen as a non-environmental strain, which is already a model bacterium for studying metabolism and adaptation. The results showed that this bacterium was able to tolerate high doses of the herbicide (10× field rate), and completely degraded mesotrione after 3 h of exposure, as determined by a High Performance Liquid Chromatography. Growth rates in the presence of mesotrione were lower than in the control, prior to the period of degradation, showing toxic effects of this herbicide on bacterial cells. Changes in the saturation of the membrane lipids reduced the damage caused by reactive oxygen species and possibly hindered the entry of xenobiotics in the cell, while activating glutathione-S-transferase enzyme in the antioxidant system and in the metabolizing process of the herbicide. Considering that *E. coli* DH5-α is a non-environmental strain and it had no previous contact with mesotrione, the defense system found in this strain could be considered non-specific. This bacterium system response may be a general adaptation mechanism by which bacterial strains resist to damage from the presence of herbicides in agricultural soils.

Editor: Marie-Joelle Virolle, University Paris South, France

Funding: The authors would like to thank the Brazilian funding agencies Coordination for the Improvement of Higher Level Personnel (www.capes.gov.br), Grant n°. 221/2007, National Council of Technological and Scientific Development (www.cnpq.br), Grant n°. 473438/2010-0, Foundation for Research Support of the State of São Paulo (www.fapesp.br), Grant n°. 2009/54676-6, and Foundation for Research Support of the State of Paraná (www.fundacaoaraucaria.org.br), Grant n°. 21724, for the financial support. The funders had no role in study design, data collection and analysis, decision to publish, or preparation of the manuscript.

Competing Interests: The authors have declared that no competing interests exist.

* E-mail: marcos.pileggi@pq.cnpq.br

Introduction

In recent years, there has been a high demand for increasing agricultural productivity and the arable land area, accompanied by the large scale use and discovery of new pesticides and fertilizers [1–3]. It was estimated that approximately 2.27 million tons of agrochemicals were released into the environment in 2001, 35% of which were herbicides [4]. Despite the fact that the use of pesticides in agriculture has had a positive impact on crop productivity, concerns have been expressed about the adverse effects of these chemicals [5], since only 0.1% of them reach their specific targets. For this reason, there is a large quantity of herbicide residues remaining in the environment, which can be metabolized by microbiota [6–8].

Herbicide application has brought damage to the soil microbiota, and may have affected the dynamics of biogeochemical cycles and soil fertility. The herbicide napropamide, for example, has been identified as harmful to soil functionality, based on the structural and functional diversity of the soil bacterial community [9]. Other studies have demonstrated the inhibition of nitrification and changes in ammonia oxidation in soils by the herbicide simazine [10]. As regard to the triketone herbicides mesotrione and sulcotrione, their toxicity level was considered equal to or higher than atrazine in studies with model organisms, such as *Tetrahymena pyriformis* and *Vibrio fischeri* [11].

Bioremediation has been the main strategy used to eliminate xenobiotics, mainly herbicides, from the environment, and this subject has been the focus of many biotechnological studies [12–13]. Degradation processes mediated by microorganisms in large part influence the persistence of herbicides in the soil [14].

Reactive oxygen species (ROS), apart from being part of normal aerobic metabolism [15], may increase in concentration as a result of exposure to toxic substances, as for example, when bacteria come into contact with herbicides [16]. An increase in the rate of

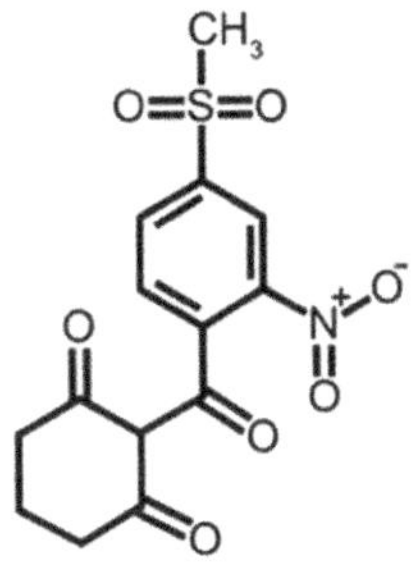

Figure 1. Molecular structure of mesotrione.

hydrogen peroxide (H_2O_2), and superoxide (O^{2-}) and hydroxyl ($OH^{\bullet}$) radicals production can cause damage to DNA, RNA, proteins and lipids [17–18]. An efficient antioxidant enzyme system is the primary line of defense for the elimination of excess ROS, such as the action of catalase, superoxide dismutases, peroxidases and glutathione reductase [15]. Glutathione-s-transferase (GST), as well as participating in the system of redox homeostasis and response to ROS [19–20], catalyzes the conjugation of glutathione with herbicides, and it can operate to degrade some herbicides, also being responsible for resistance to antibiotics [21].

Mesotrione (Fig. 1) is the active ingredient of the herbicide Callisto; it has a selective action and is recommended for the systematic control of weeds in maize cultivation, both in pre and post-planting applications. This molecule is derived from a phytotoxin that is produced by the plant *Callistemon citrinus*, and it inhibits the enzyme 4-hydroxyphenylpyruvate dioxygenase (HPPD), acting in the conversion of tyrosine to plastoquinone and α-tocopherol. The inhibition of plastoquinone leads to an interruption of the carotenoid synthesis pathway, causing the death of leaf tissues [22–23]. In mesotrione-tolerant plants, the metabolism of the herbicide is mainly carried out by the enzyme P450 [24–25]. A P450-like, synthesized by *cysj* gene, was reported in reducing paraquat herbicide in bacteria [29]. But glutathione S-transferase (GST) is also involved in the detoxification of herbicides of the triketone family [26], and in the metabolism of herbicides on microorganisms [27–28].

The degradation products of mesotrione by a *Bacillus* sp. strain were found to be 4-methylsulfonyl-2-nitrobenzoic acid (MNBA) and 2-amino-4-methylsulfonylbenzoic acid (AMBA) [30-31]. AMBA has been characterized as being more toxic than the original molecule [11]. Studies with *Pantoea ananatis* have demonstrated the degradation of mesotrione by a different route, with metabolic products probably less toxic than the original molecule ($C_{13}H_{10}NO_7S$, $C_{11}H_{13}O_8S$ and $C_{11}H_{11}O_7S$) [32]. The presence of MNBA and AMBA was reported in soils treated with high concentrations of mesotrione [33].

Strains of *E. coli* are currently used to determine the toxic effects of xenobiotics [34]. This model organism [35] has been widely used in recombinant DNA technology [36]. So far, no studies of degradation capacity have been carried out without a prior modification of *E. coli* strains.

Given the toxic effect of herbicides on bacteria, we aimed to assess whether a laboratory, non-environmental strain *E. coli* (DH5-α) possessed any mechanism of adaptation to mesotrione, even with no previous contact with this herbicide.

Materials and Methods

Bacterial strain

The *Escherichia coli* strain DH5-α [(genotype: F^- φ80dlacZΔM15 Δ(lacZYA-argF) U169 deoR, recA1 endA1 hsdR17 (r_k^- m_k^+) phoA supE44λ^- thi-1 gyrA96 relA1)] was used to evaluate the degradation of mesotrione herbicide.

Herbicides

Mesotrione (99% purity) was provided by Syngenta Crop Protection, Greensboro, NC (USA). For HPLC (High Performance Liquid Chromatography) experiments, the analytical standard Pestanal (99% pure) (Sigma-Aldrich) was used.

Evaluation of herbicide degradation by HPLC

In order to determine the mesotrione degrading ability of *E. coli*, the strain was grown in 100 mL of LB (Luria Broth: 10 g L^{-1} tryptone; 5 g L^{-1} yeast extract; 10 g L^{-1} NaCl) in 250 mL flasks, and incubated at 37°C, at 200 rpm. The cells were collected after 10 h and centrifuged at 8,000×*g* for 10 min. at 4°C. The pellet was washed twice with PBS (pH 6.8, phosphate buffered saline: 8 g L^{-1} NaCl; 0.2 g L^{-1} KCl; 1.44 g L^{-1} Na_2HPO_4; 0.24 g L^{-1} KH_2PO_4), and cells were re-suspended in 10 mL of MMM (Mineral Medium with Mesotrione: 3 g L^{-1} $NaNO_3$; 0.5 g L^{-1} $MgSO_4$; 0.5 g L^{-1} KCl; 0.01 g L^{-1} $FeSO_4$; 0.04 g L^{-1} $CaCl_2$; 0.001 g L^{-1} $MnSO_4$; 0.4 g L^{-1} glucose, 0.04 mM mesotrione; 10 mM potassium phosphate buffer, pH 7.0;). This corresponded to 1×, or normal field levels according to manufacturer's instruction. Cells were placed in mineral medium without carbon source (-CM), with five repetitions, and incubated at 37°C, at 200 rpm. For analysis of herbicide adsorption by bacteria, cells were boiled prior to being added to the treatments already described. As a negative control, a flask with MMM was incubated, in the same conditions, without *E. coli* DH5-α. Aliquots (1 mL) were collected from the culture medium every hour of incubation (from 0 to 12 h) and centrifuged at 13,000 *g* for 5 min. The supernatant (0.9 mL) was frozen for further HPLC analysis.

Samples were filtered with 0.22 μm syringe filters and HPLC analysis was performed using a Waters Alliance e2695 and a photodiode detector (Waters 2998 PDA), adjusted at a wavelength of 254 nm. An Eclipse XDB-C18 column was used, with dimensions of 4.6 mm×150, at 3.5 μm for separation, at 20°C. The gradient of the mobile phase started with 70% water (0.1% phosphoric acid) (A): 30% acetonitrile (B); 30% B for 3 min.; 55% B for 15 min.; 100% B for 17 min.; 100% B for 18 min.; 30% B for 19 min. and 30% B for 29 min., for conditioning the column to a new injection). The flow rate was 1 mL min^{-1}. The injection volume of the samples was 50 μL. The HPLC method was developed, validated and applied to analysis of mesotrione degradation rate.

Cell viability

Bacteria were grown for 10 h at 37°C in 1.2 L of LB. Cells were centrifuged at 8,000×*g* for 10 min., washed twice with PBS, and divided into vials containing 50 mL of MM (control), MMM (mineral medium with mesotrione), -C (MM without carbon source) and -CM (MMM with mesotrione as sole carbon source), with three repetitions. The flasks were incubated at 37°C (at 200 rpm), and 100 mL aliquots were withdrawn after 30 min., 3 h and 6 h of incubation. Samples were diluted to 10^{-8}, plated on LB medium and incubated at 37°C. After 12 h, the colony forming units (CFU) counting was determined.

Lipid peroxidation

Lipid peroxidation was determined by the levels of malondialdehyde (MDA) (substances reactive to thiobarbituric acid) [37]. The concentrations of MDA were monitored at 535 and 600 nm, and their concentrations were calculated using an extinction coefficient of 155 mM cm^{-1}.

Protein extraction for analysis of oxidative stress

Bacteria were grown for 10 h at 37°C (at 200 rpm) in 3.6 L of LB, to get enough cells for the following steps of protein extraction. The cells were centrifuged at 8,000 ×*g* for 10 min., washed twice with PBS, and divided into vials with 50 mL of MM (control), MMM, -C (MM without carbon source), and -CM (MMM with mesotrione as sole carbon source), with three replications. All the flasks were incubated at 37°C, and extractions were performed at 30 min., 3 h, and 6 h.

For enzyme extraction, the cultures were centrifuged at 8,000 ×*g* for 10 min., and the pellet was macerated with liquid nitrogen, and homogenized (10:1 w/v) in 100 mM potassium phosphate buffer (pH 7.5), containing 1 mM ethylenediaminetetraacetic acid (EDTA), 3 mM DL-dithiothreitol and 5% (w/v) polyvinylpolypyrrolidone [38], always kept at 4°C. The homogenate was centrifuged at 10,000 ×*g* for 30 min., and the supernatant was divided into aliquots, and frozen at −80°C for subsequent enzyme analysis. The protein concentration was determined by the Bradford method [39], using BSA as standard.

Protein analysis by polyacrylamide gel electrophoresis (PAGE)

Electrophoresis was performed in gels containing 12% polyacrylamide with 4% stacking gel. For SOD-PAGE, SDS was eliminated. A current of 15 mA gel^{-1} was applied for 3 h (SOD activity gel) at 4°C or 2 h (protein profile gel) at room temperature. Equal amounts of protein (20 µg) were applied to the gels. For SDS-PAGE, the gel was washed with distilled water, and incubated overnight in 0.05% Coomassie blue R-250 solution, at a ratio of 40:7:53 of methanol:acetic acid:water (v/v/v), and decolorized by successive washings with a solution, at a ratio of 40:7:53 of methanol:acetic acid:water (v/v/v) [40].

The SOD-PAGE activity was performed according to Beauchamp and Fridovich [41] and modified by Medici et al. [42], in which the gels were washed in distilled water, and incubated in the dark for 30 min. in 50 mM potassium phosphate buffer (pH 7.8), containing 1 mM EDTA, 0.005 mM riboflavin, 0.1 mM nitroblue tetrazolium, and 0.3% N,N,N,N-tetramethylethylenediamine. To control the reaction, a unit of bovine liver SOD (Sigma) was used. The gels were exposed to white light and immersed in water until the development of the SOD bands.

Catalase (CAT) activity

CAT activity was determined according to Kraus et al. [43] in a solution containing 1 mL of potassium phosphate buffer 100 mM (pH 7.5) and 2.5 µL H_2O_2 (30% solution), and quantified in a spectrophotometer at 25°C. The reaction was initiated with the addition of 25 µL of protein extract, and the activity was determined by following the decomposition of H_2O_2 at 240 nm for 1 min.

GST activity

GST activity was measured in a solution containing 900 µL of potassium phosphate buffer 100 mM (pH 6.8), adding 25 µL of 1-chloro-2,4-dinitrobenzene (CDNB) 40 mM and 50 µL of reduced glutathione (GSH) 0.1 M, and incubated at 30°C [44]. The reaction was initiated with the addition of 25 µL of protein extract, and was monitored for 2 min. at 340 nm.

Analysis of lipids saturation

Bacterial strain was grown in 800 mL of LB and incubated at 37°C, at 200 rpm. After 12 h, the samples were centrifuged at 8,000 ×*g* for 5 min. at 4°C. The pellet was washed twice with PBS and divided into vials containing 50 mL with LB and LB plus 0.04 mM mesotrione, in triplicate, and the cultures were incubated at 37°C, at 200 rpm. After 12 h, lipid extraction was performed, as described by Bligh and Dyer [45], with modifications. The membrane lipids were analyzed by FTIR (Fourier Transform Infrared Spectroscopy) with transmittance at wavelengths from 400 to 4,000 cm^{-1}.

Experimental design and statistical analysis

To assess the saturation of lipids, the baseline of spectrums were corrected, and then processed by PCA (Principal Component Analyzes) implemented in the Pirouette v. 4.0 software (Infometrix, Bothell, WA, USA). PCA was applied to separate the samples according to their FTIR spectra (1,400 to 3,200 cm-1). Therefore, the results obtained for each wavelength were plotted as columns and the samples as rows. Mean-Center was used as pre-treatment of the results.

Statistical analysis were conducted with three repetitions of each treatment for cell viability, MDA, GST and CAT experiments, which were performed in a completely random design. The significance of the observed differences was verified using a one-way analysis of variance (P<0.05). Analysis were made using R software version 3.0.1.

Results and Discussion

Mesotrione degrading capacity by *E. coli* DH5-α

E. coli is considered a model bacterium for study the physiology of prokaryotes [35]. Studies have been using the *E. coli* DH5-α strain as a recipient of genetic material, and it is one of the strains most frequently used as a tool in recombinant DNA technology [36,46–47]. Thus, we considered this strain as a laboratory, non-environmental bacterium, with no prior contact with herbicides.

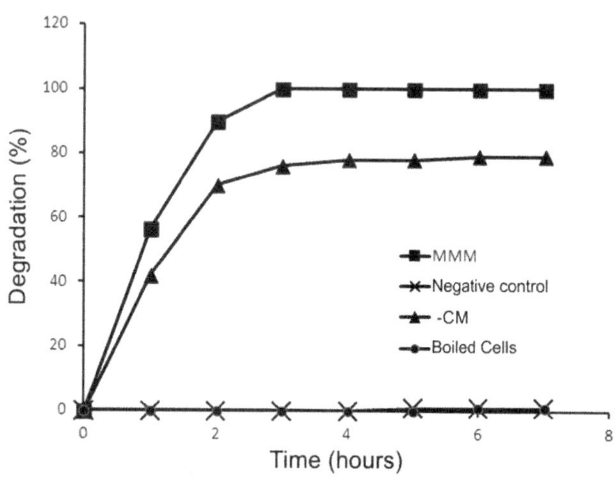

Figure 2. Degradation kinetics of mesotrione mediated by *E. coli* DH5-α. MMM (mineral medium with mesotrione), -CM (mineral medium without carbon, with mesotrione), negative control (MMM without *E. coli* DH5-α) and boiled cells.

The *in vitro* evaluation of the toxic effects of herbicides revealed a negative effect on the growth of strains of *E. coli*, particularly at higher doses [34], and also served as a model for the identification of genes and enzymatic activities involved in the antioxidant system in general [48–50]. Until this date, there are no studies involving herbicides degrading ability by non-environmental strains as *E. coli* DH5-α.

According to Batisson et al. [51], in high concentrations, mesotrione can alter the microbial community, selecting tolerant or degrading strains. *Bacillus* sp. 3B6 [52] and *P. ananatis* CCT 7673 [32] have been described as capable to degrade mesotrione in 24 h and 18 h, respectively. Similarly, *E. coli* DH5-α biotransformed mesotrione, and after 3 h of exposure to MMM, no compound was detected in the analyzed samples, and 76% in -CM (Fig. 2). The boiled cells test showed that mesotrione was not lost from the culture medium by cell adsorption. In contrast, *P. ananatis* CCT 7673 showed no ability to metabolize the herbicide without the presence of carbon [32].

In studies of mesotrione degradation in soil, it was observed that the highest rate of degradation of Callisto occurred with the application of 10× and 100× [33]. *P. ananatis* CCT 7673 (a strain isolated from water) was able to completely metabolize the herbicide mesotrione, but could not grow when in contact with the herbicide at high doses [32]. In our studies, *E. coli* DH5-α was able to tolerate and degrade the herbicide in 10× and in a shorter incubation time. The tolerance exhibited by this bacterium to the herbicide may be explained by its rapid metabolism, since the degradation process takes place as soon as the strain is exposed to the xenobiotic, reducing the time of exposure of the bacterium to the chemical.

Studies of mesotrione degradation reported difficulty in finding specific genes for the degradation of this herbicide [23,43]. According to Pileggi et al. [32], the strains *E. coli* DH5-α, TOP 10, and K-12 have the ability to metabolize mesotrione. Considering that they are related to strains developed in the laboratory, with no prior contact with the herbicide, this fact may indicate low selective pressure for specific genes to degrade mesotrione.

Characterization of mesotrione herbicide as a stress agent

In order to determine if mesotrione damaged cells of *E. coli* DH5-α, its cellular viability was assessed under the same growth conditions as for protein extraction (Fig. 3). The data obtained revealed that the herbicide mesotrione was not detected after 3 h exposure (Fig. 2) and that the metabolic process was shown to initiate in the first hour of growth; therefore, the *E. coli* DH5-α strain was probably in contact with the whole herbicide for at least the initial growing of 30 min. During this period, a decrease in cell viability was observed in the treatments with the herbicide (MMM and -CM), compared with the controls (MM and -C) (Fig. 3), indicating a toxic effect of the herbicide on the *E. coli* DH5-α. During the periods of 3 h and 6 h, similar viability rates were verified for all treatments, which is probably due to the capacity of the nutritional support from MM provided to the bacterial cells.

Botelho et al. [34] analyzed many widely used commercial formulations of herbicides, and only paraquat decreased the growth of the *E. coli* ATCC 25922 strain. In the present study, field doses (1×) of mesotrione, which is considered less toxic than the commercial herbicide [22], decreased the viability of *E. coli* DH5-α within 30 min. in treatments with the herbicide (MM and -CM) (Fig. 3).

Balagué et al. [53] used doses of 2 mM, 1 mM, 0.1 mM and 0.01 mM of the herbicide 2,4-D in spectrophotometer analysis, and verified growing inhibition of *E. coli* HB101 only at 2 mM, whilst in the present study, the growth capacity, analyzed by cellular viability, showed significant differences in *E. coli* DH5-α between control and mesotrione treatments in 30 min. evaluation (Fig. 3).

The malondialdehyde (MDA) rate has been used as an indicator of lipid peroxidation in other studies involving oxidative stress, such as that reported by Lima and Abdalla [54]. In the present study, *E. coli* DH5-α exhibited lower membrane damage, in the -CM treatment, within 30 min., while mesotrione still remained in the culture medium, (Fig. 4). Although cell viability was affected by the presence of mesotrione (Fig. 3), in the three analyzed times, the rates of MDA measured did not show an increased lipid peroxidation, in the -CM treatment. Nevertheless, in the MMM treatments, during all the periods analyzed, a statistically significant increase in MDA was observed, suggesting an imbalance in ROS production/ROS homeostasis, and consequently a toxic effect of the extra ROS was produced.

Influence of mesotrione on the level of lipids

Infrared spectrogram analysis showed changes in the structure of the membrane lipids of *E. coli* DH5-α in the presence of the herbicide mesotrione (Fig. 5). According to Böger et al. [55], an increase in the proportion of unsaturated fatty acids in bacterial cell membranes, makes them more susceptible to attack by ROS, and induces to higher rates of MDA production. These changes in membrane lipids occur from chloroacetanilide herbicides. A study

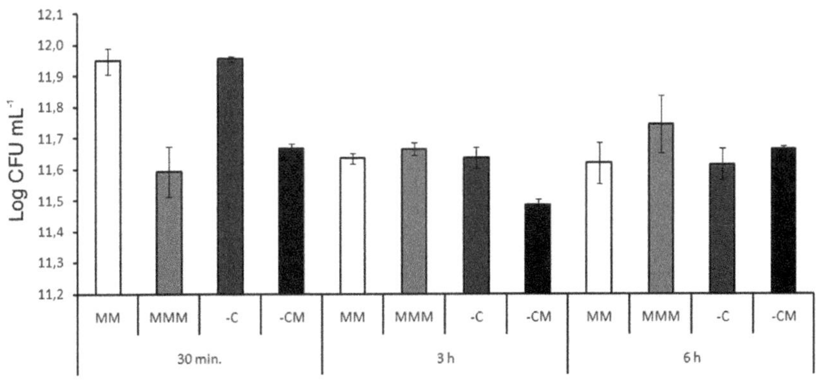

Figure 3. Cellular viability of *E. coli* DH5-α in MM and in the MMM, -C and –CM treatments, during the periods of 30 min., 3 h and 6 h. LSD = 0.13 for all pairwise comparison.

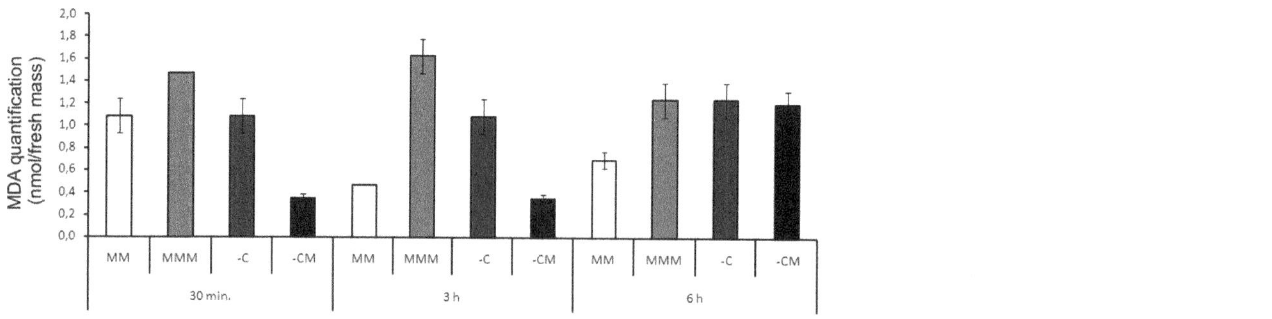

Figure 4. MDA levels in *E. coli* DH5-α in MM, and in the MMM, -C and -CM treatments, during periods of 30 min., 3 h and 6 h, respectively. LSD = 0.18 for all pairwise comparison.

by Balangué et al. [53] with *E. coli* HB101 strain demonstrated that the toxicity of 2,4-D could reduce the fluidity of bacterial membrane, and consequently alter MDA results, as a system-wide response against ROS. Sánchez et al. [56] also reported similar findings, where the strain *Klebsiella planticola* DSZ, when in contact with ethanol or the herbicide simazine, exhibited a decrease in the saturation of membrane lipids, altering the rate of selective permeability, possibly as a defense system.

In the present study, such a change in the composition of the membrane lipids of *E. coli* DH5-α may explain the lower rate of MDA in the treatments without the presence of carbon and with mesotrione (Fig. 4), as shown by data from cell viability (Fig. 3), in which the presence of mesotrione caused damage to bacteria in a time of 30 min. Apart from defense against ROS, the change in selective permeability may have prevented the entrance of mesotrione, with increased saturation of membrane lipids, characterizing a defense system against xenobiotics.

Involvement of antioxidant enzymes in the defense and degradation of mesotrione

By superoxide dismutase gel analysis, up to 6 distinct isoenzymes (Fig. 6, lane 3) were observed among the treatments used (Fig. 6). These isoenzymes were not necessarily present in all treatments, but the majority of the SOD activity detected could be accounted to SOD I and SOD IV isoenzymes. A larger number of bands were always observed when *E. coli* DH5-α was in MM (Fig. 6, lane 3) and MMM (Fig. 6, lane 6), at 6 h of exposure/growth. The higher SOD activity observed, under such conditions, and based on the higher intensity of the SOD bands, could be attributed to the increased number of visible SOD bands, and also to the higher activity of SOD I and SOD IV isoenzymes, which together clearly accounted for the majority of the SOD activity in *E. coli* DH5-α. Yet, it is likely that this difference on band intensity was due to the incubation time of *E. coli* DH5-α, and consequently occurred an increase of superoxide radical production. As shown in the treatment with carbon (MM and MMM), as longer the growing time, the higher was SOD activity. However, this was not observed in the treatment without carbon (-CM). Ongoing research is classifying the distinct SOD isoenzymes, as proposed by Azevedo et al. [57], detecting which shall be important in future studies, since they may be located in distinct cell compartments, and consequently allow to link an increase in superoxide in a specific organelle to an specific increase of one or another particular SOD isoenzyme.

An increase in CAT activity was observed, depending on the exposure time in the culture media with the presence of carbon (MMM and MM) (Fig. 7), which was probably due to the adaptation and growth of the strain in the mineral medium. However, there were no differences in relation to the presence of mesotrione. In strains of *E. coli*, depending on the growth phase, different isoforms can act, which have regulatory pathways that are activated independently [58]. This may have influenced the

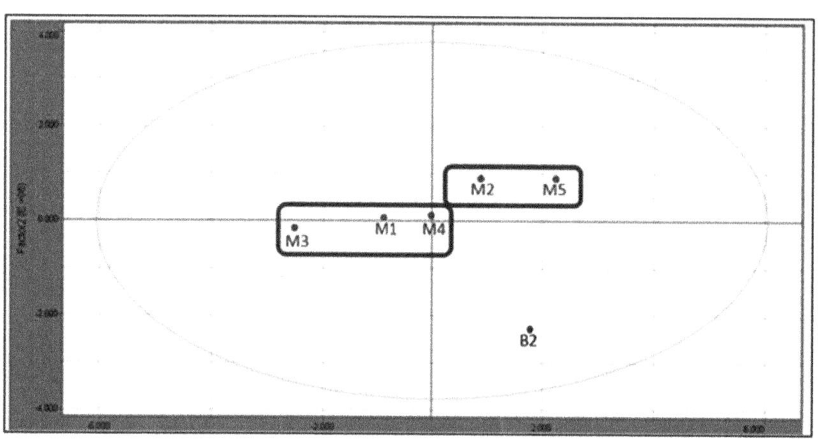

Figure 5. Analysis of lipids of *E. coli* DH5-α using FTIR and principal component analysis. Points M2 and M5: control without mesotrione (LB), in duplicate. Points M1, M3 and M4: treatment with mesotrione (LB+0.04 mM mesotrione) in triplicate. Point B2: negative control, *Bacillus* sp. in the absence of mesotrione.

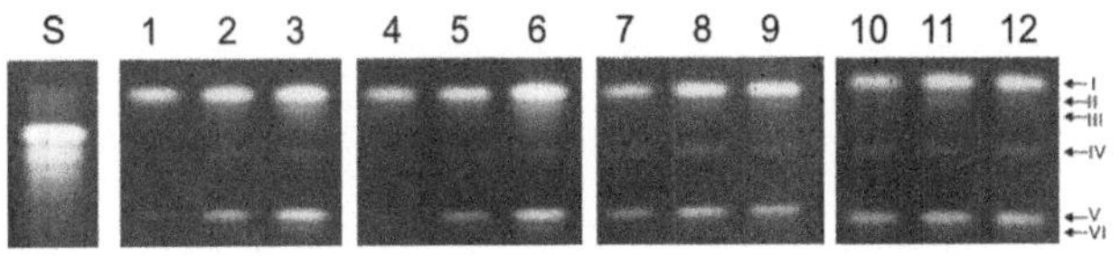

Figure 6. Non-denaturing-PAGE for SOD activity. Patterns presented by *E. coli* DH5-α in MM (lines 1, 2 and 3) and in the MMM (lines 4, 5 and 6), -C (lines 7, 8 and 9), and -CM (lines 10, 11 and 12) treatments, at periods of 30 min., 3 h and 6 h, respectively.

CAT activity response and also the potential participation of other peroxidases in the stress response.

In the culture media without carbon (-C and -CM), despite showing a significantly lower total CAT enzymatic activity, in comparison with the media with carbon, the activity of CAT was higher in the media with mesotrione, except in 3 h exposure period (Fig. 7). In the period of 30 min., during which the complete degradation of the herbicide had not yet occurred, and mesotrione appeared as the only carbon source, the amount of hydrogen peroxide may have increased in response to the xenobiotic itself, stimulating CAT activity, which is notable when compared to the control without carbon source (-C).

In relation to *E. coli* DH5-α, GST activity was higher in the 30 min. exposure treatments in the presence of mesotrione, compared to medium without the herbicide. However, in the 3 h period, no differences occurred in the activities of GST (Fig. 8). There are reports about metabolization of triketone herbicides (such as mesotrione) by influence of cytochrome P450 in plants [24–25]. According to Barrett [24], P450 is essential for the biodegradation of at least six families of herbicides used in maize culture. Furthermore, both the P450 as GST are involved in cellular defense and detoxification of these herbicides in maize and rice [26].

The involvement of the cytochrome P450 enzyme system in degradation of atrazine and herbicides EPTC (*S*-Ethyl Dipropylthiocarbamate) was observed in bacteria [59–60], but at a very slow rate [61]. A P450 like enzyme,synthesized by the *cysj* gene, is involved only in the reduction pathway of paraquat in *E. coli*, not in the degradation[29], thereby changing the mode of action of this herbicide [62].

On the other hand, GST enzymes have been characterized in plants and microorganisms, as mechanisms aiding the metabolism of herbicides and other toxic products, catalyzing the conjugation of glutathione with the herbicide and marking the compound to be degraded [27–28]. However, there is a great diversity in this enzyme, with various functions still unknown [21,63–64].

E. coli DH5-α strain exhibited the capacity to transform 100% of mesotrione in only 3 h of exposure (Fig. 2). Taking into account the GST activity during the period in which the degradation rate is high (30 min.), and any other enzyme related to mesotrione degradation was reported in bacteria, there is a possibility that this enzyme is involved in the process of degradation of the herbicide, because during the period of 3 h, when the herbicide was completely degraded, the GST activity decreased (Fig. 8). Besides, acting on the first transformation of the herbicide, the GST enzyme may also be involved in the elimination of ROS and in the adaptation of the strain to the culture medium [65–67]. The evaluation of CAT (Fig. 7) and SOD enzymes (Fig. 6) provides evidence that these enzymes showed no specific changes in activity in response to the herbicide in *E. coli* DH5-α. Such a response may vary, as shown by Martins et al. [13], who also failed to see specific changes in SOD activity that could be attributed to the herbicides (acetochlor and metolachor), but observed specific changes in CAT activity by the same bacterial isolates and to the same herbicides. Thus, it appears that GST may be acting directly in defense against ROS and in the degradation of mesotrione by *E coli* DH5-α.

Conclusions

This is the first report showing that *Escherichia coli* DH5-α, which is considered a non environmental strain, was able to degrade mesotrione without previous exposure to the herbicide The process of degradation took only 3 h, being the lowest degradation time reported until now. Previous articles tried to discover a gene responsible for mesotrione degradation, without success. In this manuscript, we describe the involvement of GST in herbicide degradation, as part of a more complex response system to mesotrione. Nevertheless, it cannot be ruled out that other systems may also be involved cooperatively or independently, such as other peroxidases.

Mesotrione was identified as an oxidative stress agent by the involvement of GST in herbicide degradation, changes in membrane lipids saturation, preventing membrane cell peroxidation, and differences in cellular viability. In this case, MDA did not indicate the occurrence of cellular damage by oxidative stress, but its reduction was related to changes on lipid structure in response to the herbicide. To our knowledge, this manuscript is the first report on the characterization of mesotrione as an oxidative stress agent in bacteria. Nevertheless, we described *Escherichia coli* DH5-α as a tolerant strain and capable of growing in the presence of the herbicide at concentrations normally used in the environment by agricultural management, even without previous contact with the herbicide. We consider that *E. coli* strains can manage to adapt to

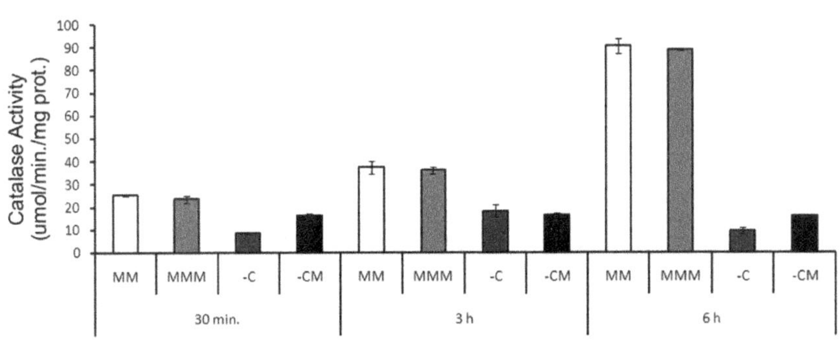

Figure 7. CAT activity in *E. coli* DH5-α in MM, and in the MMM, -C and -CM treatments, at periods of 30 min., 3 h and 6 h, respectively. LSD = 1.17 for all pairwise comparison.

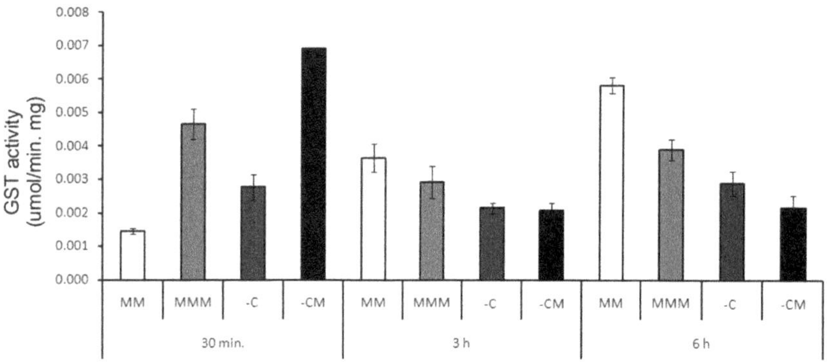

Figure 8. GST activity of *E. coli* DH5-α in MM, and in the MMM, -C, and -CM treatments, at periods of 30 min. 3 h and 6 h, respectively. LSD = 0.000398 for all pairwise comparison.

the presence of new synthesized toxic molecules in its environment through non-specific mechanisms of tolerance, regarding to non-specific anti stress enzymes involved with degradation, as GST, and changes in structure of lipid membrane, perhaps preventing the entrance of the herbicide in the cell in its toxic configuration. *E. coli* DH5-α, already a model for different studies in bacterial metabolism and adaptation, can also be used for the study of other enzymatic and structural systems related to herbicides tolerance and adaptation in contaminated environments, through phenotypic plasticity of those systems.

Acknowledgments

The authors would like to thank Bruno César do Espírito Santo for assistance in figures editing, Fernando Piotto for assistance in statistical analysis, and Jesiane Stefania da Silva Batista for manuscript review.

Author Contributions

Conceived and designed the experiments: LRO MND FLB SAVP RAA MJS MP. Performed the experiments: LRO MND FLB MP. Analyzed the data: LRO MND FLB AAFZ IMD SAVP RAA MJS MP. Contributed reagents/materials/analysis tools: FLB AAFZ IMD SAVP RAA MJS MP. Wrote the paper: LRO MND FLB AAFZ IMD SAVP RAA MJS MP.

References

1. Dayan FE, Cantrell CL, Duke SO (2009) Natural products in crop protection. Bioorg Med Chem 17: 4022–4034.
2. Armas EDD, Monteiro RTR, Antunes PM, Santos MAPFD, Camargo PBD, et al. (2007) Diagnóstico espaço-temporal da ocorrência de herbicidas nas águas superficiais e sedimentos do Rio Corumbataí e principais afluentes. Quim Nova 30: 1119–1127.
3. Sands DC, Morris CE, Dratz EA, Pilgeram A (2009) Elevating optimal human nutrition to a central goal of plant breeding and production of plant-based foods. Plant Sci 117: 377–389.
4. Kiely T, Donaldson D, Grube A (2004) Pesticides industry sales and usage: 2000 and 2001 market estimate. US Environmental Protection Agency 1–33.
5. Jiang L, Huang J, Liang L, Zheng PY, Yang H (2008) Mobility of prometryne in soil as affected by dissolved organic matter. J Agric Food Chem 56: 11933–11940.
6. Lopez-Perez GC, Arias-Estevez M, Lopez-Periago E, Soto-Gonzalez B, Cancho-Grande B, et al. (2006) Dynamics of pesticides in potato crops. J Agric Food Chem 54: 1797–1803.
7. Martins PF, Martinez CO, Carvalho G, Carneiro OIB, Azevedo RA, et al. (2007) Selection of microorganisms degrading s-metolachlor herbicide. Braz Arch Biol Technol 50: 153–159.
8. Sharma D, Nagpal A, Pakade YB, Katnoria JK (2010) Analytical methods for estimation of organophosphorus pesticide residues in fruits and vegetables: A review. Talanta 82: 1077–1089.
9. Cycon M, Markowicz A, Piotrowska-Seget Z (2013) Structural and functional diversity of bacterial community in soil treated with the herbicide napropamide estimated by the DGGE, CLPP and r/K-strategy approaches. Appl Soil Ecol 72: 242–250.
10. Hernández M, Jia Z, Conrad R, Seeger M (2011) Simazine application inhibits nitrification and changes the ammonia-oxidizing bacterial communities in a fertilized agricultural soil. FEMS Microbiol Ecol 78: 511–519.
11. Bonnet JL, Bonnemoy F, Dusser M, Bohatier J (2008) Toxicity assessment of the herbicides sulcotrione and mesotrione toward two reference environmental microorganisms: *Tetrahymena pyriformis* and *Vibrio fischeri*. Arch Environ Contam Toxicol 55: 576–583.
12. Silva TM, Stets MI, Mazzetto AM, Andrade FD, Pileggi SAV, et al. (2007) Degradation of 2,4-D herbicide by microorganisms isolated from Brazilian contaminated soil. Braz J Microbiol 38: 522–525.
13. Martins PF, Carvalho G, Gratão PL, Dourado MN, Pileggi M, et al. (2011) Effects of the herbicides acetochlor and metolachlor on antioxidant enzymes in soil bacteria. Process Biochem 46: 1186–1195.
14. Araújo ASF, Monteiro RTR, Abarkeli RB (2003) Effect of glyphosate on the microbial activity of two Brazilian soils. Chemosphere 52: 799–804.
15. Gratão PL, Polle A, Lea PJ, Azevedo RA (2005) Making the life of heavy metal-stressed plants a little easier. Funct Plant Biol 32: 481–494.
16. Zhang Y, Meng D, Wang Z, Guo H, Wang Y, et al. (2012) Oxidative stress response in atrazine-degrading bacteria exposed to atrazine. J Haz Mater 229–230: 434–438.
17. Monteiro CC, Carvalho RF, Gratão PL, Carvalho G, Tezotto T, et al. (2011) Biochemical responses of the ethylene-insensitive *Never ripe* tomato mutant subjected to cadmium and sodium stresses. Environ Exp Bot 71: 306–320.
18. Cia MC, Guimarães ACR, Medici LO, Chabregas SM, Azevedo RA (2012) Antioxidant responses to water deficit by drought-tolerant and sensitive sugarcane varieties. Ann Appl Biol 161: 313–324.
19. Allocati N, Favaloro B, Masulli M, Alexeyev MF, Di Ilio C (2003) Proteus mirabilis glutathione S-transferase B1-1 is involved in protective mechanisms against oxidative and chemical stresses. Biochem J 373: 305–311.
20. Ghelfi A, Gaziola SA, Cia MC, Chabregas SM, Falco MC, et al. (2011) Cloning, expression, molecular modelling and docking analysis of glutathione transferase from *Saccharum officinarum*. Ann Appl Biol 159: 267–280.
21. Allocati N, Federici L, Masulli M, Di Ilio C (2009) Glutathione transferases in bacteria. Febs J 276: 58–75.
22. Mitchell G, Bartlett DW, Fraser TEM, Hawkes TR, Holt DC, et al. (2001) Mesotrione: a new selective herbicide for use in maize. Pest Manag Sci 57: 120–128.
23. Norris SR, Barrette T, DellaPenna D (1995) Genetic dissection of carotenoid synthesis in *Arabidopsis* defines plastoquinone as an essential component of phytoene desaturation. Plant Cell 7: 2139–2149.
24. Barrett M (1995) Metabolism of herbicides by cytochrome P450 in corn. Drug Metabol Drug Interact 12: 299–315.
25. Nordby JN, Williams MM, Pataky JK, Riechers DE, Lutz JD (2008) A common genetic basis in sweet corn inbred Cr1 for cross sensitivity to multiple cytochrome P450-metabolized herbicides. Weed Sci 56: 376–382.
26. Hatzios KK, Burgos N (2004) Metabolism-based herbicide resistance: regulationby safeners. Weed Sci 52: 454–467.
27. Picket CB (1989) Glutathione S-transferases: gene structure, regulation, and biological function. Annu Rev Biochem 58: 743–764.
28. Oakley A (2011) Glutathione transferases: a structural perspective. Drug Metab Rev 43: 138–151.
29. Gaudu P, Fontecave M (1994) The NADPH: sulfite reductase of *Escherichia coli* is a paraquat reductase. Eur J Biochem 226: 459–463.
30. Alferness P, Wiebe L (2002) Determination of mesotrione residues and metabolites in crops, soil, and water by liquid chromatography with fluorescence detection. J Agric Food Chem 50: 3926–3934.

31. Durand S, Sancelme M, Besse-Hoggan P, Combourieu B (2010) Biodegradation pathway of mesotrione: complementarities of NMR, LC-NMR and LC-MS for qualitative and quantitative metabolic profiling. Chemosphere 81: 372–380.
32. Pileggi M, Pileggi SAV, Olchanheski LR, Silva PAG, Gonzalez AMM, et al. (2012) Isolation of mesotrione-degrading bacteria from aquatic environments in Brazil. Chemosphere 86: 1127–1132.
33. Crouzet O, Batisson I, Besse-Hoggan P, Bonnemoy F, Bardot C, et al. (2010) Response of soil microbial communities to the herbicide mesotrione: a dose-effect microcosm approach. Soil Biol Biochem 42: 193–202.
34. Botelho RG, Froes CM, Santos JB (2012) Toxicity of herbicides on *Escherichia coli* growth. Braz J Biol 72: 141–146.
35. Sondi I, Sondi BS (2004) Silver nanoparticles as antimicrobial agent: a case study on *E. coli* as a model for Gram-negative bacteria. J Coll Int Sci 275: 177–182.
36. Williams JA, Carnes AE, Hodgson CP (2009) Plasmid DNA vaccine vector design: impact on efficacy, safety and upstream production. Biotec Adv 27: 353–370.
37. Heath RL, Packer L (1968) Photoperoxidation in isolated chloroplasts. I. Kinetics and stoichiometry of fatty acid peroxidation. Arch Biochem Biophys 125: 189–198.
38. Garcia JS, Gratão PL, Azevedo RA, Arruda MAZ (2006). Metal contamination effects on sunflower (*Helianthus annuus L.*) growth and protein expression in leaves during development. J Agric Food Chem 54: 8623–8630.
39. Bradford MM (1976) A rapid and sensitive method for the quantification of microgram quantities of protein utilizing the principle of protein-dye binding. Anal Biochem 72: 248–254.
40. Gratão PL, Monteiro CC, Antunes AM, Peres LEP, Azevedo RA (2008) Acquired tolerance of tomato (*Lycopersicon esculentum* cv. Micro-Tom) plants to cadmium-induced stress. Ann Appl Biol 153: 321–333.
41. Beauchamp C, Fridovich I (1971) Superoxide Dismutase: improved assays and an assay applicable to acrylamide gels. Anal Biochem 287: 276–287.
42. Medici LO, Azevedo RA, Smith RJ, Lea PJ (2004) The influence of nitrogen supply on antioxidant enzymes in plant roots. Funct Plant Biol 31:1–9
43. Kraus TE, Mckersie BD, Fletcher RA (1995) Paclobutrazol-induced tolerance of wheat leaves to paraquat may involve increased antioxidant enzyme activity. J Plant Physiol 145: 570–576.
44. Zablotowicz RM, Hoagland RE, Locke MA, Hickey WJ (1995) Glutathione-s-transferase activity and metabolism of glutathione conjugates by rhizosphere bacteria. Appl Environ Microbiol 61: 1054–1060.
45. Bligh EG, Dyer WJ (1959) A rapid method of total lipid extraction and purification. Can J Biochem Phys 37: 911–917.
46. Souza ML, Wackett LWP, Mills KL, Mandelbaum RT, Sadowsky MJ (1995) Cloning, characterization, and expression of a gene region from *Pseudomonas* sp. Strain ADP involved in the dechlorination of atrazine. Appl Environ Microbiol 61: 3373–3378.
47. Penna TCV, Ishii M, Souza LC, Cholewa O (2004) Expression of green fluorescent protein (GFPuv) in *Escherichia coli* DH5-α, under different growth Conditions. African J Biotec 3: 105–111.
48. Greenberg JT, Manach P, Chou J, Josephy D, Demple B (1990) Positive control of a global antioxidant defense regulon activated by superoxide-generating agents in *Escherichia coli*. Proc Natl Acad Sci USA 87: 6181–6185.
49. Isarankura-Na-Ayudhya P, Isarankura-Na-Ayudhya C, Yainoy S, Thippakorn C, Singhagamol W, et al. (2010) Proteomic alterations of *Escherichia coli* by paraquat. EXCLI J 9: 108–118.
50. Lizuka M, Inoue Y, Murata K, Kimura A (1989) Purification and some properties of glutathione S-transferase from *Escherichia coli* B. J Bacteriol 171: 6039–6042.
51. Batisson I, Crouzet O, Besse-Hoggan P, Sancelme M, Mangot JF, et al. (2009) Isolation and characterization of mesotrione-degrading *Bacillus* sp. from soil. Environ Poll 157: 1195–1201.
52. Durand S, Amato P, Sancelme M, Delort AM, Combourieu B, et al. (2006) First isolation and characterization of a bacterial strain that biotransforms the herbicide mesotrione. Lett Appl Microbiol 43: 222–228.
53. Balagué C, Sturtz N, Duffard R, Duffard AME (2001) Effect of 2,4-dichlorophenoxyacetic acid herbicide on *Escherichia coli* growth, chemical composition and cellular envelope. Environ Toxicol 16: 43–53.
54. Lima ES, Abdalla DSP (2001) Peroxidação lipídica: mecanismos e avaliação em amostras biológicas. Braz J Pharm Sci 37: 293–303.
55. Böger P, Matthes B, Schmalfub J (2000) Towards the primary target of chloroacetamides - new findings pave the way. Pest Manag Sci 56: 497–508.
56. Sánchez M, Garbi C, Martínez-Álvarez R, Ortiz LT, Allende JL, et al. (2005) Klebsiella planticola strain DSZ mineralizes simazine: physiological adaptations involved in the process. Appl Microbiol Biotechnol 66: 589–596.
57. Azevedo RA, Alas RM, Smith RJ, Lea PJ (1998) Response of antioxidant enzymes to transfer from elevated carbon dioxide to air and ozone fumigation, in the leaves and roots of wild-type and a catalase-deficient mutant of barley. Physiol Plant 104: 280–292.
58. Jung IL, Kim IG (2003) Transcription of ahpC, katG, and katE genes in *Escherichia coli* is regulated by polyamines: polyamine-deficient mutant sensitive to H_2O_2-induced oxidative damage. Biochem Biophys Res Commun 301: 915–922.
59. Nagy I, Compernolle F, Ghys K, Vanderleyden J, Demot J (1995) A single cytochrome P-450 system is involved in degradation of the herbicides EPTC (*s*-ethyl dipropylthiocarbamate) and atrazine by *Rhodococcus* sp. strain NI86/21. Appl Environ Microbiol 61:2056–2060.
60. Shao ZQ, Behki R (1995) Cloning of the genes for degradation of the herbicides EPTC (*s*-ethyl dipropylthiocarbamate) and atrazine from *Rhodococcus* sp. strain TE1. Appl Environ Microbiol 61: 2061–2065.
61. Fournier J, Soulas G, Parekh N (1997) Main microbial mechanisms of pesticide degradation in soils. In: Tarradellas J (ed) Soil ecotoxicology. Lewis Publishers CRC, New York, pp 85–116.
62. Bus JS, Gibson JE (1984) Paraquat: model for oxidant-initiated toxicity. Environ Health Perspect 55: 37–46.
63. Ma XX, Jiang YL, He YX, Bao R, Chen Y, et al. (2009) Structures of yeast glutathione-S-transferase Gtt2 reveal a new catalytic type of GST family. EMBO Reps 10: 1320–1326.
64. Skopelitou K, Dhavala P, Papageorgiou AC, Labrou NE (2012) A glutathione transferase from *Agrobacterium tumefaciens* reveals a novel class of bacterial GST superfamily. PloS One 7: e34263.
65. Van Eerd LL, Haogland RE, Hall JC (2003) Pesticide metabolism in plants and microorganisms. Weed Sci 51: 472–495.
66. McGuinness MC, Mazurkiewicz V, Brennan E, Dowling DN (2007) Dechlorination of pesticide by a specific bacterial glutathione S transferase. Eng Life Sci 7: 611–615.
67. Cummins I, Andon NL, Haynes PA, Park M, Fisher WH, et al. (2011) Multiple roles for plant glutathione transferases in xenobiotic detoxification. Drug Metab Rev 43: 266–280.

17

Proteolytic Pathways Induced by Herbicides That Inhibit Amino Acid Biosynthesis

Amaia Zulet[1], Miriam Gil-Monreal[1], Joji Grace Villamor[2], Ana Zabalza[1], Renier A. L. van der Hoorn[2], Mercedes Royuela[1]*

1 Departamento de Ciencias del Medio Natural, Universidad Pública de Navarra, Pamplona, Spain, **2** Plant Chemetics Laboratory, Max-Planck Institute for Plant Breeding Research, Cologne, Germany

Abstract

Background: The herbicides glyphosate (Gly) and imazamox (Imx) inhibit the biosynthesis of aromatic and branched-chain amino acids, respectively. Although these herbicides inhibit different pathways, they have been reported to show several common physiological effects in their modes of action, such as increasing free amino acid contents and decreasing soluble protein contents. To investigate proteolytic activities upon treatment with Gly and Imx, pea plants grown in hydroponic culture were treated with Imx or Gly, and the proteolytic profile of the roots was evaluated through fluorogenic kinetic assays and activity-based protein profiling.

Results: Several common changes in proteolytic activity were detected following Gly and Imx treatment. Both herbicides induced the ubiquitin-26 S proteasome system and papain-like cysteine proteases. In contrast, the activities of vacuolar processing enzymes, cysteine proteases and metacaspase 9 were reduced following treatment with both herbicides. Moreover, the activities of several putative serine protease were similarly increased or decreased following treatment with both herbicides. In contrast, an increase in YVADase activity was observed under Imx treatment versus a decrease under Gly treatment.

Conclusion: These results suggest that several proteolytic pathways are responsible for protein degradation upon herbicide treatment, although the specific role of each proteolytic activity remains to be determined.

Editor: Abidur Rahman, Iwate University, Japan

Funding: A. Zulet and M. Gil-Monreal received funding from Spanish fellowships through the Ministerio de Educación and the Universidad Pública de Navarra, respectively. This work was financially supported through grants from the Spanish Ministry of Education and Science (AGL-2010-18621) and Max Planck Society and the COST Action CM1004 for funding. The funders had no role in study design, data collection and analysis, decision to publish, or preparation of the manuscript.

Competing Interests: The authors have declared that no competing interests exist.

* E-mail: royuela@unavarra.es

Introduction

Herbicides that inhibit amino acid biosynthesis are useful tools in weed management and have been particularly successful because of their low toxicity in mammals, as these herbicides inhibit pathways that are lacking in mammals. There are several types of herbicides whose targets or primary sites of action are associated with the specific inhibition of enzymatic activity in biosynthetic pathways for amino acids. One such group of herbicides comprises compounds that inhibit the biosynthesis of branched-chain amino acids (valine, leucine and isoleucine) through the inhibition of acetolactate synthase (ALS, EC 4.1.3.18), also referred to as acetohydroxyacid synthase. ALS inhibitors include the active ingredients of several classes of chemicals and have become one of the most widely used types of herbicides because of their wide-spectrum weed control activity, high crop selectivity, low required application rates and low toxicity in mammals [1]. Glyphosate (Gly) is another type of herbicide that inhibits amino acid biosynthesis, through inhibition of 5-enolpyruvylshikimate-3-phosphate synthase (EPSPS, EC 2.5.1.19) [2], which is involved in the biosynthesis of aromatic amino acids (tyrosine, phenylalanine and tryptophan). Gly is a wide-spectrum, non-selective post-emergence herbicide that is the most popular herbicide used worldwide, particularly since the introduction of transgenic Gly-resistant crops [3].

Although the targets (mechanisms of action) of these two types of herbicides are well known, it is not fully understood what causes plant death following the inhibition of ALS or EPSPS. Several physiological effects in the mode of action of ALS and EPSPS inhibitors have been described. Interestingly, most of these effects are common, although the target sites involved are different. A general physiological effect reported following both EPSPS and ALS inhibition is growth arrest, followed by the slow death of treated plants [4,5]. Carbon metabolism is impaired following the application of both types of herbicides, while aerobic fermentation in roots is induced [6,7], and the carbohydrate content of roots and leaves is increased upon treatment [7,8]. The increased sucrose and starch content detected in the roots following herbicide treatment triggers a decrease in sink strength, which inhibits phloem transport and causes carbohydrate accumulation in the leaves [8]. Another common physiological effect observed after the use of these classes of herbicides is accumulation of secondary metabolites, such as quinate, a compound synthesized in a lateral branch of the shikimate pathway [9].

Additional specific common biochemical effects of ALS and EPSPS inhibitors are an increase in the free amino acid pool and a decrease in the soluble protein content. Amino acid accumulation has been observed in pea plants treated with lethal doses of Gly [7,9]. Similarly, an increase in free amino acid contents following ALS inhibition have been widely reported [9–14]. Furthermore, a decrease in the amount of soluble protein after applying ALS inhibitors has been demonstrated [15]. It has been proposed that this increased free amino acid pool reflects a rise in protein turnover as a result of increased degradation and reduced synthesis rates [16]. Indeed, although protein synthesis occurs following ALS inhibitor treatment, the amino acid components of these proteins are not generated from newly incorporated nitrogen [12] but are instead primarily scavenged from protein degradation. This observation supports the hypothesis that proteases might be involved in protein degradation to release amino acids that cannot be synthesized.

Plants produce hundreds of proteases that are involved in numerous biological processes. The ubiquitin/proteasome system is a major pathway for the degradation and processing of damaged proteins. The 26 S proteasome is a large, multi-subunit protease found in the cytosol and nucleus [17]. In this proteolytic pathway, proteins are first modified through covalent conjugation with ubiquitin, which marks them for rapid hydrolysis by the 26 S proteasome. The 26 S proteasome exhibits caspase-like (peptidyl-glutamyl peptide hydrolase-like, PGPH), trypsin-like and chymotrypsin-like activities [18]. Other proteases are located in lytic vesicles or the vacuole, such as vacuolar-processing enzymes (VPEs) and papain-like cysteine proteases (PLCPs). VPEs are cysteine proteases that show YVADase activity, cleaving after Asn and Asp [19,20]. Additionally, serine proteases are involved in immune responses, stomatal density regulation, detoxification and secondary metabolism [21].

Based on the above information, we propose that there is increased protease activity upon herbicide treatment for two reasons: proteases are often induced during stress, and second, because we suspect that proteases are involved in the release of free amino acids upon herbicide treatment.

Here, we characterized the activities of different proteases in pea roots upon treatment with the ALS inhibitor imazamox (Imx) and the EPSPS inhibitor Gly. The activities of major classes of proteases were monitored using fluorescent substrates and through protease activity profiling, showing that the activities of several proteases (e.g., VPEs) were downregulated, whereas the activities of others were upregulated upon treatment with both herbicides.

Materials and Methods

Plant Material and Treatment Application

Pisum sativum L. cv. Snap Sugar Boys plants were grown in aerated hydroponic culture in a growth chamber as previously described [6]. When the plants were 12 days old, herbicide treatments were applied to the nutrient solution. The ALS-inhibiting herbicide Imx (commercial formula Pulsar®40, BASF Española SA, Barcelona, Spain) was applied at a final concentration of 0.016 mM (5 mg l^{-1}). The EPSPS-inhibiting herbicide Gly (commercial formula Roundup®Plus, MONSANTO Agricultura España SL, Madrid, Spain) was applied at a concentration of 0.234 mM (53 mg l^{-1}).

For analytical purposes, root samples were collected at different time points after treatment. At harvest, root samples were immediately frozen in liquid nitrogen and stored at −80°C until further use.

Amino Acid Content Determination

Extraction of amino acids was performed in HCl. Following protein precipitation, the amino acid concentrations in the supernatants were measured using a capillary electrophoresis apparatus equipped with a laser-induced fluorescence detector, as previously described [9].

Protease Assays and Soluble Protein Determination

Proteins were extracted from 350 mg of ground pea roots in 700 µl of extraction buffer containing 50 mM Tris-HCl (pH 7), 1 mM EDTA, 5 mM DTT, 5 mM β-mercaptoethanol and 1% PVPP following a previously described method [22] with some modifications. The samples were centrifuged for 15 min at 15,000 g at 4°C, and the supernatant was used in protease assays and subjected to protein content determination. The protein concentration was determined as previously described [23].

Protease activities were measured using specific fluorogenic substrate (see below). All fluorogenic substrates were purchased from Sigma-Aldrich (St. Louis, MO, USA), except Val-Arg-Pro-Arg-AMC (VRPR-AMC), which was purchased from Bachem (Bubendorf, Switzerland). All substrates were dissolved in DMSO and diluted in water to their final concentrations. For all examined activities, the assay mixtures contained the specific substrate and the protein extract in a reaction cocktail. Protease activities were measured at 30°C using a Sinergy™ HT microplate reader (BioTek® Instruments Inc., VT, USA) under 380 nm excitation/460 nm emission. The obtained activities were expressed as RFU min^{-1} mg^{-1} protein.

The specific substrates Leu-Leu-Val-Tyr-AMC (LLVY-AMC), Gly-Lys-Arg-AMC-HCl (GKR-AMC-HCl) and Leu-Leu-Glu-AMC (LLE-AMC) were used to evaluate the chymotrypsin- (adapted from [22]), trypsin- (adapted from [24]) and PGPH-like activities (adapted from [25]) of the proteasome, respectively. These three types of activity of the proteasome were determined as the amount of activity inhibited after incubating a crude extract with the proteasome inhibitor MG132 (100 µM final concentration) at RT for 30 min before measuring the enzymatic activities.

YVADase activity was determined using the specific substrate Tyr-Val-Ala-Asp-AMC (YVAD-AMC), which is the most common substrate of caspase 1-like activity, as previously described [26]. YVADase activity as determined in the presence or absence of a VPE inhibitor (I440), which was added at a concentration of 50 µM to the sample, followed by incubation at RT for 30 min prior to the determination of enzymatic activity. For cysteine proteases, the specific substrate Z-Gly-Gly-Arg-AMC (Z-GGR-AMC) was used, as previously described [24]. To detect metacaspase 9-like activity, the specific substrate Val-Arg-Pro-Arg-AMC (VRPR-AMC) was employed, as described elsewhere [27].

Ubiquitin Western Immunoblotting

Total protein was isolated from the roots as previously described [6]. Western blotting was performed according to standard techniques. A ubiquitin antibody (Ubq11) from Agrisera (Vännäs, Sweden) was used in this assay at a dilution of 1:5,000. Anti-Rabbit IgG Alkaline Phosphatase (Sigma-Aldrich, St. Louis, MO, USA) was employed as the secondary antibody at a dilution of 1:20,000, and of the obtained bands were visualized using the Immun-Blot Amplified AP Assay, Bio-Rad 170 6412 (Bio-Rad, Hercules, CA, USA).

Labeling, Activity-based Protein Profiling (ABPP) and Detection

A 100 mg sample of ground pea roots was labeled. The protein concentration was measured using the RC/DC Protein Assay (Bio-Rad, Hercules, CA, USA) according to the manufacturer's instructions. For ABPP of the proteasome and PLCPs, MV151 labeling was performed by incubating ~100 μg of protein in 50 μl of 50 mM PBS buffer (pH 7.5 for the proteasome) or sodium acetate (pH 6.0 for the PLCPs) containing the probe MV151 at 0.4 μM for 1 h, as previously described [28]. For ABPP of VPEs, AMS101 labeling was conducted by incubating ~100 μg of protein in 50 μl of 400 mM sodium acetate buffer (pH 5) containing 4 mM TCEP and the probe AMS101 at 1 μM for 2 h, as described elsewhere [29]. For ABPP of serine proteases, FPRh labeling was carried out by incubating ~100 μg of protein in 50 μl of 50 mM PBS buffer (pH 7.5) and the probe FPRh at 2 μM for 1 h, as previously described [21]. Equal volumes of DMSO were added to the no-probe controls. All labeling reactions were conducted in the dark at room temperature.

To achieve protease labeling in competition assays, the control extracts were pre-incubated with the inhibitor for 30 min at room temperature (RT) prior to labeling. Inhibitors and final concentrations used were: E64, cysteine proteases inhibitor 10 μM; MG132, cysteine proteases and proteasome inhibitor 10 μM; epoxomicin, specific proteasome inhibitor 20 μM; I440, VPEs inhibitor 50 μM and Ac-YVAD-CMK, YVADase inhibitor 50 μM. All inhibitors were purchased from Calbiochem (San Diego, CA, USA).

The reactions were terminated following the addition of 20 μl of 4X SDS electrophoresis gel loading buffer (280 mM SDS, 400 mM Tris, 40% glycerol, 1.4 M β-mercaptoethanol and 0.6 mM Bromophenol Blue, pH 6.8). The obtained proteins were separated in 12% SDS polyacrylamide gels. The gels were then washed three times for 15 min with distilled water, and the blots were scanned using a Typhoon 8600 scanner (GE Healthcare, Munich, Germany). The excitation wavelength was set to 532 nm, and emission was measured using a TAMRA filter (580 nm). A GS-800 densitometer (Bio-Rad, Hercules, CA, USA) was used for band quantification.

Statistical Analysis

The mean values were calculated using the samples obtained from individual plants as replicates. The results were subjected to a separate one-way ANOVA for each day of treatment (SPSS 18.0). The means were separated using the least significant difference method ($p<0.05$, Fisher protected). Significant differences between each treatment and the control plants (non-treated plants) are highlighted in the figures using a different symbol for each treatment.

Results and Discussion

Effects on Growth and the Contents of Free Amino Acids and Soluble Proteins

We previously showed that application of 0.234 mM Gly to pea roots in a hydroponic system results in slow, robust and synchronous death of this crop plant within 20 days [7]. When we tested different Imx concentrations we found that at 0.016 mM Imx have similar effects on the intensity and rate of pea growth retardation as 0.234 mM Gly (Fig. 1A). Both treatments caused arrest of root and shoot elongation, which was significant on the seventh day. In both cases, plant death occurred at 20 days.

We next evaluated the free amino acid and soluble protein contents at 1, 3 and 7 days after treatment (Fig. 1B). While no changes were detected after 24 h, the application of Imx or Gly caused an increase in the total free amino acid content in the roots and a consistent decrease in the soluble protein content from day 3 onward (Fig. 1B). These effects were consistent with the results reported in the literature [7,9–15], suggesting the existence of altered nitrogen metabolism following ALS or EPSPS inhibition. Furthermore, increased free amino acid levels and decreased protein levels suggest that proteolytic activities are increased upon herbicide treatment.

In order to find which proteolytic processes are most significant to the change of free amino acid and amount of soluble protein, it was planned to evaluate free amino acid and soluble protein in herbicide treated plants pre-incubated with proteasome and YVAD-CHO inhibitors as it has been reported before [20,24,30]. The objective was to determine if the toxicity of the herbicides was enhanced or alleviated by inhibiting the YVADase activity (with YVAD-CHO) or the proteasome (with MG132). Unfortunately, at day 3 of treatment no differences in the YVADase or the proteasome activities were found between control and pretreated plants and the results were inconclusive.

We next evaluated the activities of the main proteolytic systems found in plants with both herbicides. For these analyses, the time point of 3–4 days following herbicide treatment was selected, representing the shortest time period at which significant differences in both the amino acid and soluble protein contents were observed without any effects on growth (Fig. 1A).

Effect of Herbicide Treatments on the Proteasome

The ubiquitin-proteasome proteolytic system plays an important role in eukaryotic cell growth, development and stress responses and the environmental adaptation of plants through the degradation of short-lived and abnormal proteins [31,32]. The role of the proteasome following herbicide treatment was evaluated through the quantification of polyubiquitinated proteins and proteasome activities (Fig. 2). Ubiquitin immunoblotting of total protein extracts from plant roots was performed 1, 2, 3 and 4 days after herbicide application. Figure 2A shows that there was an increase in the amount of polyubiquitinated proteins upon treatment with both herbicides. This increase was significant at 2 days after Imx and 3 days after Gly treatment, and the signal was stronger following Imx treatment. An increase in the level of ubiquitin-protein conjugates has been reported following other types of abiotic and biotic stresses, such as heat shock [33], darkness, UV radiation, starvation, enhanced ozone levels [34] and cadmium stress [25,35].

The 26 S proteasome is a large proteolytic complex exhibiting three different types of cleavage activity: chymotrypsin-, trypsin- and caspase-like (PGPH) activities. Fluorogenic substrates were used to determine these activities of the proteasome upon herbicide treatment. Specific fluorogenic substrates are available to monitor chymotrypsin-, trypsin- and caspase-like (PGPH) activities, but these substrates are also subject to proteolytic processes that are not involved in the proteasome. Root extracts were incubated with or without MG132, a widely used proteasome inhibitor that inhibits all three enzymatic activities in proteasome at the concentration used in this experiment [36]. Then, the extracts were subsequently incubated with synthetic peptide substrates conjugated with the fluorescent reporter 7-amino-4-methylcoumarin (AMC). Therefore, the chymotrypsin-, trypsin- and caspase-like (PGPH) activities involved in the proteasome were calculated by subtracting values of activities obtained in the condition with MG132 (activities not involved in the proteasome) from that without MG132 (total activities) [25]. No general effect of Imx or Gly treatment on the chymotrypsin-, trypsin- and

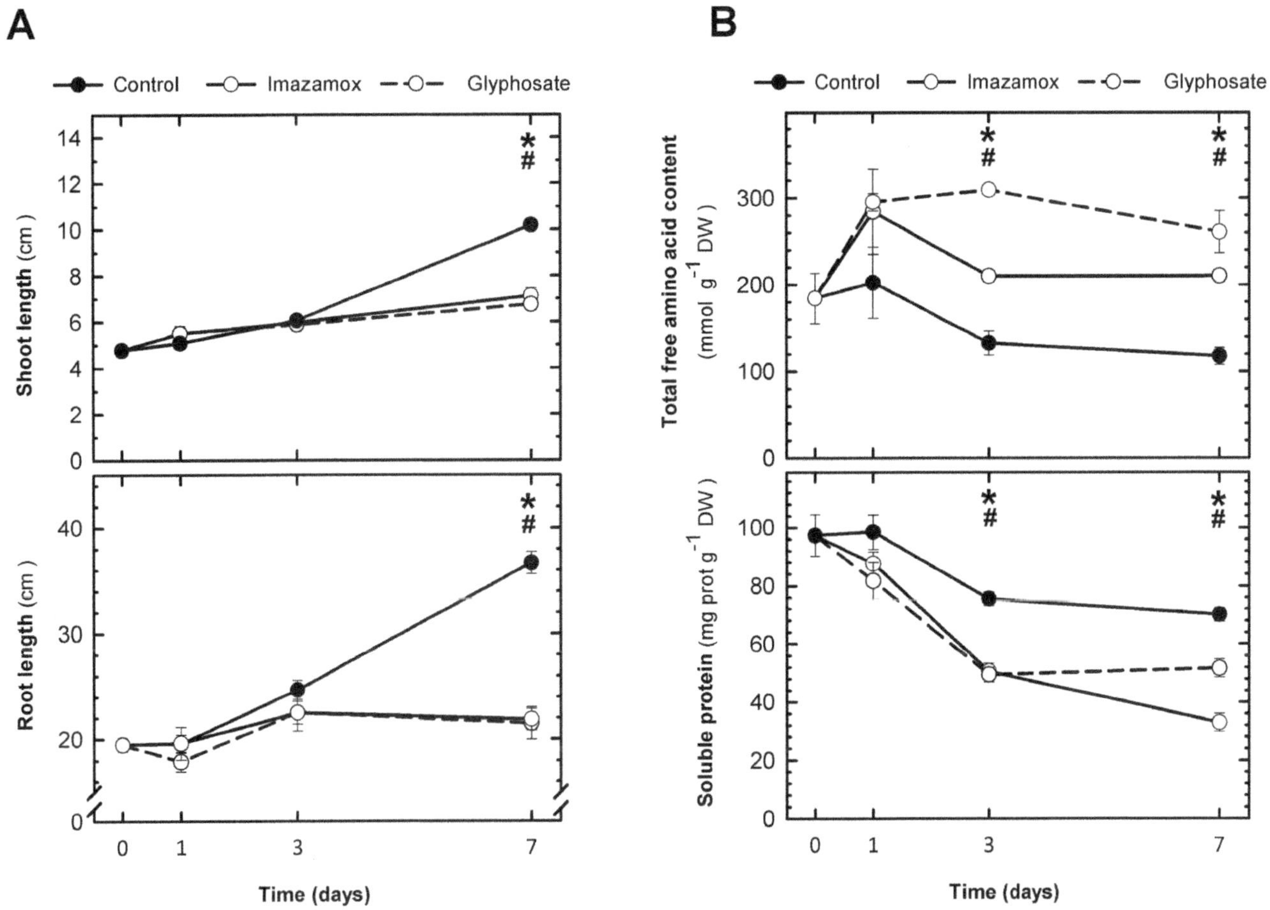

Figure 1. Effect of amino acid biosynthesis inhibitors on growth and amino acid and protein contents. (A) Shoot and root growth of control pea plants and plants treated with imazamox or glyphosate. Means ± SE (n = 8). The symbols indicate significant differences between the control and imazamox (*) or glyphosate (#) treatment on a given day ($p<0.05$). (B) Total free amino acid and soluble protein contents in the roots of control pea plants and plants treated with imazamox or glyphosate. Means ± SE (n = 4–8). The symbols indicate significant differences between the control and imazamox (*) or glyphosate (#) treatment on a given day ($p<0.05$).

caspase-like (PGPH) activities of the proteasome could be identified (Fig. 2B). However, MG132 inhibits not only the proteasome, but also cysteine proteases [37], and thus, the activities estimated following incubation with this inhibitor might not reflect proteasome activity alone.

To overcome this limitation, we used activity based protein profiling (ABPP) to directly detect the active proteasome. ABPP is based on the use of fluorescent small molecule probes that covalently react with active site residues at enzymes resulting in an irreversible labeling that can be detected and quantified form protein gels scanned for fluorescence [38]. We used MV151 [28] to monitor proteasome activity 3 days after herbicide treatment (Fig. 2C). Following labeling with the MV151, two fluorescent signals were detected at 25 kDa. Previous studies involving labeling with the MV151 in *Arabidopsis* leaves revealed three strong fluorescent signals at approximately 25 kDa, representing the three proteasome catalytic subunits [28]. However, we detected only two signals in pea roots (Fig. 2C), similar to the results of proteasome labeling in *Nicotiana benthamiana* [29]. To confirm the identities of these bands, a competitive assay was performed. In this assay, control root extracts were pre-incubated with the inhibitors E64 (which inhibits cysteine proteases, but not the proteasome [39]), MG132 (which inhibits both the proteasome and cysteine proteases) and epoxomicin (which inhibits only the proteasome [18]), followed by labeling with MV151. The observed competition of MV151 labeling with epoxomicin and MG132, but not E64 indicates that the lower signal detected at 24 kDa represents the proteasome (Fig. 2C, right). Importantly, this band showed a 2-fold greater intensity after treatment with Imx or Gly, indicating an increase in proteasome activity following herbicide treatment. Previous studies using the MV151 probe have observed an increase in proteasome activity during defense [28]. At this stage, it is unclear what the upper 25-kDa-band represents, as this band was not competed with E64 or epoxomicin.

The accumulation of ubiquitinated proteins, together with increased putative proteasome activity was observed through ABPP, indicating a role for the proteasome upon herbicide treatment. Accumulation of ubiquitinated proteins has typically been described in association with a concomitant decrease in proteasome activity [25,31,35,36]. Nevertheless, our results demonstrated increases in both proteasome substrate levels and activities. Thus, the herbicide-induced stress on the proteome might result in the accumulation of ubiquitinated proteins, despite the increased proteasome activity, or the increased availability of the substrate might induce proteasome activity. A previous study using phosphinothricin, another type of herbicide that inhibits amino acid biosynthesis, also revealed a role for the proteasome pathway in the response to herbicide treatment [40]. The authors

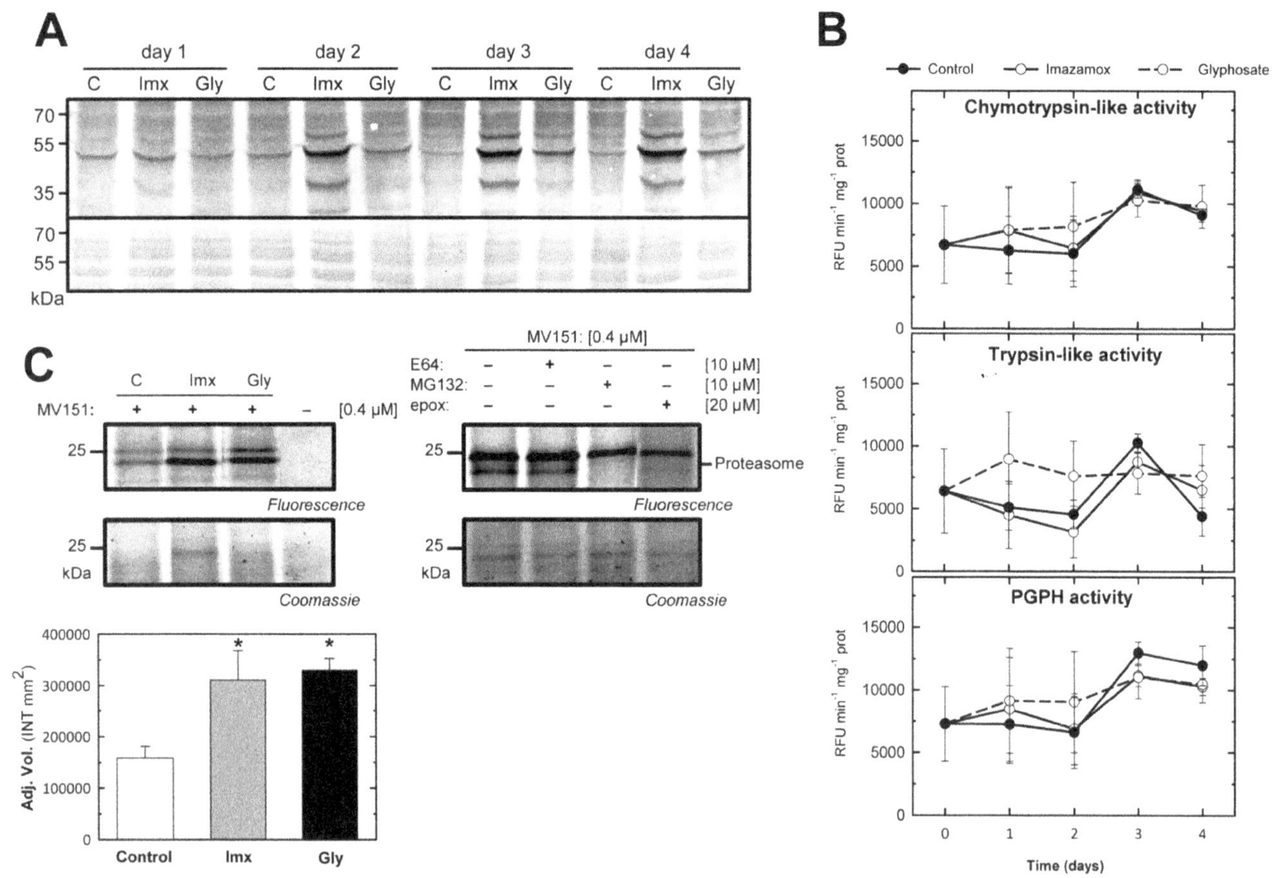

Figure 2. Effect of amino acid biosynthesis inhibitors on the proteasome. (A) Ubiquitin immunoblotting of total protein extracts from control pea plants (C) or plants treated with imazamox (Imx) or glyphosate (Gly) for 1, 2, 3 and 4 days. A total of 30 μg of protein was loaded into each well. The Coomassie-stained protein gel on the bottom shows the total amounts of input proteins. (B) Chymotrypsin-like, trypsin-like and PGPH activities of the proteasome. These activities were measured using the specific substrates LLVY-AMC, GKR-AMC-HCl and LLE-AMC, respectively. The presented data correspond to the activities after subtracting the values obtained following incubation with the proteasome inhibitor MG132 for 30 min from those recorded without MG132. Means ± SE (n = 3). (C) ABPP of the proteasome. Left, a comparison of the labeling profiles observed in pea roots treated with imazamox or glyphosate at day 3 following incubation with 0.4 μM MV151 for 1 h at pH 7.5. Right, a competitive assay in which root extracts were pre-incubated with 10 μM E64, 10 μM MG132 or 20 μM epoxomicin for 30 min before labeling. The results of the competitive assay showed that the 25 kDa signal corresponded to the proteasome. Fluorescently labeled proteins were detected from protein gels via fluorescence scanning. The signals were quantified using a densitometer, and the results are shown in the bar graph. Means ± SE (n = 3). The symbol *indicates significant differences between the control and treatments ($p<0.05$). The Coomassie-stained protein gel on the bottom shows the total amount of input protein.

observed that *Arabidopsis* mutant plants with impaired 26 S proteasome function showed increased tolerance to phosphinothricin, suggesting a relationship between the proteasome and changes in the phytotoxic response induced through herbicide treatment.

Effect of Amino Acid Biosynthesis Inhibitors on Vacuolar Processing Enzymes

VPEs, also referred to as legumains, are a family of cysteine proteases that act as asparaginyl endopeptidases [20,41]. However, Asn is not required at position P1 in the substrate cleavage site, although similar to caspase-like proteases, VPEs cleave after Asp [42,43]. Both VPE and YVADase activities are involved in programmed cell death (PCD) [20,41]. VPE activity has also been reported to increase during oomycete infection to mediate protein turnover and nutrient release [29]. The role of VPEs in the mode of action of ALS and EPSPS inhibitors was evaluated through quantification of YVADase activity, and the activity of VPEs was detected through ABPP and assessing the rate of YVADase activity associated with VPEs. A specific fluorogenic substrate was employed to determine YVADase activity upon herbicide treatment. A significant increase in YVADase activity was detected at 3 days after Imx treatment (Fig. 3A). In contrast, a significant decrease in YVADase activity was detected beginning on the first day after Gly treatment (Fig. 3A). ABPP was carried out to directly detect VPE activity, and labeling with the AMS101 probe was performed on day 3 (Fig. 3B). Similar previous labeling studies in *Arabidopsis* leaves have consistently revealed signals at 40 kDa, corresponding to VPEs [29]. Labeling pea root extracts with AMS101 revealed similar signals at 40 kDa (Fig. 3B). To determine whether both detected bands were putative VPEs, a competitive analysis was performed using the following inhibitors: I440 (which contains a Pro-Asp dipeptide and an acyloxymethylketone reactive group and inhibits VPEs), E64 (which inhibits cysteine protease activity, but not caspase-like activity) and YVAD-CMK (which inhibits YVADase activity) (Fig. 3B, right). Importantly, pre-incubation with the VPE inhibitors YVAD-

CMK and I440, but not with E64, blocked labeling, suggesting that these signals indeed represented VPEs. In addition, the lack of a reducing agent (no TCEP) also suppressed labeling, consistent with the fact that VPE activity requires reductive conditions [29]. AMS101 labeling of the extracts from herbicide-treated pea roots revealed reduced putative VPE activity upon herbicide treatment (Fig. 3B, left). The quantification of both signals indicated that there was a significant 30% reduction of VPE activity compared to the control (Fig. 3B, down). Notably, these results indicated contradictory patterns between YVADase activity and VPE labeling following herbicide treatment. The observed increase in YVADase activity and decrease in VPE activity suggests that not all YVADase activity is induced through VPEs.

To determine whether the detected YVADase activity was induced through VPEs, a fluorogenic kinetic assay was performed at day 3 with or without the VPE-specific inhibitor (Fig. 3C). Root extracts were pre-incubated with the VPE inhibitor I440 for 30 min at RT, and the YVADase activity in these extracts was subsequently measured and compared with total YVADase activity. Not all YVADase activity was inhibited following pre-incubation with I440 (Fig. 3C), indicating that not all of the observed YVADase activity could be attributed to VPEs. Consistent with the reduction of VPE labeling upon Gly treatment, the YVADase activity induced through VPE was also reduced following Gly treatment, explaining the overall decrease of YVADase activity upon Gly treatment. In the case of Imx treatment, no reduction of YVADase activity from VPEs was observed, in contrast to the decreased VPE labeling. Thus, the increased YVADase activity must reflect the activity of proteases other than VPEs.

Notably, the distinct YVADase activity observed upon herbicide treatment suggests that YVADase activity might play different roles in Imx- and Gly-treated plants. As both an increase and a decrease of YVADase activity were detected following treatment with lethal doses of herbicides, a clear role for YVADase activity in herbicide toxicity cannot be proposed. Although YVADase activity has been implicated in PCD, herbicides inhibiting amino acid biosynthesis have not been reported to induce PCD in treated plants, making it difficult to associate the increase or decrease of YVADase activity with changes in PCD.

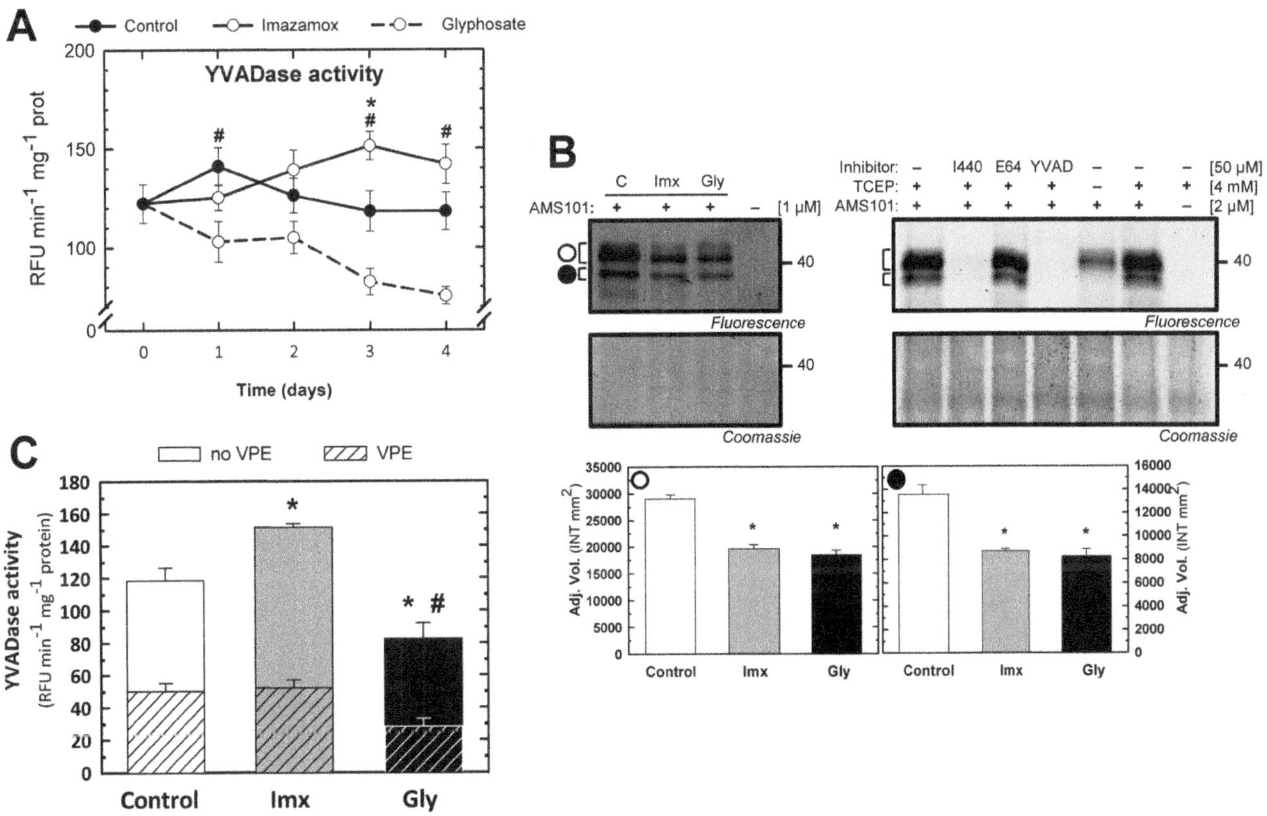

Figure 3. Effect of amino acid biosynthesis inhibitors on the vacuolar-processing enzymes. (A) Total YVADase activity determined using the specific substrate YVAD-AMC. Means ± SE (n = 7). Symbols indicate significant differences between the control and imazamox (*) or glyphosate (#) treatment on a given day ($p<0.05$). (B) Left, comparison of the VPE labeling profiles in pea roots treated with imazamox or glyphosate at day 3 following incubation 1 μM AMS101 for 2 h. Right, a competitive assay in which root extracts were pre-incubated with the inhibitors E64, Ac-YVAD-CMK (YVAD) and I440 at 10 μM, 50 μM and 50 μM, respectively, for 30 min or extracted without reducing buffer (-TCEP) prior to labeling. Fluorescently labeled proteins were detected in protein gels via fluorescence scanning, and two bands were observed (● and ○). The signals were quantified using a densitometer, and the relative values are shown in the bar graphs. Means ± SE (n = 3). The symbol * indicates significant differences between the control and the treatments ($p<0.05$). The Coomassie-stained protein gel on the bottom shows the total amounts of input proteins. (C) YVADase activity determined in the presence or absence of the VPE inhibitor I440. YVADase activity is displayed as the total YVADase activity due to VPEs ("VPE") or without VPEs ("no VPE"). The root extracts were incubated with the inhibitor at a concentration of 50 μM for 30 min at RT prior to adding the specific substrate YVAD-AMC. Means ± SE (n = 3). Symbols indicate significant differences in no-VPE YVADase activity (*) and VPE YVADase activity (#) between the control and treated plants at day 3 ($p<0.05$).

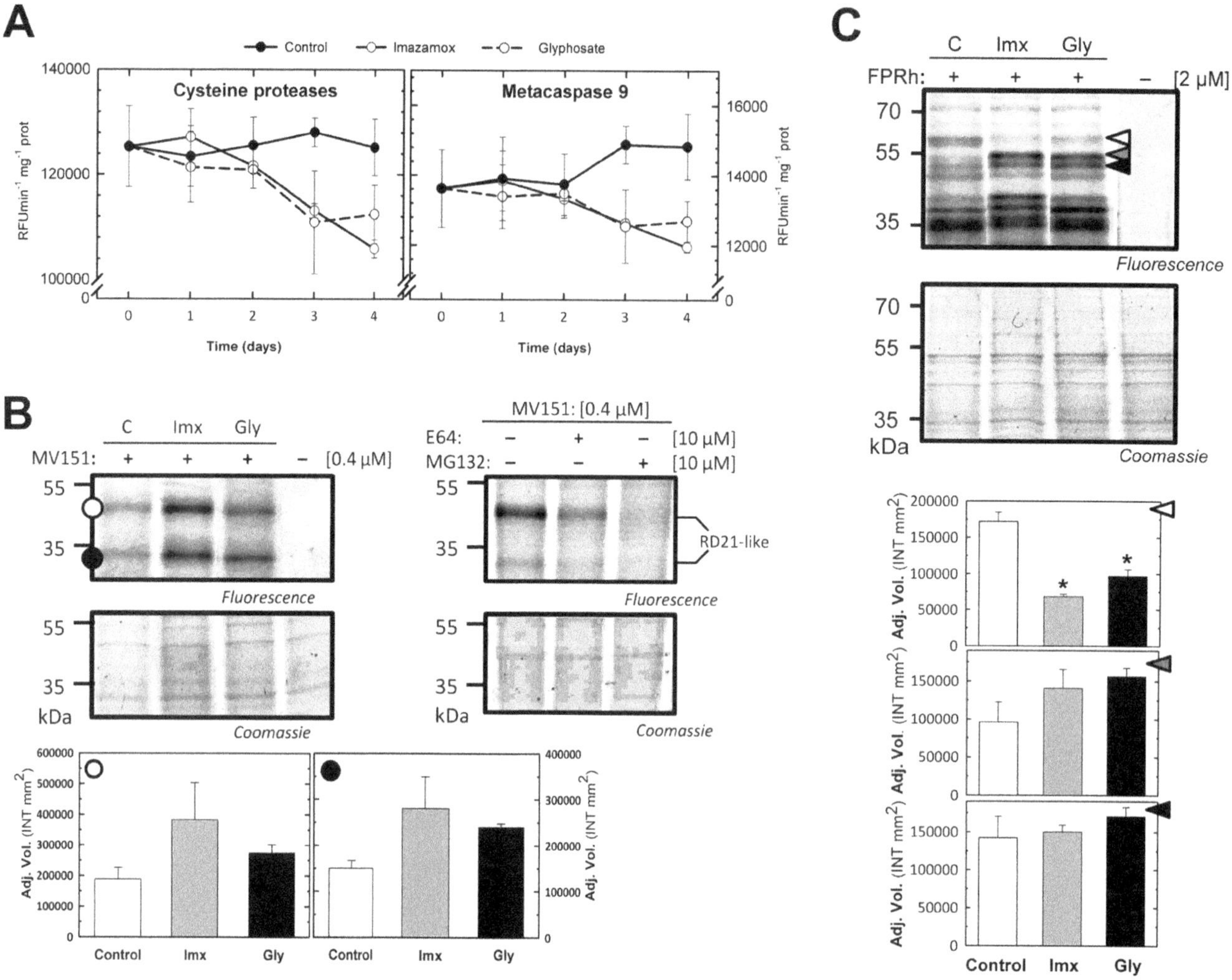

Figure 4. Effect of amino acid biosynthesis inhibitors on other proteases. (A) Cysteine protease and metacaspase 9 activities assayed using the specific substrates GGR-AMC and VRPR-AMC, respectively, in pea roots treated with herbicides that inhibit amino acid biosynthesis. Means ± SE (n = 3). (B) Left, the labeling profile of PLCPs in pea roots treated with imazamox or glyphosate at day 3 following incubation with 0.4 μM MV151 for 1 h at pH 6.0. Right, a competitive assay in which root extracts were pre-incubated with the inhibitor E64 or MG132 at a concentration of 10 μM for 30 min prior to labeling. The 30–40 kDa signals correspond to RD21 PLCPs. Fluorescently labeled proteins were detected in protein gels via fluorescence scanning, and two bands were observed (● and ○). The signals were quantified using a densitometer, and the relative values are shown in the bar graph. Means ± SE (n = 3). The symbol * indicates significant differences between the control and the treatments ($p < 0.05$). The Coomassie-stained protein gel on the bottom shows the total amounts of input proteins. (C) Labeling profile of serine proteases in pea roots treated with imazamox or glyphosate at day 3 following incubation with 2 μM FPRh for 1 h at RT in the dark. Fluorescently labeled proteins were detected in protein gels via fluorescence scanning, and three bands were detected (white, black and grey left-facing triangles). The signals were quantified using a densitometer, and the relative values are shown in the bar graph. Means ± SE (n = 3). The symbol (*) indicates significant differences between the control and the treatments ($p < 0.05$). The Coomassie-stained protein gel on the bottom shows the total amounts of input proteins.

Effect of Amino Acid Biosynthesis Inhibitors on Other Proteases

Other proteases (cysteine proteases, metacaspase 9, PLCPs and serine proteases) were evaluated to determine whether their activities were affected by herbicide treatment (Fig. 4).

Cysteine protease and metacaspase 9 activities were monitored using specific fluorogenic substrates in plants treated with Imx or Gly for 0, 1, 2, 3 and 4 days. The activity of both proteases was reduced upon treatment with both herbicides for 3 days (Fig. 4A). Several studies have shown increases in the activities of both of these proteases following stress treatment. Cysteine proteases are induced during oxidative stress and pathogen attack [44]. Metacaspases, such as metacaspase 8, are induced in response to stresses, and *Arabidopsis* metacaspase knockout mutants exhibit increased tolerance to methyl viologen [45]. It has recently been reported that different types of metacaspases act as positive or negative regulators of cell death [46].

ABPP was used to evaluate the activities of PLCPs upon herbicide treatment. Pea root extracts were labeled with the MV151 at pH 6. At this pH, MV151 labels PLCPs at 30 and 40 kDa [28]. After labeling with MV151, two bands were detected, at 30 and 40 kDa in the pea root extracts (Fig. 4B). Labeling is blocked upon pre-incubation with the PLCP inhibitors E64 and MG132, suggesting that the detected bands are PLCPs [28]. Interestingly, these bands showed stronger signals following treatment with Imx or Gly, suggesting that induction of PLCPs is another common effect of herbicide treatment.

Serine proteases are the largest class of plant proteases. These enzymes have been implicated in PCD and seed development and might play a role in the formation of plant secondary metabolites [47]. To determine whether serine protease activities were affected by herbicide treatment, an ABPP experiment using the FPRh probe was performed at day 3. Previous studies conducted using FP-based probes have identified the activities of over 50 *Arabidopsis* serine hydrolases, including subtilases, serine carboxypeptidases (SCPLs) and prolyl oligopeptidases [48]. In the present study, the major differences detected corresponded to three protein bands observed at approximately 55–70 kDa: one was significantly reduced, while the other two were increased following herbicide treatment (Fig. 4C). These different signals most likely represent subtilases (family S8) or SCPLs (family S10), as subtilases are typically 70 kDa, while SCPLs are 50 kDa in size. More experiments are required to confirm the identity of these signals. Nevertheless, while one of the putative serine protease bands was less intense following herbicide treatment, the other two bands were induced in the treated roots. Given that these putative serine proteases exhibit different patterns, it will be interesting to identify these bands through mass spectrometry.

General Conclusion

An increase in the amino acid content and decrease in the soluble protein content decrease are very well-known effects of herbicides inhibiting amino acid biosynthesis. The increased amino acid pool is thought to be derived from a rise in protein turnover, suggesting that proteases might be involved in protein degradation to provide plants with amino acids that cannot otherwise be synthesized due to herbicide inhibition. This study evaluated whether these increased free amino acid and decreased soluble protein contents were associated with major proteolytic activities. Moreover, we evaluated whether the detected changes were common to the two types of herbicides. This study is the first to evaluate and compare the effects of ALS and EPSPS inhibitors on proteolytic activities.

Theoretically, to explain the increased free amino acid pool, a general increase in proteolytic activity would be expected following herbicide treatment. We detected changes in almost all of the proteolytic systems tested following herbicide treatment. Some of the evaluated proteolytic systems were induced, while others were inhibited after treatment with Imx or Gly. Notably, most of the detected changes were common to both herbicides: proteasome and PLCP activities were induced; VPE activity was decreased; and cysteine protease and metacaspase 9 activities were reduced following both Imx and Gly treatment. Moreover, the pattern of serine proteases displayed several unidentified signals that similarly increased or decreased following treatment with both herbicides. The only observed difference between the effects of the two herbicides was in terms of YVADase activity, where an increase under Imx but a decrease under Gly treatment was observed.

New pesticides are currently being developed to replace compounds that no longer meet environmental or toxicological safety requirements. However, the battle against the development of weed resistance will require additional compounds that mimic herbicides but inhibit different biochemical targets to alleviate selection pressure [49,50]. In this context, plant proteases might represent potential new targets for herbicide development, and understanding their roles in the mode of action of herbicides could facilitate the development of new compounds with herbicidal effects [47].

The similar profiles of the proteolytic activities detected following ALS and EPSPS inhibition indicate the important role of proteolysis. Nevertheless, as some of the evaluated proteolytic systems were induced while others were reduced following the application of lethal concentrations of herbicides, the specific role of proteolytic activities in the toxicity provoked by these herbicides cannot be defined. Further pharmacological and genetic experiments are therefore required to elucidate the physiological implications of the changes in proteolytic systems observed in the presence of herbicides inhibiting amino acid biosynthesis.

Acknowledgments

The authors would like to thank Gustavo Garijo and Oscar Armendáriz for technical assistance. A. Zulet and M. Gil-Monreal received funding from fellowships through the *Ministerio de Educación* and *Universidad Pública de Navarra*, respectively.

Author Contributions

Conceived and designed the experiments: A. Zabalza RALvdH MR. Performed the experiments: A. Zulet MGM. Analyzed the data: A. Zulet A. Zabalza RALvdH MR. Contributed reagents/materials/analysis tools: A. Zulet MGM JGV A. Zabalza. Wrote the paper: A. Zulet A. Zabalza RALvdH MR.

References

1. Zhou Q, Liu W, Zhang Y, Liu K (2007) Action mechanisms of acetolactate synthase inhibiting herbicides. Pestic Biochem Physiol 89: 89–96.
2. Steinrücken H, Amrhein N (1980) The herbicide glyphosate is a potent inhibitor of 5-enolpyruvylshikimic acid-3-phosphate synthase. Biochem Biophys Res Commun 94: 1207–1212.
3. Powles S (2008) Evolved glyphosate-resistant weeds around the world: lessons to be learnt. Pest Manag Sci 64: 360–365.
4. Wittenbach V, Abell L (1999) Inhibition of valine, leucine and isoleucine biosynthesis. In: Singh B, editor. Plant amino acids: biochemistry and biotechnology. New York, Marcel Dekker. 385–416.
5. Gruys K, Sikorski J (1999) Inhibitors of tryptophan, phenylalanine, and tyrosine biosynthesis as herbicides. In: Singh B, editor. Plant amino acids: biochemistry and biotechnology. New York, Marcel Dekker. 357–384.
6. Zabalza A, González EM, Arrese-Igor C, Royuela M (2005) Fermentative metabolism is induced by inhibiting different enzymes of the branched-chain amino acid biosynthesis pathway in pea plants. J Agric Food Chem 53: 7486–7493.
7. Orcaray L, Zulet A, Zabalza A, Royuela M (2012) Impairment of carbon metabolism induced by the herbicide glyphosate. J Plant Physiol 169: 27–33.
8. Zabalza A, Orcaray L, Gaston S, Royuela M (2004) Carbohydrate accumulation in leaves of plants treated with the herbicide chlorsulfuron or imazethapyr is due to a decrease in sink strength. J Agric Food Chem 52: 7601–7606.
9. Orcaray L, Igal M, Marino D, Zabalza A, Royuela M (2010) The possible role of quinate in the mode of action of glyphosate and acetolactate synthase inhibitors. Pest Manag Sci 66: 262–269.
10. Shaner DL, Reider ML (1986) Physiological responses of corn (*Zea mays*) toAC243,997 in combination with valine, leucine, and isoleucine. Pestic Biochem Physiol 25: 248–257.
11. Anderson P, Hibberd K (1985) Evidence for the interaction of an imidazolinone herbicide with leucine, valine, and isoleucine metabolism. Weed Sci 33: 479–483.
12. Zabalza A, Gaston S, Ribas-Carbó M, Orcaray L, Igal M, et al. (2006) Nitrogen assimilation studies using 15N in soybean plants treated with imazethapyr, an inhibitor of branched-chain amino acid biosynthesis. J Agric Food Chem 54: 8818–8823.
13. Zabalza A, Orcaray L, Igal M, Schauer N, Fernie AR, et al. (2011) Unraveling the role of fermentation in the mode of action of acetolactate synthase inhibitors by metabolic profiling. J Plant Physiol 168: 1568–1575.
14. Zabalza A, Zulet A, Gil-Monreal M, Igal M, Royuela M (2013) Branched-chain amino acid biosynthesis inhibitors: Herbicide efficacy is associated with an induced carbon-nitrogen imbalance. J Plant Physiol: 170: 814–821.
15. Gaston S, Zabalza A, González EM, Arrese-Igor C, Aparicio-Tejo PM, et al. (2002) Imazethapyr, an inhibitor of the branched-chain amino acid biosynthesis, induces aerobic fermentation in pea plants. Physiol Plant 114: 524–532.

16. Rhodes D, Hogan AL, Deal L, Jamieson GC, Haworth P (1987) Amino acid metabolism of *Lemna minor* L. II. Responses to chlorsulfuron. Plant Physiol 84: 775–780.
17. Kurepa J, Smalle JA (2008) Structure, function and regulation of plant proteasomes. Biochimie 90: 324–335.
18. Dick TP, Nussbaum AK, Deeg M, Heinemeyer W, Groll M, et al. (1998) Contribution of proteasomal beta-subunits to the cleavage of peptide substrates analyzed with yeast mutants. J Biol Chem 273: 25637–25646.
19. Gruis DF, Schulze J, Jung R (2004) Storage protein accumulation in the absence of the vacuolar processing enzyme family of cysteine proteases. Plant Cell 16: 270–290.
20. Hatsugai N, Kuroyanagi M, Yamada K, Meshi T, Tsuda S, et al. (2004) A plant vacuolar protease, VPE, mediates virus-induced hypersensitive cell death. Science: 305: 855–858.
21. Kaschani F, Nickel S, Pandey B, Cravatt BF, Kaiser M, et al. (2012) Selective inhibition of plant serine hydrolases by agrochemicals revealed by competitive ABPP. Bioorg Medic Chem 20: 597–600.
22. Basset G, Raymond P, Malek L, Brouquisse R (2002) Changes in the expression and the enzymic properties of the 20 S proteasome in sugar-starved maize roots. Evidence for an *in vivo* oxidation of the proteasome. Plant Phys 128: 1149–1162.
23. Bradford MM (1976) A rapid and sensitive method for the quantitation of microgram quantities of protein utilizing the principle of protein dye-binding. Anal Biochem 72: 248–254.
24. Belenghi B, Salomon M, Levine A (2004) Caspase-like activity in the seedlings of *Pisum sativum* eliminates weaker shoots during early vegetative development by induction of cell death. J Exp Bot 55: 889–897.
25. Pena LB, Pasquini L, Tomaro ML, Gallego SM (2007) 20 S proteasome and accumulation of oxidized and ubiquitinated proteins in maize leaves subjected to cadmium stress. Phytochemistry 68: 1139–1146.
26. Bosch M, Franklin-Tong VE (2007) Temporal and spatial activation of caspase-like enzymes induced by self-incompatibility in Papaver pollen. Proc Natl Acad Sci USA 104: 18327–18332.
27. Vercammen D, Belenghi B, Van de Cotte B, Beunens T, Gavigan J-A, et al. (2006) Serpin1 of Arabidopsis thaliana is a suicide inhibitor for metacaspase 9. J Mol Biol 364: 625–636.
28. Gu C, Kolodziejek I, Misas-Villamil J, Shindo T, Colby T, et al. (2010) Proteasome activity profiling: a simple, robust and versatile method revealing subunit-selective inhibitors and cytoplasmic, defense-induced proteasome activities. Plant J 62: 160–170.
29. Misas-Villamil JC, Toenges G, Kolodziejek I, Sadaghiani AM, Kaschani F, et al. (2013) Activity profiling of vacuolar processing enzymes reveals a role for VPE during oomycete infection. Plant J 73: 689–700.
30. Cho SK, Chung HS, Ryu MY, Park MJ, Lee MM, et al. (2006) Heterologous expression and molecular and cellular characterization of CaPUB1 encoding a hot pepper U-Box E3 ubiquitin ligase homolog. Plant Phys 142: 1664–1682.
31. Vierstra RD (2003) The ubiquitin/26 S proteasome pathway, the complex last chapter in the life of many plant proteins. Trends Plant Sci 8: 135–142.
32. Callis J, Vierstra R (2000) Protein degradation in signaling. Curr Opin Plant Biol 3: 381–386.
33. Fujimuro M, Sawada H, Yokosawa H (1997) Dynamics of ubiquitin conjugation during heat-shock response revealed by using a monoclonal antibody specific to multi-ubiquitin chains. Eur J Biochem 249: 427–433.
34. Ingvardsen C, Veierskov B (2001) Ubiquitin- and proteasome-dependent proteolysis in plants. Physiol Plant 112: 451–459.
35. Pena L, Pasquini L, Tomaro M, Gallego S (2006) Proteolytic system in sunflower (*Helianthus annuus* L.) leaves under cadmium stress. Plant Sci 171: 531–537.
36. Pena LB, Zawoznik MS, Tomaro ML, Gallego SM (2008) Heavy metals effects on proteolytic system in sunflower leaves. Chemosphere 72: 741–746.
37. Kaschani F, Verhelst SHL, Van Swieten PF, Verdoes M, Wong C-S, et al. (2009) Minitags for small molecules: detecting targets of reactive small molecules in living plant tissues using "click chemistry". Plant J 57: 373–385.
38. Kołodziejek I, van der Hoorn RAL (2010) Mining the active proteome in plant science and biotechnology. Curr Opin Biotech 21: 225–233.
39. Barrett AJ, Kembhavi AA, Brown MA, Kirschke H, Knight CG, et al. (1982) L-trans-Epoxysuccinyl-leucylamido(4-guanidino)butane (E-64) and its analogues as inhibitors of cysteine proteinases including cathepsins B, H and L. Biochem J 201: 189–198.
40. Kurepa J, Karangwa C, Duke LS, Smalle JA (2010) Arabidopsis sensitivity to protein synthesis inhibitors depends on 26 S proteasome activity. Plant Cell Rep 29: 249–259.
41. Rojo E, Martín R, Carter C, Zouhar J, Pan S, et al. (2004) VPEγ exhibits a caspase-like activity that contributes to defense against pathogens. Curr Biol 14: 1897–1906.
42. Hiraiwa N, Nishimura M, Hara-Nishimura I (1999) Vacuolar processing enzyme is self-catalytically activated by sequential removal of the C-terminal and N-terminal propeptides. FEBS Lett 447: 213–216.
43. Berger AB, Witte MD, Denault JB, Sadaghiani AM, Sexton KMB, et al. (2006) Identification of early intermediates of caspase activation using selective inhibitors and activity-based probes. Mol Cell 23: 509–521.
44. Solomon M, Belenghi B, Delledonne M, Menachem E, Levine A (1999) The involvement of cysteine proteases and protease inhibitor genes in the regulation of programmed cell death in plants. Plant Cell 11: 431–444.
45. He R, Drury GE, Rotari VI, Gordon A, Willer M, et al. (2008) Metacaspase-8 modulates programmed cell death induced by ultraviolet light and H_2O_2 in Arabidopsis. J Biol Chem 283: 774–783.
46. Coll N, Vercammen D, Smidler A, Clover C, Van Breusegem F, et al. (2010) Arabidopsis Type I Metacaspases control cell death. Science 330: 1393–1397.
47. Cobb A, Reade J (2010) The inhibition of amino acid biosynthesis. In: Cobb AH, Reade JPH, editors. Herbicides and plant physiology. Oxford, UK: Wiley-Blackwell; 2010. p.176–99.
48. Kaschani F, Gu C, Niessen S, Hoover H, Cravatt BF, et al. (2009) Diversity of serine hydrolase activities of unchallenged and botrytis-infected *Arabidopsis thaliana*. Mol Cell Proteomics 8: 1082–1093.
49. Dayan FE, Cantrell CL, Duke SO (2009) Natural products in crop protection. Bioorg Med Chem 17: 4022–4034.
50. Duke S (2012) Why have no new herbicide modes of action appeared in recent years? Pest Manag Sci 68: 505–512.

18

Pesticide Acute Toxicity Is a Better Correlate of U.S. Grassland Bird Declines than Agricultural Intensification

Pierre Mineau[1]*, Mélanie Whiteside[1,2]

1 Science and Technology Branch, Environment Canada, National Wildlife Research Centre, Carleton University Campus, Ottawa, Canada, **2** Pest Management Regulatory Agency, Health Canada, Ottawa, Canada

Abstract

Common agricultural birds are in decline, both in Europe and in North America. Evidence from Europe suggests that agricultural intensification and, for some species, the indirect effects of pesticides mediated through a loss of insect food resource is in part responsible. On a state-by-state basis for the conterminous Unites States (U.S.), we looked at several agronomic variables to predict the number of grassland species increasing or declining according to breeding bird surveys conducted between 1980 and 2003. Best predictors of species declines were the lethal risk from insecticide use modeled from pesticide impact studies, followed by the loss of cropped pasture. Loss of permanent pasture or simple measures of agricultural intensification such as the proportion of land under crop or the proportion of farmland treated with herbicides did not explain bird declines as well. Because the proportion of farmland treated with insecticides, and more particularly the lethal risk to birds from the use of current insecticides feature so prominently in the best models, this suggests that, in the U.S. at least, pesticide toxicity to birds should be considered as an important factor in grassland bird declines.

Editor: Rohan H. Clarke, Monash University, Australia

Funding: The work was funded by Environment Canada. The funders had no role in study design, data collection and analysis, decision to publish, or preparation of the manuscript.

Competing Interests: The authors have declared that no competing interests exist.

* E-mail: pierre.mineau@ec.gc.ca

Introduction

Many grassland (the North American favoured term) or farmland (the European favoured term) bird species have undergone range contractions and/or population declines in recent decades in both northern Europe and North America. Indeed, North American analyses indicate that grassland birds as a group are declining faster than birds from other biomes [1,2]. Rodenhouse and colleagues [3] reported that 215 species of neotropical migrants use agricultural areas in North America. Several neotropical migrant species have, over the years, been listed as threatened or endangered or are candidates for listing, and agriculture is implicated in the decline of many. Most investigators have blamed farmland bird declines on some aspect of agricultural intensification: the shift to larger fields, row and field crop monocultures and denser more uniform crop structure, loss of native pasture and other semi-natural habitats, increase in autumn sowings, increased inputs of fertilizers and pesticides to name a few [4–10]. Increased predation as a result of habitat change has also been invoked [11]. As one component of agricultural intensification, herbicides and insecticides have been linked to population declines in some bird species in the UK, primarily via indirect, food-mediated effects [5,12]. Fox [13] reported that farmland birds in Denmark had not declined as they had in the UK and that a notable difference in crop intensification with the latter had been a clear and gradual reduction in agrochemical inputs. More recently, Geiger and colleagues [14] in a large study with study sites in 8 European countries concluded that: "*Out of the 13 studied components of agricultural intensification, use of pesticides, especially insecticides and fungicides, had the most consistent negative effects on the species diversity of plants, carabids and ground-nesting farmland birds, and on the potential for biological pest control.*" Bird diversity specifically was most highly correlated with fungicide use although the authors pointed out that fungicide use and insecticide use were highly correlated.

Historically, insecticide use in North America has been very different from that in the UK – or Denmark. Specifically, large quantities of products of very high toxicity to birds have been used for decades despite evidence that poisonings were frequent even when products were applied according to label direction [15–17]. The Avian Incident Monitoring System (AIMS), a joint project of the U.S. Environmental Protection Agency (EPA) and the American Bird Conservancy (ABC), listed 113 pesticides which have caused direct bird mortality [18]. An analysis of a large number of avian field studies [19,20] suggests that avian kills were a normal corollary of insecticide use in many crops grown in North America. For example, analyses of granular insecticide use patterns in western Canada indicated that the abundance of several common species was negatively correlated with these toxic insecticides [21]. Pimentel [22], in an oft-cited study, estimated that pesticide-induced direct mortality numbered approximately 67 million per year in the U.S. He based this estimate on the fact that 160 million ha of cropland received a very heavy dose of pesticides per year (3 kg a.i./ha on average – including a number of very toxic pesticides), a breeding density of 4.2 birds per ha (from census plot data) and a conservative kill estimate of 10% of exposed birds. This estimate ignored kills of wintering or migrant birds which could be substantial [17]. This estimate was undoubtedly quite conservative. Mineau [23] estimated, based

Table 1. Grassland species ordered by the number of total state*species declines and the number of recorded declining or increasing trends (1980–2003) for the 45 conterminous American States.

SPECIES		Total declines (significant declines)	Total increases (significant increases)	Recorded killed in pesticide field trials
Ammodramus savannarum	Grasshopper sparrow	25 (15)	3 (1)	X
Pooecetes gramineus	Vesper sparrow	18 (8)	3 (0)	X
Phasianus colchicus	Ring-necked pheasant	19 (10)	6 (2)	X
Eremophila alpestris	Horned lark	24 (16)	5 (2)	X
Passerculus sandwichensis	Savannah sparrow	12 (5)	8 (3)	X
Circus cyaneus	Northern harrier	8 (2)	8 (1)	X
Bartramia longicauda	Upland sandpiper	8 (2)	4 (1)	
Sturnella neglecta	Western meadowlark	17 (11)	4 (1)	X
Dolichonyx oryzivorus	Bobolink	12 (7)	5 (1)	X
Spiza americana	Dickcissel	8 (3)	10 (2)	X
Numenius americanus	Long-billed curlew	4 (0)	3 (1)	
Sturnella magna	Eastern meadowlark	33 (30)	0 (0)	X
Calamospiza melanocorys	Lark bunting	6 (3)	1 (1)	
Athene cunicularia	Burrowing owl	2 (0)	3 (1)	X
Cistothorus platensis	Sedge wren	2 (0)	3 (1)	
Tympanuchus phasianellus	Sharp-tailed grouse	2 (0)	1 (0)	
Ammodramus bairdii	Baird's sparrow	2 (1)	0 (0)	
Calcarius ornatus	Chestnut-collared longspur	3 (2)	0 (0)	X
Buteo regalis	Ferruginous hawk	1 (0)	4 (0)	
Tympanuchus cupido	Greater prairie chicken	1 (0)	0 (0)	
Ammodramus leconteii	Le Conte's sparrow	1 (0)	1 (1)	X
Asio flammeus	Short-eared owl	2 (1)	1 (0)	X
Anthus spragueii	Sprague's pipit	1 (0)	1 (0)	X
Aimophila cassinii	Cassin's sparrow	4 (4)	0 (0)	X
Ammodramus henslowii	Henslow's sparrow	2 (2)	0 (0)	
Charadrius montanus	Mountain plover	0 (0)	1 (1)	

Also indicated is whether the species has ever been recorded killed during pesticide field trials based on [19] augmented by similar studies on granular insecticides.

on several industry-led field studies that at its peak, a single use pattern (for corn rootworm) of a single insecticide (carbofuran), in a single crop (corn) was killing 17 to 91 million songbirds annually in the U.S. Midwest. This raises the question as to whether broader patterns of grassland bird species declines can be, in part at least, explained by the direct toxic effects of pesticides.

An analysis of pesticide use patterns in the US does suggest that the situation is improving as a result of the gradual withdrawal of the most toxic products, largely because of human health concerns [17]. The current analysis considered bird trends from 1980 to 2003; there is evidence that the acute lethal risk to birds was already dropping during the second half of that period.

Methods

Pesticide use data

We assembled State by State pesticide use data from a number of sources. The United States Department of Agriculture (USDA) National Center for Agricultural Statistics (NASS) provides crop acreage, one time pesticide application rate, and number of applications, but these data are for major crops only and are only provided for certain states. A complete matrix of State*crop*active ingredient combinations was obtained from the National Center for Food and Agriculture Policy (NCFAP). For the years circa 1992 (1990–1993) and circa 1997 (1994–1998) NCFAP reported crop acreage, the percentage of crop treated with a given active ingredient, and the cumulative yearly application rate to each treated acre based on NASS data as well as a plethora of State surveys, personal communications with extension specialists etc.... Unfortunately, one-time application rates and the annual number of repeat applications were not reported by NCFAP and thus had to be retrieved or calculated from a number of sources including the original sources cited by NCFAP where possible. The full methodology is detailed in Mineau and Whiteside [17]. For the current analysis, only the1992 data were used as they fell in the midpoint of the period for which we compiled bird trend estimates and before a number of restrictions on toxic organophosphorous (OP) insecticide use.

Bird trend data

We used data from the United States Geological Survey (USGS) Breeding Bird Survey (BBS) to calculate species trends. The BBS is an annual roadside count of standardized 39.4 km routes with stops at 0.8 km interval. Volunteer observers record all species

Table 2. Agricultural variables and BBS summarised results for each State.

State	Agricultural intensity	Proportional change in permanent pasture (1978–2002)	Proportional change in crop pasture (1978–2002)	Lethal pesticide risk	Insecticide use	Herbicide use	No. increasing species (no. significant increases)	No. declining species (no. significant declines)
ALABAMA	0.34	−0.02	−0.52	0.25	0.53	1.01	1 (1)	2 (1)
ARIZONA	0.03	−0.02	0.33	0.02	0.09	0.08	0 (0)	2 (1)
ARKANSAS	0.59	0.13	−0.42	0.06	0.22	1.23	1 (1)	1 (1)
CALIFORNIA	0.29	−0.05	−0.21	0.14	0.54	0.4	2 (1)	6 (2)
COLORADO	0.25	−0.02	0.41	0.02	0.04	0.29	5 (0)	8 (5)
CONNECTICUT	0.62	−0.36	−1.39	0.08	0.22	0.64	0 (0)	1 (1)
FLORIDA	0.26	−0.14	−0.17	0.18	0.35	0.58	0 (0)	1 (1)
GEORGIA	0.56	−0.07	−0.66	0.38	0.74	1.39	0 (0)	2 (2)
IDAHO	0.37	0.00	0.00	0.04	0.06	0.3	1 (0)	6 (3)
ILLINOIS	0.91	0.26	−1.83	0.09	0.14	1.96	1 (0)	9 (7)
INDIANA	0.83	−0.01	−1.43	0.09	0.13	1.97	1 (0)	7 (6)
KANSAS	0.56	−0.06	−0.27	0.06	0.12	0.61	2 (1)	8 (6)
KENTUCKY	0.46	0.77	−0.59	0.04	0.08	0.62	2 (0)	2 (1)
LOUISIANA	0.59	−0.08	−0.30	0.29	0.75	1.99	0 (0)	2 (1)
MAINE	0.64	−0.64	−1.04	0.1	0.21	0.35	1 (0)	2 (1)
MARYLAND	0.82	0.67	−1.26	0.06	0.25	1.34	1 (1)	4 (3)
MASSACHUSETTS	0.6	−0.28	−1.57	0.12	0.29	0.5	0 (0)	3 (1)
MICHIGAN	0.83	0.29	−0.84	0.12	0.21	1.34	1 (0)	9 (4)
MINNESOTA	0.84	0.15	−1.21	0.05	0.09	1.56	1 (0)	12 (3)
MISSISSIPPI	0.56	0.20	−1.09	0.54	0.97	1.91	0 (0)	2 (1)
MISSOURI	0.49	0.10	−0.65	0.03	0.08	0.82	2 (1)	6 (4)
MONTANA	0.24	−0.05	0.31	0.01	0.02	0.19	8 (1)	9 (3)
NEBRASKA	0.45	−0.01	−0.24	0.08	0.14	0.64	3 (0)	9 (1)
NEVADA	0.05	0.01	0.39	0	0.01	0.01	2 (1)	0 (0)
NEW HAMPSHIRE	0.64	0.47	−1.30	0.03	0.15	0.28	0 (0)	2 (2)
NEW JERSEY	0.73	−0.23	−0.46	0.11	0.26	1.24	0 (0)	2 (2)
NEW MEXICO	0.03	0.01	0.43	0	0.01	0.08	1 (0)	6 (5)
NEW YORK	0.59	−0.03	−1.34	0.09	0.18	0.65	0 (0)	9 (5)
NORTH CAROLINA	0.67	0.03	−0.48	0.29	0.56	1.5	0 (0)	2 (1)
NORTH DAKOTA	0.64	0.06	−0.20	0.02	0.03	0.87	11 (6)	8 (4)
OHIO	0.8	−0.20	−0.79	0.07	0.1	1.51	2 (1)	7 (6)
OKLAHOMA	0.32	0.00	0.19	0.05	0.11	0.32	1 (0)	6 (3)
OREGON	0.22	0.01	0.19	0.04	0.08	0.54	2 (1)	5 (4)

Table 2. Cont.

State	Agricultural intensity	Proportional change in permanent pasture (1978–2002)	Proportional change in crop pasture (1978–2002)	Lethal pesticide risk	Insecticide use	Herbicide use	No. increasing species (no. significant increases)	No. declining species (no. significant declines)
PENNSYLVANIA	0.7	0.07	−0.73	0.07	0.25	0.84	2 (0)	5 (4)
SOUTH CAROLINA	0.59	−0.19	−0.49	0.14	0.44	1.46	0 (0)	1 (0)
SOUTH DAKOTA	0.35	−0.07	0.07	0.04	0.05	0.53	4 (1)	9 (4)
TENNESSEE	0.46	−0.15	−0.41	0.1	0.26	0.8	2 (0)	1 (1)
TEXAS	0.17	0.05	0.09	0.03	0.08	0.21	1 (0)	7 (5)
UTAH	0.13	0.04	0.18	0.01	0.01	0.09	3 (0)	6 (1)
VERMONT	0.49	0.11	−2.02	0.01	0.03	0.36	1 (0)	2 (2)
VIRGINIA	0.42	−0.12	−0.36	0.09	0.18	0.69	0 (0)	3 (2)
WASHINGTON	0.44	0.12	−0.23	0.06	0.17	0.82	2 (0)	6 (2)
WEST VIRGINIA	0.28	−0.22	−0.66	0.01	0.04	0.12	0 (0)	2 (2)
WISCONSIN	0.65	0.05	−0.93	0.12	0.18	0.89	2 (1)	11 (8)
WYOMING	0.06	−0.03	0.48	0	0	0.04	6 (1)	4 (0)

seen or heard within a 0.4 km radius over a 3 min. census period. For the whole country analysis, we relied on the bird trend information provided for the 'grassland' bird species guild as generated by the USGS Migratory Bird Division for each State. We used the trends reported for route regression analyses run between 1980 and 2003 (http://www.mbr-pwrc.usgs.gov/bbs/trend/guild03.html). For any given State, a trend analysis was provided only if the species was recorded on 14 routes or more. Thus, the number of species showing positive or negative trends was taken to be a self-equilibrating index of the health of grassland birds across the conterminous U.S. given that different grassland species dropped in and out of the analysis in each State dependent on their range and abundance (Table 1). We considered restricting the analysis to only those species showing statistically significant trends but this would have greatly reduced sample size, especially the number of statistically-significant increases, making our use of logistic modeling difficult.

Agricultural variables

Agricultural Intensity and change:. We developed both 'static' and 'change' measures of agricultural intensity. Our static indicator of State agricultural intensity was developed to be analogous to the pesticide indicators. It is defined here as the percentage of total agricultural land that is allocated to active cropping. Using the NASS categories of land use, agriculture intensity was obtained by dividing the amount of 'cropland for crops' by the amount of total farmland including 'cropland for crops' which in turn included 'fallow', 'pasture', and 'grazed forest'.

The change indicators relevant to grassland birds were deemed to be changes in the land area devoted to permanent and cropped pasture. These were calculated simply as the 2002 pasture or crop pasture areas minus the 1978 pasture or crop area pastures expressed as a proportion of the 1978 area for each State.

Herbicide use:. Very few crops in North America are grown without the use of herbicides. Therefore, the acreage of crops to which herbicides have been applied divided by the area of 'total farmland' is another measure of agricultural intensification. From the pesticide dataset described earlier, areas treated were summed for all herbicide*crop combinations. Because most crops are treated with multiple herbicides, the total acreage treated with herbicides often exceeded the 'total farmland' value which resulted in a value greater than 1.0 as shown in Table 2. If that was the case, the total percentage was capped at 1.0 or 100% prior to analysis.

Insecticide use:. This variable was calculated in the same way as the herbicide use variable. There is less insecticide than herbicide use and the proportion of farmland treated never exceeded 1.0 or 100%.

Lethal pesticide risk:. This variable is the estimated lethal risk from the use of the insecticides. It was based on the logistic models developed from results of avian field studies [17,19].

The process can be summarized as follows: As a first step, a measure of acute pesticide toxicity for birds ranging from 20 to 1,000 grams (a weight range that covers most bird species found dead in farm fields and which corresponds also to the range of generic body weights considered in U.S. and Canadian risk assessments for pesticide registration) is obtained by applying species sensitivity distribution techniques [24]. A probability that birds will be killed by a given pesticide application is then derived from a model that uses logistic multiple regression with the finding of bird carcasses in a large sample of field studies as the endpoint of interest. Note that this index does not incorporate non lethal toxic effects on birds, or indirect effects such as loss of habitat or

Table 3. Correlation matrix of predictor variables. Significant (P< 0.05) correlation coefficients are shown in bold.

	Farming Intensity	Lethal pesticide risk	Insecticide Use	Herbicide Use	Change in permanent pasture	Change in cropped pasture
Farming Intensity	1.00					
Lethal pesticide risk	**0.42**	1.00				
Insecticide Use	**0.30**	**0.94**	1.00			
Herbicide Use	**0.87**	**0.59**	**0.47**	1.00		
Change in permanent pasture	**0.27**	−0.03	0.05	**0.25**	1.00	
Change in cropped pasture	**−0.80**	**−0.42**	**−0.33**	**−0.64**	**−0.30**	1.00

food source from pesticide use. Aside from the toxicity of the pesticide active ingredient, the model makes use of the application rate and physico-chemical properties of pesticides. Validation of the model for a sample of studies in field crops indicated that more than 81% of studies were correctly classified – as to whether they gave rise to mortality or not [17].

One recognized weakness of the approach is that the empirical models relating mortality to pesticide toxicity and to the other independent variables were derived entirely from foliar applications of pesticides. Because formulation-specific data (i.e. whether the material was applied as a spray or a granular formulation) were not available for many products, this problem was ignored in the current analysis.

Only insecticides were considered for the acute toxicity variable. Realistically, acute avian risk from herbicide or fungicide use is minimal. The models were run using the average application rates for each crop*pesticide*State combination (8491 unique combinations). The overall risk, expressed as the proportion of the farmland area over which avian mortality was likely, was cumulated by State. Mineau and Whiteside [17] also obtained a data set of individual insecticide applications from California (the only State where individual applications are entered into a database) and were able to ascertain that the predicted risk from a given pesticide cumulated for each separate application was similar to the risk calculated from the State application rate average.

Statistical treatment

Table 1 provides a summary of the species included in the analysis, the number of states in which trend analysis was attempted as well as the number of positive and negative trends (both significant and not) for those States having adequate coverage. The six independent indices (agricultural intensity, proportional change in permanent and cropped pasture, herbicide use, insecticide use and lethal pesticide risk) as well as the number of positive and negative bird trends for the grassland bird guild are given for each State in Table 2. The four static predictors (all proportions) were submitted to arcsin transformation; the proportional changes were left in their native state. A correlation matrix was run for all independent variables (Table 3).

In the regression analysis, each state*bird combination represents a data point. This means that States with more reported trends (with more species surveyed on more than 14 routes) are weighted more heavily in the analysis. Modeling was carried out with the GLZ module of STATISTICA (Version 10) with a logit regression model. Model strength was assessed with Akaike Information Criterion (AIC) values and Delta AIC values which are simple increases in AIC from the best model. The relative strength of the predictor variables was assessed with summed Akaike weights [25].

Results

The correlation matrix shows that most of the predictor variables were inter-correlated (Table 3). Correlations were particularly high between insecticide use and acute risk ($r = 0.94$) and between intensification and the change in cropped pasture ($r = 0.80$). The performance of all possible logit regression models is summarised in Table 4. Models are ordered by increasing delta AIC values. The proposed rule of thumb is that models with Delta AIC values of less than 2 are equally plausible [25]. Summed Akaike weights for the 6 predictor variables are provided in Table 5. On that basis, it might be tempting to conclude that 'change in cropped pasture' was the most plausible predictor followed by lethal insecticide use. However, the very high correlation between 'lethal pesticide risk' and 'insecticide use' (with the third highest summed Akaike weight) suggests that the importance of insecticides has been underestimated because of the partition of Akaike weights between those two variables. Because of the strong interdependencies of the variables (the correlation between 'change in cropped pasture' and 'farming intensity' might similarly cloud our assessment of the relative strength of the predictors), we repeated the analysis with only 'change in cropped pasture', 'change in permanent pasture' and 'lethal pesticide risk' as predictor variables. Summed Akaike weights were 0.23 for 'lethal pesticide risk', 0.17 for 'change in cropped pasture' and 0.07 for 'change in permanent pasture'.

Indeed, even in the full list of models with all 5 predictors, 'lethal pesticide risk' offered the best single-predictor model. The second best single-predictor model was with 'insectide use' and 'change in cropped pasture' came in third place. 'Lethal pesticide risk' as a predictor variable was 3.9 times more plausible than 'change in cropped pasture' based on the evidence ratio (Table 4).

Discussion

As reviewed by Murphy [9] – See citations therein), few North American studies have managed to link specific agricultural practices with bird population trends, unlike the situation that prevails in the UK and, to a lesser extent, other European countries. Murphy [9] attempted to look for correlations between changes in certain agricultural habitat types (e.g. pasture, cropland or rangeland) with BBS trends for grassland and shrub species in the Eastern and central U.S. He found that none of the variables examined exhibited a clear negative relationship with population trends. On the other hand, positive population trends were

Table 4. Summary of model results for the logit regression and 5 predictor variables.

Change in cropped pasture	Change in permanent pasture	Farming Intensity	Lethal pesticide risk	Insecticide Use	Herbicide Use	d.f.	Delta AIC	wi (Akaike weights)
1			1		1	3	0.00	0.101
1		1	1			3	0.05	0.099
1			1			2	0.70	0.071
1				1		2	1.65	0.044
			1			1	1.71	0.043
1		1	1	1		4	1.76	0.042
1		1	1		1	4	1.77	0.042
1			1	1	1	4	1.80	0.041
1	1	1	1			4	1.91	0.039
1	1		1		1	4	1.96	0.038
1		1		1		3	1.97	0.038
1				1	1	3	2.10	0.035
1	1		1			3	2.49	0.029
1			1	1		3	2.70	0.026
1	1			1		3	3.21	0.020
1		1	1	1	1	5	3.48	0.018
1	1	1		1		4	3.56	0.017
			1		1	2	3.62	0.017
			1	1		2	3.65	0.016
		1	1			2	3.66	0.016
	1		1			2	3.68	0.016
1	1	1	1		1	5	3.69	0.016
1	1	1	1	1		5	3.70	0.016
1	1		1	1	1	5	3.79	0.015
1	1			1	1	4	3.82	0.015
1		1		1	1	4	3.88	0.015
				1		1	4.47	0.011
1	1		1	1		4	4.49	0.011
1						1	4.49	0.011
		1	1		1	3	4.57	0.010
1		1				2	5.06	0.008
1	1	1	1	1	1	6	5.46	0.007
			1	1	1	3	5.51	0.006
	1		1		1	3	5.52	0.006
1	1	1		1	1	5	5.52	0.006
	1		1	1		3	5.60	0.006
		1		1		2	5.60	0.006
		1	1	1		3	5.62	0.006
	1	1	1			3	5.65	0.006
1	1					2	5.66	0.006
				1	1	2	6.22	0.005
1	1	1				3	6.23	0.004
1					1	2	6.26	0.004
	1	1	1		1	4	6.47	0.004
	1			1		2	6.47	0.004
		1	1	1	1	4	6.51	0.004
1		1			1	3	6.60	0.004
		1		1	1	3	6.96	0.003

Table 4. Cont.

Change in cropped pasture	Change in permanent pasture	Farming Intensity	Lethal pesticide risk	Insecticide Use	Herbicide Use	d.f.	Delta AIC	wi (Akaike weights)
	1		1	1	1	4	7.33	0.003
1	1				1	3	7.48	0.002
	1	1		1		3	7.55	0.002
	1	1	1	1		4	7.59	0.002
1	1	1			1	4	7.64	0.002
	1			1	1	3	8.21	0.002
	1	1	1	1	1	5	8.35	0.002
	1	1		1	1	4	8.95	0.001
		1				1	12.59	0.000
					1	1	12.98	0.000
	1	1				2	14.37	0.000
		1			1	2	14.40	0.000
	1				1	2	14.68	0.000
	1	1			1	3	16.12	0.000
	1					1	19.89	0.000

Delta AICs and Akaike weights are provided for each model.

associated with increases in rangeland, but also with increases in harvested cropland. Only when all species (whether grassland or shrub species) were combined was he able to obtain negative correlations of population trend with specific agricultural habitat types – e.g. cover crops.

When regressed one by one, all variables examined in our analysis had the expected effect on bird trends except for loss of permanent pasture. An increase in permanent pasture was associated with no trend or even a very slight trend to more declining species. It was the reverse for changes in the area under cropped pasture; i.e. increase in the area of cropped pasture equals more bird increases. An increase in any of the pesticide variables or farming intensity was associated with a higher proportion of declining species as expected.

Our analysis included both static and change-based variables. Because of the reduced temporal availability of pesticide data, we were limited to look for model fit based on variables measured at the midpoint of the bird trend period rather than estimating a change in these predictor variables over the period of interest. All of these static variables carry an implicit assumption that measurements made at the midpoint of the interval over which bird trends were calculated, are broadly representative of the entire period. We consider it meaningful that even with this handicap, the static variables 'lethal pesticide risk 'and 'insecticide use' emerged as more plausible than the pasture loss variables which we were able to measure over the entire period. Our results are all the more remarkable since any signal of acute pesticide effect in our analysis was undoubtedly 'diluted' by a gradual reduction in toxic insecticide use during the period of analysis [17].

Table 5. Summed Akaike weights for all predictor variables and logit models given in table 4.

Variable	Akaike weights
Change in cropped pasture	0.844
Lethal pesticide risk	0.775
Insecticide use	0.446
Farming intensity	0.436
Herbicide use	0.425
Change in permanent pasture	0.299

Our results suggest that the use of lethally toxic insecticides cannot be ignored when trying to identify causes of grassland population declines in North America. Indeed, they offer a more plausible explanation for overall declines than does the oft-cited 'habitat loss through agricultural intensification'. It was remarkable that loss of permanent pasture did not appear to be much of a predictor of grassland bird declines. The high correlations between our variables make it impossible to separate the direct effects of insecticides from the indirect ones and it is likely that both are operating. However, there is some indication that direct effects were most important during the period under consideration. Not only is 'lethal risk' a slightly better predictor than 'insecticide use' as judged by overall model weight but, herbicide use is a much weaker predictor. If the indirect effects of pesticides were the most important, we would expect herbicides to be contributing to grassland bird declines as noted elsewhere (e.g. [12]).

Many of the grassland species examined often have but a weak association with row or field crops where the bulk of insecticide use takes place. However, a number of factors need to be considered: 1) Mineau and colleagues showed that only a small proportion of total cropland need be treated with a dangerous pesticide to affect overall population trends [21]. Even a casual association with cropland may place birds at risk; 2) Although pastures are believed to receive much lower insecticide loads than other crops, alfalfa still carries the third highest lethal risk of any crop based on pesticide use [17]. Insecticides are used to control some rangeland pests such as grasshoppers; there are no reliable data on this but it is reasonable to assume that insecticide use on pasture may be proportional to insecticide use on field crops nearby; 3) The majority of species considered in this analysis (16/24) have been recorded killed in pesticide field trials despite the limited number of such studies (Table 1). Several of the grassland species showing the largest number of relative declines (e.g. horned lark, vesper sparrow) are species frequently picked up dead in the course of pesticide impact studies; 4) Pesticide drift is a well known phenomenon which may extend risk from a crop to an adjoining grassland or pasture.

In conclusion, it would be foolhardy for anyone to argue that habitat loss is of no importance to bird declines. However, we should be careful to consider pest control and specifically the use of highly toxic insecticides as a potential contributor to those declines. Unfortunately, information on pesticide use is often difficult to obtain or considered to be confidential, hampering any serious analysis of its true impact.

Acknowledgments

Previous versions of this paper benefitted from comments by Jude Girard and Charles Francis as well as two anonymous reviewers.

Author Contributions

Conceived and designed the experiments: PM. Analyzed the data: PM MW. Wrote the paper: PM.

References

1. Sauer JR., Hines JE, Thomas I, Fallon J, Gough G (2000) The North American breeding bird survey, results and analysis 1966–1999. Patuxent Wildl. Res. Cent., Laurel, Md.
2. Dunn E, Downes CM, Collins BT (2000) The Canadian Breeding Bird survey, 1967–1998. Canadian Wildlife Service Progress Note No. 216, Environment Canada, Gatineau, Québec.
3. Rodenhouse NL, Best LB, O'Connor RJ, Bollinger EK (1993) Effects of temperate agriculture on Neotropical migrant landbirds. In: Finch DM, Stangel PW editors. Status and management of Neotropical Migratory Birds. U.S. Dept. of Agriculture, Forest Service Gen. Tech. Rep. RM-229. Rocky Mountain and Range Experiment Station, Fort Collins, CO. pp. 280–295.
4. McLaughlin A, Mineau P (1995) The impact of agricultural practices on biodiversity. Agric Ecosyst Environ 55:201–212.
5. Campbell LH, Avery ML, Donald P, Evans AD, Green RE, et al. (1997) A Review of the Indirect Effects of Pesticides on Birds. Joint Nature Conserv. Cttee. Rep. No. 227. Peterborough, UK.
6. Fuller RJ (2000) Relationships between recent changes in lowland British agriculture and farmland bird populations: an overview. In: Aebischer, NJ, Evans AD, Grice PV, Vickery JA editors. Ecology and Conservation of Lowland Farmland Birds: 5–16. Tring: BOU.
7. Chamberlain DE, Fuller RJ, Bunce RGH, Duckworth JC, Shrubb M (2000) Changes in abundance of farmland birds in relation to the timing of agricultural intensification in England and Wales. J Appl Ecol 37:71–788.
8. Donald PF, Green RE, Heath MF (2001) Agricultural intensification and the collapse of Europe's farmland bird populations. Proc. Royal Soc. London B. 268:25–29.
9. Murphy MT (2003) Avian population trends within the evolving agricultural landscape of Eastern and Central United States. The Auk 120(1):20–34.
10. Newton I (2004) The recent declines of farmland bird populations in Britain: an appraisal of causal factors and conservation actions. Ibis 146, 579–600.
11. Evans KL (2004) The potential for interactions between predation and habitat change to cause population declines of farmland birds. Ibis 146:1–13.
12. Potts GR (1997) Cereal Farming, Pesticides and Grey Partridges. In: Pain DJ, Pienkowski MW editors. Farming and Birds in Europe. Academic Press, London.
13. Fox AD (2004) Has Danish agriculture maintained farmland bird populations? J Appl. Ecol. 41: 427–439.
14. Geiger F, Bengtsson J, Berendse F, Weisser WW, Emmerson M, et al. (2010) Persistent negative effects of pesticides on biodiversity and biological control potential on European farmland. Basic and Applied Ecology 11: 97–105.
15. Mineau P, Fletcher MR, Glazer LC, Thomas NJ, Brassard C, et al. (1999) Poisoning of raptors with organophosphorous and carbamates pesticides with emphasis on Canada, U.S. and U.K. J. Raptor Res 33(1):1–37.
16. Mineau P (2004) Birds and pesticides: Are pesticide regulatory decisions consistent with the protection afforded migratory bird species under the Migratory Bird Treaty Act? The William and Mary Environ Law Policy Rev 28(2): 313–338.
17. Mineau P, Whiteside M (2006) The lethal risk to birds from insecticide use in the U.S. – A spatial and temporal analysis. Environ Toxicol Chem 25(5):1214–1222.
18. Heath SA (2008) Birds in agricultural areas – Reducing pesticide risk to birds using a risk assessment analysis. PhD Dissertation, George Mason University, Fairfax, Virginia, USA.85 p.
19. Mineau P (2002) Estimating the probability of bird mortality from pesticide sprays on the basis of the field study record. Environ Toxicol Chem 24(7):1497–1506.
20. Anonymous (2008) Scientific Opinion of the Panel on Plant protection products and their residues on a request from the EFSA PRAPeR Unit on risk assessment for birds and mammals. The EFSA Journal 734: 1–181.
21. Mineau P, Downes CM, Kirk DA, Bayne E, Csizy M (2005) Patterns of bird species abundance in relation to granular insecticide use in the Canadian prairies. Ecoscience 12(2):267–278.
22. Pimentel D, Acquay H, Biltonen M, Rice P, Silva M, et al. (1992) Environmental and economic costs of pesticide use. Bioscience 42(10):750–760.
23. Mineau P (2005) Direct losses of birds to pesticides – Beginnings of a quantification. In: Ralph CJ, Rich TD, editors. Bird Conservation Implementation and Integration in the Americas: Proceedings of the Third International Partners in Flight Conference 2002, U.S.D.A. Forest Service, GTR-PSW-191, Albany, CA; Vol. 2, pp. 1065–1070.
24. Mineau P, Baril A, Collins BT, Duffe J, Joerman G, et al. (2001) Reference values for comparing the acute toxicity of pesticides to birds. Rev. Environ. Contam. Toxicol. 170:13–74.
25. Burnham KP, Anderson DR (2002) Model selection and multimodel inference. 2nd edition. Springer, USA. 488 p.

19

Influence of Environmental Factors on the Germination of *Urena lobata* L. and Its Response to Herbicides

Tahir Hussain Awan[1]*, Bhagirath Singh Chauhan[1], Pompe C. Sta. Cruz[2]

1 Weed Science, Crop and Environmental Sciences Division, International Rice Research Institute (IRRI), Los Baños, Laguna, Philippines, **2** Crop Science Cluster, College of Agriculture, University of Philippines Los Baños, Laguna, Philippines

Abstract

Urena lobata is becoming a noxious and invasive weed in rangelands, pastures, and undisturbed areas in the Philippines. This study determined the effects of seed scarification, light, salt and water stress, amount of rice residue, and seed burial depth on seed germination and emergence of *U. lobata*; and evaluated the weed's response to post-emergence herbicides. Germination was stimulated by both mechanical and chemical seed scarifications. The combination of the two scarification methods provided maximum (99%) seed germination. Germination was slightly stimulated when seeds were placed in light (65%) compared with when seeds were kept in the dark (46%). Sodium chloride concentrations ranging from 0 to 200 mM and osmotic potential ranging from 0 to −1.6 MPa affected the germination of *U. lobata* seeds significantly. The osmotic potential required for 50% inhibition of the maximum germination was −0.1 MPa; however, some seeds germinated at −0.8 MPa, but none germinated at −1.6 MPa. Seedling emergence and biomass increased with increase in rice residue amount up to 4 t ha^{-1}, but declined beyond this amount. Soil surface placement of weed seeds resulted in the highest seedling emergence (84%), which declined with increase in burial depth. The burial depth required for 50% inhibition of maximum emergence was 2 cm; emergence was greatly reduced (93%) at burial depth of 4 cm or more. Weed seedling biomass also decreased with increase in burial depth. Bispyribac-sodium, a commonly used herbicide in rice, sprayed at the 4-leaf stage of the weed, provided 100% control, which did not differ much with 2,4-D (98%), glyphosate (97%), and thiobencarb + 2,4-D (98%). These herbicides reduced shoot and root biomass by 99–100%.

Editor: Manuel Reigosa, University of Vigo, Spain

Funding: Co author Dr Bhagirath Singh Chauhan supervised this study and provided funding for this study. The funders had role in study design, analysis, and preparation of the manuscript.

Competing Interests: The authors have declared that no competing interests exist.

* E-mail: t.awan@irri.org

Introduction

Urena lobata L., an annual weed, belongs to the family Malvaceae. This weed is known by many local names in different countries, for example, Afulut Gad (Philippines), China pink flowered Chinese Burr or Xiao fan tian hua (China), Africain Jute (Africa), Kongojute (Germen), Bachita (India), Aguaxima (Brazil), Aramina or Cadillo (Spanish), Ke hoa dao (Vietnam), and *Urena lopastnaia* (Russia) [1]. *U. lobata* originated in Asia and has invaded most of the tropical and subtropical area of the world [2]. According to the United States Department of Agriculture (USDA) plant profile database, *U. lobata* has infested areas in Puerto Rico, Florida, Louisiana, the U.S. Virgin Islands, and New South Wales [2]. At present, *U lobata* is a common intruder in natural open areas, pastures, and gardens. It can grow over a wide altitude range; from near sea-level to about 1000-m above sea level. This weed usually grows in rainforests along roads, monsoon forests, and in disturbed areas. From these observations, the Exotic Pest Plant Council of Florida promptly placed *U. lobata* on the list of nuisance plants under Category II in 1999, which implies that the plant is increasing in number but has not yet been established as ecologically harmful [3].

During the 1970s and 1980s, the new *U. lobata* cultivar Ex-Mokwa was introduced and began to be grown as a fiber crop in Sierra Leone [4] and experimentally grown for many years [5]. This species is also used for its medicinal properties in Malaysia and Java. The leaves and barks are used as a contraceptive, and leaf and root extracts are used to prepare herbal medicines [6]. In some parts of Africa, its leaves and flowers are used as food during famine. *U. lobata* had limited success as a crop, but its cultivation resulted in its rigorous infestation as a weed in many areas. Rangelands, poorly managed disturbed areas, pastures, road sides, and eroded areas are easily infested by *U. lobata*. It can thrive on a wide range of soil types and can grow up to a height of 3 m with a basal diameter of 7 cm [7]. Throughout the year, *U. lobata* produces flowers and fruits with up to 600 seeds per plant [8]. *U. lobata* fruits are produced in about 1 cm-diameter capsules covered by straight trichomes that end in about 4–6 small hooks. These hooks cause the fruits to stick on horses' manes and tails, sheep's hair, clothes, and other similar objects that come into contact with plants.

Under field conditions, *U. lobata* seeds revealed dormancy due to an impermeable seed coat (testa) [9]. This is the main cause of poor germination under field conditions [5]. Low or irregular germination, owing to seed dormancy, has been reported by Kirkby [10]. To improve germination, mechanical scarification through removal of carpels and by rubbing the seeds in sand, have been suggested [11]. Sulfuric acid scarification has also been found successful in overcoming seed dormancy [9]. The focus of most of the research work done on *U. lobata* in the past was to improve plant characteristics including growth, seed yield, and fiber yield [5,8,12]. From research conducted on seed germination, it has been speculated that scarification of seeds is necessary to break seed dormancy [5]. The cause of poor germination is the high levels of dormancy due to an impermeable seed coat. However, the role of light, water stress, salt stress, and soil burial depth on seed germination is not well understood for the species. *U. lobata* is becoming a noteworthy weed in many parts of the world, especially in rice fields of Bangladesh, India, Indonesia, Philippines, Thailand, Vietnam [13] and Florida pastures [14]. Research is needed to better understand the seed biology of *U. lobata* and its control with herbicides. A better understanding of the environmental factors that affect the germination of *U. lobata* seeds may help in explaining its fast proliferation in various ecosystems.

Seed germination is usually influenced, directly or indirectly, by many environmental factors such as temperature, moisture, and light [15]. For example, temperature can affect germination by regulating the enzymes activities involved in germination and by promoting or inhibiting the synthesis of hormones that affect seed dormancy [15]. Germination response of seed to fluctuating temperatures is a mechanism by which seeds detect gaps in vegetation canopies and depth of burial in soil [16]. Requirement of light for germination of a weed species means that the species will germinate only when its seeds are on or near the soil surface. Similarly, the ability of a species to germinate under moisture- and salt-stressed conditions will allow that species to take advantage of conditions that suppress the germination of other species. Information on the effect of seed burial depth on seedling emergence may indicate the effectiveness of tillage operations in weed control. Crop residue on the soil surface may be helpful in weed suppression [17] and can be incorporated as a tool in integrated weed management plans. Seedling emergence may be inhibited or stimulated by crop residue, and will depend on the quantity and quality of crop residue and the biology of the weed species involved [18].

The herbicide 2,4-D is used for broadleaf weed control in agricultural and nonagricultural situations. Glyphosate is usually used as a non-selective herbicide to control all weeds before crop-sowing. However, glyphosate is marginally effective on some weeds, such as the *Ipomoea* species [19]. Little is known about the effect of herbicides on *U. lobata*, and, therefore, there is a need to evaluate the performance of available herbicides against it.

The objectives of this study were to determine the optimum time for soaking seeds in sulfuric acid to break seed dormancy; assess the effect of light, water stress, salt stress, soil burial depth, and quantity of rice straw on *U. lobata* seed germination and seedling emergence; and evaluate the performance of commonly available post-emergence herbicides on *U. lobata*.

Materials and Methods

All experiments in the laboratory and screenhouse were laid out in a randomized complete block design with four replications. Each experiment was conducted twice.

Seed collection and germination test

Seeds of *U. lobata* were collected in March 2011 at the International Rice Research Institute (IRRI) farm in Los Baños, Laguna, Philippines. The first two authors work at this institute, and thus no permission was needed to collect the weed seeds. They also confirmed that the study did not involve endangered or protected species.

Seeds collected from at least 150 plants were bulked and cleaned. Experiments were conducted at laboratory and screenhouse facilities of IRRI. Seed germination of *U. lobata* was determined by placing 25 seeds evenly spaced in a 9 cm-diameter Petri dish containing two layers of Whatman No. 1 filter paper (Whatman International Ltd., Maidstone, U.K.) moistened with 5 ml of distilled water or a treatment solution. Parafilm was used to wrap the Petri dishes to reduce water loss. Seeds were scarified with concentrated sulfuric acid (98% H_2SO_4) for 30 minutes (min). Seeds were washed with running tap water for 5 min after treating with sulfuric acid and before being subjected to various treatments. Seeds were placed in an incubator at fluctuating day/night temperatures of 30/20°C in a light/dark regime, unless otherwise specified. This temperature regime was selected because it has been found appropriate for several weed species in the Philippines. The photoperiod was set at 12 h to coincide with the high-temperature photoperiod. Fluorescent lamps were used to produce a photosynthetic photon flux density (PPFD) of 88 micro mol $m^{-2}\ s^{-1}$.

Seed germination was counted daily up to 15 d after placement in the incubator except for the continuous dark treatment, in which germination was counted once at 15 DAS. Germinated seeds (with the radicle visible) were removed from the Petri dishes at each counting. Water was added to the filter paper as needed. Non-germinated seeds were subjected to a simple pressure test to evaluate viability. White firm embryos were considered viable while brown soft embryos were considered nonviable [15].

Effect of seed scarification on germination

To establish whether germination in *U. lobata* is inhibited by an impermeable seed coat, seeds were placed in concentrated sulfuric acid (98%) for 5, 10, and 30 min. In another treatment, seeds were mechanically scarified first by using two rubber stoppers, and then soaking the seeds in sulfuric acid for 30 min. Non-treated seeds were used as a control treatment. Seeds were incubated as described above, and germination was determined.

Effect of light on germination

The effect of light on germination was determined by incubating scarified seeds of *U. lobata* at 30/20°C temperatures (alternating day/night) in both light/dark (12 h/12 h) and continuous (24 h) dark regimes. To maintain the conditions of complete darkness, Petri dishes were wrapped with a double layer of aluminum foil. Other experimental conditions were constant, as previously stated in the germination protocol. As weed seeds do not undergo continuous light and constant temperature conditions under normal conditions in nature [18], these conditions were not included in the experimental variables.

Effect of salt stress on germination

Seed germination as affected by salt stress was investigated by placing scarified seeds of *U. lobata* in dishes each containing a 5 ml of solution of 0, 25, 50, 100, and 200 mM sodium chloride (NaCl) (Mallinckrodt Baker Inc., Phillipsburg, NJ). These ranges of NaCl reflect salinity levels occurring in the tropics [20].

Effect of osmotic stress on germination

The effect of osmotic stress on seed germination was determined by incubating seeds in solutions with osmotic potentials of 0, −0.1, −0.2, −0.4, −0.8, and −1.6 MPa, which were prepared by dissolving 0, 28.9, 40.9, 57.8, 81.8, and 115.6 g of polyethylene glycol 8000 (Sigma-Aldrich Co., St. Louis, MO) in 300 ml of distilled water [21].

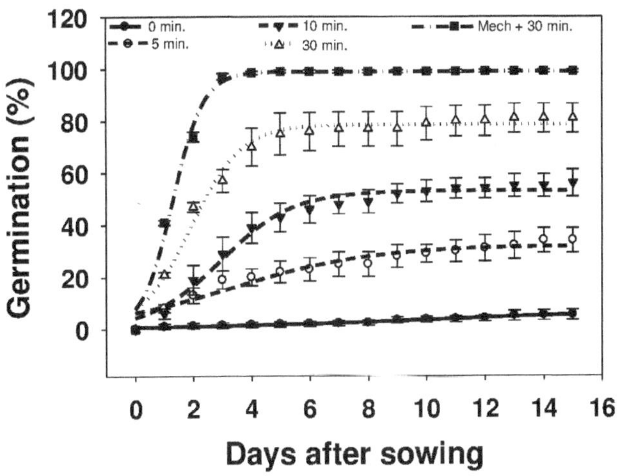

Figure 1. Effect of seed-soaking time in concentrated sulfuric acid (0, 5, 10, and 30 min) and mechanical scarification + soaking time (Mech + 30 min) on germination of *Urena lobata* seed, modeled with the use of equation $G = G_{max}/(1+e[-(x-T_{50})/G_{rate}]$. Estimated parameters are given in Table 1. Vertical bars represent standard error of the mean.

Table 1. Effect of seed-soaking time in concentrated sulfuric acid (0, 5, 10, and 30 min) and mechanical scarification + soaking time in sulfuric acid (Mech + 30 min) on germination of *Urena lobata* seed.

Soaking time in sulfuric acid (min)	G_{max}	G_{rate}	T_{50}	R^2
0	3.0 (1.5)	4.67 (0.73)	10.90 (1.96)	0.98
5	32.0 (1.4)	2.39 (0.44)	3.36 (0.43)	0.94
10	53.1 (0.9)	1.23 (0.14)	2.95 (0.15)	0.98
30	78.6 (1.1)	0.87 (0.10)	1.92 (0.11)	0.98
Mechanical + 30	99.9 (0.8)	0.53 (0.05)	1.30 (0.05)	0.99

Parameter estimates [G_{max}, maximum germination (%); T_{50}, time to reach 50% of maximum germination (days); and G_{rate}, slope] of a three-parameter sigmoid model, $G = G\max/(1+e[-(x-T50)/Grate]$, fitted to seed germination data in Figure 1. Values in parentheses represent standard error of the mean.

Effect of amount of rice residue on emergence and biomass of seedlings

Fifty scarified seeds of *U. lobata* were sown on the soil surface in plastic pots. Rice residue (leaves and stems of the rice variety NSICRc222) was spread on the soil surface at rates equivalent to 0, 1, 2, 4, and 6 t ha^{-1}. The amounts of rice straw used in this study reflect the amount of straw produced in low-yield rainfed environments and high-yield irrigated environments. The condition of the soil and the pots, emergence counting, and harvesting in this experiment were done as described above for the seed burial experiment.

Effect of seed burial depth on emergence and biomass of seedlings

The effect of seed burial depth on seedling emergence was investigated in a screenhouse. Fifty scarified seeds of *U. lobata* were covered with soil to depths of 2, 4, 6, and 8 cm or placed on the soil surface in plastic pots (15 cm in diameter). The soil had 22% sand, 38% silt, 40% clay, 35% organic carbon, 0.107% N, 0.121% Kjeldahl N, 43 mg kg^{-1} soil available P, 1.26 meq 100^{-1} g soil available K, and a pH of 6.0. Soil used for this experiment was collected from rice fields, autoclaved, and passed through a 3-mm sieve. Pots were watered initially with an overhead mist sprinkler and later sub-irrigated. Plants were watered throughout the study as needed to maintain the optimal moisture level for seed germination. Seedlings were considered emerged when a cotyledon was visible on the soil surface. Emergence was counted at 3-day intervals up to 30 days after sowing (DAS). Emerged seedlings were counted and harvested at 30 DAS. After harvesting, sample plants were oven-dried at 70°C for 72 h to obtain dry biomass. After oven-drying, the biomass of plants was measured.

Efficacy of herbicides

To evaluate the effect of recommended rates of herbicides on survival, shoot biomass, root biomass, and control of *U. lobata*, 20 scarified seeds were planted on the soil surface in plastic pots and covered with a thin layer of soil. The condition of the soil and pots used in this experiment were the same as that described for the seed burial experiment. Seedlings were thinned to 8 plants per pot at 5 DAS and later maintained at this density. Post-emergence herbicides 2,4-D ester at 0.5 kg ai ha^{-1}; glyphosate at 1 kg ai ha^{-1}; bispyribac-sodium at 0.03 kg ai ha^{-1}; thiobencarb + 2,4-D at 0.8 kg ai ha^{-1}; and fenoxaprop-p-ethyl + ethoxysulfuron at 0.045 kg ai ha^{-1} were applied using a Research Track Sprayer (DeVries Manufacturing, Hollandale, MN) that delivered 210 L ha^{-1} spray solution at a spray pressure of 140 kPa. Flat-fan nozzles (Teejet 80015) were used in the sprayer. Herbicides were sprayed on the seedlings of *U. lobata* at the 4- and 6-leaf stages. The difference between the two leaf stages was 9 days. Herbicides were not sprayed in the control treatments for each leaf stage. Seedling survival was determined at 25 days after herbicide application, with the criterion being the appearance of a new sprouting leaf. After counting the surviving plants, these were removed from the pots. Soil was washed off through a steel strainer. Plants were separated into shoots and roots and oven-dried at 70°C for 72 h to obtain dry biomass weight.

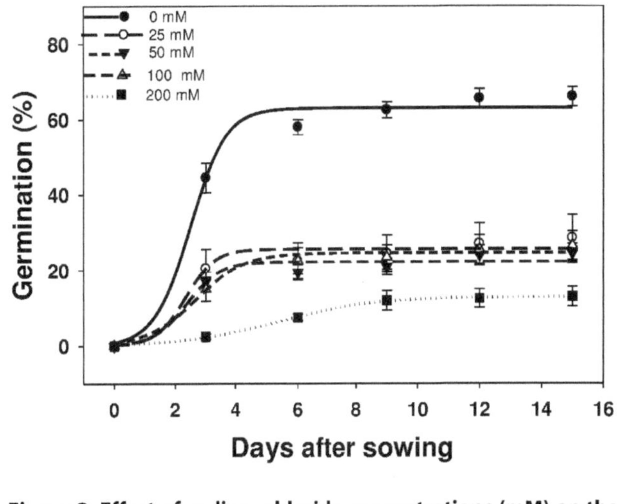

Figure 2. Effect of sodium chloride concentrations (mM) on the germination of *Urena lobata* seeds, incubated at 30/20°C in light/dark, modeled with the use of equation $G = G_{max}/(1+e[-(x-T_{50})/G_{rate}]$. Estimated parameters are given in Table 2. Vertical bars represent standard error of the mean.

Table 2. Effect of sodium chloride concentrations (mM) on the germination of *Urena lobata* seeds, incubated at 30/20°C in light/dark.

NaCl (mM)	G_{max}	G_{rate}	T_{50}	R^2
0	63.1 (1.8)	0.59 (0.47)	2.49 (0.43)	0.99
25	25.6 (1.3)	0.48 (1.03)	2.34 (1.45)	0.98
50	24.6 (0.9)	0.80 (0.44)	2.68 (0.28)	0.98
100	22.3 (1.2)	0.54 (0.86)	2.37 (1.03)	0.96
200	12.9 (0.3)	1.55 (0.15)	5.42 (0.17)	0.998

Parameter estimates [G_{max}, maximum germination (%); T_{50}, time to reach 50% of maximum germination (days); and G_{rate}, slope] of a three-parameter sigmoid model, $G = G\max/(1+e[-(x-T50)/Grate]$, fitted to seed germination data in Figure 2. Values in parentheses represent standard error of the mean.

Table 3. Effect of different osmotic potential (MPa) on germination of *U. lobata* seed, incubated at 30/20°C in light/dark conditions.

Osmotic Potential (MPa)	G_{max}	G_{rate}	T_{50}	R^2
0	63.1 (1.8)	0.59 (0.47)	2.49 (0.43)	0.988
−0.1	38.0 (1.1)	1.16 (0.25)	3.29 (0.24)	0.987
−0.2	29.0 (1.0)	1.30 (0.23)	4.19 (0.29)	0.990
−0.4	27.8 (0.5)	0.93 (0.09)	4.30 (0.15)	0.998
−0.8	19.2 (0.3)	0.57 (0.27)	5.65 (0.17)	0.998
−1.6	0	0	0	0

Parameter estimates [G_{max}, maximum germination (%); T_{50}, time to reach 50% of maximum germination (days); and G_{rate}, slope] of a three-parameter sigmoid model, $G = G\max/(1+e[-(x-T50)/Grate]$, fitted to seed germination data in Figure 3. Values in parentheses represent standard error of the mean. No germination occurred at −1.6 MPa.

Statistical analyses

In the laboratory and screenhouse, each replication was arranged on different shelves in the incubator or on different benches in the screenhouse. Different shelves in the incubator or different benches in the screenhouse may experience yet varying conditions, and therefore each replication was considered as a block. Data were subjected to ANOVA and means were separated using least significant difference (LSD) at 5% level of significance. No interaction was detected between the treatments and experimental runs; therefore, data were pooled over the runs for analysis (GenStat 8.0). Before statistical analysis, homogeneity of variance was visually confirmed by inspecting the residuals.

Germination, emergence, and biomass data were analyzed using regression analysis. Germination resulting from scarification and salt- and water-stress treatments, and emergence resulting from different burial depths and residue amounts were modeled using a functional three-parameter sigmoid function. The fitted model was:

$$G = G\max/(1+e[-(x-T50)/Grate]$$

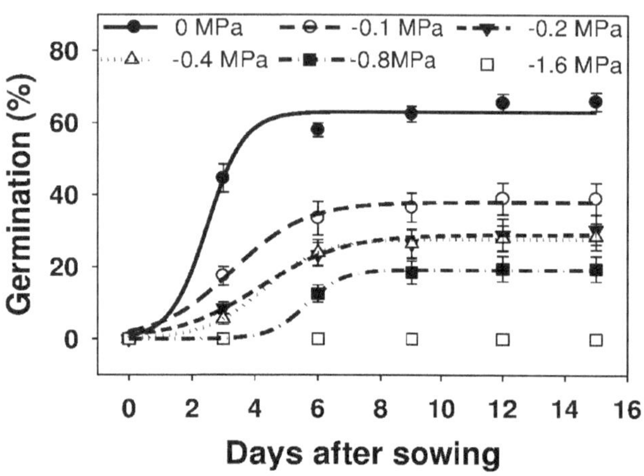

Figure 3. Effect of different osmotic potential (MPa) on germination of *U. lobata* seed, in 3-day intervals, incubated at 30/20°C in light/dark, modeled with the use of equation $G = G_{max}/(1+e\ [-(x-T_{50})/G_{rate}]$. Estimated parameters are given in Table 3. Vertical bars represent standard error of the mean.

where G is the cumulative percentage germination at time x, G_{max} is the maximum germination (%), T_{50} is the time (d) required for 50% of maximum germination, and G_{rate} indicates the slope. An exponential decay curve in the form of

$$E = E\max.e(-Erate.x)$$

was fitted to seedling emergence (%) or seedling biomass (g) obtained at different depths, where E represents cumulative emergence (%) and biomass (g) at depth x, E_{max} is the maximum emergence or biomass, and E_{rate} indicates the slope. Parameter estimates for each model were compared using their standard errors [22–24]. Survival and weed control data were transformed before statistical analysis using arcsine transformation. The transformation did not improve the results; therefore, the nontransformed values were used for analysis.

Results and Discussion

Effect of seed scarification on germination

This study included the effect of mechanical and chemical scarifications on the germination of *U. lobata*. Increased soaking time in concentrated sulfuric acid was found to improve the germination percentage (Figure 1). Maximum germination (80%) was achieved at 30 min of soaking in sulfuric acid compared to the control (3%). Mechanical scarification plus 30 min of soaking in sulfuric acid resulted in the highest germination rate (99%) (Table 1). Similar observations were reported in an earlier study [14], in which mechanical and chemical scarification enhanced germination rate of *U. lobata* by 42 and 78%, respectively.

Mechanical scarification was laborious and time consuming, and may not be done uniformly on a large quantity of seeds. Thus, *U. lobata* seeds were chemically scarified (30 min of soaking) for all the subsequent experiments. Non-germinated seeds in all treatments were tested through the pressing method as described earlier [15] and were found to be viable. These results are consistent with earlier findings that sulfuric acid scarification was effective in breaking seed dormancy in *U. lobata* [17] and *Ipomoea triloba* L. [25].

Seed dormancy of *U. lobata* was mainly caused by its hard seed coat. Physical dormancy is widespread among Malvaceae species [5] and is the most ubiquitous form of dormancy among plant species [27]. Mechanical or chemical scarification can break,

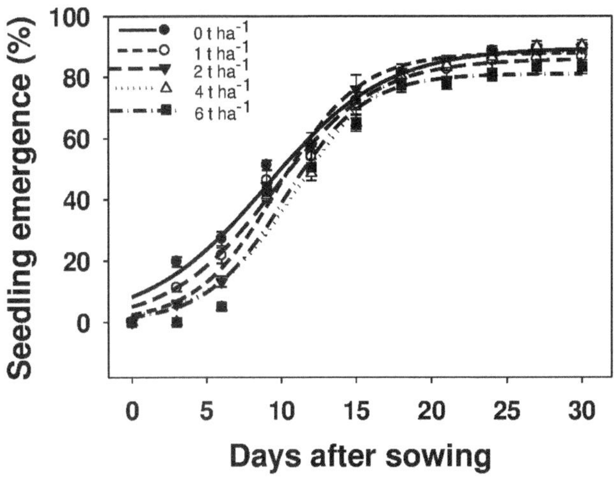

Figure 4. Effect of rice residue amount (t ha^{-1}) on emergence of *Urena lobata* at various days after sowing, modeled with the use of equation $G = G_{max}/(1{+}e\ [-(x - T_{50})/G_{rate}]$. Parameter estimates are given in Table 4. Vertical bars represent standard error of the mean.

Table 4. Effect of rice residue amount (t ha^{-1}) on seedling emergence of *Urena lobata*.

Residue quantity (t ha^{-1})	G_{max}	G_{rate}	T_{50}	R^2
0	89.5 (2.7)	3.97 (0.49)	9.04 (0.55)	0.98
1	85.9 (1.8)	3.44 (0.32)	9.47 (0.37)	0.99
2	87.9 (1.7)	2.76 (0.28)	9.79 (0.32)	0.99
4	88.5 (3.4)	3.03 (0.52)	11.17 (0.61)	0.98
6	81.0 (3.0)	2.65 (0.50)	10.25 (0.58)	0.98

Parameter estimates [G_{max}, maximum germination (%); T_{50}, time to reach 50% of maximum germination (days); and G_{rate}, slope] of a three-parameter sigmoid model, $G = Gmax/(1 + e[-(x - T50)/Grate]$, fitted to seedling emergence data in Figure 4. Values in parentheses represent standard error of the mean.

scratch, or soften seed coats, which effectively release dormancy in *U. lobata* seeds. Based on previous research on *U. lobata* [14], it was pragmatic that the germination percentage of intact seeds (without removing husk) was less than 2%. The low percentage of germination was attributed to the enclosure of the seed inside a bur, which resulted in low germination rates [28]. Removal of the husk increased germination slightly (10%), while water leaching inside resulted in a nearly 20% germination.

Research suggests that seed coat-imposed dormancy may be a significant cause for low germination. The hard seed coat helps *U. lobata* persist for a long period of time in the soil. When dormancy is released, seeds start to germinate. The results of the present study suggest that *U. lobata* seeds are less likely to germinate in the field unless scarified. It may thus persist in the soil for a long period of time, causing a setback in future crops, if scarification does not take place. Under natural environmental conditions, scarification usually occurs due to soil burial of seed [28,29], microbial attack [29], acute changes in temperature which cause contraction and expansion of the seed coat) [30,31], humid atmospheric conditions [32,33], passage through the digestive system of animals [34,35], flora and fauna actions, burning of vegetation and fire release the dormancy of hard seed especially if fires coincide with wet and hot conditions [36,37,38], and scratching by soil particles probably during tillage operations [28]. However, the effects of these factors are not yet known in the case of *U. lobata.*

Effect of light on germination

The average germination percentage of *U. lobata* seed was 65% (±2.5%) in light/dark and 46% (±6.1%) in continuous dark conditions (data not shown). The results suggest that light is not a prerequisite for the germination of *U. lobata* and that its seeds can germinate without light. This may explain why *U. lobata* is often found under tree canopies within pastures, rangelands, and natural areas. Other plant species of the Malvaceae family, such as *Sida spinosa* L., do not require light for germination [26].

Effect of salt stress on germination

Salt concentration affected the germination of *U. lobata* greatly (Figure 2). Seeds had the highest cumulative germination (63%) in the nonstressed treatment (0 mM NaCl); cumulative germination declined with increasing salt concentration. Germination of *U. lobata* was at 22% at NaCl concentration of 100 mM, and decreased drastically to 13% at 200 mM concentration. It is evident from the data that salt stresses significantly delayed the start of germination and reduce germination percentage. The time to inhibit 50% of the maximum germination (T_{50}) increased with increase in the level of salt stress (Table 2).

Similar results on the effect of varying NaCl concentrations were reported for *Amaranthus spinosus* L. and *A. viridis* L. [24], *I. triloba* [25], *Chromolaena odorata* (L.) King and H.E. Robins and *Tridax procumbens* L. [22], and *Digitaria ciliaris* (Retz.) Koel. and *D. longiflora* (Retz.) Pers. [23]. In Asia, a significant area (21.5 M ha) has been documented to have salt-affected soil [39]. Soil with more than 100 mM concentration of NaCl is considered as having a high level of salinity. Rice, a staple food of people in the tropics, is considered a salt-sensitive crop [40]. Salinity stress, even at 40 mM NaCl, can affect flooded rice [41]. Some species in the Asteraceae, Fabaceae, and Poaceae families were reported as able to adapt to saline conditions [20,22,23]. In this study, *U. lobata* germinated even at 200 mM NaCl. This characteristic may allow

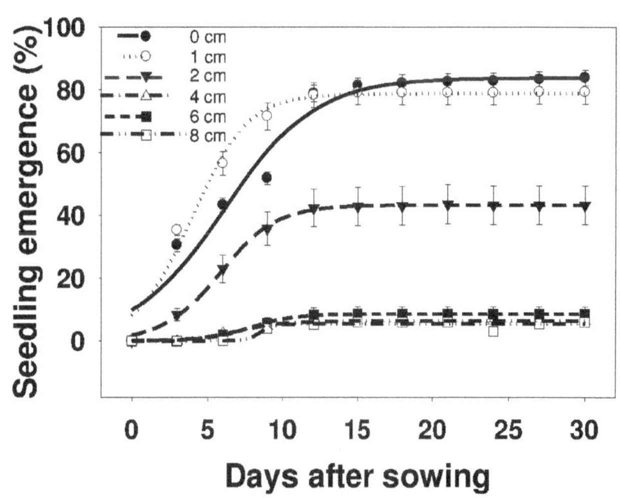

Figure 5. Effect of seed burial depth (cm) on seedling emergence of *Urena lobata* at various days after sowing, modeled with the use of equation $G = G_{max}/(1{+}e\ [-(x - T_{50})/G_{rate}]$. Estimated parameters are given in Table 5. Vertical bars represent standard error of the mean.

Table 5. Effect of seed burial depth (cm) on seedling emergence of *Urena lobata* Parameter estimates [G_{max}, maximum germination (%); T_{50}, time to reach 50% of maximum germination (days); and G_{rate}, slope] of a three-parameter sigmoid model, $G = G\max/(1 + e[-(x - T50)/Grate]$, fitted to seedling emergence data in Figure 5.

Soil depth (cm)	G_{max}	G_{rate}	T_{50}	R^2
0	83.8 (2.7)	3.02 (0.65)	6.09 (0.60)	0.97
1	78.7 (1.5)	1.84 (0.27)	3.96 (0.30)	0.98
2	43.2 (0.3)	1.87 (0.10)	5.82 (0.11)	0.99
4	6.3 (0.1)	1.53 (0.17)	7.05 (0.20)	0.99
6	8.5 (0.1)	1.45 (0.08)	7.87 (0.09)	0.99
8	5.3 (0.4)	0.45 (3.38)	8.52 (3.65)	0.89

Values in parentheses represent standard error of the mean.

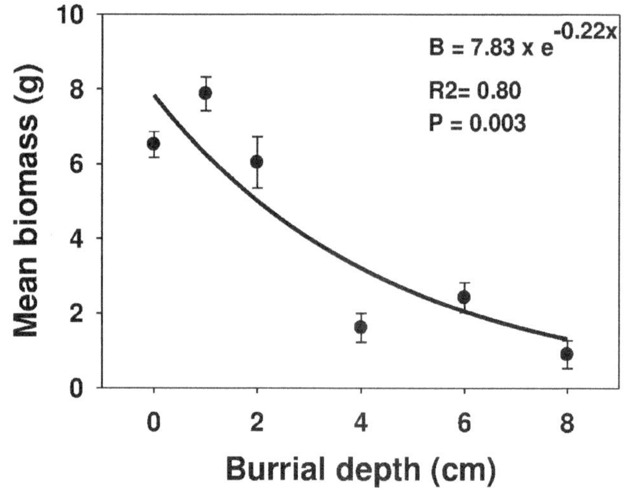

Figure 6. Effect of different seed burial depth (cm) on seedling biomass (g pot^{-1}) of *Urena lobata* at 30 days after sowing, modeled with the use of equation $E = E_{max}\ x\ e^{(-Erate.x)}$. Vertical bars represent standard error of the mean.

U. lobata to have a distinct advantage over other crops, such as rice, in these saline environments.

Effect of osmotic stress on germination

Germination of *U. lobata* was affected greatly by increasing water stress (Figure 3). The maximum germination of *U. lobata* at 0 MPa was 63%. An osmotic potential of −0.1, −0.2, −0.4, −0.8, and −1.6 MPa reduced germination of *U. lobata* to 38, 29, 28, 19, and 0%, respectively (Figure 3; Table 3). Time to 50% germination (T_{50}) increased with increase in degree of osmotic stress.

The results of our study suggest that germination of *U. lobata* is favored by a moist environment and that seeds can remain ungerminated in drought conditions. This could be a valuable tactic of *U. lobata*, allowing for an increased ability to delay germination until favorable conditions occur, which thus prolongs the existence of these seeds in the seed bank. Similar results were reported by Wang and colleagues, wherein *U. lobata* seed germination was significantly affected by increasing water stress. Germination percentage was at 86% in deionized water and, as water stress is increased, germination declined [14].

Seed germination is a process of growth of a previously dormant or inactive seed starting with the imbibitions of water [42]. Seed imbibition rate and germination normally decrease as the water potential of the surrounding decreases, and the critical hydration level for seed germination is species-specific [43].

Based on the present study, germination of *U. lobata* cannot occur when water potential is equal or lower than −1.6 MPa, which is the case for soils that are dry or have high salt levels. The tolerance level of a weed species to water stress is related to its ecological habitat. Species, such as *Eupatorium compositifolium* Walt., when in dry or well-drained terrestrial habitats, can usually germinate at low moisture or water potential [44]. Conversely, species that require moist soils for germination, such as *A. spinosus* and *A. viridis*, can germinate only when water potential is at −0.6 MPa or higher [24,45]. Our data support the findings reported for *I. triloba* [25], *A. spinosus* and *A. viridis* [24,45], in which germination decreased with decrease in water potential. At 30/20°C alternating temperatures, no seeds germinated at water potential of −0.6 MPa and below. Germination of *A. retroflexus* L. was also sensitive to water availability, with final germination being reduced by 50% at −0.4 MPa compared to 0 MPa [46]. *U. lobata* seeds are sensitive to water stress and may not germinate during the dry season in well-drained soils. Rainfall is typically low during April and May in the Philippines, which may delay the germination of *U. lobata* until the rainy season starts in June.

Effect of amount of rice residue on emergence and biomass of seedlings

Seedling emergence continued until 27 DAS and subsequently became constant (Figure 4). The emergence of *U. lobata* was not influenced by the amount of residue tested, and it varied from 81 to 90% (Figure 4; Table 4). Seedling biomass was not influenced by the amount of rice residue, and was in the range of 5.4–6.2 g pot^{-1} (data not shown). Similar results were reported for *A. viridis* [24]. Contrary to our findings on *U. Lobata*, however, the emergence of several weed species (e.g., *C. odorata*, *Tridax procumbens*, *A. spinosus*, and *A. viridis*) were reduced by the addition of crop residue [22,24].

Effect of seed burial depth on emergence and biomass of seedlings

The emergence of *U. lobata* was markedly affected by seed burial depth (Figure 5). Seeds placed on the soil surface had an emergence percentage of 84% that decreased with increase in burial depth, i.e., 79% at 1 cm and 50% at 2 cm. Increase in burial depth to 6 and 8 cm resulted in 9 and 5% emergence, respectively. With increase in soil burial depth, there was a resulting increase in time (d) to reach 50% of the maximum emergence, except in the treatment where seeds were placed on the soil surface. For example, time to reach 50% emergence was 4, 6, 7, 8, and 9 days at seed burial depths of 1, 2, 4, 6, and 8 cm, respectively (Table 5).

Biomass was lower (6.5 g pot^{-1}) for seedlings emerging from the soil surface compared with those emerging from 1 cm burial depth (7.9 g pot^{-1}) (Figure 6). However, seedling biomass decreased with burial depth beyond 1 cm. Biomass was only 0.9 g pot^{-1} for seedlings emerging from a depth of 8 cm.

These results are in agreement with reported findings on *I. triloba* [25]. *U. lobata* seeds are most likely to germinate within 2 cm of the soil surface, but low germination rate (6%) was observed at

Table 6. Effect of post-emergence herbicides on survival (%), shoot and root biomass (g pot^{-1}), and control (%) of *Urena lobata* when sprayed at 4- and 6-leaf stages of the weed.

Herbicide	Rate	4-leaf stage				6-leaf stage			
		Survival	Shoot biomass	Root biomass	Control	Survival	Shoot biomass	Root biomass	Control
	kg ha^{-1}	%	g pot^{-1}	g pot^{-1}	%	%	g pot^{-1}	g pot^{-1}	%
Control	-	100	4.342	1.647	0	100	6.971	2.053	0
2,4-D ester	0.50	1.6	0.006	0.001	98.4	17.2	0.229	0.051	82.8
Bispyribac-sodium	0.03	0	0	0	100	89.1	1.336	0.329	10.9
Fenoxaprop+ ethoxysulfuron	0.045	90.6	1.014	0.307	9.4	100	2.447	0.809	0
Glyphosate	1.00	3.1	0.014	0.003	96.9	12.5	0.117	0.021	87.5
Thiobencarb + 2,4-D	0.80	1.6	0.008	0.005	98.4	18.8	0.218	0.072	81.2
$LSD_{0.05}$		7.88	0.1727	0.0748	7.88	10.74	0.3376	0.1309	10.74
P		<0.001	<0.001	<0.001	<0.001	<0.001	<0.001	<0.001	<0.001

burial depth of 8 cm. A similar germination percentage (84%) was reported by Wang and workers when *U. lobata* seeds were placed on the soil surface [14]. Probably at deeper depths, light and seed size are usually the limiting factors for seedling emergence. Light penetration is generally limited to the first few millimeters below the soil surface [47]. However, the present study found that light is not a prerequisite for *U. lobata* seed germination. In previous studies, reduced emergence of several weed species due to increased burial depth were reported [22,24].

Fluctuating temperatures are more likely to promote germination of small-seeded species than of large-seeded species [16]. Seeds of small weed species are more likely to be buried [48,49], and their seedlings are less likely to emerge from deeper soil, from where seedlings of large seeds can emerge [50]. Thus, depth sensing is much more imperative for small seeds than for large seeds. Small-seeded species, such as *A. spinosus* and *A. viridis*, may have limited reserves to support germination and seedling emergence, thus limiting the depth from which their seedlings can emerge [24]. Conversely, larger seeds usually have higher reserves of carbohydrate and are capable of emerging from greater burial depths [17], such as in the case of *Sicyos angulatus* L. that could emerge even from a depth of 16 cm [51]. The seed diameter of *U. lobata* is about 2 mm, compared to that of *S. angulatus* that is about 5 mm. *U. lobata* seedlings thus failed to emerge when buried too deep. *U. lobata* seedlings were able to emerge from a depth of 6 cm, suggesting the possibility of an escape mechanism when treated with pre-emergence herbicides. Another possible reason for germination inhibition caused by burial depth may be the secondary dormancy imposed by the interaction between soil gases and seed metabolism [52,53]. The lack of emergence from deeply buried seeds could be attributed to hypoxia and low rates of gaseous diffusion deep in the soil [47,54].

The results of the present study suggest that farming practices that enable shallow burial of weed seeds will only suppress the emergence of *U. lobata* seedlings but may not provide complete control of *U. lobata* emergence. On the one hand, deep-tillage operations may be needed to overturn the soil and bury *U. lobata* seeds very deep, from where its seedlings are unable to emerge. On the other hand, succeeding tillage operations may bring viable weed seeds near the soil surface and cause these to germinate and re-infest the succeeding crop. To cope with weed infestation, it is necessary to understand seed viability in the soil seed bank.

Efficacy of herbicides

Urena lobata seedlings did not survive when sprayed with bispyribac-sodium at the 4-leaf stage, resulting in 100% weed control. Similar weed control was achieved with 2,4-D, glyphosate, and thiobencarb + 2,4-D (Table 6). These herbicides reduced the shoot and root biomass by 99–100%. At the 6-leaf stage, the above-mentioned herbicides, except bispyribac-sodium, had similar effects but with little decline in weed control (80%). Bispyribac-sodium at this leaf stage provided poor control (11%). As far as root and shoot biomass is concerned, other herbicides reduced biomass by 97% while bispyribac-sodium reduced biomass by 82%. Similar results from the use of 2,4-D and glyphosate were reported for *I. triloba* [25]. Percentage control and survival data suggest that fenoxaprop-p-ethyl + ethoxysulfuron provided poor control (9 and 0% at 4- and 6-leaf stages, respectively).

Manual control of a weed plant that has a fibrous root system, such as *U. lobata*, is very difficult, time consuming, and expensive. Therefore, farmers have to rely on chemicals to control *U. lobata.* Knowledge of herbicides that are effective against *U. lobata* is very important for its control. Based on the results of this study, it can be concluded that if *U. lobata* infests upland rice, farmers can easily control this weed through the application of 2,4-D or thiobencarb + 2,4-D. These herbicides can provide excellent control of *U. lobata* and the control is comparable to glyphosate application. Glyphosate application remains as the ideal control option for *U. lobata* when there is no rice crop in the field.

Author Contributions

Conceived and designed the experiments: THA BSC PCSC. Performed the experiments: THA BSC. Analyzed the data: THA BSC. Contributed reagents/materials/analysis tools: THA BSC. Wrote the paper: THA BSC PCSC.

References

1. Philippine Medicinal plants. Available: http://www.stuartxchange.com/Dalupang.html. Accessed 2013 Nov 27.
2. Leisure services department exotic management plan 2009. Available: http://seminolecountyfl.gov/parksrec/pdf/scnl_exoticmgmtplan_FINAL.pdf. Accessed 2013 Sept 23.
3. FLERPC (2008) Florida Exotic Pest Plant Council's 2008 list of invasive plant species. Wildland Weeds 10: 13–16.
4. Dasgupta DK (1973) Studies on the crop physiology and cultural practices of *Urena lobata* L. for fibre production in Sierra Leone. 2 Cultural practices. Sols Afr 18: 33–46.
5. Harris PJC (1981) Seed viability, dormancy, and field emergence of *Urena lobata* L. in Sierra Leone. Trop Agric 58: 205–213.
6. Forest Research Institute of Malaysia (2003) Plant Information: *Urena lobata* Griff., pulut-pulut, Malvaceae. Forest Research Institute of Malaysia, Kuala Lumpur, Malaysia. http://www.frim.gov.my/tu/Urena.htm.
7. Francis JK (2003) *Urena lobata* L. Malvaceae. US Department of Agriculture, Forest Service, International Institute of Tropical Forestry, Jardin Botanico Sur, 1201 Calle Ceiba, San Juan PR 00926-1119, in cooperation with the University of Puerto Rico, Rio Piedras, PR 00936-4984. http://www.fs.fed.us/global/iitf/pdf/shrubs/Urena%20lobata.pdf.
8. Harris PJC, Brewah Y (1986) Seed production of *Urena lobata* in Sierra Leone: effect of sowing date. Trop Agric 63: 30–32.
9. Horn CL, Natal Colon JE (1942) Acid scarification of the seed of two Cuban fiber plants. J Am Soc Agron 34: 1137–1138.
10. Kirby RH (1963) Vegetable fibres, pp. 135–136. London: Leonard Hill (Books) Ltd.
11. Dempsey JM (1975) Fiber crops, p 382. Gainsville: University Presses of Florida.
12. Harris PJC (1985) Seed production of *Urena lobata* in Sierra Leone: effect of harvest date on yield. Trop Agric 62: 229–232.
13. Moody K (1989) Weeds Reported In Rice in South And Southeast Asia. International Rice Research Institute, Los Baños, Laguna, Philippines. 83 pp.
14. Wang J, Jason F, Gregory MD, Brent S (2009) Factors affecting seed germination of Cadillo (*Urena lobata*). Weed Sci 57: 31–35.
15. Baskin CC, Baskin JM (1998) Seeds: ecology, biogeography, and evolution of dormancy and germination. San Diego, California, USA: Academic Press. 666 p.
16. Liu K, Baskin JM, Baskin CC, Bu H, Du G, et al. (2013) Effect of diurnal fluctuating versus constant temperatures on germination of 445 species from the Eastern Tibet Plateau. PLOS ONE 8(7): 1–7. e69364. doi:10.1371/journal.pone.0069364
17. Chauhan BS, Johnson DE (2010) The role of seed ecology in improving weed management strategies in the tropics. Adv Agron. 105: 221–262.
18. Bolfrey AGEK, Chauhan BS, Johnson DE (2011) Seed germination ecology of itchgrass (*Rottboellia cochinchinensis*). Weed Sci 59: 182–187.
19. Culpepper AS, Gimenez AE, York AC, Batts RB, Wilcut JW (2001) Morningglory (*Ipomoea* spp.) and large crabgrass (*Digitaria sanguinalis*) control with glyphosate and 2,4-DB mixtures in glyphosate-resistant soybean (*Glycine max*). Weed Technol 15: 56–61.
20. Chauhan BS, Johnson DE (2008a) Influence of environmental factors on seed germination and seedling emergence of eclipta (*Eclipta prostrata*) in a tropical environment. Weed Sci 56: 383–388.
21. Michel BE (1983) Evaluation of the water potentials of solutions of polyethylene glycol 8000 both in the absence and presence of other solutes. Plant Physiol 72: 66–70.
22. Chauhan BS, Johnson DE (2008b) Germination ecology of two troublesome Asteraceae species of rainfed rice: Siam weed (*Chromolaena odorata*) and coat buttons (*Tridax procumbens*). Weed Sci 56: 567–573.
23. Chauhan BS, Johnson DE (2008c) Germination ecology of southern crabgrass (*Digitaria ciliaris*) and India crabgrass (*Digitaria longiflora*): two important weeds of rice in tropics. Weed Sci 56: 722–728.
24. Chauhan BS, Johnson DE (2009) Germination ecology of spiny (*Amaranthus spinosus*) and Slender Amaranth (*A. viridis*): troublesome weeds of direct-seeded rice. Weed Sci 57: 379–385.
25. Chauhan BS, Abugho SB (2012) Threelobe morningglory (*Ipomoea triloba*) germination and response to herbicides. Weed Sci 60: 199–204.
26. Smith CA, Shaw DR, Newsom LJ (1992) Arrow leaf sida (*Sida rhombifolia*) and prickly sida (*Sida spinosa*): germination and emergence. Weed Res 32: 103–109.
27. Foley ME (2001) Seed dormancy: an update on terminology, physiological genetics, and quantitative trait loci regulating germinability. Weed Sci 49: 305–317.
28. Crocker W (1906) Role of seed coats in delayed germination. Bot Gaz 44: 375–380.
29. Taylor GB (1985) Effect of tillage practices on the fate of hard seeds of subterranean clover in a ley farming system. Aust J Exp Agri. 25: 568–573.
30. Taylor GB, Ewing MA (1988) Effect of depth of seed burial on the longevity of hard seeds of subterranean clover and annual medics. Aust J Exp Agri 28: 77–81.
31. Hagon MW (1971) The action of temperature fluctuating on hard seed of subterranean clover Aust J Exp Agri Ani Husb 11: 440–443.
32. Scott KA (2006) Seed coat morphology of *Parkinsonia aculeata* L. & pattern of heat-induced fractures. In: Proceedings of the 15th Australian Weeds Conference (Adelaide, SA, Australia, 24–28 September 2006). Weed Management Society of South Australia, Adelaide, SA, p. 268–271.
33. Taylor GB (1982) Longevity of subterranean clover. Annual Report, CSIRO Division of Land Resources management, CSIRO.
34. Fairbrother TE (1991) Effect of fluctuating temperature and humidity on the softening rate of hard seed of subterranean clover (*Trifolium subterraneum* L.). Seed Sci Tech 19: 93–105.
35. Carter Ed, Challis S, Knowles RC, Bahrani J (1989) Pasture legume seed survival following ingestion by sheep. In 'Proceedings of the 5th Australian Agronomy Conference'. Perth, WA, p. 437. (Australian Society of Agronomy: Carlton, Vic.).
36. Squella F, Carter ED (1993) Significance of seed size and level of hard seededness on survival of medic seeds ingested by sheep. In 'Proceedings of the XVII International Grassland Congress Palmerston North, New Zealand Vol. 1. Pp. 419–420. (Keely and Mundy: Palmerston North, New Zealand). Perth, WA, p. 437. (Australian Society of Agronomy: Carlton, Vic.).
37. Auld TD, O'Connel MA (1991) Predicting patterns of post-fire germination in eastern Australian Fabaceae. Aust J Ecol 16: 53–70.
38. Tieu A, Dixon KW, Meney KA, Sivasitijamparam K (2001) The interaction of heat and smoke in the release of seed dormancy in seven species from southwestern Western Australia. Ann Bot 88: 259–265.
39. Lafitte HR, Ismail A, Bennett J (2006) Abiotic stress tolerance in tropical rice: progress and future prospects. Oryza 43: 171–186.
40. Tanji KK, Kielen NC (2002) Agricultural Drainage Water Management in Arid and Semi-Arid Areas. FAO Irrigation and Drainage Paper 61. Rome: Food and Agriculture Organization of the United Nations. 202 p.
41. Sanchez PA (1976) Soil management in rice cultivation systems. Pages 413–477 in Properties and Management of Soils in the Tropics. Raleigh, NC: John Wiley and Sons.
42. Bradford KJ (1990) A water relations analysis of seed germination rates. Plant Physiol 94: 840–849.
43. Evans CE, Eitherington JR (1990) The effect of soil water potential on seed germination of some British plants. New Phytol 115: 539–548.
44. MacDonald GE, Brecke BJ, Shilling DG (1992) Factors affecting germination of dogfennel (*Euparotium capillifolium*) and yankeeweed (*Eupatorium compositifolium*). Weed Sci 40: 424–428.
45. Thomas WE, Burke IC, Spears JF, Wilcut JW (2006) Influence of environmental factors on slender amaranth (*Amaranthus viridis*) germination. Weed Sci 54: 316–320.
46. Ghorbani R, Seel W, Leifert C (1999) Effects of environmental factors on germination and emergence of *Amaranthus retroflexus*. Weed Sci 47: 505–510.
47. Woolley JT, Stoller E (1978) Light penetration and light-induced seed germination in soil. Plant Physiol 61: 597–600.
48. Peart MH (1984) The effects of morphology, orientation and position of grass diaspores on seedling survival. J Ecol 72: 437–453. doi:10.2307/2260057
49. Thompson K, Green A, Jewels AM (1994) Seeds in soil and worm casts from a neutral grassland. Funct Ecol 8: 29–35. doi: 10.2307/2390108
50. Pearson TRH, Burslem DFRP, Mullins CE, Dalling JW (2002) Germination ecology of neotropical pioneers: interacting effects of environmental conditions and seed size. Ecology 83: 2798–2807.
51. Mann RK, Rieck CE, Witt WW (1981) Germination and emergence of burcucumber (*Sicyos angulatus*). Weed Sci 29: 83–86.
52. Benvenuti S, Macchia M (1995) Effects of hypoxia on buried weed seeds germination. Weed Res. 35: 343–351.
53. Benvenuti S, Macchia M (1997) Phytochrome-meditated germination of *Datura stramonium* L. seeds after seed burial. Weed Res 38: 199–205.
54. Benvenuti S (2003) Soil texture involvement in germination and emergence of buried weed seeds. Agron J 95: 191–198.

20

Seed Germination Ecology of *Echinochloa glabrescens* and Its Implication for Management in Rice (*Oryza sativa* L.)

Jhoana L. Opeña[1], Bhagirath S. Chauhan[1]*, Aurora M. Baltazar[2]

1 Crop and Environmental Sciences Division, International Rice Research Institute, Los Baños, Laguna, Philippines, **2** University of the Philippines Los Baños, Los Baños, Laguna, Philippines

Abstract

Echinochloa glabrescens is a C_4 grass weed that is very competitive with rice when left uncontrolled. The competitive ability of weeds is intensified in direct-seeded rice production systems. A better understanding is needed of factors affecting weed seed germination, which can be used as a component of integrated weed management in direct-seeded rice. This study was conducted to determine the effects of temperature, light, salt and osmotic stress, burial depth, crop residue, time and depth of flooding, and herbicide application on the emergence, survival, and growth of two populations [Nueva Ecija (NE) and Los Baños (IR)] of *E. glabrescens*. Seeds from both populations germinated at all temperatures. The NE population had a higher germination rate (88%) from light stimulation than did the IR population (34%). The salt concentration and osmotic potential required to inhibit 50% of germination were 313 mM and −0.24 MPa, respectively, for the NE population and 254 mM and −0.33 MPa, respectively, for the IR population. Emergence in the NE population was totally inhibited at 4-cm burial depth in the soil, whereas that of the IR population was inhibited at 8 cm. Compared with zero residue, the addition of 5 t ha^{-1} of rice residue reduced emergence in the NE and IR populations by 38% and 9%, respectively. Early flooding (within 2 days after sowing) at 2-cm depth reduced shoot growth by 50% compared with non-flooded conditions. Pretilachlor applied at 0.075 kg ai ha^{-1} followed by shallow flooding (2-cm depth) reduced seedling emergence by 94−96% compared with the nontreated flooded treatment. Application of postemergence herbicides at 4-leaf stage provided 85−100% control in both populations. Results suggest that integration of different strategies may enable sustainable management of this weed and of weeds with similar germination responses.

Editor: Jauhar Ali, International Rice Research Institute, Philippines

Funding: The authors have no support or funding to report.

Competing Interests: The authors have declared that no competing interests exist.

* E-mail: b.chauhan@irri.org

Introduction

The looming water crisis, increasing labor cost, and shortage of labor have led to a shift from transplanted to direct-seeded rice production systems in many Asian countries [1,2]. In the Philippines, the availability of water for irrigation has been declining since the last decade. Recently, significant yield loss (34−44%) in transplanted rice was reported at Cavite Province in the Philippines and it was mainly due to water scarcity [3]. Rapid adoption of direct-seeded rice has since been observed in the Philippines. In 1979, only 10% of rice farmers at Nueva Ecija Province in the Philippines used direct-seeding. By 1986, however, 27% of farmers who used transplanting as a method of rice establishment forcibly switched to direct-seeded rice in the dry season (broadcast seeding) and farmers who were practicing a combination of transplanting and wet seeding switched entirely to direct-seeded rice [4]. The moisture status of a field, the introduction of more herbicides, and the lack of timely labor supply for transplanting encouraged this shift to direct-seeded rice [5,6].

Weed infestation is a serious problem in direct-seeded rice because of the absence of standing water, which has a suppressive effect on weed growth during rice emergence and seedling size advantage of rice seedlings over the weeds [1,7]. Manual weeding is commonly done, but it is very labor-intensive, is becoming expensive, and getting scarce because of the migration of farm labor to urban areas [1,8]. Herbicide use is becoming popular as an alternative to manual weeding in many countries. But the intensive use of herbicides poses hazards to human health and the environment, causes evolution of herbicide resistance and shifts in weed population, and increases production costs.

Echinochloa glabrescens Munro ex Hook. F., *E. crus-galli* (L.) Beauv., and *E. colona* (L.) Link. count among the world's most serious grass weeds [9]. All *Echinochloa* species have a C_4 photosynthetic pathway and show a great advantage when grown with C_3 crops such as rice (*Oryza sativa* L.). These weeds look very similar to rice plants at the seedling stage, causing farmers to sometimes unknowingly transplant these weeds onto the field [10]. By the time the weeds can be easily distinguished from the rice plants, crop yield losses have already been incurred [9−11]. *E. glabrescens*, one of the most common species in lowland rice fields [12], is highly competitive with rice. Uncontrolled *E. glabrescens* transplanted with rice reduced yield by 7% and 87% at weed infestation levels of 5% and 50%, respectively [13]. In another study, yield loss of 90% occurred when all rice hills were infested with *E. glabrescens* seedlings [14].

Better understanding is needed of the factors that affect germination. Such understanding will help in the development of an effective cultural management practice for this problematic weed through either inhibiting its germination or encouraging germination when weed seedlings can be easily controlled [1]. Weed seed germination is influenced by various ecological factors such as temperature, light, soil salinity, soil moisture, seed burial depth in the soil through tillage, and the use of rice residue as mulch. Germination response to these factors is species-specific [15] and such information on *E. glabrescens* is limited in literature.

Weed species, such as *Ipomeoa triloba*, can germinate equally in both light and dark [7], while germination in other species, such as *Leptochloa chinensis*, is completely inhibited in the dark [16]. *E. crus-galli* and *E. colona* do not require light for germination, but light can stimulate their germination [10,11]. Different weed species also have different responses to salt and osmotic stress. Seed germination was reduced by 50% in *E. colona* at 106 mM NaCl [10] and in *Cyperus difformis* at 23 mM [15]. The osmotic potential required for 50% inhibition of germination in *E. colona* was -0.46 MPa [10]. Seed burial experiments for other related *Echinochloa* species revealed lower germination with increase in seed burial depth. Germination in *E. colona* (97%) and *E. crus-galli* (92%) was highest at the soil surface, and no emergence was seen at 6- and 8-cm depths, respectively [10,11]. The addition of high amounts (4−6 tons (t) ha^{-1}) of rice straw as mulch reduced seedling emergence in *E. crus-galli* and *E. colona* [10,11].

Different weed ecotypes or weed populations may also have differential responses to the factors that affect germination [17]. In a previous study, for example, two populations of *Trianthema portulacastrum* had differential germination responses to light and seed burial depth [17]. In another study, however, germination response to light, osmotic stress, and burial depth was similar between two populations of *Rottboellia cochinchinensis* [18].

Flooding has long been used as an effective means of controlling weeds in rice. Due to shortage in irrigation water, however, flooding as a method of control may not be possible in the future. In addition, a previous study suggests that flooding alone may not be able to effectively control the *Echinochloa* species. One study suggested that, in order to control *E. glabrescens* in lowland rice, shallow flooding (3 cm) must be combined with suitable tillage systems to lessen the need for manual weeding and herbicide application [12]. In another study, emergence in *C. difformis*, *Ludwigia hyssopifolia*, and *E. colona* was suppressed with the application of a preemergence (PRE) herbicide alongside shallow flooding [8]. The use of postemergence (POST) herbicides is also considered effective in controlling weeds in rice, particularly in direct-seeded rice.

The efficacy of herbicides, however, is dependent on the growth stage of the weed [19,20]. The efficacy of bispyribac-sodium, penoxsulam + cyhalofop-butyl, and fenoxaprop + ethoxysulfuron on *E. crus-galli*, *E. colona*, and *Digitaria ciliaris*, for example, was reduced when the herbicides were applied at 8-leaf stage compared to when these were applied at 4-leaf stage [19].

A study on the seed ecology of *E. glabrescens* will provide information on the factors that affect its germination and growth, which can serve as a basis in the development of cost-effective control measures of this and other weeds with a similar germination requirement. This study was therefore conducted to determine the effects of light, temperature, salt, osmotic stress, burial depth, rice residue, time and depth of flooding, and flooding depth combined with different rates and application timing of PRE and POST herbicides on the emergence and growth of two populations of *E. glabrescens*.

Materials and Methods

Experiments were conducted from October 2011 to August 2012 at the International Rice Research Institute in Los Baños, Laguna, Philippines. Seeds of mature *E. glabrescens* were harvested in lowland rice fields of Muñoz City, Nueva Ecija (NE) (latitude 15°43′N, longitude 120°54′E), and Los Baños, Laguna (IR) (latitude 14°10′ N, longitude 121°13′ E), both in the Philippines, in March 2011. The distance between the two sites is approximately 175 km. Seeds were bulked, cleaned, and stored in plastic containers at room temperature until use. The authors confirm that no permission was needed to collect the weed seeds, as well as that the field studies did not involve endangered or protected species.

Germination test procedure

All experiments were conducted using seeds from both NE and IR populations. To avoid fungal contamination, seeds were soaked in sodium hypochlorite (5%) for 5 minutes and rinsed with running tap water for 3−5 minutes prior to the germination test.

Freshly harvested seeds were tested for germination by placing 25 of the *E. glabrescens* seeds in a 9-cm diameter Petri dish lined with two pieces of Whatman no. 1 filter paper (Whatman International Ltd., Maidstone, U.K.) and 5 ml of distilled water. The Petri dishes were sealed with parafilm and placed inside incubators at fluctuating day/night temperatures of 30/20 °C in light/dark conditions. The photoperiod was set at 12 h to coincide with the higher temperature interval. In all germination tests, seeds with a visible protrusion of the radicle were considered to have germinated. The number of germinated seeds was counted at 15 days after sowing (DAS), or until there was no further germination.

Effect of temperature and light on germination

Germination in the two populations was determined by placing 25 seeds of *E. glabrescens* in 9-cm diameter Petri dishes. These were then incubated in growth chambers under fluctuating day/night temperatures (35/25, 30/20, and 25/15 °C) in both light/dark and dark regimes. These temperature regimes were selected to reflect temperature variations in the Philippines and other areas in the tropics. For the dark treatment, the dishes were wrapped in three layers of aluminum foil to prevent any light penetration.

The number of germinated seeds was counted at 3-day (d) intervals for up to 15 d for those grown in the light/dark regime and only after 15 d for those grown under dark conditions. Seeds that failed to germinate in the dark after 15 d were moved to the light/dark regime with 5 ml of distilled water added. The number of germinated seeds was counted after 15 d more.

Effect of salt and osmotic stress on germination

Twenty-five seeds of *E. glabrescens* were placed in dishes containing 5 ml solutions of 0, 25, 50, 100, 150, 200, and 250 mM sodium chloride (NaCl, Mallinckrodt Baker inc., Phillipsburg, NJ) to determine the effect of salt on germination. The solutions were prepared by dissolving 0, 1.46, 2.42, 5.84, 8.77, 11.69, and 14.61 g of NaCl per 1 L of distilled water.

The effect of osmotic stress was studied by placing 25 seeds in solutions with osmotic potentials of 0, −0.1, −0.2, −0.4, −0.6, −0.8, and −1.0 MPa, prepared by dissolving polyethylene glycol 8000 (Sigma Aldrich Co., St. Louis, MO) in distilled water [21].

These ranges of NaCl and osmotic potentials reflect salinity and water-stress levels in the tropics. The Petri dishes were placed inside growth chambers under fluctuating day/night temperatures of 30/20 °C to test germination. Seeds that did not germinate at 250 mM (highest salt concentration) and −1.0 MPa (highest water-stress condition) were rinsed with running water for 5 minutes and placed in the incubator again after adding 5 ml of distilled water. The number of germinated seeds was counted after 15 d.

Effect of burial depth on seedling emergence

To determine the effect of burial depth, a pot experiment was conducted inside a screenhouse (a chamber of size 10 m×20 m, framed with 2-mm iron mesh and covered overhead with a transparent plastic cover to prevent rain damage). The pots (15.0 cm diameter×15.0 cm height) used in the experiment had holes at the bottom; hence, a piece of paper was placed at the bottom of each pot to prevent soil from leaking out.

The soil used in all pot experiments was collected from upland rice fields; had a pH of 6.6; and was composed of 31% sand, 37% silt, and 32% clay. Soil was autoclaved and passed through a 3-mm sieve. Fifty seeds of *E. glabrescens* from each of the two populations (NE and IR) were sown on the soil surface and then covered with the same soil to depths of 0, 0.5, 1, 2, 4, and 8 cm. Pots were irrigated with an overhead sprinkler initially and, later, subirrigated. In all pot experiments, the visibility of the coleoptile on the soil surface indicated emergence.

The number of seeds that emerged was recorded at 3-d intervals until no further emergence was observed, at 24 DAS.

Effect of rice residue amount on seedling emergence and biomass

Fifty seeds of *E. glabrescens* from each of the the two populations were sown on the soil surface in plastic pots containing finely chopped rice straw (leaves and stems) of the variety NSIC Rc222. Residue was spread on the soil surface at rates equivalent to 0, 1, 2, 4, and 6 t ha^{-1}. The pots and soil used in this experiment were the same as those described in the burial depth experiment.

The number of germinated seeds was recorded at 3-d intervals until no further emergence was observed. At 24 DAS, shoots (leaf and stem) were oven-dried at 70 °C for 72 h and biomass was recorded.

Effect of depth and time of flooding on seedling emergence and biomass

Fifty seeds from the two populations were sown on the soil surface in small plastic trays (8.0 cm×8.0 cm×5.5 cm). Soil that was used in this experiment was the same as that described in the previous experiment. The pots containing the seeds were placed inside larger plastic pots (15 cm diameter×12 cm height) to retain and maintain water. The pots were marked with 0, 2, 4, and 6 cm to indicate flooding depth. Pots were kept at saturated conditions until flooding was introduced, i.e., at 0, 2, 4, and 8 DAS. Pots were kept flooded at the aforementioned depths for 14 d.

The number of seedlings that emerged was counted at 14 d after start of flooding. The shoot (leaf and stem) and root were separated and oven-dried to constant weights at 70 °C for 72 h to measure biomass.

Effect of flooding depth and pretilachlor rates on seedling emergence and biomass

Fifty seeds from each of the two populations were sown at 0.5-cm depth in small plastic trays (8.0 cm×8.0 cm×5.5 cm) containing the same soil described in the previous experiments. The pots with the seeds were placed inside larger plastic pots (15 cm diameter×12.0 cm height) to retain and maintain water. The pots were placed on benches inside the screenhouse. These pots were sprayed with pretilachlor (Sofit 300 EC, Syngenta, Fort Bonifacio, Taguig City, Philippines) at 75, 150, 255, and 300 g ai ha^{-1} at 2 DAS. Spraying was done with a research track sprayer (DeVries Manufacturing, Hollandale, MN) with 210 L ha^{-1} spray volume delivery and spray pressure of 140 kPa, fitted with flat fan nozzles (Teejet 80015E, Spraying Systems Co., North Ave. at Schmale Rd, Wheaton, IL 60189). Unsprayed plants were included as the control (nontreated) treatment. Pots were kept at saturated conditions until 5 DAS. Afterwards, flooding was introduced at 0-, 2-, and 4-cm depth and kept at these depths until 14 d.

The number of seedlings that emerged was counted at 14 d after start of flooding. Shoots (leaf and stem) and roots were separated and oven-dried at 70 °C for 72 h and biomass was recorded.

Effect of POST herbicides on seedling survival and growth

Twenty-five seeds from each of the two populations were sown in small plastic trays (8.0 cm×8.0 cm×5.5 cm). The pots were filled with the same soil as described above. Seedlings were thinned to 6 plants per tray at 3 DAS. The POST herbicides were: bispyribac-sodium (Bayer Crop Science, Canlubang City, Laguna, Philippines) at 30 g ai ha^{-1}, fenoxaprop + ethoxysulfuron (Bayer Crop Science, Canlubang City, Laguna, Philippines) at 45 g ai ha^{-1}, and penoxsulam + cyhalofop-butyl (Dow Agro Sciences, Ayala Alabang, Muntinlupa City, Philippines) at 72 g ai ha^{-1}. Herbicides were sprayed at 4-, 6-, and 8-leaf stages using the research track sprayer described above. A nontreated control was included for comparison.

The number of seedlings that survived (at least one green leaf on the plant) the treatments was counted at 14 d after herbicide application. Shoots (leaf and stem) and roots were separated and oven-dried at 70 °C for 72 h and biomass was recorded.

Statistical analyses

Experiments were arranged in a randomized complete block design with four replications. Each replication was considered as a block and arranged on different shelves or benches in the incubators or screenhouse. All experiments were conducted twice. The data were combined for analysis, as there were no interaction effects of treatment and experiments. Data were analyzed using regression analysis to determine relationships among different temperature regimes, salt and osmotic concentrations, burial depths, and residue amounts. These were best fitted to a functional three-parameter sigmoid model using SigmaPlot 10.0. The model fitted to the fluctuating temperatures was:

$$G = G\max/\{1 + e[-(T - T50)/Grate]\}$$

where G is the total germination (%) at time T, G_{max} is the maximum germination (%), T_{50} is the time required for 50% of maximum germination, and G_{rate} indicates the slope. The model fitted to varying salt and osmotic concentrations was:

$$G = G\max/\{1 + e[-(x - x50)/Grate]\}$$

where G is the total germination (%) at salt concentration or osmotic potential x, G_{max} is the maximum germination (%), x_{50} is the salt concentration or osmotic potential required for 50% inhibition of maximum germination, and G_{rate} indicates the slope. The model fitted to varying burial depth and amount of rice residue was:

$$E = E\max/\{1 + e[-(T - T50)/Erate]\}$$

where E is the total emergence (%) at time T, E_{max} is the maximum emergence (%), T_{50} is the time required for 50% of maximum emergence, and G_{rate} indicates the slope. An exponential decay curve of the form:

$$E = E\max.e(-Erate.x)$$

was fitted to seedling emergence (%) obtained at different seed burial depths, where E represents cumulative emergence (%) at burial depth x, E_{max} is the maximum emergence, and E_{rate} indicates the slope. A linear model of the form

$$E = ax + b$$

was fitted to seedling emergence (%) obtained at different residue amounts, where E represents emergence (%) at residue amount x, and a and b are constants. Parameter estimates were compared using the standard error. Treatment means were compared using least significant difference (LSD) at P = 0.05 (GenStat 13th Edition VSN International Ltd., U.K.), when regression analysis was not possible. Analysis was done separately for the two populations.

Results and Discussion

Effect of temperature and light

Germination of *E. glabrescens* seeds from the IR population was influenced by the interaction between light and temperature ($P<0.001$) while seeds of the NE population were stimulated by light but not by the alternating day/night temperatures. In the continuous dark conditions, seed germination in the IR population was lower at 35/25 °C (41.5±1.7%) than at 25/15 °C (72.5±5.4%) and 30/20 °C (67.5±2.7%), whereas in the alternating light/dark conditions, germination was similar at all temperatures (Figure 1).

In a previous study, germination in *Eclipta prostrata* at 30/20 °C under light/dark conditions was significantly higher than at 25/15 and 35/25 °C temperature regimes [22]. Germination in *Eleusine indica* was significantly higher in light/dark conditions (68–72%) than in dark conditions (17–25%) at 30/20 and 35/25 °C [23].

Germination of the NE population was more sensitive to dark, as it resulted in an 86–92% reduction in germination. In the IR population, germination decreased by only 25–52%. The differential response of these two populations might be due to the differential conditions—such as water and fertilizer management, nutrient status of the soil, and weather conditions (temperature and sunlight)—during the growth and reproduction of their mother plants.

In our study, we were not able to record these observations at the seed collection areas. A previous study showed contrasting germination responses to heat, salinity, and drought of two populations of *Chenopodium album*, and that the differential germination responses were due to either adaptation via natural selection or maternal effects, or combinations of both [24]. In contrast, a similar germination response was observed between two populations of *R. cochinchinensis* [18].

In the light/dark conditions, seed germination in both populations was similar at all temperatures (Figure 1). The ability of *E. glabrescens* to germinate at all temperatures suggests that this species could emerge throughout the year at low altitudes in tropical countries. Similar results were reported for *E. prostrata*, in which seeds germinated at all test temperatures (i.e., 25/15, 30/20, and 35/25 °C) [22]. Authors suggested a similar speculation regarding germination in *E. prostrata* at all temperatures.

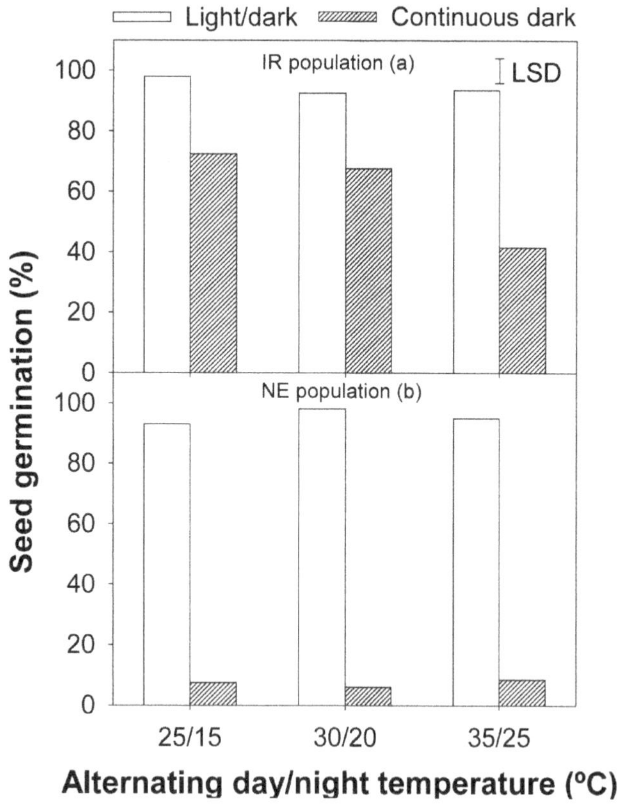

Figure 1. Cumulative germination in two populations (NE and IR) of *Echinochloa glabrescens* incubated at alternating day/night temperatures (25/15, 30/20, and 35/25 °C) and light (light/dark and dark).

Seeds of both populations, when exposed to light/dark conditions at the lowest temperature regime, took longer to reach 50% germination (T_{50}) (Figure 2). In the NE population, for example, the time it took for 50% germination (of the maximum germination) at 25/15, 30/20, and 35/25 °C was 5.4±2.8, 3.8±0.01, and 2.7±0.15 d. However, as mentioned earlier, cumulative germination (%) at 15 DAS was similar among test temperatures. Although light was not an absolute requirement for germination, it did stimulate seed germination of the NE and IR populations by 86–92 and 25–52%, respectively.

Weed species vary in germination response to light. Seeds of some species can germinate only with light and others may germinate equally in light or dark conditions. In previous studies, germination in *E. colona* and *E. crus-galli* was stimulated by 39 and 19%, respectively, when exposed to a light/dark regime [10,11].

When nongerminated seeds from the dark treatment were transferred to light/dark conditions at 30/20 °C, 72–100% of the seeds germinated (data not shown), indicating that the seeds were not adversely affected by incubation in the dark. Similarly, seed germination in *E. prostrata* reached 92–96% when transferred from dark to light/dark conditions [22].

The ecological significance attributed to light response in this species is that amount of light acts as a soil depth indicator, allowing greater germination of seeds on the surface than those buried deeper in soil [25]. The light stimulation germination

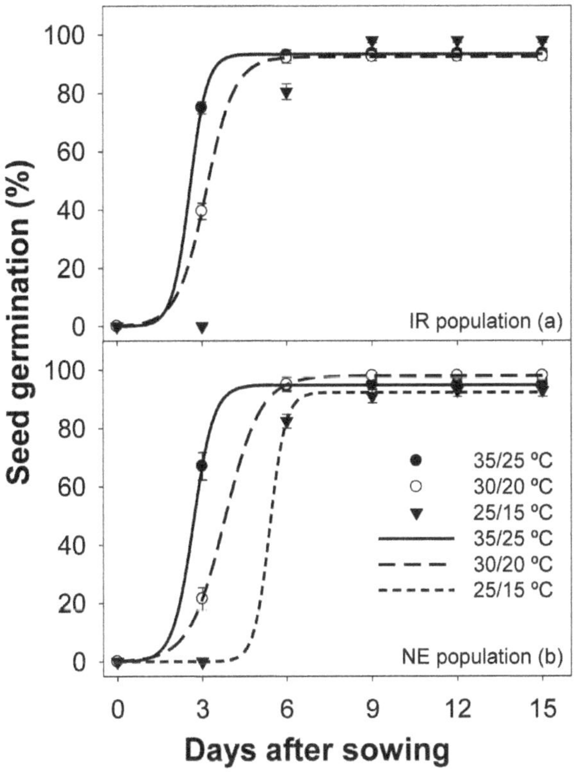

IR Population
25/15 °C; did not fit
30/20 °C; $G\% = 92.5/[1 + e^{-(x-3.1)/0.5}]$; $R^2 = 0.99$
35/25 °C; $G\% = 93.4/[1 + e^{-(x-2.6)/0.3}]$; $R^2 = 0.99$
NE Population
25/15 °C; $G\% = 92.3/[1 + e^{-(x-5.4)/0.3}]$; $R^2 = 0.99$
30/20 °C; $G\% = 98.0/[1 + e^{-(x-3.8)/0.6}]$; $R^2 = 0.99$
35/25 °C; $G\% = 94.9/[1 + e^{-(x-2.7)/0.4}]$; $R^2 = 0.99$

Figure 2. Seed germination in two populations (IR and NE) of *Echinochloa glabrescens* incubated at alternating day/night temperatures (25/15, 30/20, and 35/25 °C) and light (light/dark and dark). The lines represent a three-parameter sigmoid model fitted to the data.

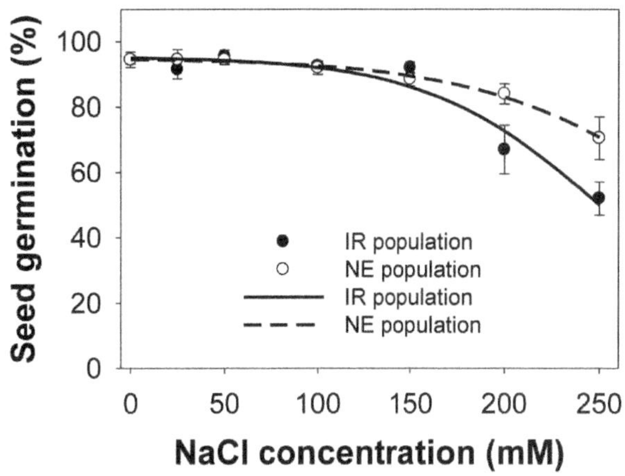

IR Population
$G\% = 95.5/[1 + e^{-(x - 254.43)/-47.0}]$; $R^2 = 0.98$
NE Population
$G\% = 95.0/[1 + e^{-(x - 313.0)/-58.3}]$; $R^2 = 0.99$

Figure 3. Seed germination in two populations (IR and NE) of *Echinochloa glabrescens*, in response to sodium chloride (NaCl) concentrations when incubated in a growth chamber at 30/20 °C day/night temperature over a 12-h photoperiod for 15 d. The lines represent a three-parameter sigmoid model fitted to the data.

response of *E. glabrescens* suggests that the species could be a problematic weed in no-till systems, wherein much of the weed seed bank remains on the surface, exposed to light [26]. On the other hand, the exposure of previously buried *E. glabrescens* seeds to light may trigger germination, and that some seeds may still germinate even when buried.

Effect of salt and osmotic stress

Seed germination in *E. glabrescens* from both populations decreased with increasing salt concentration (Figure 3). Maximum germination (95%) in both populations was observed at 0 mM. Germination in both populations was more than 60% up to a concentration of 200 mM NaCl. Seed germination in the IR and NE populations at 250 mM was still 52% and 71%, respectively. The salt concentration required to inhibit 50% germination in the IR and NE populations was 254 and 313 mM NaCl, respectively. In an earlier study, seed germination in *E. colona* was more than 60% up to a concentration of 50 mM NaCl, and that 106 mM of NaCl was needed to reduce seed germination by 50% [10]. Germination in *E. colona* was completely inhibited at 200 mM NaCl. *L. chinensis*, on the other hand, showed sensitivity to saline stress [16].

When nongerminated seeds after 15 d at the highest tested salinity concentration (250 mM NaCl) were rinsed with distilled water and incubated again at 30/20 °C, germination increased 60–100%, indicating that the seeds were not adversely affected by the saline condition (data not shown). Such a result suggests that seeds in saline conditions may wait until favorable conditions prevail before germination occurs. The longevity of *E. glabrescens* seeds in saline conditions is still unclear, however, and further investigation is needed. Results suggest that *E. glabrescens* will still germinate at high soil salinity, which indicates that the species may be able to grow well in saline areas. Thus, crop cultivation in such areas will be affected not only by salinity but also by competition with this weed species.

Seed germination decreased with reduced osmotic potential in both populations of *E. glabrescens* (Figure 4). Maximum seed germination (98%) in the IR and NE populations was observed at 0 MPa. At osmotic potential of −0.2 MPa, seed germination in the IR and NE populations was 77 and 58%, respectively. An osmotic potential of −0.4 MPa reduced seed germination in the IR and NE populations by 60 and 66%, respectively, compared with the control (i.e., 0 MPa). Some seeds from both populations germinated (4%) at −0.8 MPa, but germination was totally inhibited at −1.0 MPa. The osmotic potential required to inhibit 50% germination in the IR and NE populations was −0.33 and −0.24 mM, respectively.

A similar germination response was observed in *E. colona*, in which germination decreased from 80 to 1% as osmotic potential decreased from 0 to −0.8 MPa and was completely inhibited at osmotic potential of −1.0 MPa [10]. In *E. crus-galli*, increase in osmotic potential allowed increase in germination when seeds were exposed to light [27].

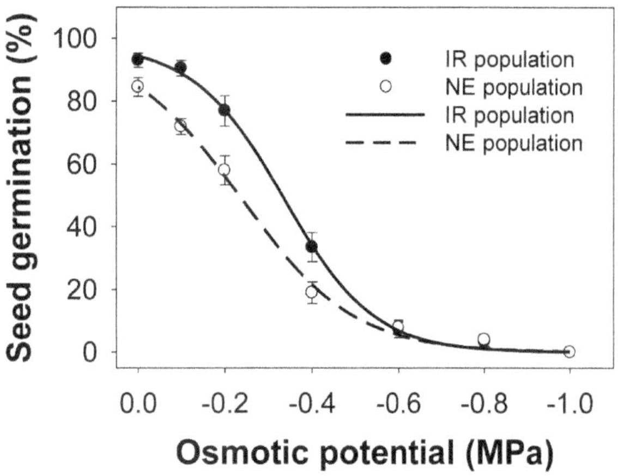

IR Population
$G\% = 97.9/[1 + e^{-(x - -0.34)/0.10}]$; $R^2 = 0.99$
NE Population
$G\% = 97.6/[1 + e^{-(x - -0.24)/-0.13}]$; $R^2 = 0.99$

Figure 4. Seed germination in two populations (IR and NE) of *Echinochloa glabrescens*, in response to osmotic potentials (MPa) when incubated in a growth chamber at 30/20 °C day/night temperature over a 12-h photoperiod for 15 d. The lines represent a three-parameter sigmoid model fitted to the data.

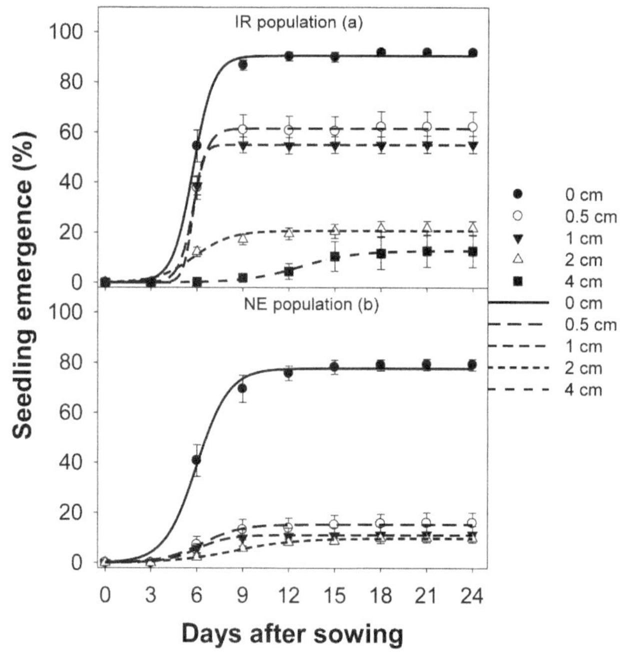

IR population
0 cm; $E\% = 90.5/[1 + e^{-(x-5.7)/0.7}]$; $R^2 = 0.99$
0.5 cm; $E\% = 61.5/[1 + e^{-(x-5.8)/0.5}]$; $R^2 = 0.99$
1 cm; $E\% = 55.0/[1 + e^{-(x-5.7)/0.3}]$; $R^2 = 0.99$
2 cm; $E\% = 20.7/[1 + e^{-(x-5.8)/1.9}]$; $R^2 = 0.99$
4 cm; $E\% = 12.7/[1 + e^{-(x-12.7)/1.8}]$; $R^2 = 0.99$
NE population
0 cm; $E\% = 77.7/[1 + e^{-(x-6.0)/1.0}]$; $R^2 = 0.99$
0.5 cm; $E\% = 15.4/[1 + e^{-(x-6.3)/1.3}]$; $R^2 = 0.99$
1 cm; $E\% = 11.1/[1 + e^{-(x-6.0)/1.0}]$; $R^2 = 0.99$
2 cm; $E\% = 9.8/[1 + e^{-(x-8.5)/2.0}]$; $R^2 = 0.99$
4 cm; did not fit

Figure 5. Seedling emergence in two populations (IR and NE) of *Echinochloa glabrescens*, in response to burial depth (cm) when grown in screenhouse conditions for 24 d. The lines represent a three-parameter sigmoid model fitted to the data.

When nongerminated seeds from −1.0 MPa were rinsed and placed in distilled water, seed germinated by 60–80% for the IR population and 40–72% for the NE population. These results indicate that the seeds of *E. glabrescens* were not adversely affected by exposure to low osmotic potentials for up to 15 d. These results also suggest that most of the seeds will germinate in moist conditions, while seeds in dry conditions may wait until moisture conditions are favorable before germination occurs. The ability of *E. glabrescens* seeds to germinate at low osmotic potentials would allow this weed to have a competitive advantage over crops that are susceptible to drought stress. This ability also suggests that germination will be favored immediately after a rainfall, which may coincide with the onset of the wet cropping season in rainfed areas.

Effect of burial depth

Seedling emergence in both populations of *E. glabrescens* was reduced by seed burial depth ($P<0.001$). Estimates from the three-parameter sigmoid model indicated that 50% seedling emergence (T_{50}) was achieved at 5.7 d in the IR population and 6.0 d in the NE population (Figure 5). Fifty percent of the seeds that were buried from 0.5 to 4 cm for the IR population and from 0.5 to 2 cm for the NE population emerged at the same time as seeds that were sown on the soil surface.

On the other hand, the time needed for 50% emergence was delayed by 13 d at 4-cm depth in the IR population, and by 9 d at 2-cm depth in the NE population, compared with seeds sown on the surface.

In both populations, the cumulative seedling emergence at 24 DAS declined with increasing burial depth (Figure 6). The maximum emergence (91 and 79% for the IR and NE populations, respectively) was observed at the soil surface. Seedling emergence in the NE population declined more rapidly with increasing burial depth, compared with that in the IR population. At 0.5-cm burial depth, seedling emergence decreased by 63% in the NE population and 30% in the IR population. No emergence was observed at 4-cm depth in the NE population and at 8-cm depth in the IR population.

The slopes (E_{rate}) were not similar (0.6 and 2.8) between the IR and NE populations, suggesting that the NE population was more affected by increasing burial depth than was the IR population.

Greater germination on the soil surface in *E. glabrescens* is consistent with stimulation of germination by light. Reduced emergence with increase in burial depth in *E. glabrescens* could be due to the absence of light to signal germination or a limitation on soil gas diffusion, or both [28]. Seeds buried more than 2 mm below the soil surface receive less than 1% of incident light [29]. Another possible reason for reduced emergence with increasing depth could be physical limitations of the seedling, i.e., insufficient seed reserves to enable it to reach the soil surface [18]. The seed mass of *E. glabrescens* was 3.3 mg seed^{-1}. Larger seeds with more carbohydrate reserves are more able to emerge from greater burial depths compared with seeds that have lower reserves [30].

It has been reported that the most number of weed species germinated when seeds were placed on the soil surface, although burial depth response varied with the species [15]. Emergence in *Cyperus iria*, *C. difformis*, *Fimbristylis miliacea*, and *L. chinensis* at depths greater than 0.5 cm was completely inhibited [15]. Emergence in

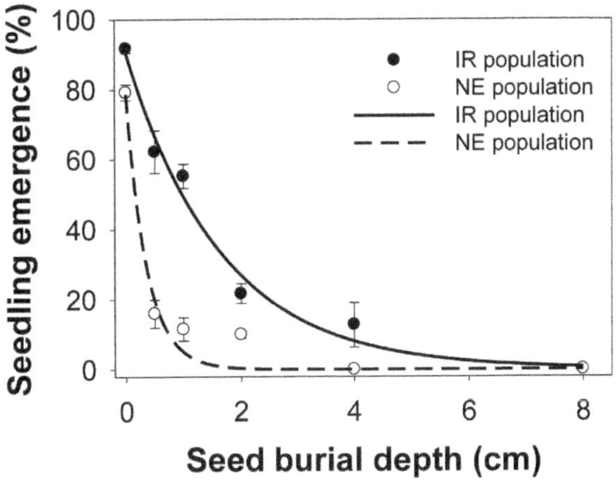

IR population
$E\ \% = 90.4\ x \cdot e^{\ 0.6x};\ R^2 = 0.99$
NE population
$E\ \% = 78.7\ x \cdot e^{\ 2.7x};\ R^2 = 0.98$

Figure 6. Cumulative seedling emergence in two populations (IR and NE) of *Echinochloa glabrescens*, in response to burial depth (cm) when grown in screenhouse conditions after 24 d. The lines represent an exponential curve fitted to the data.

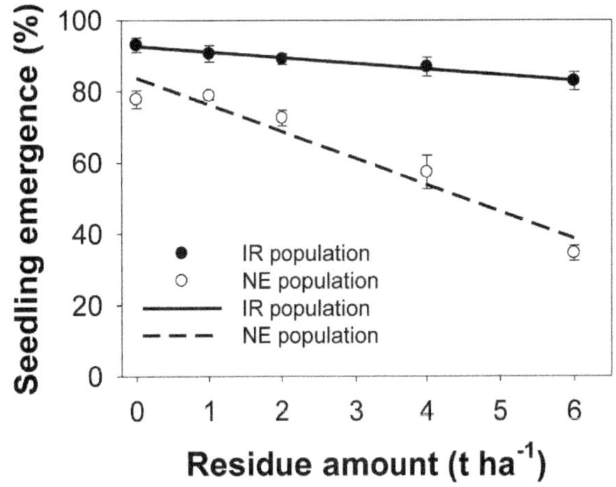

IR population
$E\ \% = 92.57 - 1.60x;\ R^2 = 0.99$
NE population
$E\ \% = 83.67 - 7.51x,\ R^2 = 0.96$

Figure 7. Seedling emergence in two populations (IR and NE) of *Echinochloa glabrescens*, in response to residue amount (t ha^{-1}) when grown in screenhouse conditions after 24 d. The lines represent a linear model fitted to the data.

Digitaria longiflora, *Heliotropium indicum*, and *Portulaca oleracea* was completely inhibited at 2-cm burial depth [15]. Emergence in *E. colona* decreased by 70–88% at 0.5-cm burial depth compared with that among seeds sown on the soil surface [10], while 11–21% of *E. indica* seedlings emerged at 6-cm burial depth [23].

Stimulation by light and greater emergence on the soil surface in *E. glabrescens* suggests that no-till systems may favor its emergence in the field. This weed species and others with a similar germination response could pose the same problem, particularly in continuous no-till systems.

On the other hand, seeds that remain on the soil surface may be prone to faster desiccation and insect predation [31]. Since emergence is greatly enhanced in these systems, control of *E. glabrescens* before planting is necessary so as to avoid having the crop compete with it. One method is the stale seedbed technique, in which a nonselective herbicide is applied when most of the weeds have emerged before planting. In some areas in the Philippines, farmers practice the rice-maize cropping system, in which glyphosate-resistant maize is grown under a no-till system, thus favoring the use of the broad-spectrum herbicide glyphosate to kill most of the weeds. In order to reduce the build-up of seed banks of this weed species, these tillage systems must undergo rotation. Deep tillage using a moldboard plow that would bury the seeds of *E. glabrescens* below 8 cm in the soil may discourage the weed's emergence. The longevity of buried *E. glabrescens* seeds is still unknown and needs further investigation.

Effect of residue amount

Seedling emergence in the NE ($P<0.001$) and IR ($P<0.03$) populations decreased with increasing rates of residues (Figure 7). Maximum seedling emergence in *E. glabrescens* was 93 and 84% in the IR and NE populations, respectively, when no residue was applied. The NE population was more affected than the IR population by increasing amount of residue. Emergence in the NE population increased significantly with the addition of 4 t ha^{-1} residue, although there was still no reduction in emergence in the IR population at this level. The addition of 5 t ha^{-1} of rice residue caused a 55% and 9% reduction in maximum emergence in the NE and IR populations, respectively, compared with when no residue was added.

The difference in response between the two populations is consistent with observations from the seed burial experiment, in which emergence in the NE population was affected more than that in the IR population by increase in burial depth.

Weed suppression by mulch is attributed to various physical and chemical factors. The physical factors include lower soil temperatures, shading, and physical obstruction provided by the mulch itself [32,33]. No study has yet recorded the rice variety used in our experiments (NSICRc222) as having allelopathic effects. Thus, it is not clear whether the effect of residue on emergence was a result of biochemical or physical factors.

Emergence in many weed species has been reported to decrease with the addition of crop residue. *E. colona* and *E. crus-galli*, for example, were reported to have a marked reduction in seedling emergence with increased amount of residue [10,11].

The addition of residue at 1–2 and 1–4 t ha^{-1} increased the biomass of *E. glabrescens* seedlings in the NE and IR populations, respectively (Table 1). The addition of rice residue of 4–6 t ha^{-1} in the NE population and 6 t ha^{-1} in the IR population resulted in similar biomass compared with those without rice residue. The increase in biomass with the addition of residue could be a result of higher availability of soil moisture due to better soil-seed contact. Seeds placed directly on the soil surface could have less soil-seed contact, which resulted in slow growth. A similar response was observed in *E. crus-galli*, in which biomass increased with the addition of residue of up to 4 t ha^{-1} [11]. On the other hand, the biomass of *E. prostrata* and *E. indica* decreased as amount of residue increased [22,23].

Our results suggest that high amounts of residue were needed to reduce emergence and growth of *E. glabrescens*. These high amounts of residue may not be feasible in the Philippines, as the common practice of majority (95%) of farmers is to burn the rice straw after harvest [34]. In other areas in the Philippines—those

Table 1. Shoot biomass (g pot^{-1}) of two populations (IR and NE) of *E. glabrescens* at different residue amounts (t ha^{-1}).

Amount of residue (t ha^{-1})	Shoot biomass (g pot^{-1})	
	NE population	*IR population*
0	5.35	4.97
1	7.04	8.32
2	7.13	7.57
4	6.68	8.38
6	3.93	6.59
$LSD_{0.05}$	1.50	2.24

where rice hull burning is banned because of air pollution concerns, and in areas with low soil organic matter but which practice organic farming—the addition of high amounts of rice residue mulch may be considered.

Effect of depth and time of flooding

Seedling emergence and shoot and root biomass in the IR population was influenced ($P<0.001$) by the interaction between flooding depth and timing. On the other hand, seedling emergence and shoot and root growth in the NE population was influenced ($P<0.001$) by flooding depth and timing. Regardless of flooding depth, flooding did not kill any seedlings that had emerged (Figures 8a, b). Flooding had a more pronounced effect on shoot and root growth than on seedling emergence in *E. glabrescens* (Figures 8c–f). Timing of flooding, more than depth, affected emergence and growth. On the other hand, depth of flooding became more critical when flooding was delayed.

Although flooding is an important tool for controlling weeds, the response of weeds to flooding varies among species. *L. hyssopifolia* can be easily controlled with flooding [35], but some weeds, such as *Monochoria vaginalis* and *Sphenoclea zeylanica*, are very difficult to control because they are well-adapted to submerged conditions [36,37]. Similarly, emergence in *E. crus-galli* seedlings can be reduced by flooding at 2 DAS but not at 15 DAS [11].

Reduction in emergence and biomass of weeds in flooded conditions may be a result of several factors that include reduced O_2 levels; accumulation of CO_2 and toxic gaseous products of anaerobic decomposition; and the presence of reduced forms of chemical radicals and gases such as methane, nitrogen, nitrous oxides, and sulfides [38]. In addition, the depth-sensing fluctuation in weed seeds might be related to the amplitude of temperature fluctuations. Deeper water could have smaller temperature fluctuations, which might allow less germination [37].

Results showed that, in both populations, flooding at an earlier stage (0–2 DAS) reduced the emergence and shoot and root growth of *E. glabrescens*. Delaying flooding to 4 DAS reduced shoot and root growth but not emergence. At late flooding, reduction occurred only with increased water depth (4–6 cm). Flooding at 8 DAS did not reduce emergence nor growth, regardless of increase in water depth. A similar response was observed in *C. difformis*, *C. iria*, *F. miliacea*, and *L. chinensis*, where flooding depth to as low as 2 cm, compared with 0 cm, greatly reduced emergence when the soil had been flooded continuously for 28 d [15].

These results suggest that, to achieve about 50% reduction in shoot growth, flooding at 2 cm should be introduced not later than 2 DAS. If flooding is delayed up to 4 DAS, then the depth of flooding must be increased. Flooding at 8 DAS did not control emergence and growth in *E. glabrescens*. Seeding depth and flooding reduces germination, survival, and growth in *E. glabrescens* [12]. Thus, in order to control *E. glabrescens* in lowland rice, suitable tillage systems must be combined with shallow flooding (3 cm) to reduce the need for other weed control methods such as hand-weeding and herbicides [12].

In the future, many areas will experience water scarcity and that continuous flooding for weed control will be restricted [39].

The results of our study showed that shallow flooding (2 cm) at 0–2 DAS may significantly reduce emergence and growth in *E. glabrescens* and other similar species. Thus, in areas with limited irrigation water, early flooding at shallow depths rather than late flooding will be a better water management strategy for the control of *E. glabrescens*.

In direct-seeded rice, early flooding could be possible with the use of anaerobic rice cultivars that can emerge despite submerged conditions, and using pre-germinated and/or high-vigor seeds or varieties that can emerge in a short time. By the time the canopy closes, weed growth can be suppressed by shading. However, flooding alone will not completely suppress *E. glabrescens* and still needs to be combined with other weed management strategies. Further investigation of the effect of flooding of different depths and timing under field conditions should be done in the future.

Effect of flooding depth and pretilachlor rates

Seedling emergence and shoot biomass of both populations and the root biomass of the IR population was influenced ($P<0.001$) by the interaction between the pretilachlor dose and flooding depth. However, root growth in the NE population was not influenced by the interaction of flooding depth and the pretilachlor dose. It decreased with increase in flooding depth ($P<0.003$) and increase in the pretilachlor dose ($P<0.001$).

Flooding alone did not suppress the emergence and growth of *E. glabrescens* (Figure 9). Pretilachlor applied alone also did not completely suppress the emergence of *E. glabrescens* seeds, although it did reduce shoot and root biomass. The IR population showed more sensitivity to increasing pretilachlor dose than the NE population. The emergence and growth of both populations were reduced more by pretilachlor when application was followed by flooding (Figure 9). In flooded conditions, the emergence and growth of both populations were similar at all pretilachlor doses. In saturated conditions, on the other hand, as pretilachlor dose increased, emergence in the IR population and growth in both populations declined.

Results suggest that the application of pretilachlor at 0.075 kg ai ha^{-1} (1/4 of the recommended dose) followed by flooding at shallow depth (2 cm) suppressed the emergence and growth of *E. glabrescens*. This supports earlier observations that pretilachlor combined with flooding controlled *E. glabrescens* in direct-seeded rice [40]. Similar results were also observed in *E. colona*, in which pretilachlor application with subsequent flooding greatly reduced its emergence and growth [8].

Application of pretilachlor alone reduced emergence and growth, but the dose required for such reduction was high (i.e., 0.300 kg ai ha^{-1}). Consequently, increasing flooding depth also reduced emergence and growth of *E. glabrescens* but the suppression was not as effective as that with the application of pretilachlor. In *C. difformis*, growth was not affected by flooding alone, but was completely controlled by pretilachlor [8].

In the Philippines, flooding depths in farmers' fields may vary because of poorly leveled fields and limited farm resources [8]. Many rice farmers, now and in the near future, will experience limited water availability that will then restrict their capacity to use continuous flooding as a weed control strategy [39,41]. Conse-

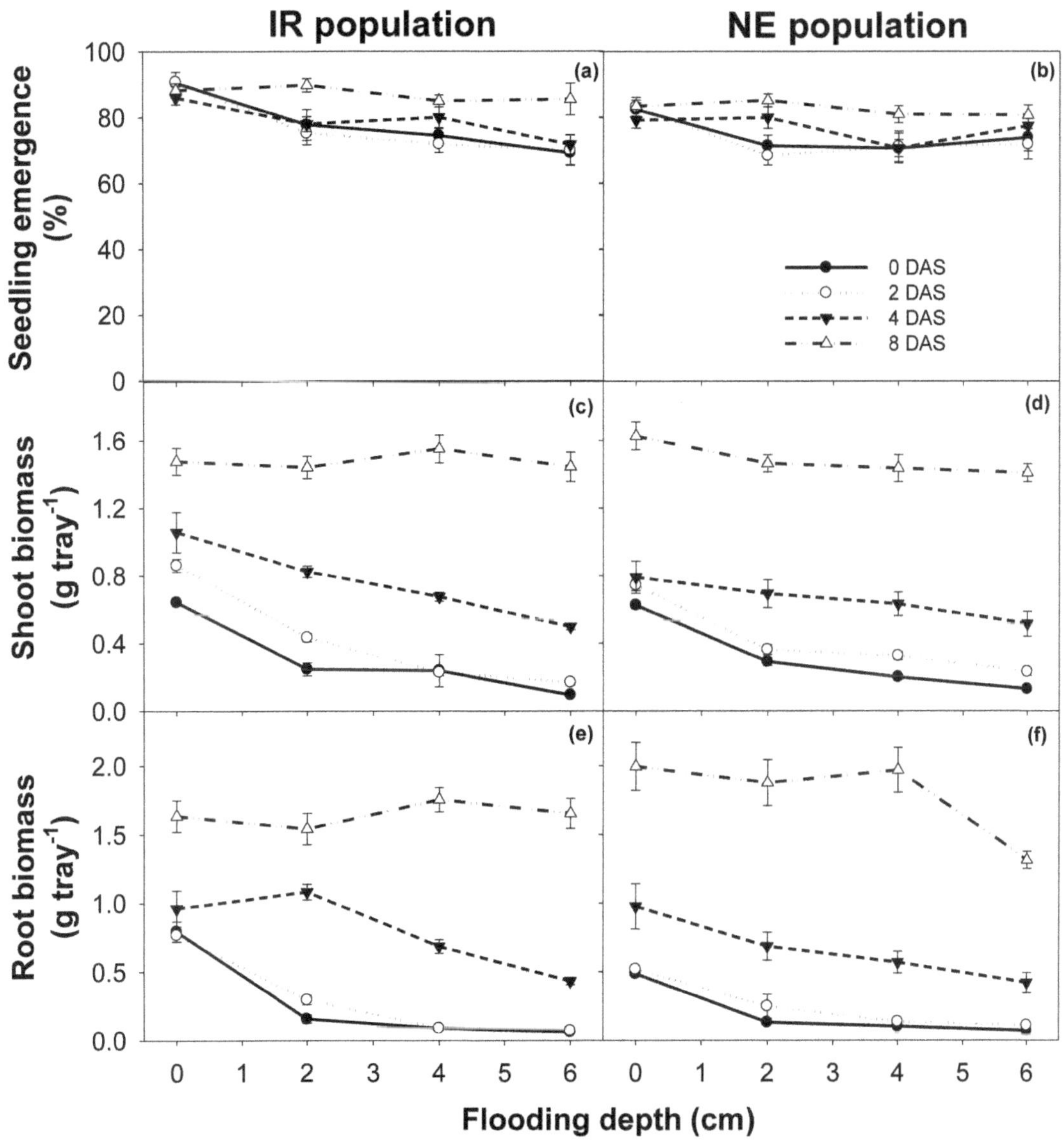

Figure 8. Seedling emergence, shoot biomass (g tray^{-1}), and root biomass (g tray^{-1}) of two populations (IR and NE) of *Echinochloa glabrescens*, in response to various depth (cm) and time of flooding (0, 2, 4, and 8 days after sowing) when grown in screenhouse conditions. Error bars represent standard error of means.

quently, a shift to direct-seeded rice is anticipated to increase in the future as the availability of water and labor decreases and input costs increase. The results of our study suggest that the use of a lower dose of pretilachlor combined with shallow flooding may provide an effective weed management strategy in direct-seeded rice systems for the control of *E. glabrescens* and weeds with a similar response.

Effect of POST herbicides

Application of POST herbicides at the 4-leaf stage of *E. glabrescens* reduced seedling survival in both populations. Bispyribac-sodium, fenoxaprop + ethoxysulfuron, and penoxsulam + cyhalofop provided 98–100% reduction in seedling survival in the IR population at the 4-leaf stage (Figure 10a). At the same leaf stage, penoxsulan + cyhalofop provided only a 35% reduction in seedling survival in the NE population, while fenoxaprop + ethoxysulfuron provided an 85–94% reduction (Figure 10b). Efficacy of the POST herbicides on both populations was reduced when applied at the 6-leaf stage (Figure 10c and d), at which only fenoxaprop + ethoxysulfuron reduced seedling survival in both populations. Further delay in herbicide application to the 8-leaf stage no longer reduced seedling survival in the NE population (Figure 10f). At 8-leaf stage, however, fenoxaprop + ethoxysulfuron still caused reduction in seedling survival in the IR population, by 19% (Figure 10e).

The survival of plants treated with POST herbicides increased with age; reduction in surviving seedlings decreased as the application was delayed from the 4-leaf to the 8-leaf stage. Likewise, shoot biomass increased when POST herbicides were applied at later stages. Shoot biomass in both populations was greatly reduced by the application of all POST herbicides at the 4-leaf stage (Table 2), at which bispyribac-sodium, fenoxaprop + ethoxysulfuron, and penoxsulam + cyhalofop were equally effective in reducing shoot biomass in the IR population.

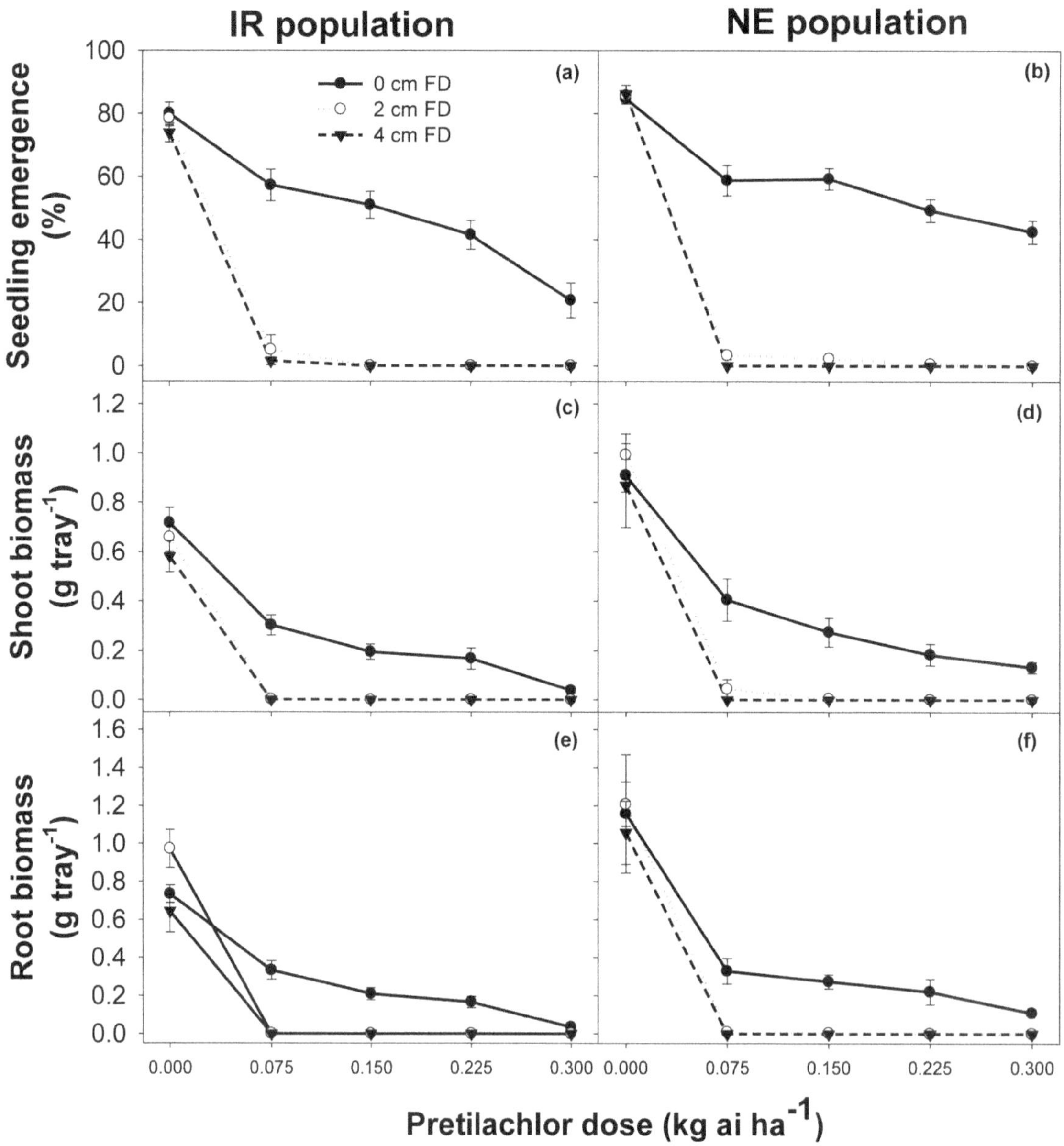

Figure 9. Seedling emergence, shoot biomass (g tray^{-1}), and root biomass (g tray^{-1}) of two populations (IR and NE) of *Echinochloa glabrescens*, in response to various depth of flooding (cm) and pretilachlor application rates (0, 0.075, 0.150, 0.225, and 0.300 kg ai ha^{-1}) when grown in screenhouse conditions. Error bars represent standard error of means.

However, the commercial mixture of penoxsulam + cyhalofop was less effective compared with bispyribac-sodium and fenoxaprop + ethoxysulfuron in reducing shoot biomass in the NE population at this stage. All herbicides reduced shoot biomass in the IR population by 96–100%. Bispyribac-sodium and fenoxaprop + ethoxysulfuron reduced shoot biomass in the NE population by 67–87%, while penoxsulam + cyhalofop provided only a 41% reduction in shoot biomass.

Efficacy of the POST herbicides was reduced in both populations when applied at the 6- and 8-leaf stages (Table 2). At the 6-leaf stage, fenoxaprop + ethoxysulfuron reduced shoot biomass in the IR population by only 48%. Application of all herbicides at the 6-leaf stage did not reduce shoot biomass in the NE population. Application of these POST herbicides at the 8-leaf stage also did not reduce shoot biomass in both populations (Table 2).

Root biomass of both populations was greatly reduced by all herbicides when applied at the 4-leaf stage (Table 3). Reduction in root biomass when herbicides were applied at this stage was 99–100% and 85–95% in the IR and NE populations, respectively. Application of these herbicides at the 6-leaf stage reduced root biomass in both populations but to a lesser degree compared with when application was made at the 4-leaf stage. All herbicides were equally effective in reducing root biomass in both populations. At the 4-leaf stage, root biomass decreased by 83–87% and 63–76% in the IR and NE populations, respectively. Application of herbicides at the 8-leaf stage reduced root biomass in the IR population by 41–55%, but no reduction was observed in root biomass in the NE population (Table 3).

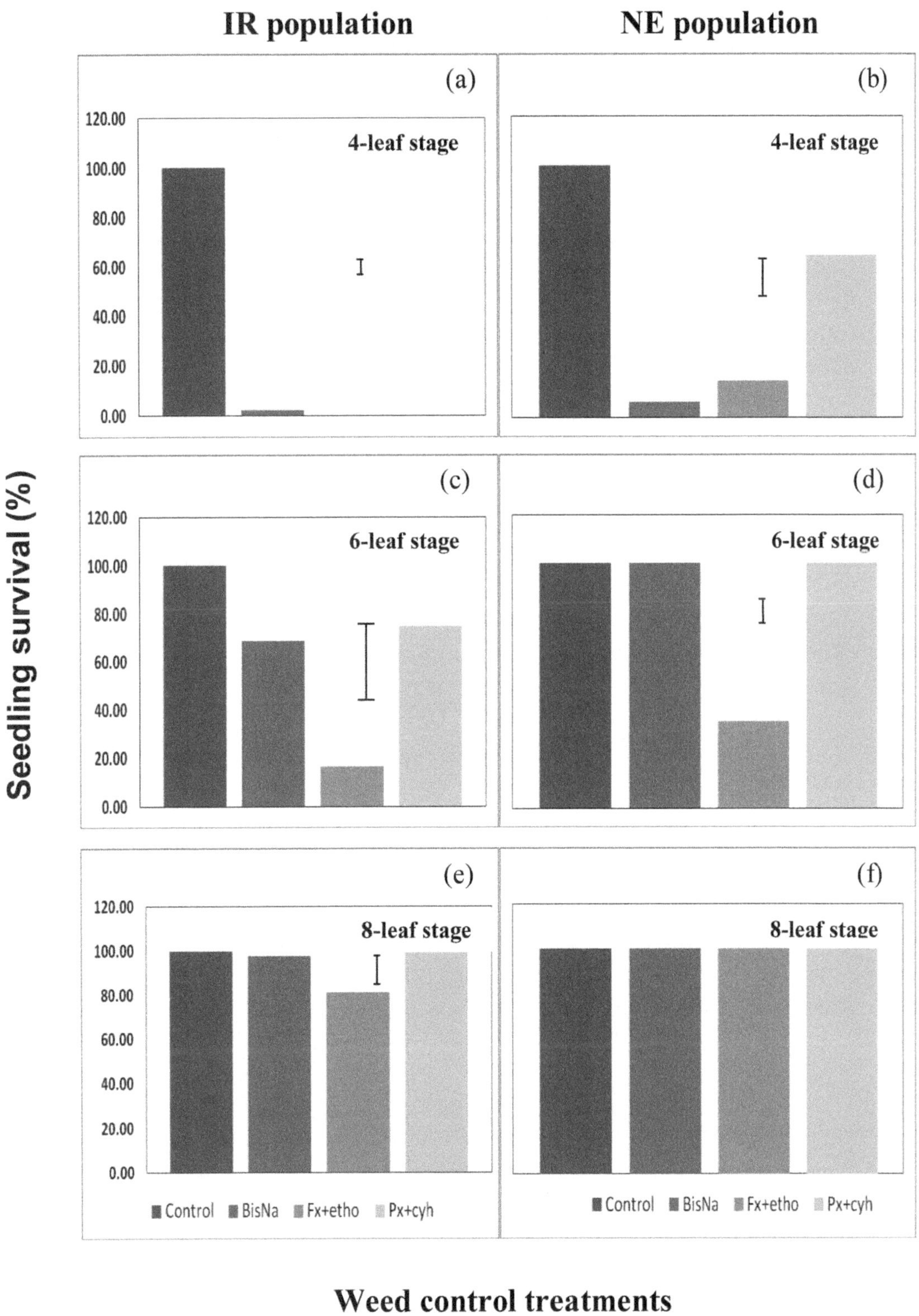

Figure 10. Seedling survival in two populations (IR and NE) of *Echinochloa glabrescens* when treated with postemergence herbicides (BisNa, bispyribac-sodium; Fx + etho, fenoxaprop + ethoxysulfuron; Px + cyh, penoxsulam + cyhalofop) at four-, six-, and eight-leaf stages. Error bars represent least significant difference values at 5% level of significance.

Reduction in the root-shoot ratio was observed in both populations and could be a result of the application of POST herbicides at 4- to 6-leaf stages (Table 4). At the 4-leaf stage, bispyribac-sodium, fenoxaprop + ethoxysulfuron, and penoxsulam + cyhalofop all caused similar levels reductions in the root-shoot ratio of the IR population. At the same stage, bispyribac-sodium and penoxsulam + cyhalofop caused similar levels of reduction in the root-shoot ratio of the NE population. There was less

Table 2. Shoot biomass (g tray^{-1}) of two populations (IR and NE) of *E. glabrescens* applied with postemergence herbicides (bispyribac-sodium, fenoxaprop + ethoxysulfuron, and penoxsulam + cyhalofop) at four-, six-, and eight-leaf stages.

Weed control treatments	Rate (g ha^{-1})	Shoot biomass (g tray^{-1})		
		Four-leaf	Six-leaf	Eight-leaf
IR population				
Nontreated control	0	1.30	1.50	1.48
Bispyribac-sodium	30	0.05	1.20	1.50
Fenoxaprop + ethoxysulfuron	45	0.00	0.78	1.46
Penoxsulam + cyhalofop	72	0.00	1.41	1.49
$LSD_{0.05}$		0.20	0.37	NS
NE population				
Nontreated control	0	1.10	1.26	1.76
Bispyribac-sodium	30	0.14	1.34	1.93
Fenoxaprop + ethoxysulfuron	45	0.36	1.18	2.20
Penoxsulam + cyhalofop	72	0.65	1.46	1.76
$LSD_{0.05}$		0.23	NS	NS

Abbreviations: NS = not significant.

Table 3. Root biomass (g tray^{-1}) of two populations (IR and NE) of *E. glabrescens* applied with postemergence herbicides (bispyribac-sodium, fenoxaprop + ethoxysulfuron, and penoxsulam + cyhalofop) at four-, six-, and eight-leaf stages.

Weed control treatments	Rate (g ha^{-1})	Root biomass (g tray^{-1})		
		Four-leaf	Six-leaf	Eight-leaf
IR population				
Nontreated control	0	0.80	4.08	1.80
Bispyribac-sodium	30	0.01	0.68	1.00
Fenoxaprop + ethoxysulfuron	45	0.00	0.70	0.82
Penoxsulam + cyhalofop	72	0.00	0.52	1.06
$LSD_{0.05}$		0.06	1.47	0.43
NE population				
Nontreated control	0	1.10	2.61	3.41
Bispyribac-sodium	30	0.05	0.97	2.21
Fenoxaprop + ethoxysulfuron	45	0.17	0.62	3.72
Penoxsulam + cyhalofop	72	0.17	0.74	3.39
$LSD_{0.05}$		0.16	0.37	NS

Abbreviations: NS = not significant.

reduction in the root-shoot ratio among plants treated with fenoxaprop + ethoxysulfuron (Table 4). When herbicides were applied at the 6-leaf stage, the decrease in the root-shoot ratio was similar in both populations. The decrease in the root-shoot ratio could mean allocation of more biomass to the shoots at the expense of the roots. This may be a survival mechanism of the plant to enable the shoot to recover through photosynthesis. Application of POST herbicides at the 8-leaf stage did not reduce root-shoot ratio in both populations.

These results suggest that the timing of the application of POST herbicides is critical to achieving maximum efficacy. Early herbicide application (4-leaf stage) provided more effective control of both populations of *E. glabrescens*.

The importance of early application of POST herbicides was also observed in a previous study [19], the authors of which stated that the application of all herbicides at 4-leaf stage provided a 89–98% control of *E. crus-galli* and a more than 95% control of *E. colona*.

In our study, the herbicides provided reductions of 85–100% in seedling survival, 41–81% in shoot biomass, and 75–100% in root biomass in both populations when applied at the 4-leaf stage.

Since all herbicides were equally effective against *E. glabrescens*, these can be used in rotation to prevent a possible evolution of resistance to these herbicides when these are continuously used against the weeds. Fenoxaprop + ethoxysulfuron applied at the 6-leaf stage controlled *L. chinensis* and *D. ciliaris*, while penoxsulam + cyhalofop controlled only *L. chinensis* and bispyribac-sodium controlled only *E. colona* [19].

In our study, we observed that at the 6-leaf stage, only fenoxaprop + ethoxysulfuron reduced seedling survival (65–77%) in *E. glabrescens*. This observation suggests that fenoxaprop + ethoxysulfuron can be used in situations in which herbicides cannot be applied at the early stages due to rain or other reasons.

Failure of all treatments to control *E. glabrescens* at the 8-leaf stage could be the result of higher tolerance of herbicides among bigger and older plants [20]. In a similar study, the efficacy of bispyribac-sodium, fenoxaprop + ethoxysulfuron, and penoxsulam + cyhalofop in *E. crus-galli* and *E. colona* decreased when applied at the 8-leaf stage [19].

The results of our study suggest that for a more effective, season-long, and sustainable weed control in direct-seeded rice systems, weed management methods must be used in combination with one another as a strategy. In no-till systems, the use of a stale seedbed combined with the retention of large amounts of residue, shallow flooding at early stage (0–4 DAS), and application of POST

Table 4. Root-shoot ratio of two populations (IR and NE) of *E. glabrescens* applied with postemergence herbicides (bispyribac-sodium, fenoxaprop + ethoxysulfuron, and penoxsulam + cyhalofop) at four-, six-, and eight-leaf stages.

Weed control treatments	Rate (g ha^{-1})	Root-shoot ratio		
		Four-leaf	Six-leaf	Eight-leaf
IR population				
Nontreated control	0	0.66	2.84	0.61
Bispyribac-sodium	30	0.06	0.55	0.67
Fenoxaprop + ethoxysulfuron	45	0.00	0.94	0.70
Penoxsulam + cyhalofop	72	0.00	0.40	0.93
$LSD_{0.05}$		0.15	1.10	0.46
NE population				
Nontreated control	0	1.00	2.06	1.99
Bispyribac-sodium	30	0.08	0.74	1.18
Fenoxaprop + ethoxysulfuron	45	0.31	0.53	1.77
Penoxsulam + cyhalofop	72	0.26	0.52	1.97
$LSD_{0.05}$		0.19	0.30	0.89

Abbreviations: NS = not significant.

herbicide at the 4-leaf stage of the weed can reduce its density. Since *E. glabrescens* is expected to proliferate in continuous no-till systems, there is also a need to rotate tillage systems. Deep cultivation that would bury weeds below 8 cm is thus suggested. Succeeding cultivation should be shallower to prevent previously buried weed seeds from coming back onto the soil surface. In conventional systems, the stale seedbed method followed by the application of PRE herbicides at low dose followed by flooding at shallow depths may also be an option in controlling *E. glabrescens*. In the absence of PRE herbicides, POST herbicides can be applied at the early stages.

Acknowledgments

The authors would like to thank Ms. Maria Leah Cruz for providing comments on this manuscript.

Author Contributions

Conceived and designed the experiments: JLO BSC. Performed the experiments: JLO BSC. Analyzed the data: BSC. Contributed reagents/materials/analysis tools: JLO BSC. Wrote the paper: JLO BSC AMB.

References

1. Chauhan BS (2012) Weed ecology and weed management strategies for dry-seeded rice in Asia. Weed Technol 26:1–13.
2. Pandey S, Velasco L (2005) Trends in crop establishment methods in Asia and research issues, in *Rice Is Life: Scientific Perspectives for the 21st Century*, K Toriyama, Heong KL, Hardy B, Editors. Los Baños, Philippines: International Rice Research Institute and Tsukuba, Japan: Japan International Research Center for Agricultural Sciences, p 178–181.
3. Bureau of Agricultural Statistics (2011) *Monthly Agricultural Situation Report*. 2011. Available: http://www.bas.gov.ph/?ids = masr&id = 20&mon = 2011-04. Accessed 2013 Nov 15.
4. Erguiza A, Duff B, Khan C (1990) Choice of rice crop establishment technique: transplanting vs wet seeding, in IRRI Reserach Paper Series 139. 10 pp.
5. Abilay WP, Garcia FV, Alcantara JM, De Datta SK (1984) *Changes in input use and grain yields in lowland rice farms in three Philippine provinces*, in *IRRI Research Paper Series No. 100*. International Rice Research Institute: Los Baños, Philippines.
6. Mandac AM, Kalirajan KP, Flinn JC (1982) Economic limitations to increasing shallow rainfed rice productivity in Bicol, Philippines, in IRRI Research Paper Series 80. 21 pp.
7. Chauhan BS, Abugho SB (2012) Threelobe morningglory (*Ipomoea triloba*) germination and response to herbicides. Weed Sci 60:199–204.
8. Dorji S, Chauhan BS, Baltazar AM, Johnson D (2013) Effect of flooding depth and pretilachlor rate on emergence and growth of three rice weeds: junglerice (*Echinochloa colona*), smallflower umbrella sedge (*Cyperus difformis*), and ludwigia (*Ludwigia hyssopifolia*). Can J Plant Prot 1:43–48.
9. Holm LG, Plucknett DL, Pancho JV, Herberger JP (1991) *The World's Worst Weeds: Distribution and Biology*. Malabar, Florida: The University Press of Hawaii. 609 p.
10. Chauhan BS, Johnson DE (2009) Seed germination ecology of junglerice (*Echinochloa colona*): a major weed of rice. Weed Sci 57:235–240.
11. Chauhan BS, Johnson DE (2011) Ecological studies on *Echinochloa crus-galli* and the implications for weed management in direct-seeded rice. Crop Prot 30:1385–1391.
12. Diop AM, Moody K (1984) Effect of seeding depth and flooding regime on germination and growth of *Echinochloa glabrescens*. Phil J Weed Sci 11:65–69.
13. Rao AN, Moody K (1987) Weed seedlings transplanted with rice seedlings reduce grain yield. Int Rice Res Newsletter 12:51.
14. Rao AN, Moody K (1992) Competition between *Echinochloa glabrescens* and rice (*Oryza sativa*). Trop Pest Manage 38:25–29.
15. Chauhan BS, Johnson DE (2010) The role of seed ecology in improving weed management strategies in the tropics. Adv Agron 105:221–262.
16. Chauhan BS, Johnson DE (2008) Germination ecology of Chinese sprangletop (*Leptochloa chinensis*) in the Philippines. Weed Sci 56:820–825.
17. Lee J, Chauhan BS, Johnson DE (2011) Germination of fresh horse purslane (*Trianthema portulacastrum*) seeds in response to different environmental factors. Weed Sci 59:495–499.
18. Bolfrey-Arku GE-K, Chauhan BS, Johnson DE (2011) Seed germination ecology of itchgrass (*Rottboellia cochinchinensis*). Weed Sci 59:182–187.
19. Chauhan BS, Abugho SB (2012) Effect of growth stage on the efficacy of postemergence herbicides on four weed species of direct-seeded rice. Scientific World J, Article ID 123071, 7 pages.
20. Singh S, Singh M (2004) Effect of growth stage on trifloxysulfuron and glyphosate efficacy in twelve weed species of citrus groves. Weed Technol 18:1031–1036.
21. Michel BE (1983) Evaluation of the water potentials of solutions of polyethylene glycol 8000 both in the absence and presence of other solutes. Plant Physiol 72:66–70.
22. Chauhan BS, Johnson DE (2008) Influence of environmental factors on seed germination and seedling emergence of eclipta (*Eclipta prostrata*) in a tropical environment. Weed Sci 56:383–388.
23. Chauhan BS, Johnson DE (2008) Germination ecology of goosegrass (*Eleusine indica*): an important grass weed of rainfed rice. Weed Sci 56:699–706.
24. Eslami SV (2011) Comparative Germination and emergence ecology of two populations of common lambsquarters (*Chenopodium album*) from Iran and Denmark. Weed Sci 59:90–97.
25. Schütz W, Milberg P, Lamont BB (2002) Seed dormancy, after-ripening and light requirements of four annual Asteracea in south-western Australia. Ann Bot 90:707–714.
26. Chauhan BS, Johnson DE (2009) Influence of tillage systems on weed seedling emergence pattern in rainfed rice. Soil Till Res 106:15–21.
27. Boyd NS, Van Acker RC (2004) Seed germination of common weed species as affected by oxygen concentration, light, and osmotic potential. Weed Sci 52:589–596.
28. Benvenuti S, Macchia M (1995) Hypoxia effect on buried weed seed germination. Weed Res 35:343–351.
29. Egley GH (1986) Stimulation of weed seed germination in soil. *Reviews in* Weed Sci 2:67–89.
30. Baskin CC, Baskin JM (1998) Seeds: Ecology, Biogeography, and Evolution of Dormancy and Germination. San Diego, CA: Academic. 666 p.
31. Chauhan BS, Migo T, Westerman PR, Johnson DE (2010) Post-dispersal predation of weed seeds in rice fields. Weed Res 50:553–560.
32. Crutchfield DA, Wicks GA, Burnside OC (1985) Effect of winter wheat (*Triticum aestivum*) straw mulch level on weed control. Weed Sci 34:110–114.
33. Facelli JM, Pickett STA (1991) Plant litter: its dynamics and effects on plant community structure. Bot Rev 57:1–32.
34. Gadde B, Bonnet S, Menke C, Garivait S (2009) Air pollutant emissions from rice straw open field burning in India, Thailand and the Philippines. Environ Pollution 157:1554–1558.
35. Chauhan BS, Johnson DE (2009) *Ludwigia hyssopifolia* emergence and growth as affected by light, burial depth and water management. Crop Prot 28:887–890.
36. Kent RJ, Johnson DE (2001) Influence of flood depth and duration on growth of lowland rice weeds, Cote d'Ivoire. Crop Prot 20:691–694.
37. Pons TL (1982) Factors affecting weed seed germination and seedling growth in lowland rice in Indonesia. Weed Res 22:155–161.
38. Smith RJJ, Fox WT (1973) Soil water and growth of rice and weeds. Weed Sci 21:61–63.
39. Tuong TP, Bouman BAM, Mortimer M (2005) More rice, less water: integrated approaches for increasing water productivity in irrigated rice-based systems in Asia. Plant Prod Sci 8:231–241.
40. Pablico PP, Moody K (1993) Effect of flooding duration on the performance of pretilachlor + fenclorim in wet-seeded and transplanted rice. Phil J Weed Sci 20:56–61.
41. Chauhan BS, Prabhjyot-Kaur, Mahajan G, Randhawa RK, Singh H, et al. (2014) Global warming and its possible impact on agriculture in India. Adv Agron 123:65–121.

21

Impact of Biotic and Abiotic Stresses on the Competitive Ability of Multiple Herbicide Resistant Wild Oat (*Avena fatua*)

Erik A. Lehnhoff[1]*, Barbara K. Keith[2], William E. Dyer[2], Fabian D. Menalled[1]

1 Department of Land Resources and Environmental Sciences, Montana State University, Bozeman, Montana, United States of America, **2** Department of Plant Sciences and Plant Pathology, Montana State University, Bozeman, Montana, United States of America

Abstract

Ecological theory predicts that fitness costs of herbicide resistance should lead to the reduced relative abundance of resistant populations upon the cessation of herbicide use. This greenhouse research investigated the potential fitness costs of two multiple herbicide resistant (MHR) wild oat (*Avena fatua*) populations, an economically important weed that affects cereal and pulse crop production in the Northern Great Plains of North America. We compared the competitive ability of two MHR and two herbicide susceptible (HS) *A. fatua* populations along a gradient of biotic and abiotic stresses The biotic stress was imposed by three levels of wheat (*Triticum aestivum*) competition (0, 4, and 8 individuals pot^{-1}) and an abiotic stress by three nitrogen (N) fertilization rates (0, 50 and 100 kg N ha^{-1}). Data were analyzed with linear mixed-effects models and results showed that the biomass of all *A. fatua* populations decreased with increasing *T. aestivum* competition at all N rates. Similarly, *A. fatua* relative growth rate (RGR) decreased with increasing *T. aestivum* competition at the medium and high N rates but there was no response with 0 N. There were no differences between the levels of biomass or RGR of HS and MHR populations in response to *T. aestivum* competition. Overall, the results indicate that MHR does not confer growth-related fitness costs in these *A. fatua* populations, and that their relative abundance will not be diminished with respect to HS populations in the absence of herbicide treatment.

Editor: Randall P. Niedz, United States Department of Agriculture, United States of America

Funding: This research was funded through grants by the U.S. Environmental Protection Agency, http://www.epa.gov, (Grant # X8-97873401-0); the Montana Wheat and Barley Committee, http://wbc.agr.mt.gov, (Grant # 408-1655); and the Montana Noxious Weed Trust Fund, http://agr.mt.gov/agr/Programs/Weeds/TrustFund, (Grant# 2011-033). The funders had no role in study design, data collection and analysis, decision to publish, or preparation of the manuscript.

Competing Interests: The authors have declared that no competing interests exist.

* E-mail: erik.lehnhoff@montana.edu

Introduction

The reliance on herbicides for weed control has posed strong selection pressure for resistant populations, and there are now nearly 400 unique cases (plant species × site of action) of herbicide resistance in 217 plant species [1]. Most of these cases involve target site mutations that confer resistance to a single herbicide or related herbicides with the same mechanism of action. However, non-target-site-based resistance has recently become more common, and in some cases the use of one herbicide mode of action may substantially increase selection for non-target-site-based resistance genes that confer resistance to other unrelated herbicides [2,3]. The physiological mechanisms of non-target-site-based resistance are usually based on enhanced herbicide metabolism or detoxification as mediated by cytochrome P450 monooxygenases [4] (hereafter P450s), glutathione *S*-transferases (GSTs) [5], and other enzymes of Phase II metabolism [6].

Ecological theory predicts that individuals with heritable resistance to an environmental stress may have an ecological disadvantage as compared to susceptible individuals in the absence of the stress [7,8]. For example, herbicide resistant biotypes are predicted to experience a fitness cost as resources are shifted to resistance mechanism(s) rather than to growth and reproduction. Such fitness costs have been associated with a number of specific gene mutations conferring resistance to herbicides (see [9] for a review). For example, fitness costs have been demonstrated for the Pro197His and Trp547Leu mutations that confer resistance to acetolactate synthase (ALS)-inhibiting herbicides in prickly lettuce (*Lactuca serriola*) [10,11] and Powell's amaranth (*Amaranthus powellii*) [12]. However, while Menchari et al. [13] determined fitness costs for the resistance-conferring Asp2078Gly and Ile2041Asn mutations in acetyl CoA carboxylase (ACCase) in slender meadow foxtail (*Alopecuris myosuroides*), they did not find fitness costs associated with the Ile1781Leu mutation. Similarly, Vila-Aiub et al. [9] found resistance costs for the Cys2088Arg mutation associated with resistance to ACCase herbicides in Wimmera ryegrass (*Lolium rigidum*), yet fitness costs were not demonstrated for the Ile1781Leu mutation. Thus, particular target site mutations may or may not be associated with fitness costs in resistant populations.

Weed populations with herbicide resistance conferred by enhanced metabolic rates may be more likely to exhibit fitness costs, due to the constitutive and/or inducible overexpression of genes involved in energetically expensive pathways like those involving P450- and GST-mediated metabolism [14]. For example, a *L. rigidum* biotype (isolated from one population) with suspected P450-mediated herbicide metabolism produced less aboveground biomass and had a lower relative growth rate (RGR)

Table 1. Soil nutrient concentrations.

Greenhouse	Nitrate (kg ha^{-1})	Phosphorus (mg kg^{-1})	Potassium (mg kg^{-1})	Organic Matter (%)
1	39.7±3.5^{a}	13±0.0^{a}	261.3±8.0^{a}	7.2±0.7^{a}
2	129.3±5.0^{b}	14.7±0.6^{b}	302.0±13.9^{b}	7.6±0.6^{a}

Initial nutrient and organic matter concentrations in greenhouse soil (mean ± SD, n = 3 per greenhouse) used in experiments to assess the impact of environmental and biological stressors on *Avena fatua* growth. Significant differences across greenhouses are indicated by different letters (P<0.05).

[15] and was a weaker competitor with *T. aestivum* [4] in the greenhouse than susceptible biotypes. Similarly, Park and Mallory-Smith [16] found that plants from a metabolically-based resistant downy brome (*Bromus tectorum*) population produced less shoot biomass, leaf area and seeds, and was less competitive than plants from an adjacent susceptible population.

A. fatua is one of the most economically important weeds across the Northern Great Plains of North America, where it competes with and reduces yields of cereal and pulse crops [17,18]. ACCase and ALS-inhibiting herbicides have been used to control *A. fatua* since the 1970s and 1980s, respectively. However, repeated use of these herbicides, as well as others such as triallate and difenzoquat, led to the evolution of herbicide resistant *A. fatua* populations in the 1980s and 1990s [19,20,21]. Recently, two *A. fatua* populations with resistance to multiple herbicides including triallate (emergence inhibitor), flucarbazone (ALS-inhibitor), imazamethabenz (ACCase-inhibitor), paraquat (membrane disruptor), and difenzoquat (membrane disruptor) were reported [22]. Resistance to some of these modes of action may be conferred by P450-enhanced metabolism [3,23,24] or by protection against oxidative stress via GST [6]. While controlling resistant weeds with herbicides is difficult, resistance via enhanced herbicide metabolism is especially problematic because weeds do not need to be exposed to a herbicide in order to be resistant to it [6].

Few studies have evaluated fitness costs for herbicide resistance in *A. fatua*. O'Donovan et al. [25] showed that shoot weight in triallate/difenzoquat-resistant *A. fatua* populations was generally greater than in susceptible ones, but this difference did not translate into competitive advantages. Similarly, Lehnhoff et al. [22] demonstrated that in non-competitive conditions, the MHR *A. fatua* populations utilized in this study had photosynthetic capacities and relative growth rates similar to susceptible populations, and while the MHR populations reached anthesis quicker, they ultimately produced fewer tillers and seeds.

Understanding the effects of herbicide resistance on individual fitness is crucial to the effective management of MHR weed biotypes. Specifically, assessment of fitness costs is important for the prediction of resistant biotype population demographics after the cessation of herbicide use [26,27]. Three issues highlight the importance of assessing fitness costs of the MHR *A. fatua* populations. First, the observations that herbicide resistance fitness costs are not universal [9] indicates that each case should be examined individually. Second, the increasing incidence of MHR populations and their spread has profound implications for their prevention and management [28]. And finally, the fact that the MHR populations studied here are the first with confirmed constitutively elevated and inducible P450 expression levels (Keith et al. unpublished data) provides an important genetic resource in which to test the resource-based allocation theory.

The purpose of this research was to evaluate the competitive performance of two MHR and two herbicide susceptible (HS) *A. fatua* populations along a gradient of biotic and abiotic stresses. The biotic stress was provided by growing *A. fatua* plants in competition with three planting densities of *T. aestivum*, and the abiotic stress was a gradient of resource availability using three rates of N fertilization. Our objectives were to compare (1) total above-ground biomass production, and (2) relative growth rates of the HS and MHR *A. fatua* populations across a factorial combination of these abiotic and biotic gradients.

Materials and Methods

Plant Materials

Two herbicide susceptible (HS1 and HS2) *A. fatua* populations were used as controls. Population HS1 was obtained originally from seeds collected from a field adjacent to where MHR populations were collected, and subsequently confirmed to be 100% HS. All seeds were collected with private landowner permission, and no endangered or protected species were involved in this research. The second susceptible population (HS2, technically a biotype) was the nondormant inbred SH430 biotype used for seed dormancy research [29,30]. This biotype exhibited similar growth to HS1 and the MHR populations in previous research [22].

MHR populations (MHR3 and MHR4) used for this research were derived from seeds collected from two *A. fatua* populations not controlled by 60 g a.i. ha^{-1} pinoxaden (Axial, Syngenta Crop Protection, Inc., Greensboro, NC, USA; ACCase inhibitor) from two fields in Teton County, Montana, USA in 2006. To ensure a 100% resistant population, these seeds (initially about 90% resistant to 60 g a.i. ha^{-1} pinoxaden) were subjected to two generations of recurrent group selection with 50 plants per generation by spraying with the same pinoxaden dose. Because we maintained some genetic diversity by using 50 random seeds in each generation, we refer to MHR3 and MHR4 as populations rather than biotypes. Therefore this work assessed the population level effects of MHR [22] and did not explicitly assess fitness costs of resistance as would be done via isogenic lines of MHR biotypes [31].

Prior to conducting this study, MHR3 and MH4 populations were determined to be resistant to field use rates of difenzoquat, flucarbazone, imazamethabenz, and tralkoxydim as compared to HS1 and HS2 [22]. Resistance to triallate and paraquat was subsequently determined for both MHR populations (Keith et al. unpublished data). We use the MHR acronym to describe these populations because they are resistant to members of five different mode of action families, and we suspect the presence of different physiological mechanisms (Keith et al. unpublished data).

Plant Growth

This study was conducted as a complete randomized block design with four blocks of four *A. fatua* populations, three N fertilization rates, three levels of *T. aestivum* (Reeder hard red spring wheat) competition, and two harvest times, for a total of 288 experimental units (plastic pots, 17.8 cm dia. ×15.2 cm deep) per trial. It was conducted twice, simultaneously in two different

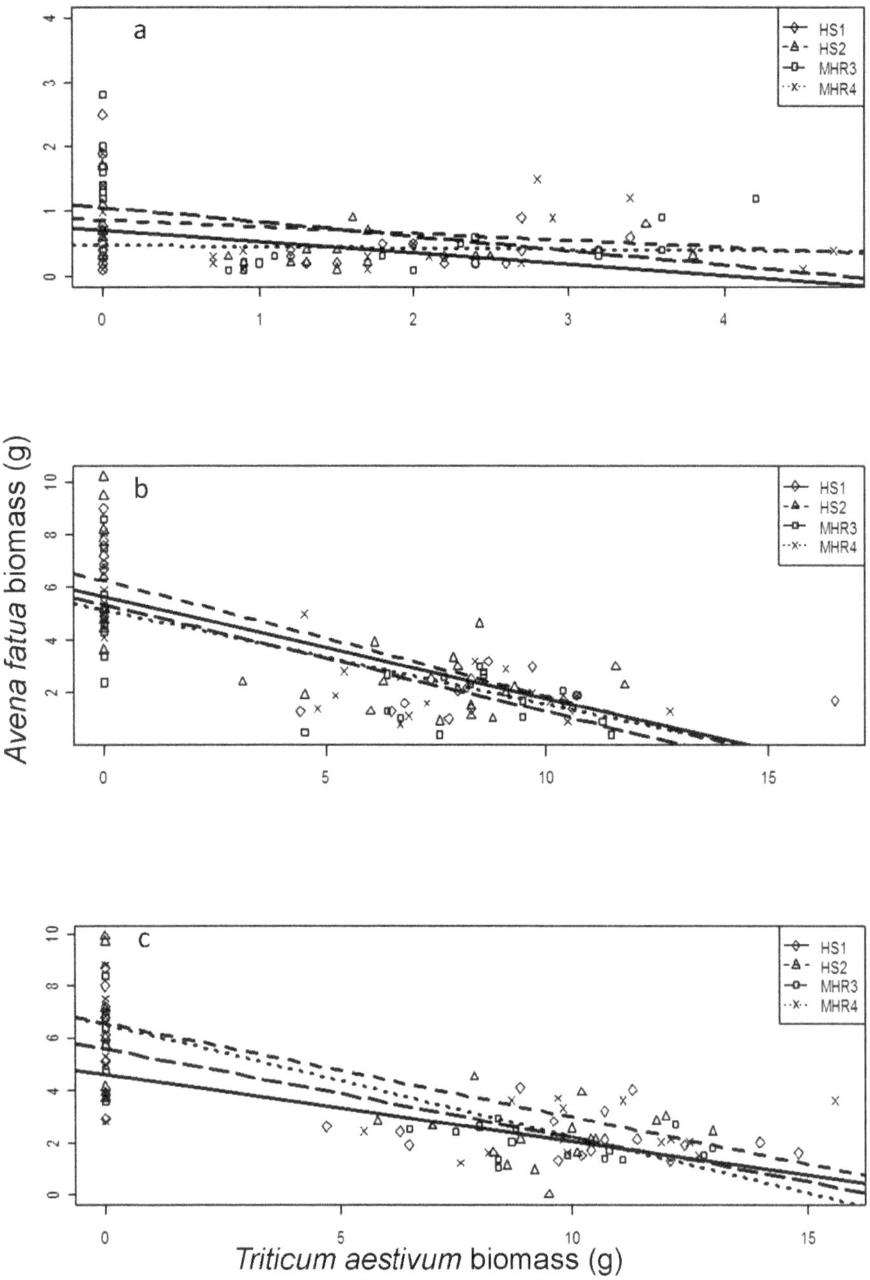

Figure 1. Effects of *Triticum aestivum* competition and nitrogen on *Avena fatua* biomass. Nitrogen fertilizer rates (N) are (a) 0, (b) 50 and (c) 100 kg N ha^{-1}. HS1 and HS2 are herbicide susceptible *A. fatua* populations and MHR3 and MHR4 are multiple herbicide resistant populations. n = 8 for each N, *T. aestivum* competition, and *A. fatua* combination.

greenhouses at the Montana State University Plant Growth Center under a 16-hr photoperiod of natural sunlight supplemented with mercury vapor lamps (165 $\mu E\ m^{-2}\ sec^{-1}$) at 25±4 C. Pots were filled with a mixture of 1:1:1 [by vol] sphagnum moss, sand, and Sunshine Mix #1 (Sun Gro Horticulture, Inc., Bellevue, WA) and leached to reduce background N concentrations by draining four pot volumes of water through the soil over two days. Because of changes in greenhouse soil availability, different batches of soil were used in each greenhouse. Three soil samples, each composited from ten pots, were collected from each greenhouse and analyzed to determine initial concentrations of nitrate (NO_3^-), Olsen phosphorus (P), potassium (K^+), and organic matter. The pots were then fertilized with ammonium sulfate at rates of 0, 50, or 100 kg N ha^{-1}. In each pot, two seeds of one of the four *A. fatua* populations (HS1, HS2, MHR3, or MHR4) were planted 2 cm deep in the center of the pot, and *T. aestivum* seeds (0, 5, or 9) were planted evenly spaced around the circumference of the pot approximately 2 cm from the edge. *A. fatua* seedlings were thinned to the single largest individual per pot and *T. aestivum* seedlings were thinned to create densities of 0, 4, and 8 individuals per pot at 7 days after planting (DAP).

At 32 DAP, one half of the pots from each treatment were randomly chosen and *A. fatua* and *T. aestivum* plants were harvested separately by cutting them at the root crown, followed by drying at

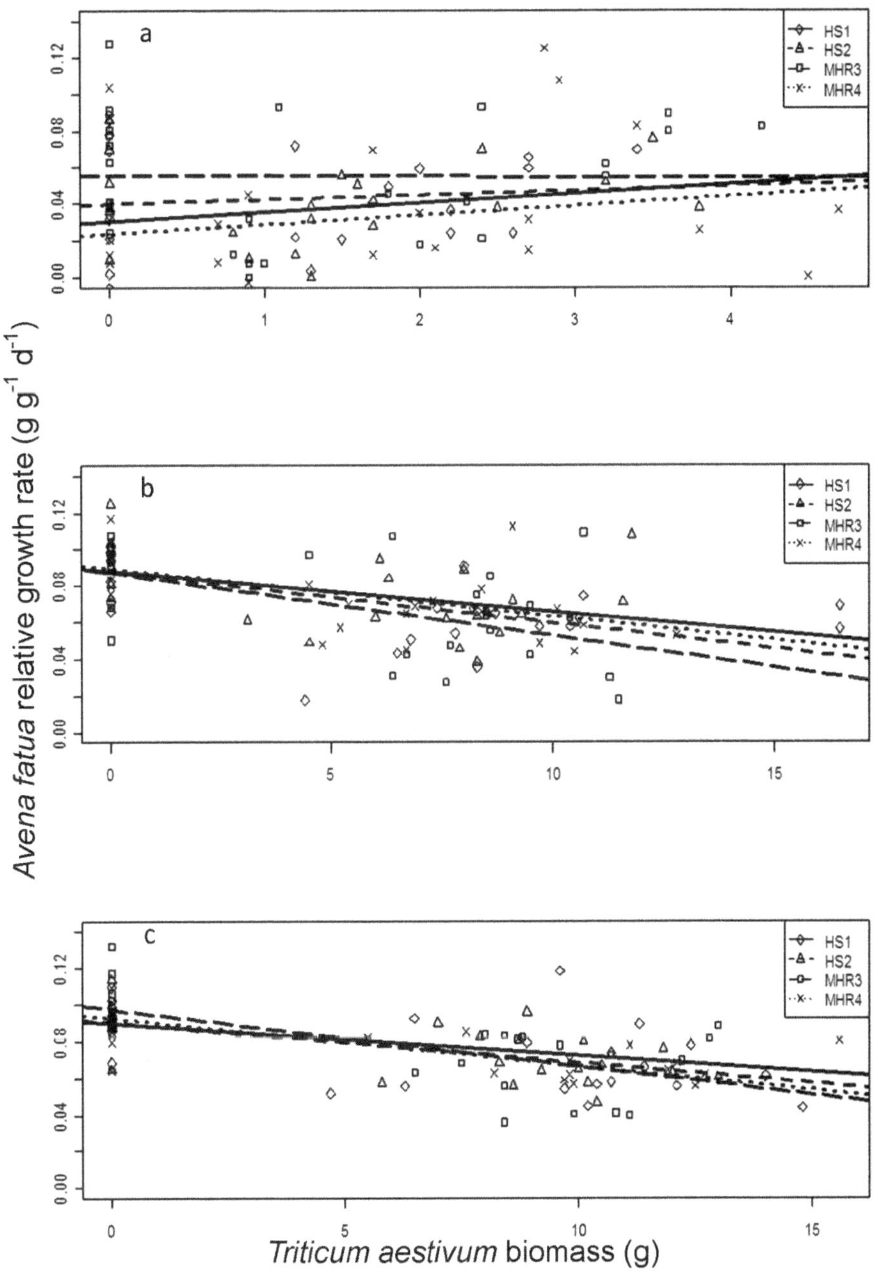

Figure 2. Effects of *Triticum aestivum* competition on *Avena fatua* relative growth rate. Nitrogen fertilizer rates (N) are (a) 0, (b) 50 and (c) 100 kg N ha^{-1}. HS1 and HS2 are herbicide susceptible *A. fatua* populations and MHR3 and MHR4 are multiple herbicide resistant populations. n = 8 for each N, *T. aestivum* competition, and *A. fatua* combination.

40 C for 10 days, and weighing. The pots receiving 50 and 100 kg N ha^{-1} were then fertilized at these rates bi-weekly and weekly, respectively, for the next four weeks. The remaining plants were harvested at 61 DAP, and plant materials were dried and weighed as described above.

Data Analysis

Differences in initial soil nutrient contents between greenhouse soil sources were evaluated by Student's t-test and were found to be different for concentrations of NO_3^- ($P<0.001$), P ($P = 0.038$), and K^+ ($P = 0.019$), but there were no differences in percent organic matter ($P = 0.484$) (Table 1). The effects of initial nutrient content on *A. fatua* biomass and RGR were evaluated via ANOVA with greenhouse as a random error term, and initial nutrient differences between greenhouses did not translate into differences in final *A. fatua* biomass ($F_{1,283} = 1.09$, $P = 0.297$) or relative growth rate ($F_{1,283} = 3.03$, $P = 0.083$), so data were combined for analysis.

A. fatua RGRs were calculated following Hunt [32] (equation 1):

$$RGR = \frac{(\ln W_2 - \ln W_1)}{(t_2 - t_1)} \quad (1)$$

where ln is the natural log, W_2 and W_1 are final and first harvest

Table 2. Models predicting Avena fatua biomass and growth rate.

Model variables	*k*	Biomass AIC	Biomass Δ (AIC)	Biomass *w*(AIC)	Relative growth rate AIC	Relative growth rate Δ (AIC)	Relative growth rate *w*(AIC)
Intercept only	3	1296	237	0	−1158	62	0
N	4	1215	156	0	−1210	10	0.01
T. aestivum biomass	4	1287	228	0	−1143	77	0
A. fatua population	4	1300	241	0	−1145	75	0
N + *T. aestivum* biomass	5	1076	17	0	**−1220**	**0**	**0.97**
N + *A. fatua* population	5	1219	160	0	−1197	23	0
T. aestivum biomass + *A. fatua* population	5	1291	232	0	−1130	90	0
N × *T. aestivum* biomass	6	**1059**	**0**	**1**	−1212	8	0.02
N × *A. fatua* population	6	1231	172	0	−1177	43	0
T. aestivum biomass × *A. fatua* population	6	1298	239	0	−1114	106	0
N + *T. aestivum* biomass + *A. fatua* population	6	1081	22	0	−1206	14	0
N × *T. aestivum* biomass × *A. fatua* population	10	1096	37	0	−1139	81	0

Akaike information criterion (AIC) scores for mixed-effects models of *Avena fatua* biomass and relative growth rate, where *k* is the number of parameters predicted by intercept only, nitrogen fertilization rate (N), competition from *Triticum aestivum* (*T. aestivum* biomass), *A. fatua* population, and their interactions. Bold values indicate the best-fit models with the lowest AIC scores. Δ(AIC) is the change in AIC with respect to the best candidate model and *w*AIC is the AIC weight. N rate, *T. aestivum* biomass, and *A. fatua* population were fixed effects and greenhouse was a random effect included in every model. n = 8 for each N, *T. aestivum* competition, and *A. fatua* combination.

biomasses, respectively, and t_2 and t_1 are final and first harvest times in days. The biomass and RGR responses of the *A. fatua* populations to N rates and *T. aestivum* competition were analyzed via linear mixed effects-models, with N rate, *T. aestivum* biomass, and *A. fatua* population as fixed effects, and the greenhouse in which experiments were conducted as a random effect to account for initial soil differences, using the 'lme4' package in R (version 2.12.1, The R Foundation for Statistical Computing). Best models predicting *A. fatua* responses were determined based on the lowest Akaike Information Criterion (AIC) values. Change in AIC (ΔAIC) values between the best candidate model and all other models, and the models' AIC weights (*w*AIC, higher weight indicates better model), were also assessed for model comparison. For individual N rates, the response of each *A. fatua* population to *T. aestivum* was modeled via linear regression in R.

Results and Discussion

The best candidate models predicting *A. fatua* biomass and RGR based on AIC scores and weights were N fertilization rate × *T. aestivum* biomass and N fertilization rate + *T. aestivum* biomass, respectively (Table 2). The second best candidate models for *A. fatua* biomass and RGR based on ΔAIC were N + *T. aestivum* biomass and N × *T. aestivum* biomass, respectively. The best models that included *A. fatua* population as a predictor had ΔAIC values of 22 and 14 for *A. fatua* biomass and RGR, respectively; however, these models had *w*(AIC) values of zero, indicating a poor fit to the data. At all N rates, the biomass of *A. fatua* decreased as *T. aestivum* biomass increased ($P<0.05$) (Figure 1) while the RGR of all *A. fatua* populations decreased with increasing *T. aestivum* biomass for the two higher N fertilization rates, but not at the zero N rate ($P = 0.213$) (Figure 2).

As expected, *A. fatua* biomass declined with increasing *T. aestivum* competition, although this relationship was not significant in the zero N treatments, where both *A. fatua* and *T. aestivum* biomasses were quite low compared to those in the 50 and 100 kg N ha^{-1} treatments. The minimal decrease of *A. fatua* biomass in the unfertilized pots can be explained by the apparent lack of competition from *T. aestivum*, as both species were stunted in these treatments (Lehnhoff, personal observation). The similar pattern observed with respect to *A. fatua* RGR can also be attributed to the low *T. aestivum* biomass in the unfertilized pots compared to fertilized ones. These results are generally in accordance with those reported by Blackshaw and Brandt [33] who showed that *A. fatua* competition with *T. aestivum* was unaffected by fertilizer rate.

More importantly, there were no differences in the response of HS and MHR populations to *T. aestivum* competition or N stress, suggesting that there were no growth-related fitness costs for multiple herbicide resistance. This is similar to our previous findings [22], where MHR populations did not experience growth-related fitness costs in the absence of competition. Because these MHR populations exhibit constitutively higher levels of P450 expression (Keith et al., unpublished data), the resource allocation theory would predict that the resulting resource diversion would be associated with a fitness cost. These results based on *A. fatua* growth therefore contrast with previous studies reporting various fitness costs associated with enhanced rates of herbicide metabolism [4,15,16].

The current results suggest that if competitiveness is positively related to reproductive fitness, the frequency of resistant alleles in the *A. fatua* MHR populations should not decline due to fitness costs. However, we previously showed that, in the absence of competition, one MHR population produced fewer seeds than a HS population in the greenhouse [22]. Thus, it is possible that if the selection pressure is interrupted, the proportion of MHR to HS seeds in the seedbank will decline, ultimately leading to a reduction in MHR populations, *sensu* Maxwell et al. [34] who predicted a decrease in resistant populations after the cessation of herbicide use. The relationship between *A. fatua* competitiveness (i.e., biomass production and RGR) and reproductive fitness is unknown, and will need to be examined under realistic field conditions to determine the actual implications for field management of MHR *A. fatua*.

While competitive ability provides a potential indication of the fitness costs associated with herbicide resistance [4], other traits that contribute to individual success through the entire life cycle, such as seed germination, seedling survival, and seed production should be included in a comprehensive analysis [14,35]. In this context, this study provides useful information that can be integrated with future field research on MHR and HS demography. Such investigations will be required in order to develop successful alternative management options for herbicide resistant weed populations.

Finally, it should be noted that our study examined a small number of populations (two HS and two MHR). As Cousens et al. [36] discussed, a non-significant result from a small number of populations may not be enough to support a hypothesis that there is no fitness cost of resistance. Additional studies comparing the MHR populations to more HS populations could provide valuable insight into potential fitness costs. Alternatively, fitness costs could be assessed through the generation and testing of HS and MHR isogenic biotypes, and this process is currently underway in our laboratory.

Acknowledgments

We thank Mark Boyd and Ethan Mayes for assistance with greenhouse work.

Author Contributions

Conceived and designed the experiments: EAL FDM WED. Performed the experiments: EAL BKK. Analyzed the data: EAL. Wrote the paper: EAL BKK WED FDM.

References

1. Heap IM (2013) International survey of herbicide resistant weeds. Weed Science Society of America website. Available: http://www.weedscience.org. Accessed 2013 Mar 1.
2. Delye C, Gardin JAC, Boucansaud K, Chauvel B, Petit C (2011) Non-target-site-based resistance should be the centre of attention for herbicide resistance research: *Alopecurus myosuroides* as an illustration. Weed Research 51: 433–437.
3. Beckie HJ, Tardif FJ (2012) Herbicide cross resistance in weeds. Crop Protection 35: 15–28.
4. Vila-Aiub MM, Neve P, Powles SB (2009) Evidence for an ecological cost of enhanced herbicide metabolism in *Lolium rigidum*. Journal of Ecology 97: 772–780.
5. Reade JPH, Milner LJ, Cobb AH (2004) A role for glutathione S-transferases in resistance to herbicides in grasses. Weed Science 52: 468–474.
6. Powles SB, Yu Q (2010) Evolution in Action: Plants Resistant to Herbicides. In: Merchant SBWROD, editor. Annual Review of Plant Biology, Vol 61. 317–347.
7. Herms DA, Mattson WJ (1992) The dilemna of plants – to grow or defend. Quarterly Review of Biology 67: 283–335.
8. Bazzaz FA, Chiariello NR, Coley PD, Pitelka LF (1987) Allocating resources to reproduction and defense. Bioscience 37: 58–67.
9. Vila-Aiub MM, Neve P, Powles SB (2009) Fitness costs associated with evolved herbicide resistance alleles in plants. New Phytologist 184: 751–767.
10. Alcocer-Ruthling M, Thill DC, Mallorysmith C (1992) Monitoring the occurrence of sulfonylurea-resistant prickly lettuce (*Lactuca serriola*). Weed Technology 6: 437–440.
11. Alcocer-Ruthling M, Thill DC, Shafii B (1992) Differential competitiveness of sulfonylurea resistant and susceptible prickly lettice (*Lactuca serriola*). Weed Technology 6: 303–309.
12. Tardif FJ, Rajcan I, Costea M (2006) A mutation in the herbicide target site acetohydroxyacid synthase produces morphological and structural alterations and reduces fitness in *Amaranthus powellii*. New Phytologist 169: 251–264.
13. Menchari Y, Chauvel B, Darmency H, Delye C (2008) Fitness costs associated with three mutant acetyl-coenzyme A carboxylase alleles endowing herbicide resistance in black-grass *Alopecurus myosuroides*. Journal of Applied Ecology 45: 939–947.
14. Vila-Aiub MM, Neve P, Steadman KJ, Powles SB (2005) Ecological fitness of a multiple herbicide-resistant *Lolium rigidum* population: dynamics of seed germination and seedling emergence of resistant and susceptible phenotypes. Journal of Applied Ecology 42: 288–298.
15. Vila-Aiub MM, Neve P, Powles SB (2005) Resistance cost of a cytochrome P450 herbicide metabolism mechanism but not an ACCase target site mutation in a multiple resistant *Lolium rigidum* population. New Phytologist 167: 787–796.
16. Park KW, Mallory-Smith CA (2005) Multiple herbicide resistance in downy brome (*Bromus tectorum*) and its impact on fitness. Weed Science 53: 780–786.
17. Evans RM, Thill DC, Tapia L, Shafii B, Lish JM (1991) Wild oat (*Avena fatua*) and spring barley (*Hordeum vulgare*) density affect spring barley grain yield. Weed Technology 5: 33–39.
18. Beckie HJ, Francis A, Hall LM (2012) The Biology of Canadian Weeds. 27. *Avena fatua* L. (updated). Canadian Journal of Plant Science 92: 1329–1357.
19. O'Donovan JT, Sharma MP, Harker KN, Maurice D, Baig MN, et al. (1994) Wild oat (*Avena fatua*) populations resistant to triallate are also resistant to difenzoquat. Weed Science 42: 195–199.
20. Heap IM, Murray BG, Loeppky HA, Morrison IN (1993) Resistance to aryloxyphenoxypropionate and cyclohexanedione herbicides in wild oat (*Avena fatua*). Weed Science 41: 232–238.
21. Somody CN, Nalewaja JD, Miller SD (1984) Wild oat (*Avena fatua*) and *Avena sterilis* morphological characteristics and response to herbicides. Weed Science 32: 353–359.
22. Lehnhoff EA, Keith BK, Dyer WE, Peterson RKD, Menalled FD (*in press*) Multiple herbicide resistance in wild oat (*Avena fatua*) and its impacts on physiology, germinability, and seed production. Agronomy Journal.
23. Beckie HJ, Warwick SI, Sauder CA (2012) Basis for herbicide resistance in Canadian populations of wild oat (*Avena fatua*). Weed Science 60: 10–18.
24. Maneechote C, Preston C, Powles SB (1997) A diclofop-methyl-resistant *Avena sterilis* biotype with a herbicide-resistant acetyl-coenzyme A carboxylase and enhanced metabolism of diclofop-methyl. Pesticide Science 49: 105–114.
25. O'Donovan JT, Newman JC, Blackshaw RE, Harker KN, Derksen DA, et al. (1999) Growth, competitiveness, and seed germination of triallate/difenzoquat-susceptible and -resistant wild oat populations. Canadian Journal of Plant Science 79: 303–312.
26. Holt J, Thill DC (1994) Growth and productivity of resistant plants. In: J.A.M PSBaH, editor. Herbicide Resistance in Plants: Biology and Biochemistry. Boca Raton, FL: Lewis. 299–316.
27. Maxwell BD, Mortimer AM (1994) Selection for Herbicide Resistance. In: Powles SB, Holtum JAM, editors. Herbicide Resistance in Plants: Biology and Chemistry. Boca Raton, Florida: CRC Press, Inc. 353.
28. Mortensen DA, Egan JF, Maxwell BD, Ryan MR, Smith RG (2012) Navigating a Critical Juncture for Sustainable Weed Management. Bioscience 62: 75–84.
29. Naylor JM, Jana S (1976) Genetic adaptation for seed dormancy in *Avena fatua*. Canadian Journal of Botany-Revue Canadienne De Botanique 54: 306–312.
30. Johnson RR, Cranston HJ, Chaverra ME, Dyer WE (1995) Characterization of CDNA clones for differentially expressed genes in embryos of dormant and nondormant *Avana fatua* caryopses. Plant Molecular Biology 28: 113–122.
31. Vila-Aiub MM, Neve P, Roux F (2011) A unified approach to the estimation and interpretation of resistance costs in plants. Heredity 107: 386–394.
32. Hunt R (1982) Plant Growth Curves: The functional approach to plant growth analysis. London: Edward Arnond. 248 p.
33. Blackshaw RE, Brandt RN (2008) Nitrogen fertilizer rate effects on weed competitiveness is species dependent. Weed Science 56: 743–747.
34. Maxwell BD, Roush ML, Radosevich SR (1990) Predicting the evolution and dynamics of herbicide resistance in weed populations. Weed Technology 4: 2–13.
35. Délye C, Menchari Y, Michel S, Cadet É, Corre VL (2013) A new insight into arable weed adaptive evolution: mutations endowing herbicide resistance also affect germination dynamics and seedling emergence. Annals of Botany 111: 481–691.
36. Cousens RD, Gill GS, Speijers EJ (1997) Comment: Number of sample populations required to determine the effects of herbicide resistance on plant growth and fitness. Weed Research 37: 1–4.

Chronic Exposure to the Herbicide, Atrazine, Causes Mitochondrial Dysfunction and Insulin Resistance

Soo Lim[1], Sun Young Ahn[4], In Chan Song[2], Myung Hee Chung[3], Hak Chul Jang[1], Kyong Soo Park[1], Ki-Up Lee[5], Youngmi Kim Pak[4]*, Hong Kyu Lee[1]*

1 Department of Internal Medicine, Seoul National University College of Medicine, Seoul, Korea, **2** Department of Radiology, Seoul National University College of Medicine, Seoul, Korea, **3** Department of Pharmacology, Seoul National University College of Medicine, Seoul, Korea, **4** Age-Related and Brain Diseases Research Center, Department of Nanopharmaceutical and Life Sciences, Department of Physiology, Kyung Hee University College of Medicine, Seoul, Korea, **5** Department of Internal Medicine, University of Ulsan College of Medicine, Seoul, Korea

Abstract

There is an apparent overlap between areas in the USA where the herbicide, atrazine (ATZ), is heavily used and obesity-prevalence maps of people with a BMI over 30. Given that herbicides act on photosystem II of the thylakoid membrane of chloroplasts, which have a functional structure similar to mitochondria, we investigated whether chronic exposure to low concentrations of ATZ might cause obesity or insulin resistance by damaging mitochondrial function. Sprague-Dawley rats ($n=48$) were treated for 5 months with low concentrations (30 or 300 $\mu g\ kg^{-1}\ day^{-1}$) of ATZ provided in drinking water. One group of animals was fed a regular diet for the entire period, and another group of animals was fed a high-fat diet (40% fat) for 2 months after 3 months of regular diet. Various parameters of insulin resistance were measured. Morphology and functional activities of mitochondria were evaluated in tissues of ATZ-exposed animals and in isolated mitochondria. Chronic administration of ATZ decreased basal metabolic rate, and increased body weight, intra-abdominal fat and insulin resistance without changing food intake or physical activity level. A high-fat diet further exacerbated insulin resistance and obesity. Mitochondria in skeletal muscle and liver of ATZ-treated rats were swollen with disrupted cristae. ATZ blocked the activities of oxidative phosphorylation complexes I and III, resulting in decreased oxygen consumption. It also suppressed the insulin-mediated phosphorylation of Akt. These results suggest that long-term exposure to the herbicide ATZ might contribute to the development of insulin resistance and obesity, particularly where a high-fat diet is prevalent.

Editor: German Malaga, Universidad Peruana Cayetano Heredia, Peru

Funding: This work is supported by the IT R&D program of MIC/IITA [2006-S075-01 to HK Lee, Development of an early diagnostic system of metabolic syndrome based on nanosensor integrated network computing], a Korea Science and Engineering Foundation (KOSEF) grant (M10642140004-06N4214-00410 to YK Pak) and a grant from the 21C Frontier Functional Proteomics Project (FPR08A1-070 to YK Pak) from the Ministry of Education, Science & Technology, Korea. The funders had no role in study design, data collection and analysis, decision to publish, or preparation of the manuscript.

Competing Interests: The authors have declared that no competing interests exist.

* E-mail: ykpak@khu.ac.kr (YKP); hkleemd@snu.ac.kr (HKL)

Introduction

A close association between mitochondrial dysfunction and insulin resistance is well established [1–3]. In *in vitro* studies, we found that artificial induction of mitochondrial dysfunction induced insulin resistance [4,5]. However, there are no *in vivo* studies showing that exposure to an environmental mitochondrial toxin causes insulin resistance.

Persistent organic pollutants (POPs) that contaminate ground and water may accumulate in the tissues of animals and be passed up the food chain, leading to human exposure. Some POPs have recently been associated with the prevalence of diabetes in a serum concentration-dependent manner [6]. The triazine herbicide, atrazine (ATZ, 2-chloro-4-ethylamine-6-isopropylamino-S-triazine), has been extensively used in the USA since the early 1960s, a time frame that corresponds to the beginning of the present obesity epidemic [7,8]. Because it is moderately persistent under normal soil condition and has low to moderate water-solubility, ATZ is routinely found as a contaminant in many surface and ground waters [9,10]. Maps of ATZ usage show that the Corn Belt region of the Midwest USA has the heaviest application (http://water.usgs.gov/GIS/browse/herbicide1.gif) (supplementary Figure S1A). Interestingly, the Behavior Risk Factor Surveillance Survey (BRFSS) from 1985 to 2005 by the Center for Disease Control and Prevention revealed a high concentrations of individuals with a body mass index (BMI) over 30 kg/m^2 in the Corn Belt and surroundings connected via water sources [11] (http://www.cdc.gov/nccdphp/dnpa/obesity/trend/maps/) (supplementary Figure S1B). ATZ-usage and obesity maps show striking overlaps, suggesting that heavy usage of ATZ may be associated with the risk of obesity.

ATZ binds irreversibly to the plastoquinone binding sites of photosystem complex II on thylakoid membranes in chloroplasts, thereby inhibiting electron transport [12]. As mitochondrial electron transfer chain (ETC) complexes I and III also have similar Q binding sites, we hypothesized that ATZ might bind to these mitochondrial sites, resulting in the suppression of mitochondrial oxidative phosphorylation. Previous studies have shown that exposure to ATZ reduces metabolic activity in the gills of fish [13] and induces cellular DNA damage [14–18], tumorigenesis [19–22], and hermaphroditism of exposed male frogs [23]. In the present study, we found that chronic exposure to low concentrations of ATZ induced abdominal obesity and insulin resistance in rats by impairing mitochondrial function.

Materials and Methods

Animals

Male eight-week-old Sprague-Dawley rats (n = 48) were treated for 5 months with vehicle or ATZ (30 or 300 μg kg^{-1} day^{-1}) supplied in drinking water. One group of animals was fed a regular diet for the entire period, and another group of animals was fed a high-fat diet (40% fat) for 2 months after 3 months of a regular diet. Initial body weights were the same in both control (187.1±9.4 g) and ATZ (187.5±14.0 g) groups. All rats were fed regular chow (Han Sam R&D, Seoul, Korea) *ad libitum* for three months. Then, half of each group was fed a high-fat diet (high-fat diet group) and the other half was fed regular chow (regular-diet group) for another two months. Regular chow consisted of 16.0% fat, 63.0% carbohydrate and 20.0% protein (by calories), and 7.0% corn oil, 10.0% sucrose, 13.2% dextrose, 40.0% cornstarch, 5.0% cellulose and 20.0% casein (by weight). The high-fat diet consisted of 64.0% fat, 20.0% carbohydrate and 14.0% protein (by calories), and 33.0% shortening, 7.0% corn oil, 10.0% sucrose, 13.2% dextrose, 5.0% cornstarch, 5.0% cellulose and 20.0% casein (by weight). The remaining percentages of the two diets consisted of vitamins and minerals. All rats were maintained in plastic cages in an air-conditioned room at 22±2°C and 55±10% humidity. All procedures were approved by the Institutional Animal Care and Use Committee of the Seoul National University Hospital. The ATZ dose was calculated by multiplying ATZ concentrations by volume of water consumed. Diet and water consumption amounts were measured twice a month.

Measurement of activity levels

Movement was evaluated using a spontaneous motor activity analyzer (IW-800CT, modular test chamber, Seoul, Korea). Horizontal locomotion and rearing activity were evaluated for 4 hours per month.

Measurements of obesity

1) Body weight. The rats were weighed twice a month.

2) Visceral fat measurement by high resolution computed tomography (CT). Visceral fat areas were quantified using non-contrast CT scans (conditions: 120 kVp, 150 mA, 3 mm slice thickness, 3 mm reconstruction interval) using a Somatom Sensation 16 (Siemens, Munich, Germany). With the rats in a supine position, a 3 mm CT slice scan was acquired at the upper margin of the L3 vertebra to measure the amount of abdominal and visceral fat at a single level, and over L1 to L5 for the whole abdomen. Adipose tissue attenuation was determined by measuring the mean value of all pixels within the range of −250 to −50 Hounsfield units. Visceral fat amounts were measured by one radiologist using the computer software, Rapidia (INFINITT, Seoul, Korea).

3) Intramuscular and intrahepatic lipid content by proton magnetic resonance spectroscopy (^{1}H-MRS). Intramuscular lipid (IML) and intrahepatic lipid (IHL) content was measured *in vivo* by ^{1}H-MRI, using a 3-Tesla clinical unit (Signa Excite, GE, Milwaukee, WI, USA) with an 8-channel head coil. Voxels of 10 mm^3 were located in the anterior thigh muscles and liver, avoiding vascular structures and gross adipose tissue deposits. A probe-p sequence (TR/TE = 2,000 ms/35 ms) was used for MRS. Magnetic resonance imaging with the T2-weighted fast spin echo sequences (3,000–3,500/100–120, 18-cm field of view, 256×160 matrix, 3 mm slice thickness) in the axial or coronal planes preceded 1H-MR spectra in order to define the volume of interest. All spectra were processed using Mrdx (CAD Impact, Inc., Seoul, Korea), based on Interactive Data Language (Research Systems, Inc., Boulder, CO, USA). The water signal was suppressed by a frequency-selective saturation pulse at the water resonance value. A sweep width of 5,000 Hz was used with a data size of 2,048 points. Only the second half of the echo was acquired. Following the zero-filling of 8,192 points in all the free induction-decay data, an exponential line-broadening (center 0 ms; half time 150 m) was done before Fourier transformation. Zero-order phase correction was applied to all spectra. The integral of the IML signal (1.3 ppm) was related to that of total creatine (tCr; 3.05 ppm). The IML/tCr ratio corresponded to the total muscle IML value. Fat content was expressed as the ratio of the fat-to-water signal as a percentage. The reliability of the method was assessed by performing repeated measurements on the same individual on different study days and found to be <15%. IML and IHL content was compared between ATZ-treated and normal groups after adjusting for weight.

Analysis of blood samples for insulin resistance and lipid profiles

Plasma glucose levels were measured using a glucose-oxidase method (YSI 2300-STAT; Yellow Springs Instrument Co., Inc.) immediately after blood was drawn. Serum insulin was measured using insulin-specific radioimmunoassay kits for rats (Linco Research). The homeostasis model assessment-insulin resistance (HOMA-IR) index and beta cell function (HOMA-Beta) were obtained to evaluate insulin resistance and beta cell function. Total cholesterol and triglyceride concentrations were determined by enzymatic procedures (Hitachi 747 chemistry analyzer, Hitachi, Tokyo, Japan).

Intravenous glucose tolerance test (IVGTT)

IVGTTs were performed as described [24]. All experiments were performed during the light period between 1:00 and 2:00 pm. Animals were food-deprived for 5 hours before starting the experiment. To facilitate stress-free blood sampling, two infusion catheters (PE-10, Intramedic, Clay Adams, Parsippany, NJ, USA) were placed in the tail veins of rats on the evening before the experiment. A bolus dose of 0.5 g glucose/kg body weight was injected into the right tail vein immediately after blood sampling from the left tail vein for measurement of serum concentrations of glucose and insulin (t_0). Blood samples were collected again from the left tail vein at 2, 4, 6, 10, 20, 30 and 60 minutes for measurement of serum glucose and insulin concentrations. Glucose levels were measured using a YSI 2300 (Yellow Springs Instrument Co., Inc., Yellow Springs, OH, USA). Blood samples were stored at −70°C prior to insulin measurements. Plasma insulin levels were measured using commercial radioimmunoassay kits (Linco Research, St. Charles, MO, USA). Areas under the curve for glucose ($AUC_{glucose}$) and insulin ($AUC_{insulin}$) were calculated using the trapezoid rule for insulin data collected from 0 to 60 minutes.

Hyperinsulinemic-euglycemic clamp study

Hyperinsulinemic-euglycemic clamping was also performed at the end of the study (5 months) to determine whether ATZ exposure impaired the ability of rats to maintain glucose levels when challenged by a high insulin load, as described [25]. Two tail-vein infusion catheters (PE-10, Intramedic, Clay Adams) were placed in the rats on the evening before the experiment, and one tail-artery blood sampling catheter was emplaced 6 hours before the start of insulin infusion. Whole-body glucose kinetics were estimated in awake, unstressed rats 6 hours after food removal. Patency of the arterial catheter was maintained by a slow infusion

(15 μL/minute) of saline. Blood samples (200 μL) were drawn from the tail artery prior to the clamp study to measure baseline fasting glucose and insulin levels. Human insulin (Novolin-R, Novo Nordisk, Bagsvaerd, Denmark) was infused into one of the tail vein microcatheters at a rate of 72 pmol kg^{-1} $minute^{-1}$. Aliquots of venous blood (50 μL) were collected from one tail vein catheter to measure blood glucose at 10-minute intervals, and 25% glucose was infused into the other tail vein catheter at variable rates to maintain plasma glucose at basal concentrations. Steady-state plasma glucose concentrations were reached after 50 to 60 minutes. Blood samples were drawn at 90 and 120 minutes for the measurement of insulin. The steady-state glucose infusion rate (GIR) is defined as the amount of glucose required to maintain euglycemia between 90 and 120 minutes of hyperinsulinemia. The insulin sensitivity index (ISI) was calculated by dividing the GIR by the mean insulin concentration during the 90–120-minute clamping window and expressed as mg glucose kg^{-1} $minute^{-1}$ of for each ng/mL of insulin.

Energy expenditure

Energy expenditure in rats was measured following a 5-hour fasting period using an Oxymax apparatus (Columbus Instruments, Columbus, OH, USA) two days before each hyperinsulinemic euglycemic clamp experiment [26]. After baseline O_2, CO_2 and flow were measured, each rat from each group was situated in metabolic monitor cage for 10 minutes and then O_2 and CO_2 were measured again after the values had stabilized. Energy expenditure was calculated according to the formula provided by the manufacturer.

Electron microscopic images of mitochondria

At the end of the clamp, rats were euthanized with an intravenous injection of pentobarbital, and liver and muscle tissues were collected. The tissue samples were stored frozen at −70°C for later analysis. Soleus muscle and liver tissues were dissected and immersed in fixatives. Mitochondrial morphology was examined by electron microscopy (20,000× magnification).

Western blot analysis

Cells were washed twice with Dulbecco's phosphate buffered saline (DPBS) and harvested with lysis buffer (50 mM Tris HCl, pH 7.5, 0.1 M NaCl, 1 mM EDTA, 1% Triton X-100, 10 μg/mL each aprotinin and leupeptin, 1 mM PMSF). In some cases, frozen tissue was homogenized in lysis buffer. A portion of cell or tissue lysate (20 μg) was separated by SDS-PAGE on 10–12% gels, transferred to a nitrocellulose membrane (Schleicher & Schnell, Inc., Keene, NH, USA) and analyzed by western blotting using specific antibodies against oxidative phosphorylation (OXPHOS) complex I (39 kDa α subcomplex 9, NDUFA9), complex II (70 kDa flavoprotein, SDHA), complex III (core II, UQCRC2), complex IV-subunit I (COXI, MTCO1), complex IV-subunit IV (COXIV, COX4) or complex V (F1 complex α, ATP5A1). All OXPHOS complex antibodies were purchased from Molecular Probes (Invitrogen, Eugene, OR, USA). The antibodies against Akt and phospho-Akt (Thr308 or Ser473) were purchased from Cell Signaling Technology, Inc (MA, USA). Hsp60 and/or β-actin antibodies were utilized to control for equal loading of protein. The immunoblots were developed using an enhanced chemiluminescence system (ECL, Amersham Pharmacia Biotech., Arlington Heights, IL, USA).

Endogenous cellular oxygen consumption

Endogenous cellular respiration was measured as described [27], with modifications. Briefly, isolated L6 rat skeletal muscle cells (5×10^6) in DMEM containing 0.5% FBS were cultured in the presence of ATZ (100 μg/mL) or DMSO (vehicle control) for 48 hours, washed with DPBS (pH 7.4) and collected by trypsinization. After resuspending in 1 mL complete phenol red-free DMEM, the cells were transferred to the chamber of an Oxygraph-2K apparatus (Oroboros, Innsbruck, Austria). Coupled and uncoupled OCRs were measured before and after adding 2.5 μM carbonyl cyanide p-(trifluoromethoxy)phenylhydrazone). KCN-insensitive non-mitochondrial respiration was measured by adding 2.5 mM KCN.

OXPHOS complex activity was measured by placing cells (5×10^6) in 1 mL of mitochondria respiration buffer (MiRO5, 20 mM HEPES, 110 mM sucrose, 3 mM $MgCl_2$, 0.5 mM EGTA, 10 mM KH_2PO_4, 60 mM K-lactobionate, 20 mM taurine, 0.1% BSA, pH 7.1 at 30°C) into the continuously stirred chamber of an Oxygraph-2K apparatus (Oroboros, Austria). After recording basal respiration rate, the cells were permeabilized by adding digitonin (25 μg/mL) and adenylate was blocked by adding 100 μM p1,p5-di(adenosine-5′)-pentaphosphate. The activity of each OXPHOS complex was determined by sequential addition of the following inhibitors and substrates: 2 mM ADP, 8 mM malate and 20 mM glutamate for complex I; 1 μM rotenone, 10 mM succinate and 2.5 mM glycerol-3-phosphate for complexes II and III; 25 μM antimycin A, 80 μM ascorbate and 0.42 mM N,N,N,N-tetramethyl-p-phenylenediamine (TMPD) for complex IV; and 2.5 mM KCN for KCN-insensitive respiration. Oxygen consumption rate (OCR) was expressed as picomoles of oxygen consumed per second per milligram of mitochondrial protein.

Assays for mitochondrial respiratory chain activities

The OCR of complex I, complexes II and III and complex IV of soleus muscle or liver mitochondria were measured using an Oxygraph-2K. Mitochondria were isolated by differential centrifugation, as described previously [28]. Approximately 400 μg mitochondrial protein was suspended in 1 mL MiRO5 buffer. In some cases, isolated mitochondria were incubated with different concentrations (0–300 μg/mL) of ATZ for 30 minutes before OCR measurement. After recording a basal respiration rate, the OCR of each OXPHOS complex was measured by sequential addition of the substrates and inhibitors, as described above.

Because it is not possible to determine the activities of complex II and complex III separately by OCR measurements, the enzyme activities of succinate dehydrogenase (SDH, complex II) and cytochrome bc_1 complex (complex III) of the tissue or cell lysates were assayed spectrophotometrically as described previously [29], with slight modifications. Briefly, tissue or L6 muscle cells were lysed in 100 mM Tris HCl (pH 7.4) containing 250 mM sucrose and 2 mM EDTA. SDH activity was determined by reduction of 2,6-dichlorophenolindophenol (DCPIP) at 600 minus 520 nm (extinction coefficient 19.1 mM/cm) in a mixture of 50 mM potassium phosphate, pH 7.4, 20 mM succinate, 2 μg/mL antimycin A, 2 μg/mL rotenone, 2 mM KCN and 50 μM DCPIP. Complex III activity was measured by reduction of cytochrome *c* at 550 minus 540 nm (extinction coefficient 19.0 mM/cm) in 50 mM Tris-HCl, pH 7.4, 250 mM sucrose, 1 mM EDTA, 0.2 mM KCN, 1 mg/mL antimycin A, 100 μM decylubiquinol and 200 μM oxidized cytochrome *c*. Decylubiquinol was synthesized in the laboratory from decylubiquinone by reduction with potassium borohydride.

Complex I (NADH dehydrogenase) activity was determined in 50 mM potassium phosphate, pH 7.4, 3.5 mg/mL BSA, 0.2 mM NADH, 1 μM antimycin A, 2 mM KCN, 70 μM decylubiquinone and 180 μM DCPIP by measuring the reduction of DCPIP at 520–600 nm (extinction coefficient 19.1 mM/cm) for 1 minute [30]. Complex IV (cytochrome *c* oxidase) activity was determined in 10 mM potassium phosphate (pH 7.4) with 0.1% reduced cytochrome *c* (reduced by sodium hydrosulfate) by measuring oxidation of reduced cytochrome *c* at 550 nm (extinction coefficient 19.0 mM/cm) [29]. Data are expressed as percentages of control activity.

Statistical analysis

Data are expressed as the mean±SE. Significant differences between groups were evaluated using Student's *t* tests and ANOVA with a post hoc test. Correlations between variables were analyzed using Pearson's correlation coefficient. Differences were considered statistically significant when *P* values were <0.05.

Results

ATZ treatments led to weight gain

After being fed a normal diet for 3 months, control and the ATZ-treated rats exhibited no significant differences in mean body weight. From month 4 onward, mean body weight in the regular-diet groups was higher for ATZ-treated rats than for control rats. At the end of the study, the body weights of the regular-diet rats were 582.3±9.3 g for the 30-μg kg^{-1} day^{-1} ATZ group (ATZ30), 585.0±22.7 g for the 300-μg kg^{-1} day^{-1} ATZ group (ATZ300) and 554.6±13.4 g for controls; overall, the mean body weight of ATZ-treated rats was 5.5% higher than that of the controls ($p<0.05$; **Fig. 1A and 1C**). In the high-fat diet groups, the body weights of ATZ30 and ATZ300 groups were 621.1±18.8 g and 626.9±16.9 g, respectively, or 9.8% higher overall than control rats (570.8±13.6 g) ($p<0.01$, **Fig. 1B and 1C**). No treatment-related differences in food/water intake or horizontal/spontaneous locomotor activities were observed at any point. Thus, chronic exposure to low concentrations of ATZ resulted in weight gain, particularly when combined with a high-fat diet.

ATZ increased visceral fat and intracellular fat content

CT scans were utilized to determine whether the weight gain induced by ATZ was related to the amount of visceral fat. In high-fat diet groups, ATZ-treated rats had a greater amount of visceral fat at the L3 level and between L1 and L5 than controls ($p<0.05$; n = 8/group; **Fig. 1D and 1E**). Fat deposition in the muscle and liver is positively correlated with measures of obesity and negatively with insulin sensitivity [24]. Non-invasive ^{1}H-MRS revealed that, even in regular diet groups, ATZ treatment increased intramuscular and intrahepatic fat accumulation, with borderline significance after adjusting for body weight (**Fig. 1F and 1G**). Thus, ATZ treatment for 5 months enhanced visceral fat accumulation in the groups fed a high-fat diet and induced lipid accumulation in both muscle and liver in rats fed a regular diet.

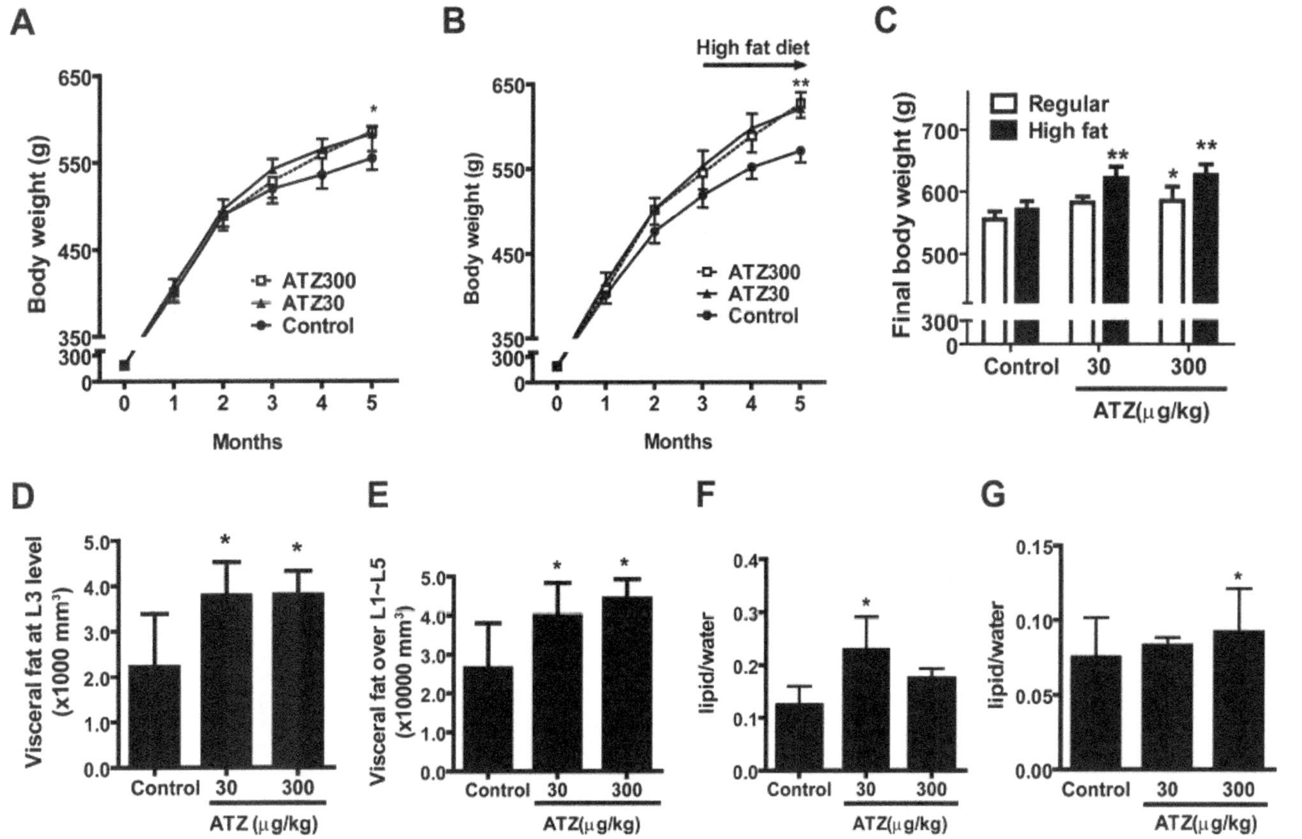

Figure 1. Induction of obesity in rats by ATZ treatment. (A and B) Changes in body weight of ATZ-treated rats (ATZ30, 30 μg kg^{-1} day^{-1}; ATZ300, 300 μg kg^{-1} day^{-1}) versus control rats over time. **(A)** Regular-diet group. Rats were fed a regular diet for 5 months during treatment with ATZ, provided in drinking water. **(B)** High-fat-diet group. Rats were fed a regular diet for 3 months and then fed a high-fat diet for another 2 months. **(C)** End-of-study comparison of body weights between two different diet-treated rats. (D and E) Increase in visceral fat by ATZ. The amount of visceral fat in the high-fat diet group was measured by horizontal CT scan. Abdominal fat area at the L3 level **(D)** and over L1 to L5 **(E)** were calculated from the scanned image using a Hounsfield unit. (E and F) Intracellular fat deposition by ATZ. The amount of intrahepatic **(F)** and intramuscular **(G)** fat in rats on a regular diet was measured by non-invasive ^{1}H-MRS and adjusted for body weight. (*$p<0.05$, **$p<0.01$ vs. control).

ATZ treatment worsened insulin levels and HOMA-IR values

For rats fed a regular diet, ATZ treatment tended to increase triglyceride and total cholesterol levels, but the differences did not reach statistical significance (data not shown). However, ATZ accelerated the increase in triglycerides induced by the high-fat diet ($p<0.05$ vs. control), and also worsened insulin levels and HOMA-IR index values. Specifically, in high-fat diet groups, insulin levels were increased to 2.39±0.91 and 2.21±0.86 ng/mL in the ATZ300 and ATZ30 groups, respectively compared to 1.70±0.57 ng/mL in controls ($p<0.01$). The corresponding HOMA-IR values for ATZ300, ATZ30, and control groups were 0.60±0.24, 0.51±0.20 and 0.37±0.11, respectively ($p<0.05$).

ATZ induced insulin resistance

To determine if ATZ treatment induced glucose intolerance and insulin resistance, we performed IVGTTs and a hyperinsulinemic-euglycemic clamp study. Fasting plasma glucose levels of ATZ-treated rats on the regular diet were higher than those in controls (86.5±9.8, 91.1±8.5, and 78.5±8.3 mg/dL in ATZ30, ATZ300 and controls, respectively; $p<0.05$), although they were within the normal range. The same pattern was found in the high-fat diet group: fasting glucose levels were 97.8±6.5 and 102±8.3 in ATZ30 and ATZ300 groups, respectively, compared to 90.0±6.7 mg/dL in controls ($p<0.05$). Interestingly, ATZ exposure impaired glucose tolerance as monitored by IVGTT. Plasma glucose and insulin concentrations in both ATZ-treated groups were significantly higher than those in controls at early points (<10 min) after intravenous loads of glucose ($p<0.05$ vs. control) (**Fig. 2A and 2B**).

The hyperinsulinemic-euglycemic clamp study demonstrates an animal's ability to maintain glucose levels when challenged by a high insulin load. Steady-state plasma insulin concentrations were measured as plasma glucose reached steady state during clamping. For the ATZ-treated rats on a regular diet, insulin levels increased

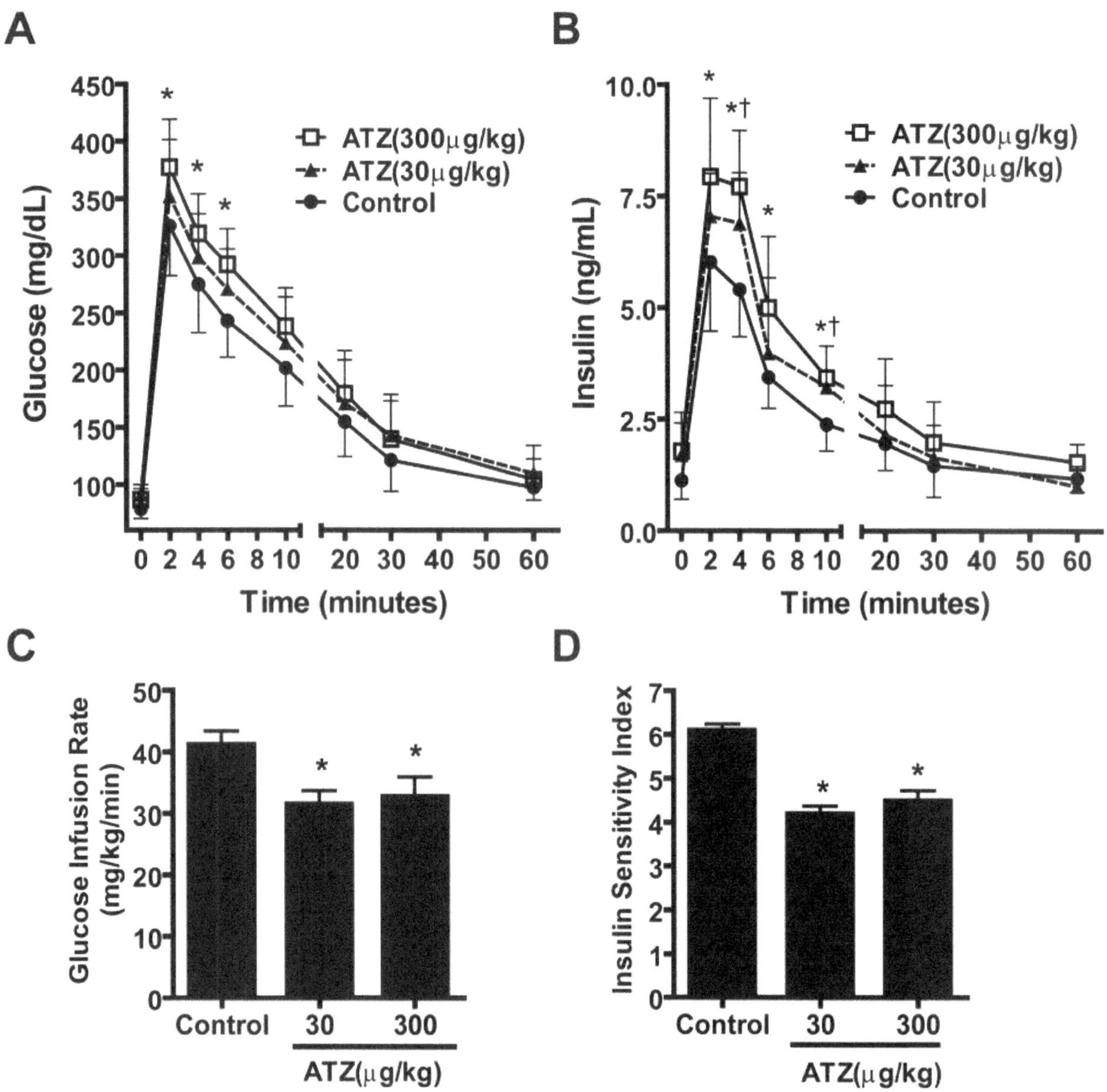

Figure 2. Impairment of IVGTT and insulin sensitivity by ATZ exposure in regular-diet rats. (A) Changes in plasma glucose during IVGTT. **(B)** Changes in plasma insulin during IVGTT. Plasma glucose or insulin concentrations were determined at the indicated times after i.v. glucose load. **(C)** Steady-state glucose infusion rate (GIR) and **(D)** insulin sensitivity index (ISI) during clamp study. (*$p<0.05$ vs. control; n = 8 per group).

to 7.20±0.70 ng/mL in ATZ30 and 7.50±0.50 ng/mL in ATZ300 compared with 4.90±0.60 ng/mL glucose infusion rate for controls ($p<0.05$ for both). AZT significantly decreased the steady-state glucose infusion rate (GIR) during the clamp to 31.5±2.22 and 32.75±3.18 mg kg^{-1} $minute^{-1}$ for ATZ30 and ATZ300, respectively compared to 41.13±2.28 mg kg^{-1} $minute^{-1}$ for controls ($p<0.05$ for both) (**Fig. 2C**). Thus, the ATZ-exposed rats showed an impaired response to insulin loading, even under a regular diet. The insulin sensitivity index (ISI) was also significantly reduced to 4.2±0.17 for ATZ30 and 4.5±0.23 for ATZ300 compared with 6.1±0.15 for controls ($p<0.05$ for both; **Fig. 2D**). Collectively, these data support the conclusion that rats chronically exposed to ATZ developed insulin resistance without being fed a high-fat diet.

ATZ changed the ultrastructure of mitochondria

The soleus muscles of ATZ-treated rats were not discernibly different from those of control rats by light microscopy. However, electron microscopy showed that soleus muscle mitochondria in these animals on a regular diet were swollen and their cristae were partially destroyed (**Fig. 3A**). There was also prominent accumulation of lipid droplets in the livers of ATZ-treated rats, and electron microscopy revealed that some liver mitochondria from the ATZ treated group showed partially disrupted cristae. Despite the fact that mitochondrial morphology was altered in muscle and liver by ATZ administration, protein expression levels of mitochondrial OXPHOS complex subunits in liver and muscle tissues were not changed significantly (**Fig. 3B**). Therefore, the AZT-induced disruption of mitochondrial morphology might not be associated with changes in mitochondrial complex protein expression.

ATZ exposure decreased energy metabolism

Insulin resistance and obesity are affected by caloric intake, physical activity and energy expenditure [31,32]. Since no treatment-related changes in food or water intake or physical activity were observed at any point during the study, the development of insulin resistance by ATZ might be related to energy metabolism. To determine if ATZ affected energy metabolism, we measured energy expenditure and OCR using indirect calorimetry and oxygen sensors, respectively. The energy expenditure of ATZ-treated rats on regular diet was reduced in a dose-dependent manner (**Fig. 4A**). The OCR of each OXPHOS complex was measured using mitochondria isolated from the soleus muscle of ATZ-treated rats on regular diet. The combined OCR measured for complexes II plus III, both of which use ubiquinone (Q) as an electron transfer intermediate, was significantly reduced by ATZ administration. The activities of other complexes were not altered (**Fig. 4B**), although the OCR of complex I showed a tendency toward a decrease that did not reach statistical significance. There were some analytical limitations in measuring OCR where the changes were modest, especially when using animal tissues.

To distinguish which of these complexes (II or III) were affected by ATZ, we measured the enzymatic activities of SDH (complex II) and cytochrome bc_1 (complex III) in gastrocnemius muscle tissues by spectrophotometry. As expected, SDH activity was not altered by ATZ treatment while the activity of the cytochrome bc_1 complex was decreased by 10% in the muscles of ATZ-treated animals ($p<0.05$, **Fig. 4C**). These results suggest that ATZ may bind to the Q site of the cytochrome bc_1 complex, a counterpart of the plastoquinone of photosynthesis in plants. Consistent with tissue lysate western blotting results, the amounts of OXPHOS complex proteins in mitochondria were not significantly altered. The western blot results imply that ATZ might interfere with the electron transfer from complex I or II to complex III without impacting mitochondrial OXPHOS protein expression levels (**Fig. 4D**). However, we cannot exclude the possibility that the expression of other proteins that were not examined by western blotting might be involved in changes in mitochondria respiration activity.

ATZ impairs mitochondrial OXHPOS through direct action

Changes in cellular respiration and differences in the respiratory effects of inhibitors are important indicators of mitochondrial functional defects that result from damaged mitochondrial proteins or DNA (mtDNA), or substantial alterations to mitochondrial signaling cascades. To test if ATZ impaired mitochondrial function directly or indirectly, we monitored both cellular and mitochondrial respiration levels in ATZ-treated, cultured L6 muscle cells and isolated mitochondria. Incubation of L6 muscle cells with ATZ decreased both endogenous coupled and FCCP-uncoupled oxygen consumption by 30% and 40%, respectively (**Fig. 5A**). ATZ had no effect on the levels of nuclear DNA (nDNA)- or mtDNA-encoded OXPHOS complex proteins (**Fig. 5B**). When oxygen consumption was measured using digitonin-permeabilized cells, respiration levels in complex II plus III were inhibited by ATZ (**Fig. 5C**), similar to the *in vivo* results (**see Fig. 4C**). Enzyme activities of complex III (cytochrome bc_1) were also decreased by ATZ treatment (**Fig. 5D**). Direct treatment of isolated mouse liver mitochondria with ATZ also reduced the OCR of complex I and complex II plus III by 49% and 37%, respectively (**Fig. 6A**); the magnitude of these ATZ effects were greater in isolated mitochondria than in cells. Complex IV activities were not significantly changed. Again, consistent with the results of the *in vivo* study, SDH enzymatic activity was not changed, but the activity of the cytochrome bc_1 complex was decreased significantly by ATZ treatment (**Fig. 6B**). These results strongly suggest that ATZ itself, not ATZ metabolites, intervenes directly at Q binding sites between complex I and III, or II and III.

ATZ treatment blocked the insulin-Akt signaling pathway

To investigate the mechanism by which ATZ-induced mitochondrial dysfunction caused insulin resistance, we analyzed insulin-stimulated Akt phosphorylation in ATZ-treated L6 muscle cells. Pretreatment with ATZ (100 µg/mL) for 24–48 hours abolished insulin-mediated Akt phosphorylation at both Thr308 and Ser473 residues (**Fig. 7**). Park *et al.* demonstrated that mitochondrial dysfunction induced by mtDNA depletion suppressed IRS-1 expression, resulting in diminished downstream signaling and glucose transport, and insulin resistance [5]. The suppression of insulin-mediated Akt phosphorylation is clearly linked to the development of insulin resistance *in vitro* [33] and *in vivo* [34]. Thus, the reduction in Akt phosphorylation that results from ATZ-mediated inhibition of mitochondrial OXPHOS complex III may help to explain why chronic exposure to ATZ induced weight gain and insulin resistance.

Discussion

In this study, we found that long-term treatment with low concentrations of the herbicide, ATZ, induced insulin resistance and weight gain in Sprague Dawley rats. We acknowledge that, unlike the current results, previous studies have reported that the body weights of ATZ-treated animals were generally decreased or unchanged [35–37]. However, the doses of ATZ used in these other studies were 10–100 fold higher (2.7 to 50 mg kg^{-1} day^{-1})

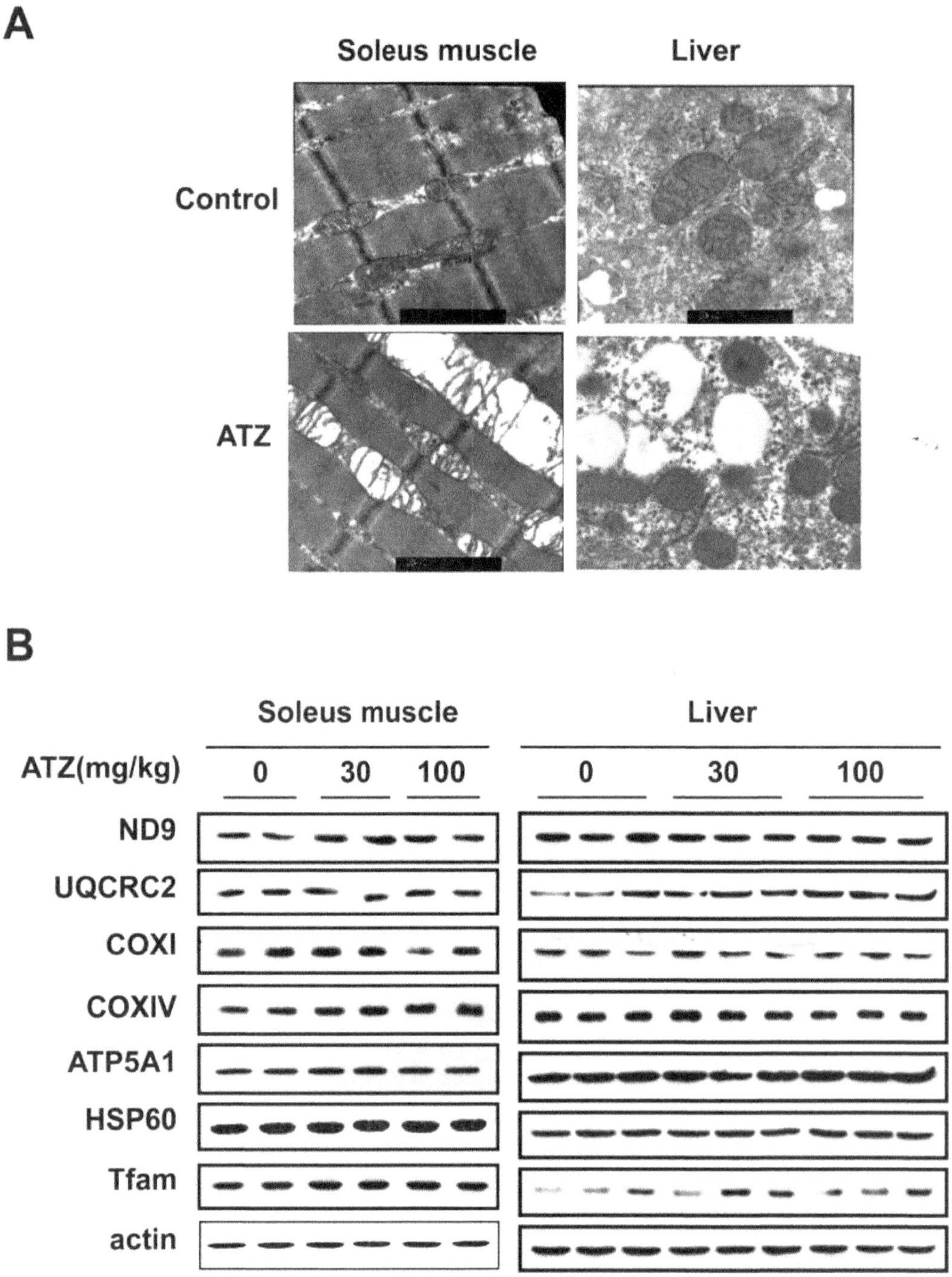

Figure 3. Mitochondrial morphology and protein expression in ATZ-treated rat muscle and liver. (**A**) Mitochondrial morphology by electron microscopy. Rats on a regular diet were treated with or without ATZ300 for 5 months. Rat soleus muscle and liver were isolated and observed by electron microscopy (magnification ×20,000). Enlarged, swollen or cristae-disrupted mitochondria are indicated by arrows. Lipid droplets were observed in ATZ-treated liver. (**B**) Expression of mitochondrial proteins in rat tissues. Total lysates of soleus muscle or liver of rats on a regular diet with or without ATZ treatment were analyzed by western blotting using the indicated antibodies.

than those used in our study. Thus, our interpretation is that acute exposure to high concentrations of ATZ is toxic, and thus prevents weight gain and possibly causes weight loss. In contrast, chronic low-dose ATZ exposure might lead to mild mitochondrial damage that mimics the characteristic of the insulin-resistance state, and hence, leads to weight gain.

Because ATZ treatment in our study induced obesity without changing food intake or physical activity, ATZ presumably lowered energy metabolism. Indeed, indirect calorimetry measurements revealed that the ATZ-induced weight gain was associated with decreased energy metabolism. Furthermore, *in vitro* experiments provided evidence that ATZ interferes with electron transfer through OXPHOS complex at Q sites in mitochondria, resulting in reduced oxygen consumption. We have found that treatment with rotenone, a complex I inhibitor, or dideoxydytidine, an mtDNA-replication inhibitor, directly inhibits insulin-mediated

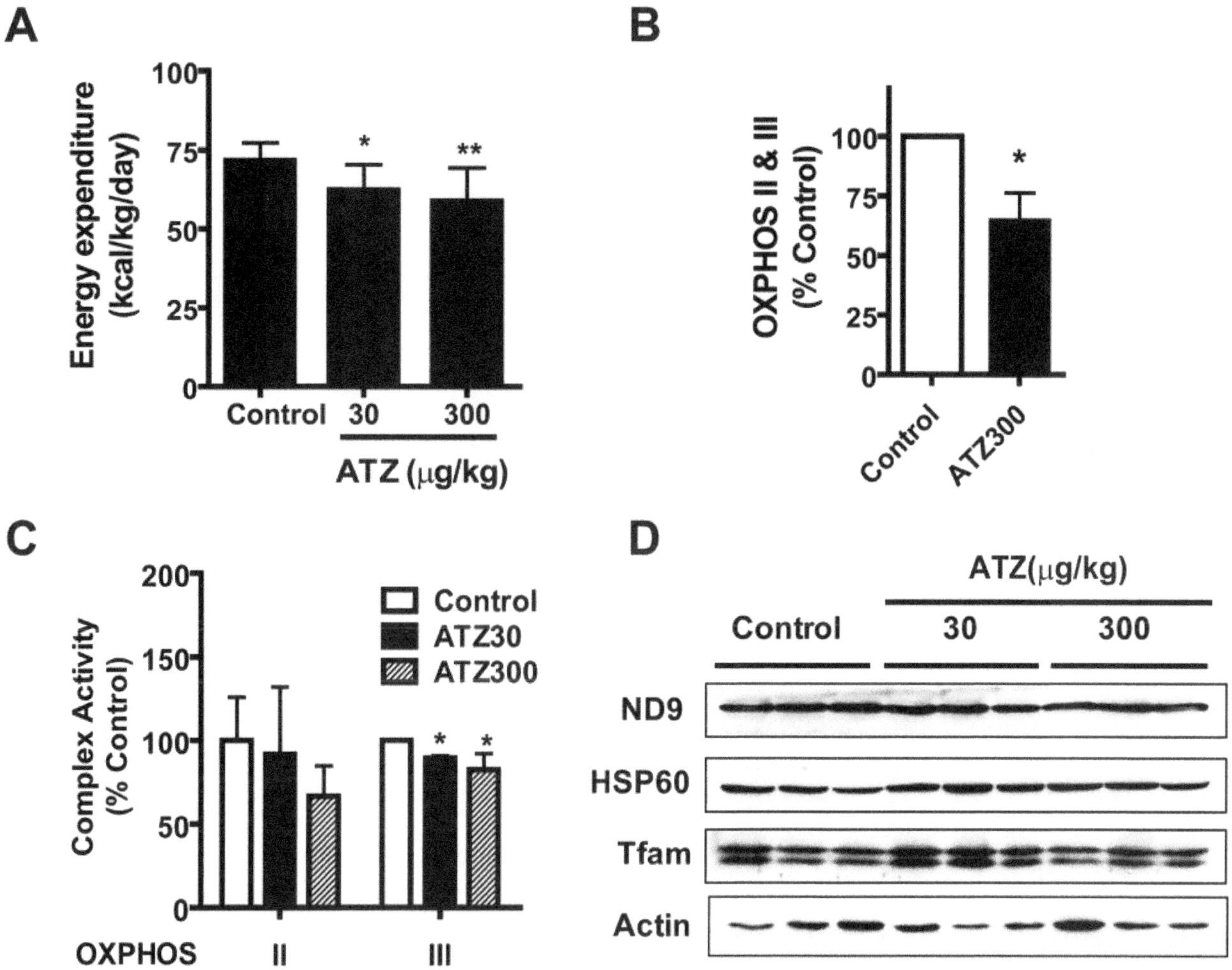

Figure 4. Mitochondrial OXPHOS activity of ATZ-treated rat skeletal muscle mitochondria. Rats on a regular diet were treated with or without ATZ for 5 months. **(A)** Dose-dependent decrease in energy expenditure in ATZ-treated rats, monitored using indirect calorimetry. **(B)** Decrease in OCR of complex II plus III in ATZ300-treated skeletal muscle mitochondria (n = 5). **(C)** Decrease in the activity of complex III enzyme in ATZ-treated liver lysates, determined by spectrophotometry (*p<0.05, **p<0.01; n = 5). **(D)** Western blot of muscle mitochondria proteins. Total skeletal muscle mitochondrial lysates from rats on a regular diet were subjected to western blot analysis using the indicated antibodies.

phosphorylation of Akt without affecting upstream signaling molecules (data not shown). In addition, overexpression of dominant-positive Akt (myr-Akt) completely reverses downstream Akt function. For example, Akt-mediated phosphorylation of Foxo1, which is blocked by mitochondrial dysfunction, is abrogated by myr-Akt overexpression, implying that Akt is the focal point of the mitochondrial dysfunction (manuscript in preparation). As part of this study, we tested whether ATZ also blocked Akt function, and found, as expected, that ATZ blocked insulin-mediated Akt phosphorylation in skeletal muscle cells.

Treatment with ATZ changed the ultrastructure of mitochondria in liver and muscle of rats, producing morphological alterations such as ring- and cup-shaped mitochondria without significantly altering the expression of mitochondrial OXPHOS complex proteins. This observation is in agreement with the findings of others that ATZ induced mitochondrial damage in fresh water mussels [13] and rainbow trout [38]. In our study, we also found that ATZ decreased the membrane potential of mitochondria and reduced intracellular ATP content in various cells (data not shown). Collectively, these data led us to conclude that ATZ treatment damaged both mitochondrial respiratory function and morphology.

Intramuscular lipid accumulation is known to be closely related to insulin resistance [39,40], and is considered as an early phenomenon in the development of obesity. Increased lipid content in muscle is also associated with decreased ATP synthesis and diminished mitochondrial function [1]. The observed increase in intracellular lipid content in ATZ-treated rats also supports the interpretation that ATZ-induced mitochondrial damage affects the insulin-signaling pathway, and consequently induces insulin resistance and fat accumulation in metabolically active tissue, such as muscle.

There is epidemiological evidence that human exposure to POPs such as TCDD may also disturb glucose metabolism and induce insulin resistance [6,41–47], and it has been reported that exposure to herbicides or pesticides, including ATZ, is associated with an increased risk of gestational diabetes [47]. However, there is scant information in the literature to indicate the level of human exposure to ATZ or similar herbicides. Obviously, there is no direct evidence of ATZ accumulation in diabetic or obese human subjects. However, a toxicological report has shown that acute occupational or dietary exposure of humans to ATZ was in the range of 0.2–90 μg kg^{-1} day^{-1}, and the annual average exposure (chronic) was between 0.046 and 0.286 μg kg^{-1} day^{-1} [48].

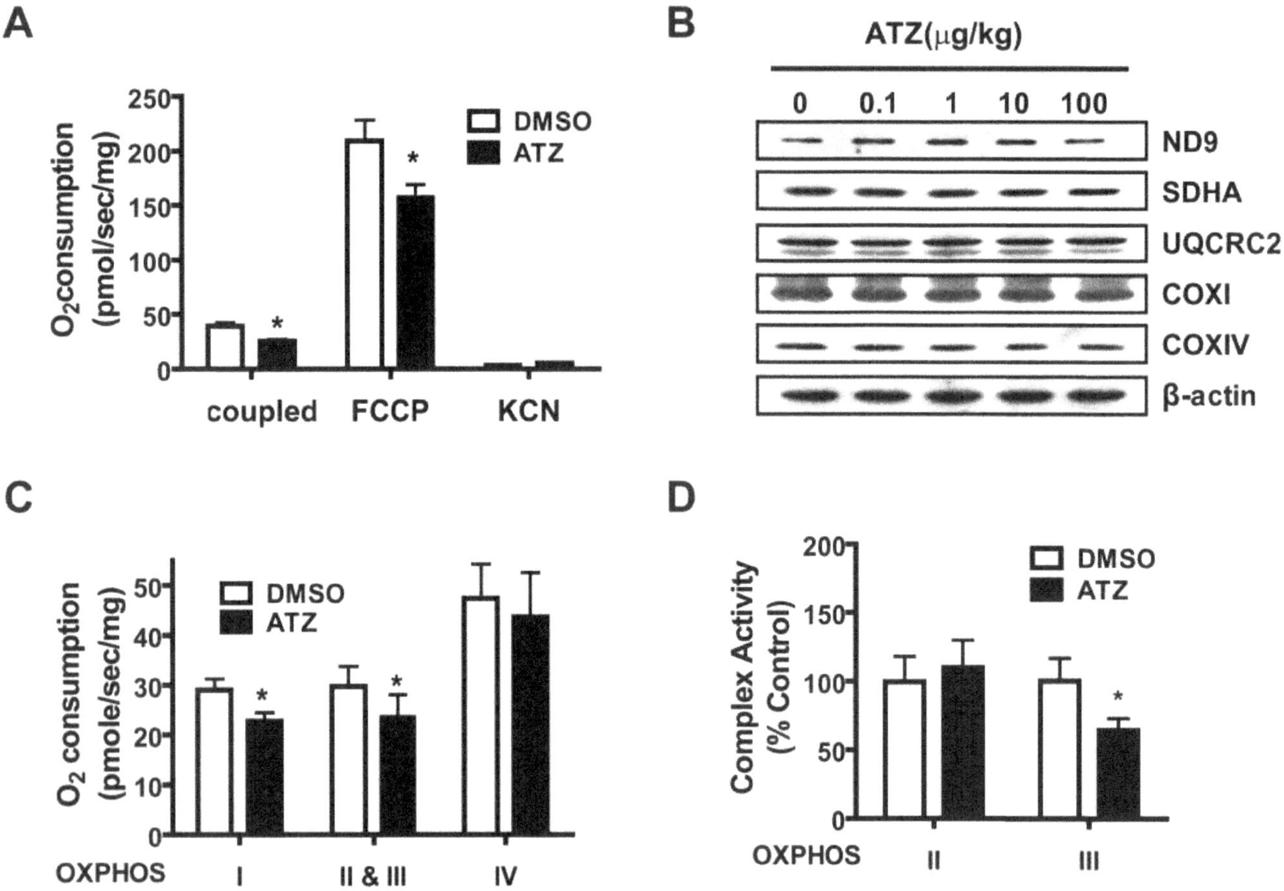

Figure 5. Decrease in endogenous oxygen consumption by ATZ in L6 cells. L6 rat skeletal muscle cells were treated for 48 hours with or without ATZ (100 μg/mL). **(A)** Endogenous cellular (coupled), FCCP-uncoupled (FCCP) and KCN-insensitive respiration (KCN) of trypsinized intact cells were measured in phenol red-free media using an Oxygraph-2K apparatus (*$P<0.05$; n = 3). **(B)** Western analysis of nDNA- and mtDNA-encoded OXPHOS complex subunit proteins. Complex I (ND9), complex II (SDHA), complex III (UQCRC2) and complex IV (COXI, mtDNA-encoded, COXIV, nDNA-encoded) were examined. β-actin was used as an equal loading control. **(C)** Oxygen consumption by each complex in digitonin-permeabilized cells was measured using an Oxygraph-2K apparatus ($P<0.05$; n = 4). **(D)** Enzyme activities of complex II or III were determined by spectrophotometry in ATZ-treated L6 muscle cells (*$P<0.05$; n = 3).

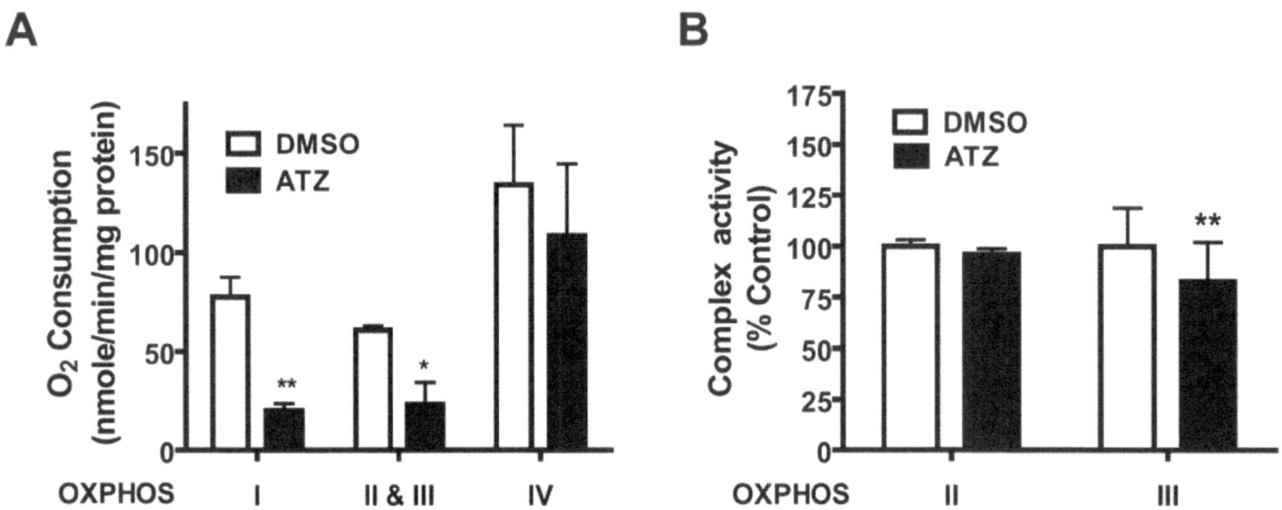

Figure 6. Inhibition of complex I and II plus III in isolated mitochondria by ATZ. (A) OCR. Mouse liver mitochondria (40 μg/100 μL assay) isolated by differential centrifugation were incubated with ATZ (100 μg/mL) for 30 minutes and the OCR of each complex was measured. **(B)** Enzyme activities of complex II (SDH) or complex III (cytochrome *bc*1 complex). Liver tissue lysates of ATZ-treated mitochondria were assayed as described (*$p<0.05$, **$p<0.01$; n = 4).

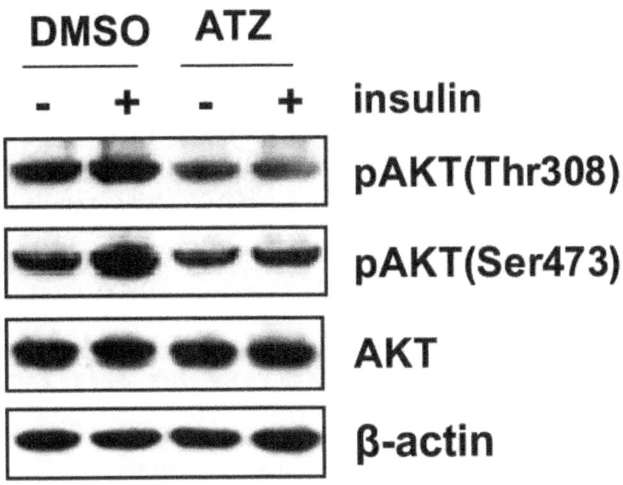

Figure 7. Inhibition of insulin signaling by ATZ. L6 cells were treated with ATZ (100 μg/mL) in DMEM containing 0.5% FBS for 24 hours and stimulated with insulin (100 ng/mL) for 30 minutes. The cells were harvested and analyzed by western blotting using anti-Akt or -pAkt antibodies. Akt Thr308 and Ser473 phosphorylation, both important in the insulin-signaling pathway, were blocked by ATZ treatment.

We believe that ATZ or its metabolites may be introduced to humans through air, water and and/or corn products as contaminants, and accumulate in tissues. One such pathway by which ATZ or its metabolites might be introduced into humans is through corn-derived foods (e.g., high fructose corn syrup or corn oil). Since corn syrup [49] and fast foods served in the USA [50,51] are suspected of causing an obesity epidemic, this seems a reasonable supposition. Recently, it was reported that of 160 food products purchased at a fast food restaurant throughout the USA, not a single item could be traced back to a non-corn source [52]. This work also identified corn as the overwhelmingly predominant animal feed for the beef and chicken served at fast food restaurants.

Considering that the process of corn wet milling requires a huge amount of fresh water for steeping (digesting), which in turn generates 1,300 to 1,600 L of light steepwater per ton of corn that is then evaporated or dried to make intermediate products, it is conceivable that ATZ and related herbicides are present in steepwater and may be concentrated during the process [53]. Because herbicides like ATZ and their metabolites are present in streams and groundwaters throughout USA and are at highest concentrations in agricultural areas where wet corn milling plants are located [54], there is a substantial possibility of ATZ contamination of corn products. However, in 2006 the U.S. Environmental Protection Agency issued a cumulative risk assessment of triazine herbicides, concluding that they posed "no harm that would result to the general U.S. population, infants, children, or other consumers" [55].

Based on evidence presented in this study, we conclude that environmental ATZ might be an important contributing factor to the obesity epidemic in the United States. It damages mitochondrial function, affects insulin signaling, and induces insulin resistance and obesity, especially when exposure is associated with a high-fat diet. Sanguine official assessments notwithstanding, further studies are definitely needed to clarify issues related to human exposure to ATZ and related herbicides in the US and elsewhere around the world.

Acknowledgments

We thank Jeong Hoon Kim, PhD for locomotor activity measurements, Min-Seon Kim, MD for energy expenditure measurements, Kyu Ri Son, MD for measuring fat amounts using CT and MRI, and Dong-Young Lee and Sunny Lim for their technical assistance.

Author Contributions

Conceived and designed the experiments: KSP YKP HKL. Performed the experiments: SL SYA ICS MHC HCJ KUL YKP. Analyzed the data: SL SYA ICS MHC HCJ KUL YKP HKL. Contributed reagents/materials/ analysis tools: KSP YKP HKL. Wrote the paper: SL YKP HKL.

References

1. Petersen KF, Befroy D, Dufour S, Dziura J, Ariyan C, et al. (2003) Mitochondrial dysfunction in the elderly: possible role in insulin resistance. Science 300: 1140–2.
2. Poulton J, Brown MS, Cooper A, Marchington DR, Phillips DI (1998) A common mitochondrial DNA variant is associated with insulin resistance in adult life. Diabetologia 41: 54–8.
3. Lowell BB, Shulman GI (2005) Mitochondrial dysfunction and type 2 diabetes. Science 307: 384–7.
4. Park KS, Nam KJ, Kim JW, Lee YB, Han CY, et al. (2001) Depletion of mitochondrial DNA alters glucose metabolism in SK-Hep1 cells. Am J Physiol Endocrinol Metab 280: E1007–E1014.
5. Park SY, Choi GH, Choi HI, Ryu J, Jung CY, et al. (2005) Depletion of mitochondrial DNA causes impaired glucose utilization and insulin resistance in L6 GLUT4myc myocytes. J Biol Chem 280: 9855–64.
6. Lee DH, Lee IK, Song K, Steffes M, Toscano W, et al. (2006) A strong dose-response relation between serum concentrations of persistent organic pollutants and diabetes: results from the National Health and Examination Survey 1999–2002. Diabetes Care 29: 1638–44.
7. Flegal KM, Carroll MD, Ogden CL, Johnson CL (2002) Prevalence and Trends in Obesity Among US Adults, 1999–2000. JAMA: The Journal of the American Medical Association 288: 1723–7.
8. Kuczmarski RJ, Flegal KM, Campbell SM, Johnson CL (1994) Increasing prevalence of overweight among US adults. The National Health and Nutrition Examination Surveys, 1960 to 1991. JAMA: The Journal of the American Medical Association 272: 205–11.
9. Kolpin DW, Sneck-Fahrer DA, Hallberg GR, Libra RD (1996) Temporal trends of selected agricultural chemicals in Iowa's ground-water, 1982–95: Are things getting better? J Environ Qual 26: 1007–17.
10. Schottler SP, Eisenreich SJ (1994) Herbicides in the Great Lakes. Environmental Science & Technology 28: 2228–32.
11. Mokdad AH, Bowman BA, Ford ES, Vinicor F, Marks JS, et al. (2001) The continuing epidemics of obesity and diabetes in the United States. JAMA: The Journal of the American Medical Association 286: 1195–200.
12. Soloman KR, Baker DB, Richards RP, Dixon KR, Klaine SJ, et al. (1996) Ecological risk assessment of atrzine in North American surface waters. Environ Toxicol Chem 15: 31–76.
13. Cheney MA, Fiorillo R, Criddle RS (1997) Herbicide and estrogen effects on the metabolic activity of Elliptio complanata measured by calorespirometry. Comp Biochem Physiol C Pharmacol Toxicol Endocrinol 118: 159–64.
14. Clements C, Ralph S, Petras M (1997) Genotoxicity of select herbicides in Rana catesbeiana tadpoles using the alkaline single-cell gel DNA electrophoresis (comet) assay. Environ Mol Mutagen 29: 277–88.
15. Zeljezic D, Garaj-Vrhovac V, Perkovic P (2006) Evaluation of DNA damage induced by atrazine and atrazine-based herbicide in human lymphocytes in vitro using a comet and DNA diffusion assay. Toxicol In Vitro 20: 923–35.
16. Ribas G, Frenzilli G, Barale R, Marcos R (1995) Herbicide-induced DNA damage in human lymphocytes evaluated by the single-cell gel electrophoresis (SCGE) assay. Mutat Res 344: 41–54.
17. Pino A, Maura A, Grillo P (1988) DNA damage in stomach, kidney, liver and lung of rats treated with atrazine. Mutat Res 209: 145–7.
18. Della CC, Morichetti E, Intorre L, Soldani G, Bertini S, et al. (1996) Biochemical and genetic interactions of two commercial pesticides with the monooxygenase system and chlorophyllin. J Environ Pathol Toxicol Oncol 15: 21–8.
19. De Roos AJ, Zahm SH, Cantor KP, Weisenburger DD, Holmes FF, et al. (2003) Integrative assessment of multiple pesticides as risk factors for non-Hodgkin's lymphoma among men. Occupational and Environmental Medicine 60: 11e.

20. McElroy JA, Gangnon RE, Newcomb PA, Kanarek MS, Anderson HA, et al. (2006) Risk of breast cancer for women living in rural areas from adult exposure to atrazine from well water in Wisconsin. J Expo Sci Environ Epidemiol.
21. Young HA, Mills PK, Riordan DG, Cress RD (2005) Triazine herbicides and epithelial ovarian cancer risk in central California. J Occup Environ Med 47: 1148–56.
22. Hessel PA, Kalmes R, Smith TJ, Lau E, Mink PJ, et al. (2004) A nested case-control study of prostate cancer and atrazine exposure. J Occup Environ Med 46: 379–85.
23. Hayes T, Haston K, Tsui M, Hoang A, Haeffele C, et al. (2002) Herbicides: feminization of male frogs in the wild. Nature 419: 895–6.
24. Lim S, Son KR, Song IC, Park HS, Jin CJ, et al. (2009) Fat in liver/muscle correlates more strongly with insulin sensitivity in rats than abdominal fat. Obesity (Silver Spring) 17: 188–95.
25. Jin CJ, Park HK, Cho YM, Pak YK, Lee KU, et al. (2007) S-adenosyl-L-methionine increases skeletal muscle mitochondrial DNA density and whole body insulin sensitivity in OLETF rats. J Nutr 137: 339–44.
26. Kim MS, Pak YK, Jang PG, Namkoong C, Choi YS, et al. (2006) Role of hypothalamic Foxo1 in the regulation of food intake and energy homeostasis. Nat Neurosci 9: 901–6.
27. Yoon YS, Byun HO, Cho H, Kim BK, Yoon G (2003) Complex II defect via down-regulation of iron-sulfur subunit induces mitochondrial dysfunction and cell cycle delay in iron chelation-induced senescence-associated growth arrest. J Biol Chem 278: 51577–86.
28. Pchelintseva O, Pak YK, Weiner H (1995) Identification of protein-receptor components required for the import of prealdehyde dehydrogenase into rat liver mitochondria. Arch Biochem Biophys 323: 54–62.
29. Attardi G, Chomyn A (1996) Mitochondrial biogenesis and genetics. Part B. Mitochondrial biogenesis and genetics. Part B. San Diego: Academic Press. 621 p.
30. Janssen AJ, Trijbels FJ, Sengers RC, Smeitink JA, van den Heuvel LP, et al. (2007) Spectrophotometric assay for complex I of the respiratory chain in tissue samples and cultured fibroblasts. Clin Chem 53: 729–34.
31. Noland RC, Thyfault JP, Henes ST, Whitfield BR, Woodlief TL, et al. (2007) Artificial selection for high-capacity endurance running is protective against high-fat diet-induced insulin resistance. Am J Physiol Endocrinol Metab 293: E31–E41.
32. Takahashi N, Li F, Hua K, Deng J, Wang CH, et al. (2007) Increased energy expenditure, dietary fat wasting, and resistance to diet-induced obesity in mice lacking renin. Cell Metab 6: 506–12.
33. Nelson BA, Robinson KA, Buse MG (2002) Defective Akt activation is associated with glucose- but not glucosamine-induced insulin resistance. Am J Physiol Endocrinol Metab 282: E497–E506.
34. Cho H, Thorvaldsen JL, Chu Q, Feng F, Birnbaum MJ (2001) Akt1/PKBalpha is required for normal growth but dispensable for maintenance of glucose homeostasis in mice. J Biol Chem 276: 38349–52.
35. Cantemir C, Cozmei C, Scutaru B, Nicoara S, Carasevici E (1997) p53 protein expression in peripheral lymphocytes from atrazine chronically intoxicated rats. Toxicol Lett 93: 87–94.
36. Kandori H, Suzuki S, Asamoto M, Murasaki T, Mingxi T, et al. (2005) Influence of atrazine administration and reduction of calorie intake on prostate carcinogenesis in probasin/SV40 T antigen transgenic rats. Cancer Sci 96: 221–6.
37. Fukamachi K, Han BS, Kim CK, Takasuka N, Matsuoka Y, et al. (2004) Possible enhancing effects of atrazine and nonylphenol on 7,12-dimethylbenz[a]anthracene-induced mammary tumor development in human c-Ha-ras proto-oncogene transgenic rats. Cancer Sci 95: 404–10.
38. Oulmi Y, Negele RD, Braunbeck T (1995) Segment specificity of the cytological response in rainbow trout (Oncorhynchus mykiss) renal tubules following prolonged exposure to sublethal concentrations of atrazine. Ecotoxicol Environ Saf 32: 39–50.
39. Machann J, Haring H, Schick F, Stumvoll M (2004) Intramyocellular lipids and insulin resistance. Diabetes Obes Metab 6: 239–48.
40. Sinha R, Dufour S, Petersen KF, LeBon V, Enoksson S, et al. (2002) Assessment of skeletal muscle triglyceride content by (1)H nuclear magnetic resonance spectroscopy in lean and obese adolescents: relationships to insulin sensitivity, total body fat, and central adiposity. Diabetes 51: 1022–7.
41. Fisher MT, Nagarkatti M, Nagarkatti PS (2005) Aryl hydrocarbon receptor-dependent induction of loss of mitochondrial membrane potential in epididydimal spermatozoa by 2,3,7,8-tetrachlorodibenzo-p-dioxin (TCDD). Toxicol Lett 157: 99–107.
42. Lee DH, Jacobs DR Jr, Porta M (2006) Could low-level background exposure to persistent organic pollutants contribute to the social burden of type 2 diabetes? J Epidemiol Community Health 60: 1006–8.
43. Lee DH, Lee IK, Jin SH, Steffes M, Jacobs DR Jr (2007) Association between serum concentrations of persistent organic pollutants and insulin resistance among nondiabetic adults: results from the National Health and Nutrition Examination Survey 1999–2002. Diabetes Care 30: 622–8.
44. Lind PM, Orberg J, Edlund UB, Sjoblom L, Lind L (2004) The dioxin-like pollutant PCB 126 (3,3′,4,4′,5-pentachlorobiphenyl) affects risk factors for cardiovascular disease in female rats. Toxicol Lett 150: 293–9.
45. Pelletier C, Imbeault P, Tremblay A (2003) Energy balance and pollution by organochlorines and polychlorinated biphenyls. Obes Rev 4: 17–24.
46. Shertzer HG, Genter MB, Shen D, Nebert DW, Chen Y, et al. (2006) TCDD decreases ATP levels and increases reactive oxygen production through changes in mitochondrial F(0)F(1)-ATP synthase and ubiquinone. Toxicol Appl Pharmacol 217: 363–74.
47. Saldana TM, Basso O, Hoppin JA, Baird DD, Knott C, et al. (2007) Pesticide exposure and self-reported gestational diabetes mellitus in the Agricultural Health Study. Diabetes Care 30: 529–34.
48. Gammon DW, Aldous CN, Carr WC Jr, Sanborn JR, Pfeifer KF (2005) A risk assessment of atrazine use in California: human health and ecological aspects. Pest Manag Sci 61: 331–55.
49. Bray GA, Nielsen SJ, Popkin BM (2004) Consumption of high-fructose corn syrup in beverages may play a role in the epidemic of obesity. Am J Clin Nutr 79: 537–43.
50. Maddock J (2004) The relationship between obesity and the prevalence of fast food restaurants: state-level analysis. Am J Health Promot 19: 137–43.
51. Satia JA, Galanko JA, Siega-Riz AM (2004) Eating at fast-food restaurants is associated with dietary intake, demographic, psychosocial and behavioural factors among African Americans in North Carolina. Public Health Nutr 7: 1089–96.
52. Jahren AH, Kraft RA (2008) Carbon and nitrogen stable isotopes in fast food: signatures of corn and confinement. Proc Natl Acad Sci U S A 105: 17855–60.
53. May JB (1987) Wet milling: process and products. In: Watson SA, Ramstad PE, eds. Corn: chemistry and technology. St. Paul, MN: American Association of Cereal Chemists, Inc.
54. Robert J, Gillion RJ (2002) Pesticides in the Nation's Streams and Ground Water, 1992–2001. USGS Circular 1291.
55. EPA (2006) Triazine cumulative risk assessment and atrazine, simazine, and propazine decisions. EPA.

A New F131V Mutation in Chlamydomonas Phytoene Desaturase Locates a Cluster of Norflurazon Resistance Mutations near the FAD-Binding Site in 3D Protein Models

Julio V. Suarez[1], **Stephen Banks**[1¤a], **Paul G. Thomas**[2¤b], **Anil Day**[1]*

1 Faculty of Life Sciences, The University of Manchester, Manchester, United Kingdom, **2** Syngenta, Bracknell, Berkshire, United Kingdom

Abstract

The green alga *Chlamydomonas reinhardtii* provides a tractable genetic model to study herbicide mode of action using forward genetics. The herbicide norflurazon inhibits phytoene desaturase, which is required for carotenoid synthesis. Locating amino acid substitutions in mutant phytoene desaturases conferring norflurazon resistance provides a genetic approach to map the herbicide binding site. We isolated a UV-induced mutant able to grow in very high concentrations of norflurazon (150 μM). The phytoene desaturase gene in the mutant strain contained the first resistance mutation to be localised to the dinucleotide-binding Rossmann-likedomain. A highly conserved phenylalanine amino acid at position 131 of the 564 amino acid precursor protein was changed to a valine in the mutant protein. F131, and two other amino acids whose substitution confers norflurazon resistance in homologous phytoene desaturase proteins, map to distant regions in the primary sequence of the *C. reinhardtii* protein (V472, L505) but in tertiary models these residues cluster together to a region close to the predicted FAD binding site. The mutant gene allowed direct 5 μM norflurazon based selection of transformants, which were tolerant to other bleaching herbicides including fluridone, flurtamone, and diflufenican but were more sensitive to beflubutamid than wild type cells. Norflurazon resistance and beflubutamid sensitivity allow either positive or negative selection against transformants expressing the mutant phytoene desaturase gene.

Editor: Vladimir N. Uversky, University of South Florida College of Medicine, United States of America

Funding: JVS was supported by CONACyT scholarship 209303 (Mexico), SB was the recipient of a BBSRC CASE studentship 95/A2/P/00330 with Syngenta (formerly Zeneca Agrochemicals). Zeneca Agrochemicals provided support in the form of salaries for author PGT and a stipend supplement to SB, but did not have any additional role in the study design, data collection and analysis, decision to publish, or preparation of the manuscript. The specific roles of these authors are articulated in the 'author contributions' section.

Competing Interests: PGT was employed by Zeneca Agrochemicals (now Syngenta), whose company provided funding as a BBSRC CASE studentship. The work on the mutant C. reinhardtii PDS gene disclosed in this publication is not the subject of any patent applications by the authors. The mutant C. reinhardtii PDS gene disclosed in this publication is not the subject of commercial products in development by the authors or marketed products by the authors. This does not alter the authors' adherence to all the PLOS ONE policies on sharing data and materials.

* E-mail: anil.day@manchester.ac.uk

¤a Current address: University of Otago, Dunedin, New Zealand
¤b Current address: Cambridge Enterprise Limited, University of Cambridge, Cambridge, United Kingdom

Introduction

Chlamydomonas reinhardtii provides an attractive model organism to dissect processes unique to photosynthetic eukaryotes [1]. Forward genetic screens for gain-of-function mutations that confer herbicide resistance provide a powerful approach to investigate herbicide mode of action [2]. These studies proceed rapidly using algal and plant models containing herbicide target sites, which are often absent in animals and fungi. Resistance resulting from a mutation in a target protein can identify an herbicide's binding site, which is normally revealed by identifying and sequencing the corresponding mutant gene. Identifying genes responsible for herbicide resistance is facilitated by the availability of complete sequences of the *C. reinhardtii* nuclear and organelle genomes [3,4]. Forward genetics provided a powerful tool to locate the binding sites of triazine herbicides in the D1 protein of photosystem II by sequencing mutant *psbA* genes encoding D1 from herbicide tolerant *C. reinhardtii* strains [5,6,7]. The locations of the amino acid substitutions in the D1 protein identified the herbicide target site which overlapped with the Q_B quinone binding site in photosystem II [8,9].

Norflurazon (CID 33775) is a bleaching herbicide that blocks the synthesis of carotenoids by inhibiting the activity of phytoene desaturase (PDS) [10]. The first committed step of the carotenoid synthesis pathway produces phytoene in a reaction catalysed by phytoene synthase. PDS catalyses the next step of the pathway involving the two-step dehydrogenation of phytoene to produce zeta-carotene via a phytofluene intermediate [11,12]. Carotenoids protect chloroplasts from excess light energy [13,14] and their absence results in chlorophyll bleaching and eventual cell death [15]. In *C. reinhardtii*, PDS is encoded by the single nuclear *PDS1* gene (Cre12.g509650) located on chromosome 12 [4]. The *C. reinhardtii* and *A. thaliana* PDS3 (At4G14210.1) proteins are homologous and share 66% amino acid identity.

PDS mutations conferring norflurazon resistance have been described in algae, plants and cyanobacteria. A *C. reinhardtii* mutant tolerant to 2 μM norflurazon has been described [16] but the mutation (s) responsible remain uncharacterised. Substitutions at arginine 304 (R304S, R304T, R304H, R304C) of the 580 amino acid PDS protein from the aquatic flowering plant *Hydrilla verticillata* conferred resistance to norflurazon and the related bleaching herbicide fluridone [17,18]. A L516F substitution in the 558 amino acid PDS protein of the green alga *Chlorella zofingiensis* prevented bleaching in media containing 0.25 μM norflurazon; wild type (WT) cells were bleached by 0.05 μM norflurazon [19]. In the cyanobacterium *Synechococcus* PCC 7942, mutant PDS proteins conferring resistance to norflurazon contain R195P, L320P, V403G and L436R substitutions [11]. The herbicide lethal dose was 40 μM and 70 μM for strains harbouring the R195P and L436R substitutions, respectively, and 0.5 μM for WT *Synechococcus* [11]. The 474 amino acid *Synechococcus* and *C. reinhardtii* PDS proteins share 62% identity over 448 amino acids. In Synechocystis PCC 6803, R195P and R195S substitutions in PDS gave higher levels of norflurazon resistance than a R195C substitution [20]. The relationship between these resistance mutations located at distant sites of the plant, algal and cyanobacterial PDS proteins and the norflurazon target site in the folded protein is not known.

C. reinhardtii mutants tolerant to norflurazon have been previously isolated [16,21] but the mutant genes responsible for herbicide tolerance have not been isolated and sequenced. Down regulation of PDS RNA levels by about 80% did not appear to affect carotenoid levels in *C. reinhardtii* [22]. Whilst there are no previous reports of using forward genetics to isolate and sequence mutant PDS alleles conferring norflurazon resistance in *C. reinhardtii*, substitutions that impair PDS function resulting in light-induced bleaching have been identified [23]. Selection with relatively high concentrations of herbicide has the potential to isolate a new *pds1* allele encoding an extremely resistant mutant protein that could act as a marker gene for nuclear transformation [24]. New missense mutations located in regions of the primary sequence not previously associated with herbicide resistance are particularly valuable for mapping the norflurazon binding site. Here we describe the isolation of a new mutant *pds1* allele containing the first herbicide-resistance mutation to be localised to the dinucleotide-binding Rossmann-like domain. The mutant allele confers resistance to 150 μM norflurazon allowing direct herbicide-based selection of nuclear transformants. We mapped this new mutation, and other amino acid substitutions conferring norflurazon resistance in homologous proteins, from algae, cyanobacteria and plants, on 3D models of *C. reinhardtii* PDS to identify the target site of norflurazon.

Results

Isolation of a Mutant Strain Highly Resistant to Norflurazon

After exposure of *C. reinhardtii nic7 arg7 y1* mt+ cells to high energy UV light (254 nm), colonies were selected by their ability to grow on TAP medium containing 33 μM norflurazon. Norflurazon resistant mutant strain 1 (NFR1) was chosen for further analysis because it grew much more strongly under 33 μM herbicide selection. The parent *nic7 arg7 y1* mt+ strain giving rise to NFR1 was sensitive to norflurazon and henceforth will be referred to as wild type (WT). The lowest concentration of norflurazon tested (1 μM) reduced WT cell densities by about 70% after four days of growth in liquid TAP medium (Fig. 1A). Growth of the WT strain was prevented by 5 μM norflurazon. The effective concentration of norflurazon inhibiting WT growth by 50% (Ec50) after a four day period was 0.7±0.1 μM. In contrast, strain NFR1 grew to similar densities in the presence or absence of 5 μM norflurazon after four days growth (Fig. 1B). The NFR1 culture densities were highest at four days when exposed to herbicide and then gradually declined. The Ec50 for the NFR1 strain was calculated to be 108 μM norflurazon at four days, which is over 150-fold higher than the WT Ec50. NFR1 grew in media containing 150 μM norflurazon, the highest concentration tested, and after four days reached cell densities that were about 30% of those reached in the absence of herbicide (Fig. 1B). Coloured carotenoid content was measured by absorbance at 470 nm. The coloured carotenoid content of WT cells was reduced by 79% and 99% following exposure to 2 μM and 10 μM norflurazon, respectively (Fig. 2A). By comparison 2 μM and 10 μM norflurazon reduced coloured carotenoids in NFR1 cells by 13% and 44%, respectively (Fig. 2B).

Genetic Analysis of Norflurazon Resistance

Segregation analysis of resistance to norflurazon was performed by crossing NFR1 mt+ with WT norflurazon-sensitive strains S1D2 mt- or CC621 mt- and scoring haploid progeny dissected from the meiotic products of diploid zygotes. Of the 179 colonies derived from random haploid spores dissected from tetrads, 81 were norflurazon resistant. This data fits a 1:1 segregation pattern for a single Mendelian locus better than the 3:1 segregation ratio predicted for a resistance trait encoded by two unlinked genes, where only 25% of progeny would be resistant to norflurazon. The mutation conferring herbicide resistance was localised to a *pds1-nfr1d* allele encoding a mutant PDS (see below).

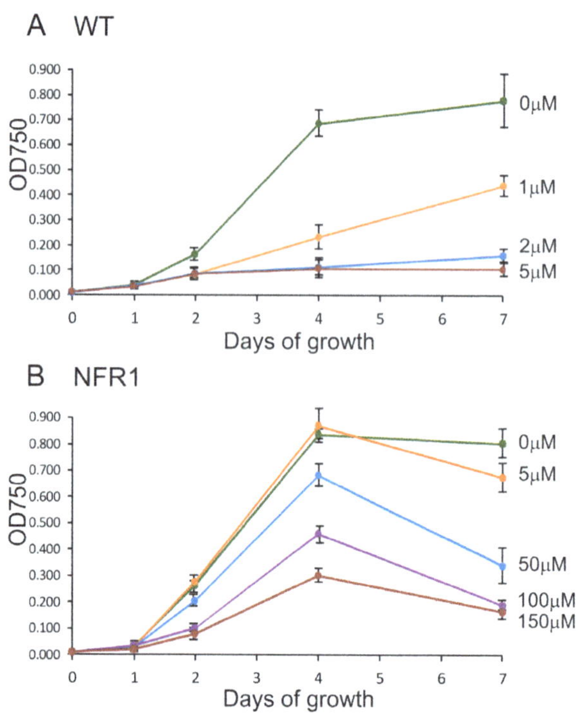

Figure 1. The influence of increasing norflurazon concentration on growth of *C. reinhardtii*. Cultures were inoculated to a density of 1×10^5 cells/ml on day 0 and optical density at 750 nm used to estimate cell density. (**A**) WT progenitor *PDS1* strain. (**B**) NFR1 mutant *pds1-nfr1d* strain. The mean values of three experiments were plotted. Standard error bars are shown.

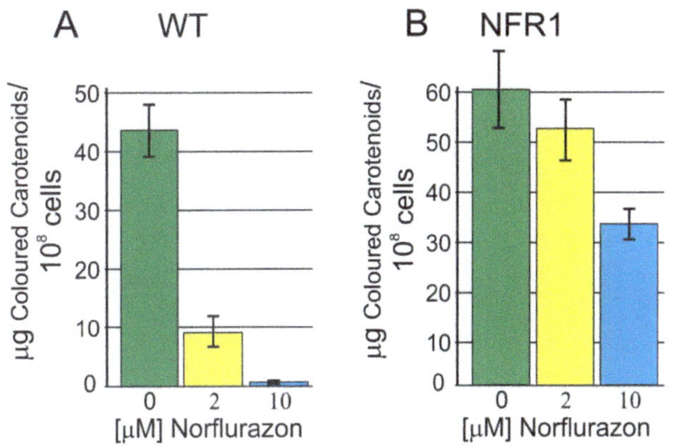

Figure 2. Impact of norflurazon concentration on coloured carotenoid content. (**A**) WT progenitor *PDS1* strain, (**B**) NFR1 mutant *pds1-nfr1d* strain. Data obtained after seven days of growth in a 12 h light/12 h dark cycle from a starting cell density of 10^5 cells/ml. Norflurazon was added at day 0. Average values of three replicate cultures were plotted. Standard error bars are shown.

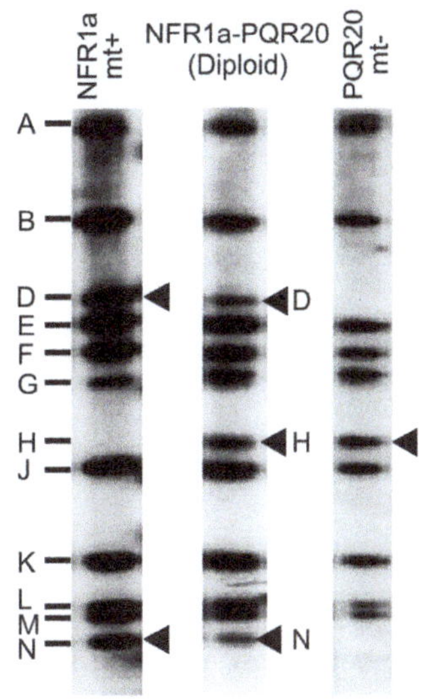

Figure 3. DNA blot showing Gulliver transposon bands in Hind III digests of total DNA from parental NFR1a and PQR20 strains and a diploid strain. Letters on the left represent Gulliver transposons [26]. Arrowheads indicate bands unique to either parent strain.

Diploid (2 n) cells [25] were used to determine whether the mutant allele in the NFR1 strain was dominant or recessive to the corresponding WT gene. Norflurazon resistant strain NFR1a (*pds1-nfr1d, arg7* mt+) was mated to a paraquat (PQR) resistant strain PQR20 (*PDS1, nic7, pqr* mt−) previously isolated in our laboratory (unpublished) to produce diploid cells. Complementation of the *arg7* and *nic7 loci* in diploid cells resulted in prototrophic growth. All diploid colonies isolated were tolerant to norflurazon and paraquat. This indicated that the mutant genes responsible for norflurazon and paraquat resistance were dominant or semi-dominant to their corresponding WT alleles. Dominance or semi-dominance of the *pds1-nfr1d* allele over the WT *PDS1* gene was confirmed by its use in transformation experiments (see below). DNA blot analysis was used to show that the cells tolerant to both herbicides were indeed diploid rather than recombinant cells. Hind III digests of DNA from diploid strains gave rise to a pattern of Gulliver transposable elements [26] representative of all the elements present in the parental PQR20 and NFR1a strains. An example is shown in Figure 3. NFR1a mt+ possesses two Gulliver bands (D and N in Figure 3) not present in the PQR20 mt− parent, whereas PQR20 contains Gulliver H, which is not present in DNA digests from strain NFR1a. These monosomic D, H and N single copy Gulliver bands unique to one of the parents were less intense in the diploid cells relative to the duplicated Gulliver bands that were common to both parents. This additive pattern of Gulliver elements in diploid strains was consistent with their derivation from the combined genomes of the two parental strains.

A Base Substitution in the *pds1-nfr1d* Allele Results in a F131V Missense Mutation

The PDS enzyme is a well characterised target of norflurazon [10,11,18,27]. The *PDS1* gene (Cre12.g509650) located on chromosome 12 is comprised of six exons (Fig. 4A) which are spliced together into a continuous coding region of 1692 nucleotides terminated by a TAA stop codon. All the *pds1-nfr1d* exons from strain NFR1 genomic DNA were amplified into PCR products and sequenced directly. Two mutations were found in the *pds1-nfr1d* allele compared to the *C. reinhardtii* WT *PDS1* sequence in GenBank (Acc. No. XM_001690807). A guanine to adenine transition in exon three (base 987 of the CDS) was a silent mutation that does not change glycine 329 in the protein; base one is the adenine in the start ATG codon. A thymine to guanine transversion in exon 2, located at base 476 of the gene (base 391 of the coding region), changed the phenylalanine TTC codon at amino acid position 131 to a valine GTC codon (Fig. 4A). This F131V missense mutation resides within the Rossmann-like domain (see below) in a region showing a high degree of conservation between cyanobacterial, algal and plant PDS proteins (Fig. 4B).

Use of the Mutant *pds1-nfr1d* Allele as a Selectable Marker for Transformation

To show that the F131V mutation conferred norflurazon resistance, the mutant *pds1-nfr1d* gene was cloned in an *E. coli* plasmid vector and used to transform *C. reinhardtii* cw92, which is a norflurazon sensitive cell-wall deficient strain [28]. The 5′ region of the *pds1-nfr1d* gene was isolated as a 3.2 kbp PCR product (Fig. 4A, primers 1017 and 1114) and joined to the 3′ region of the WT gene amplified as a 3.0 kbp PCR product (Fig. 4A, primers 1113 and 1065). The two cloned PCR products were combined at a *ClaI* site inserted in the third intron and ligated to pBluescript to assemble the 9121 bp pNFR1 plasmid (Fig. 5A). The *ClaI* site was created by inserting an ATC before the GAT trinucleotide at base 130 of the third 612 bp intron. In order to include 5′ and 3′ expression elements, the coding region was flanked by 1468 bp of upstream and 1390 bp of downstream sequences. As a control construct, the equivalent 5′ region was cloned from the $PDS1^{WT}$ gene and fused to the 3′ end of the $PDS1^{WT}$ gene to produce the pWTPDS1 plasmid. The only difference between the pNFR1 and pWTPDS1 vectors was the thymine to guanine transversion in exon 2 of the *pds1-nfr1d* gene (Fig. 4A) in the pNFR1 vector, which was confirmed by sequencing both plasmids.

pNFR1 and pWTPDS1 were used to transform the cell wall deficient and norflurazon sensitive strain cw92 using the glass-bead vortexing method [29]. Herbicide resistant colonies were isolated by direct selection on 5 µM norflurazon medium, followed by several rounds of continued selection on 5 µM norflurazon media. We found selection on 2.5 µM norflurazon gave rise to a high

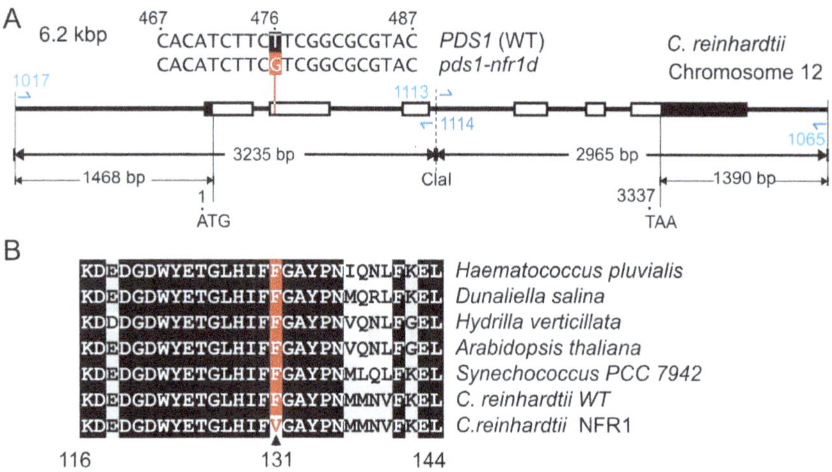

Figure 4. Localisation of the norflurazon resistance mutation in the NFR1 strain. (**A**) Section of Chromosome 12 of *C. reinhardtii* containing the *PDS1* gene (Cre12.g509650). Exons are shown as boxes; white boxes represent coding regions and black boxes UTRs. Base one is the adenine in the ATG start codon. At base 476 in exon 2, the WT thymine is replaced by guanine in the *pds1-nfr1d* allele resulting in a phenylalanine to valine missense mutation. Primers used to amplify the gene and an inserted *Cla I* site are shown. (**B**) The F131V substitution (red) is found within a highly conserved region of the PDS protein. Coordinates correspond to the *C. reinhardtii* PDS sequence, where amino acid one is the initiator methionine at the start of the transit peptide.

background of surviving non-transformed cells. For pNFR1, 45 to 60 herbicide resistant colonies were isolated using 2 μg of supercoiled plasmid with $\sim 1.5 \times 10^7$ cells. Sixty percent of the colonies isolated on primary selection plates were PCR positive for the pNFR1 plasmid (not shown). In pWTPDS1 or control no-plasmid transformation experiments, typically 0–15 herbicide resistant colonies were obtained with $\sim 1.5 \times 10^7$ cells on primary selection plates. PCR positive herbicide resistant colonies were only obtained following transformation with pNFR1 and none were isolated using pWTPDS1. The majority of PCR negative colonies from pNFR1 transformation experiments, and colonies from the pWTPDS1 and no plasmid DNA transformation experiments isolated on primary selection plates, grew relatively slowly and died following continued rounds of selection. Four norflurazon resistant colonies transformed with pNFR1 were chosen for further analysis.

Molecular Analyses of Transgenic pNFR1 Strains

DNA blots hybridized with a *PDS1* cDNA probe are shown in Fig. 5C. The pNFR1 vector contains two *Apa* I sites and one *Bam* HI site (Fig. 5A). The *Apa* I site downstream of the *Eco* R1 site is located in the vector polylinker. *Apa* I digests of non-transformed cw92 DNA gave rise to a broad band comprised of closely migrating 3.6 and 3.8 kbp fragments (Fig. 5C lane 6, see map Fig. 5B). Transformant lanes 2–5 (Fig. 5C) contained additional *Apa* I bands consistent with integration of pNFR1 into the genome. A 5.3 kbp *ApaI* band common to three transformant lanes (Fig. 5C) represents a pNFR1 derived fragment (Fig. 5A). Additional *Apa I* bands in transformant lanes not corresponding to the pNFR1 bands of 3.8 kbp and 5.3 kbp are likely to represent junction fragments. *Bam HI* digests of DNA from all four transformants (Fig. 5C, lanes 7–10) contain the 5.8 kbp WT band (Fig. 5C, lane 11). The transformant lanes contained additional higher MW bands corresponding to integration of pNFR1 into their genomes. Three transformant lanes (A2, A10, 1A1) contained one additional higher MW Bam HI band of similar intensity to the 5.8 kbp BamH1 band corresponding to the single copy endogenous *PDS1* gene. This is consistent with single copy insertions. Transformant 2A1 appeared to contain multiple insertions giving rise to three additional bands in *Bam* HI digests (Fig. 5C, lane 10). Transforming DNA integrates by illegitimate recombination into the nucleus of *C. reinhardtii* and would account for differences in the banding pattern between lanes.

Accumulation of the PDS protein in transformants was compared to non-transgenic strains cw92 and cc621 by protein blot analysis (Fig. 5D). A PDS-specific antibody showed variable accumulation of a PDS band in the four transgenic lines. Transgenic strains A10 and 2A1 accumulated over two-fold higher levels of PDS protein (Fig. 5D, lanes 3, 5 and 6) than the untransformed cw92 and cc621 strains (Fig. 5D, lanes 7–8). Transgenic strains A2 and 1A1 (Fig. 5D, lanes 2 and 4) accumulated similar levels of PDS compared to the non-transgenic cw92 and cc621 strains. A~50 kDa band present in all lanes serves as a loading control and indicates similar amounts of protein were loaded per lane. Differences in PDS levels between transformants can be explained by position effects resulting from random integration of foreign genes into the *C. reinhardtii* nuclear genome by illegitimate recombination.

Transgenic pNFR1 Strains were Highly Resistant to Norflurazon

Growth of pNFR1 transformants and the recipient non-transgenic cw92 strain were compared in TAP media containing increasing concentrations of norflurazon (Fig. 6A). Growth of cw92 was severely reduced by 1 μM norflurazon (67% reduction) and prevented by 5 μM norflurazon after seven days growth. The growth of all transformants showed marked resistance to norflurazon. The highest levels of resistance were shown by strains 2A1 and A10 (Fig. 6A), which accumulated more PDS protein (Fig. 5D). All transgenic strains grew in media containing 150 μM norflurazon. When grown in TAP media without herbicide, the transformants grew at similar rates to WT (Fig. 6B). This indicates that the expression of *pds1-nfr1d* in transgenic strains does not have a marked impact on cell growth. Growth of transformant A2 plateaued earlier than the other strains and reached a final density that was 70% of the WT density at day six. The reduced cell density observed for one transformant

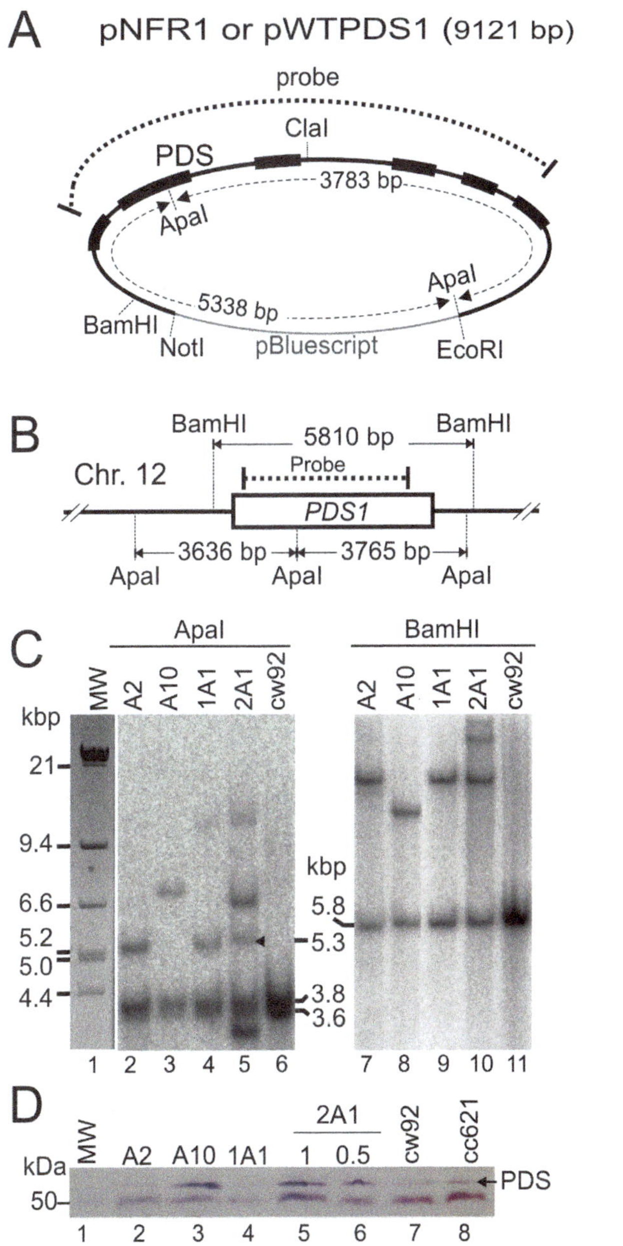

Figure 5. Nuclear transformation vector and analysis of pNFR1 transformants. (**A**) pNFR1 and pWTPDS1 transformation vectors. The two vectors are distinguished by a single base substitution conferring norflurazon resistance in pNFR1. Shown are coding regions (black boxes), pBluescript vector backbone (thin grey line), extent of cDNA probe used for DNA hybridization and restriction enzyme sites. (**B**) Map of the *PDS1* gene (white box) located on chromosome 12 (Chr12). Restriction sites and overlap region of the cDNA probe are shown. (**C**) Blots of digested DNA from pNFR1 transformants and untransformed cw92 strain hybridized with a *PDS1* cDNA probe. Restriction enzyme used, band sizes and MW standards are indicated. (**D**) Protein blot of fractionated total cell protein from pNFR1 cw92 transformants, the cc621 WT strain and the non-transformed cw92 strain incubated with an affinity-purified polyclonal antibody raised against a PDS peptide. The PDS band is arrowed.

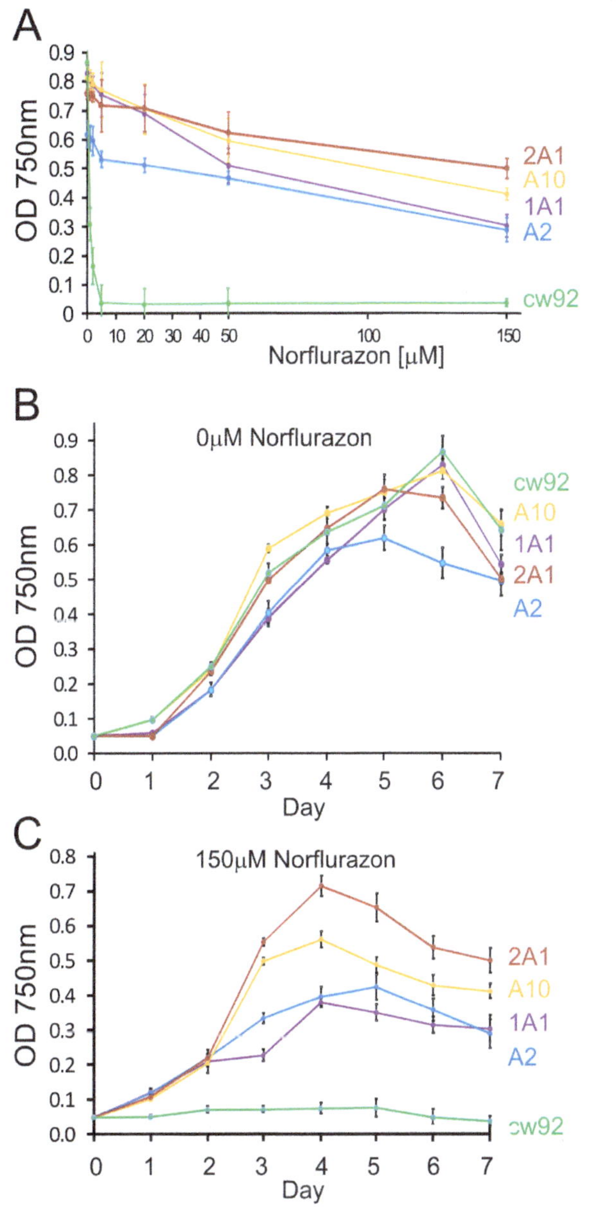

Figure 6. Resistance levels of pNFR1 transformants to norflurazon. Cells were inoculated at an initial OD750 of 0.05 and were grown in TAP media with varying concentrations of norflurazon. (**A**) Norflurazon resistance of cw92 (norflurazon sensitive non-transformed strain) and pNFR1 A2, A10, 1A1 and 2A1 transformants. Optical density was measured after 7 days of growth. (**B–C**) Growth curves of NFR1 cw92 transformants and non-transformed cw92 strain in liquid TAP media without norflurazon (**B**) or with 150 μM norflurazon (**C**). Green line: cw92; blue line: A2; yellow line: A10; purple line 1A1; red line: 2A1. The mean values of three cultures were plotted. Standard error bars are shown.

(transformant A2) relative to cw92 is likely to reflect factors that are not directly related to the presence of transgenic DNA containing *pds1-nfr1d*, such as the transgene integration site. In the presence of 150 μM norflurazon, transformant 2A1 grew to the highest densities (Fig. 6C).

F131 Maps to a Rossmann-like Domain and Clusters with other Norflurazon-resistant Mutations in 3D Protein Models

A linear scheme of the 564 amino acid precursor PDS protein from *C. reinhardtii* containing the chloroplast transit peptide is shown in Figure 7A. F131V and previously described resistance mutations in homologous PDS proteins identify amino acids that are likely to affect norflurazon binding. The high conservation amongst PDS proteins allowed us to map substitutions conferring norflurazon resistance in homologous proteins (Fig. 8) onto the *C. reinhardtii* PDS sequence (Fig. 7A). The F131V mutation reported here lies in the N-terminal quarter, whereas previously reported amino acids substitutions conferring norflurazon-resistant mutations map closer to the middle (R268, L388) or C-terminus (V472, L505, L517) of the 564 amino acid protein (Fig. 7A). The F131V is the first herbicide resistance mutation localising to the dinucleotide binding Rossmann-like domain [30,31,32] spanning amino acids Y91 to L144 (Fig. 7A). This organisation of Rossmann-like domain amino acids contributing to a FAD-binding domain is characteristic of pyridine nucleotide-disulphide oxidoreductases [31]. A previously described substitution (E143K) in this domain reduces function and is suppressed by intragenic suppressor mutation K90M [23].

The crystal structure of a plant PDS has not been determined. Whilst the structure of a CRTI-type PDS enzyme present in bacteria and fungi is available [33], it is distantly related to the plant PDS enzyme and may have an independent evolutionary origin [34]. In order to map the F131V mutation and the positions of other norflurazon resistance mutations onto the structure of PDS, 3D models of the *C. reinhardtii* protein were constructed using four programs: 3DLigandSite [35] which uses the Phyre prediction system [36], MODELLER [37,38] via the BioSerf service [39], SwissModel [40] (Fig. 7B), and 3D-Jigsaw [41]. The predicted PDS structures are shown in Figure 7B and Figure S1. Each program used the crystal structures of different proteins to predict the folded structure of PDS. Phyre used human monoamine oxidase type B [42], MODELLER used an oxidoreductase from *Methanosarcina mazei* [43] and SwissModel used putrescine oxidase from *Arthrobacter aurescens* [44]. *C.reinhardtii* PDS shares: 15% identity (32% similarity) with human monoamine oxidase chain B over aligned PDS amino acids 76–537; 16% identity (30% similarity) with *M. mazei* oxidoreductase for PDS amino acids 76–526; 18% identity (32% similarity) with *A. aurescens* putrescine oxidase for PDS amino acids 76–553.

In all four models of the folded protein, three amino acids implicated in norflurazon resistance F131 (red), V472 (pink) and L505 (green), cluster together in the protein (Fig. 7B, Fig. S1). R268 (blue), whose substitution confers norflurazon resistance in cyanobacteria and *Hydrilla* (Fig. 8), is also part of the cluster in the 3D structures predicted by MODELLER, SwissModel (Fig. 7B) and 3-DJigsaw (Fig. S1). In the 3D structure predicted by Phyre, R268 lies outside the cluster in close proximity to F131 (Figs. 7B and 7D). Mutations of amino acids L388 (purple) and L517 (orange) also confer norflurazon resistance (Fig. 8). L388 and L517 are located on opposite sides of the cluster (Fig. 7B). Previous work using 3D–LigandSite has suggested E143 (black) and K90 (cyan) are juxtaposed in the folded protein [23]. Consistent results were obtained with Bioserf/MODELLER, SwissModel (Fig. 7B) and 3D–Jigsaw [41,45] (Fig. S1), where E143 and K90 lie close to each other in the predicted protein structures. Amino acids in the Rossmann-like domain (Y91-L144) are shown as a yellow ribbon in Figure 7B. Whilst both F131 (red) and E143 (black) are located in the Rossmann-like domain (Fig. 7A), they are separated in the folded protein models (Fig. 7B).

Norflurazon inhibits PDS activity by competing with the enzyme's cofactors rather than the phytoene substrate [46]. The purified plant enzyme is a flavoprotein containing FAD [47]. In bacterial CRT-I type PDS, FAD appears to be the only co-factor required for phytoene desaturation [33]. Following the desaturation reaction the reduced $FADH_2$ produced is re-oxidised to FAD by plastoquinone either directly or indirectly [48]. The 3DLigandSite program [35] predicts the location of the FAD binding domain [32,49,50] in PDS. The amino acids contributing to the predicted FAD binding domain in the *C. reinhardtii* PDS enzyme are shown in Fig. 7C. These include amino acids that are also part of the Rossmann-like domain (Fig. 7C, shaded yellow). Of the amino acids whose substitution confers norflurazon resistance, V472 (pink) is part of the predicted FAD-binding domain, whilst F131 (red) and L505 (green) lie adjacent to FAD-binding domain residues F130 and A506 (Fig. 7C). A magnified view of the relative positions of F131, V472 and L505 and FAD in the 3D-ligand model of PDS is shown in Figure 7D. The close proximity of F131, V472 and L505 to the predicted FAD binding site is consistent with norflurazon inhibiting the function of this co-factor.

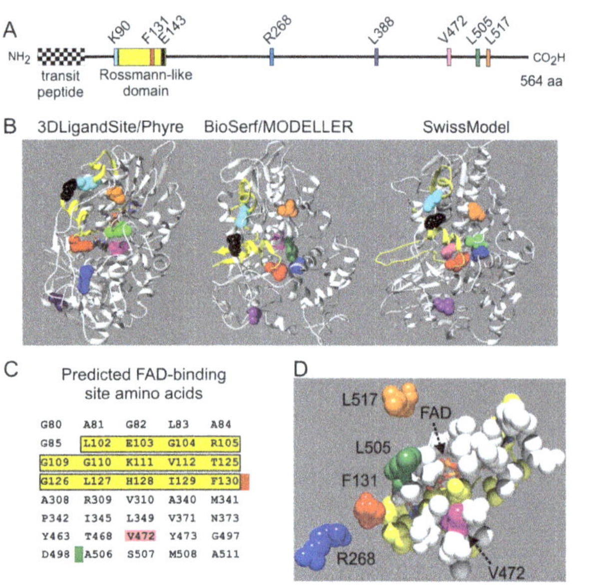

Figure 7. Location of norflurazon resistance mutations on the PDS protein. (**A**) Primary structure of *C. reinhardtii* prePDS protein showing plastid transit peptide, dinucleotide binding Rossmann-like domain, and amino acid substitutions affecting PDS function (see Fig. 8). The same colour coding is used throughout Figs 7 and 8. (**B**) Ribbon representation of the predicted three-dimensional structure of *C. reinhardtii* PDS. Left: Phyre-based 3DLigandSite model; middle: Modeller-based BioSerf predicted structure; right: SwissModel predicted structure. Amino acids shown in A are indicated. The conserved Rossmann-like domain is shown as a yellow ribbon. (**C**) Residues in the predicted FAD-binding domain. Amino acids in the Rossmann-like domain are shaded yellow. Amino acids conferring norflurazon resistance lie adjacent to (F131/red, L505/green) or are part of the predicted domain (V472/pink). (**D**) 3D structure of the predicted FAD-binding region. Bound FAD is shown in the centre of the structure as a ball stick model. Yellow atoms are part of the Rossmann-like domain. F131, R268, V472, L505 and L517 are shown.

Cross-tolerance Conferred by the *pds1-nfr1d* Gene to other Bleaching Herbicides

The highest levels of norflurazon resistance were exhibited by pNFR1 transformants 2A1 and A10. Four different bleaching

C. reinhardtii PDS residue	species	amino acid substitution	phenotype	reference
K90	*C.reinhardtii*	K90M	suppresses E143K	[23]
F131	*C.reinhardtii*	F131V	norflurazon resistance	this work
E143	*C.reinhardtii*	E143K	impaired function	[23]
R268	*Synechococcus* *Synechocystis* *Hydrilla verticillata*	R195P R195C/P/S R304C/H/S/T		[11] [20] [18]
L388	*Synechococcus*	L320P		[11]
V472	*Synechococcus*	V403G	norflurazon resistance (R268–L517)	[10]
L505	*Synechococcus*	L436R		[11]
L517	*Chlorella zofingiensis*	L516F		[19]

Figure 8. Location of mutations on the *C. reinhardtii* PDS sequence. Shown are amino acid residue numbers in *C.reinhardtii* PDS, species in which the missense mutation was isolated, numbering of the original substitutions in homologous PDS proteins and phenotype of the substitutions. Synechococcus PCC7942 and Synechocystis PCC 6803 are listed. Colour coding of amino acids is the same as in Figure 7.

herbicides were tested on these transformants: fluridone, flurtamone, beflubutamid and diflufenican (Figs. 9A–D). Wild-type cells showed a marked reduction in growth when placed in TAP medium containing 50 nM fluridone (Ec50 = 38 nM), 25 nM flurtamone (Ec50 = 16 nM) and 100 nM diflufenican (Ec50 = 68 nM). The WT strain was relatively tolerant to beflubutamid and grew at all concentrations tested (Ec50> 100 nM). Transformants 2A1 and A10 showed greater tolerance to fluridone, flurtamone and diflufenican at all the concentrations tested. The average Ec50 values for the two transformants were: 110 nM for fluridone, 56 nM for flurtamone, and 170 nM for diflufenican. In contrast, pNFR1 transformants exhibited enhanced sensitivity to beflubutamid (Ec50 = 21 nM) with resistance levels 4.8 fold lower than wild type cells (Ec50>100 nm).

Discussion

We used forward genetics to isolate a new F131V substitution in the *C. reinhardtii* PDS protein that confers a high level of resistance to norflurazon. Whilst growth of WT cells was prevented by 5 µM norflurazon, cells containing the mutant *pds1-nfr1d* gene were able to grow in media containing 150 µM norflurazon. The Ec50 values to norflurazon were 0.7 µM and around 100 µM, for WT and mutant NFR1 cells, respectively. The norflurazon-resistance mutation conferred cross-tolerance to the bleaching herbicides fluridone, flurtamone, and diflufenican but enhanced sensitivity to beflubutamid, relative to WT cells. The mutant *pds1-nfr1d* allele is dominant or semi-dominant to the WT gene allowing the cloned *pds1-nfr1d* gene to be used as a marker for nuclear transformation with direct norflurazon selection for resistant transgenic cells.

The F131V substitution is the first norflurazon resistance mutation to be mapped to the dinucleotide binding Rossmann-like domain of PDS. F131V and three other resistance mutations, which correspond to substitutions at R268, V472 or L505 in the *C. reinhardtii* protein (Fig. 8), are located at distant sites of the primary sequence but cluster together in the predicted tertiary structure. Moreover, three of these amino acids lie adjacent to (F131, L505) or are part of (V472) the predicted FAD binding domain (Fig. 7C). These results, based on locating amino acid substitutions conferring norflurazon resistance on 3D-models of PDS, predict the herbicide's target region on the protein and suggest the herbicide's mode of action involves inhibition of FAD function. Substitutions of amino acids L388 and L517 also confer norflurazon resistance but these residues do not map close to the F131-R268-V472-L505 cluster of resistance mutations on 3D models. To resolve whether L388 and L517 interact with norflurazon directly or affect binding of norflurazon indirectly by affecting the folding of the protein will become clearer if a crystal structure becomes available.

Use of the *pds1-nfr1d* allele as a positive selection marker for nuclear transformation of *C. reinhardtii* avoids the use of antibiotic resistance genes [51,52,53,54,55], which may raise regulatory concerns for commercial applications of transgenic algae. The transformation procedure used the mutant *pds1-nfr1d* allele with its own promoter and terminator and 5 µM norflurazon, which was substantially higher than the concentration required to kill the majority of non-transformed cells (Ec50 of 0.7 µM). In a recent study, 0.5 µM norflurazon was used to select *C. reinhardtii* transformants expressing a cDNA encoding a mutant PDS protein in which the L505F substitution was introduced by site-directed mutagenesis [24]. The RbcS2 promoter [52] was used to express the L505F mutant gene resulting in elevated PDS transcripts [24]. Here we found that 5 µM norflurazon was required for effective selection of transformants in the light regimes used. Use of 2.5 µM norflurazon on selection plates led to a high background of non-transformed colonies. Previous herbicide resistant marker genes [24,56] and a mutant CRY1 gene providing resistance to the translational inhibitors emetine and cryptopleurine [57] have used 5′ and 3′ regulatory elements from the RbcS2 gene to allow the isolation of nuclear *C. reinhardtii* transformants [24,56,57]. The use of herbicide resistance genes with native regulatory elements [58] including the mutant *pds1-nfr1d* gene (this work) as marker genes overcomes the possibility of *RbcS* gene silencing and titration of

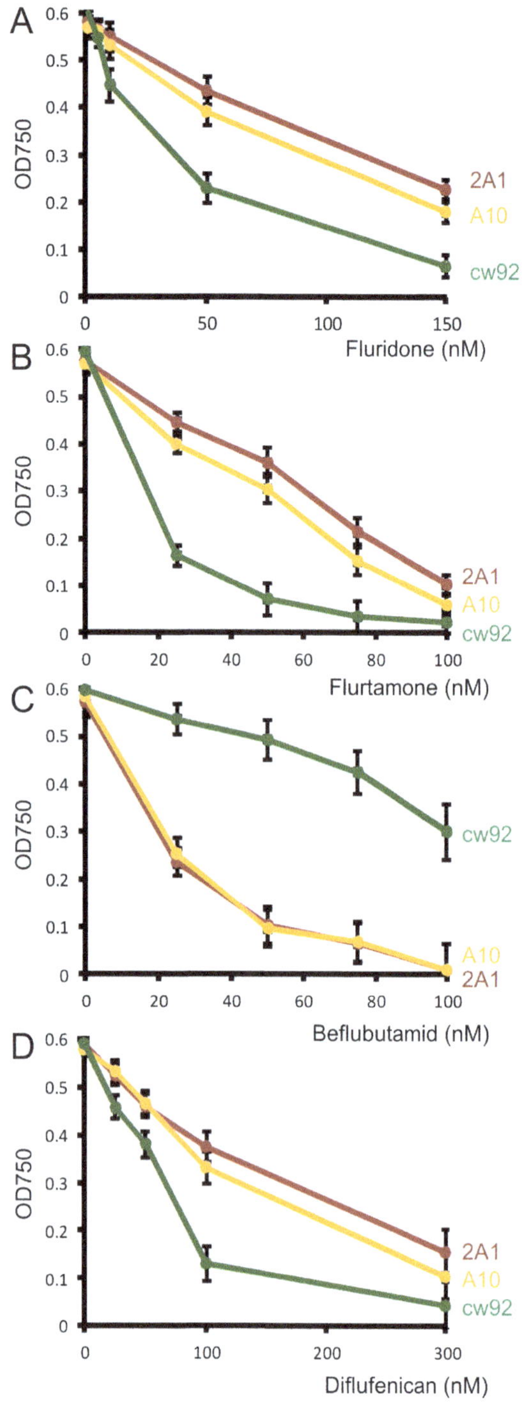

Figure 9. The effect of bleaching herbicides on growth of cw92 (non-transformed strain) and pNFR1 transgenic lines. (A) Fluridone, **(B)** Flurtamone, **(C)** Beflubutamid, and **(D)** Diflufenican. Cells were inoculated at an OD750 of 0.05 at day 0 in TAP media and cell density estimated by OD750 on day 7. Values represent the average of three replicate cultures. Standard error bars are shown. Red line: strain 2A1; yellow line: strain A10; green line: cw92.

transcription or translation factors needed to express the native *RbcS* genes [59] in *C. reinhardtii*.

Transformants containing the *pds1-nfr1d* gene were 4.8 fold less resistant to beflubutamid than WT cells. By comparison, transformation of the WT *Hydrilla* PDS protein into Arabidopsis increased seedling sensitivity to diflufenican [18]. Arabidopsis seedlings expressing a *Hydrilla* norflurazon resistant PDS protein with a R304C (corresponds to position R268 in the *C. reinhardtii* protein) mutation, were more sensitive to both diflufenican and beflubutamid [18]. These results show that changes in the region identified by the cluster of amino acid substitutions conferring norflurazon resistance (Figs. 7B, 7D) are critical in determining the action of herbicides and provide a step towards the rational design of new herbicides targeting this protein. The enhanced susceptibility of *pds1-nfr1d* transformants to beflubutamid allows negative selection. This enables *pds1-nfr1d*, which is a native algal gene, to be used for either positive or negative selection, which is conditional on the selection agent used. Norflurazon enables positive selection whereas beflubutamid allows negative selection. Because the transformants contain both the WT and norflurazon-resistant PDS enzymes, this suggests that beflubutamid sensitivity is dominant or semi-dominant to resistance conferred by the WT enzyme. Raised PDS levels in the A10 and 2A1 transformants relative to WT cells was consistent with higher levels of the beflubutamid-sensitive mutant PDS enzyme in transgenic cells.

Dual and conditional positive/negative selection markers are powerful tools for manipulating genomes and controlling the growth of transgenic cells [60,61,62]. The high norflurazon resistance levels exhibited by *pds1-nfr1d* transformants suggests the coding region flanked by appropriate regulatory elements may serve as a marker for transformation of a wide range of algae and plants susceptible to norflurazon and related bleaching herbicides. The PDS protein is localised to chloroplasts, which raises the possibility of using a modified coding sequence with chloroplast regulatory elements as a new marker gene for the chloroplast transformation toolbox [63]. Negative selection can be used to restrict the spread of transgenic cells, influence recombination events such as marker excision, and in mutant screens. The *pds1-nfr1d* allele confers sensitivity to beflutamid and is a candidate for a new negatively selectable marker for algal genetics.

Conclusions

A phenylalanine to valine substitution at position 131 in the dinucleotide binding Rossmann-like domain of phytoene desaturase increased the 5 μM norflurazon concentration needed to restrict growth of WT *C. reinhardtii* cells to over 150 μM in mutant cells. The F131V substitution clusters with two other substitutions conferring norflurazon resistance close to the predicted FAD-binding site in 3D-protein models.

Materials and Methods

C. reinhardtii Strains, Media, Mutagenesis and Genetic Crosses

All *C. reinhardtii* strains were from the Chlamydomonas Center (University of Minnesota). Strains were maintained at 20°C to 25°C on TAP medium [1]. Transformants were propagated on TAP medium with 5 μM norflurazon (Sigma-Aldrich, Poole, UK). Large 25 ml–200 ml liquid cultures were grown at 25°C in a New Brunswick Scientific Innova 4340 illuminated orbital shaker at 120 rpm. Arginine (Sigma-Aldrich) was used at a final concentration of 50 mg L^{-1} and nicotinic acid (Sigma-Aldrich) at 25 mg L^{-1}.

For mutagenesis 5 ml of 1×10^7 cells ml^{-1} of *C. reinhardtii* strain *nic7 arg7 y1* mt+ were irradiated for 20 seconds with a 254 nm XX-15 BlackRay germicidal UV lamp (UVP, Upland, California) at 2.42 mW cm^{-2} in 9 cm petri dishes with gentle rocking.

Irradiated cells were kept in the dark for 12 h, and then concentrated by centrifugation at 600×g at room temperature. Three norflurazon resistant cell lines (NFR1-3) were selected on media containing 33 μM norflurazon at 25°C with a light intensity of 60 μmol m^{-2} s^{-1} in a 12 h light cycle. Only NFR1 survived several rounds of serial propagation on 33 μM norflurazon selective plates. Two additional resistant mutants were isolated using UV irradiation of 2.86 and 3.62 mW cm^{-2} for 20 seconds but these grew very slowly and could not be propagated.

Genetic crosses to isolate vegetative diploid cells or zygotes giving rise to meiotic tetrads for segregation analyses were performed as described [1,64]. Norflurazon resistant strain NFR1a (*arg7, NIC7, pds1-nfr1d*, mt+) was isolated from a meiotic product of the cross NFR1 (*nic7, arg7, pds1-nfr1d*, mt+) with an *arg2* mt− strain. NFR1a was sensitive to paraquat. Diploids were made by crossing paraquat-resistant strain PQR20 (ARG7, *nic7*, *PDS1*, *pqr* mt−) with NFR1a; the *pqr* locus responsible for paraquat resistance has not been identified. PQR20 was sensitive to norflurazon. Eighty one diploid colonies capable of growing on media without arginine or nicotinic acid were resistant to both norflurazon (33 μM) and paraquat (39 μM). NFR1 (*nic7, arg7, pds1-nfr1d, y1* mt+) was crossed with either S1D2 mt− [65] or CC621 mt− to provide tetrads for segregation analysis.

Nucleic Acid Methods

Total DNA from *C. reinhardtii* was prepared according to Day et al 1988 [66], except an additional DNA precipitation was included by adding 0.6 volumes of 20% PEG 8000, 1.5 M NaCl (Sigma-Aldrich) to the total DNA solution. RNA was purified using the RNeasy plant mini kit (Qiagen) according to the manufacturer's instructions. Genomic DNA for PCR screening was isolated according to Cao *et al.* (2009) [67]. Hybridisation probes were comprised of a Gulliver [26] internal fragment and a 1.5 kbp *PDS1* cDNA from *C. reinhardtii* strain CC-503 cw92 mt+ prepared using nested RT-PCR with primers 1297 and 1298 followed by primers 1328 and 1329. Probes were labelled with [α-^{32}P] dCTP (PerkinElmerNEN, Seer Green, UK) using the High Prime labelling kit (Roche, Mannheim, Germany). Digests of 2 μg of total DNA digests were fractionated on 0.7% W/V agarose gels for DNA blot analysis [66]. Membranes were washed at 65°C in 0.1×SSC buffer and 0.1% w/v SDS. Table S1 lists the oligonucleotides used. PCR screening of transformants was performed using primer pairs 1322+1323, and 1324+1325, which amplified the regions spanning the PDS-vector junctions in pNFR1 and pWTPDS1 transformants. PCR reactions with actin gene primers 1127 and 1351 were used as positive controls to verify the quality of genomic DNA. PCR screening was performed using BioMix PCR master mix (Bioline, London). BigDye terminator v3.1 (Applied Biosystems) was used for cycle sequencing and the ladders fractionated using a 48 capillary ABI 3730 DNA Analyser.

Isolation and Sequence Analysis of the *pds1-nfr1d* Gene

PCR products from NFR1 genomic DNA using Vent DNA polymerase (New England Biolabs, Massachusetts, USA) were sequenced directly or following cloning into pGEM-T Easy (Promega, Madison, USA). Direct sequencing was on: (i) a 1.3 kbp product containing exons 1 and 2 (PCR primers 889 and 891; sequencing primers 870, 871, 890 and 892); (ii) a 470 bp product containing exon 3 (PCR primers 872 and 873; sequencing primers 873 and 874); (iii) a 462 bp product containing exon 4 (PCR primers 875 and 876; sequencing primers 876 and 877); (iv) a 662 bp product containing exons 5 and 6 (PCR primers 878 and 879; sequencing primers 879 and 880). A 3235 bp PCR product containing exons 1–3 was amplified with primers 1017+1114, cloned into pGEM-T and sequenced.

Construction of pNFR1 and pWTPDS1

The pWTPDS1 plasmid was constructed by joining 5′ (3235 bp; primers 1017+1114) and 3′ (2965 bp; 1113+1065) PCR products (cloned in pGEM-T Easy) amplified from CC503 cw92 mt+ total DNA via a *Cla* I site located in primers 1113 and 1114. The WTPDS1 sequences in pGEM-T Easy were excised with *Not I* and *Cla I* (3235 bp sequence) and *Cla I* and *EcoR1* (2965 bp sequence) and cloned into pBluescript KS- [68]. To construct pNFR1, the 3.2 kbp 5′ WT PDS1 *Not* I - Cla I fragment was replaced with the equivalent 5′ *pds1-nfr1d* fragment. The complete inserts in both plasmids were sequenced to exclude PCR errors.

Nuclear Transformation of *C. reinhardtii*

Chlamydomonas cell wall deficient strain CC-503 cw92 mt+ was transformed according to Kindle, 1990 [29]. 25 ml of cell culture (~5×10^6 cells/ml) was sedimented at 500×g. The cell pellet was resuspended in 2.5 ml of TAP media and 300 μl mixed with 0.3 g of sterile 0.4 mm glass beads (BDH, Poole, UK), 2 μg circular (mainly supercoiled) plasmid DNA and 100 μl of 20% (W/V) polyethylene glycol (M_W 8000). The mixture was vortex-mixed (Rotamixer, *Hook and Tucker Instruments* Ltd., Croydon, UK) at maximum setting for 30 seconds, 700 μl of TAP media was added, and the cells spread onto three 9 cm selection plates containing solid (1% W/V agar) TAP media with 5 μM norflurazon. Cells were allowed to recover for 36 h in dim light (10 μmol m^{-2} sec^{-1}) and then transferred to high-light (250 μmol m^{-2} sec^{-1}). Colonies became visible after 21 days and were transferred to fresh selection media.

The Effects of Herbicides on Cell Cultures

Growth curves for the NFR1 strain were prepared by inoculating cells to a concentration of 1×10^5 cells ml^{-1}, growing them for 7 days in 50 ml of TAP media in a 12 h light cycle (60 μmol m^{-2} s^{-1}) at 25°C with shaking (120 rpm) with increasing concentrations of norflurazon and measuring optical density (OD) at 750 nm. Transformant herbicide growth assays were carried out in continuous light (260 μmol m^{-2} s^{-1}) at 25°C without shaking and culture ODs measured daily in a microplate reader (Biotek Synergy, Potton, UK). Transformant cells were inoculated in 1 ml of TAP media to an OD750 nm of 0.05 containing increasing concentrations of norflurazon in 24-well plates and grown for 7 days. Resistance curves for other herbicides were obtained by growing cells in 200 μl TAP media with increasing concentrations of herbicides in 96-well plates. Fluridone was used at 5, 10, 50 and 150 nM concentrations. Befublutamid and flurtamone were utilised at 25, 50, 75 and 100 nM concentrations. Diflufenican was used at 25, 50, 100 and 300 nM concentrations. All experiments were performed in triplicate. Herbicides were from Sigma-Aldrich.

For estimation of coloured carotenoids, cultures were grown for 7 days in TAP media in a 12 h day (60 μmol m^{-2} s^{-1}) with shaking (120 rpm) at a starting density of 1×10^5 cells ml^{-1}. Carotenoids in sedimented cells (1.1×10^8 cells) were extracted twice with 4 ml of 80% acetone for 20 minutes. Carotenoids were partitioned into 8 ml of 50% diethyl ether with the addition of 2 M NaCl to promote phase separation. The organic phase was washed twice with 5 ml of 4 M NaCl and then dried under nitrogen. The dried carotenoids were dissolved in diethyl ether and coloured carotenoids were estimated from the absorbance at 470 nm [1].

Protein Blot Analyses

Cells were sedimented from 50 ml cultures at 1×10^7 cells ml^{-1}, frozen, resuspended in 300 µl of 1×SSC (0.15 M NaCl, 0.015 M Trisodium citrate) and 100 µl 4×gel sample loading buffer added (0.25 M Tris-HCl, pH 6.8, 40% V/V glycerol, 12% W/V SDS, 20% V/V β-mercaptoethanol, 0.4% W/V bromophenol blue). The mixture was placed in a boiling water bath for 5 minutes and then sedimented for 2 min (2000 rpm, in an Eppendorf 5415C microfuge). For SDS-PAGE, 6 µl of the supernatant was loaded per lane on 10 well, 12% (W/V) polyacrylamide gels with 4% stacking gels and electrophoresed in Tris-Glycine SDS buffer using the MiniProtean II system (BioRad, Hemel Hempsted, UK). Proteins were electroblotted onto Hybond ECL nitrocellulose (GE Healthcare) and incubated with antibody as previously described [69]. An affinity-purified polyclonal antibody raised in rabbits against peptide VAAWKDEDGDWYETG (Eurogentec, Belgium) was used to detect PDS. The primary antibody was detected with an anti-rabbit IgG-alkaline phosphatase conjugate (Sigma-Aldrich, Poole, UK) and visualised using 5-bromo-4chloro-3-indolyl phosphate/nitro blue tetrazolium (NBT/BCIP) reagent (Sigma-Aldrich).

Models of Protein Tertiary Structure

The *C. reinhardtii* PDS sequence (GI 158279545) was submitted for structural prediction using default parameters to the 3DLigandSite [35], BioSerf [39], SWISS MODEL [40] and 3D-JIGSAW [41,45]. Protein structure files were visualised and annotated using UCSF Chimera [70]. Images were rendered using POV-Ray within UCSF Chimera.

Supporting Information

Figure S1 Space-filling model of *C.reinhardtii* phytoene desaturase predicted by 3D Jigsaw [41,45]**.** Amino acids F131, R268, L388, V472, L505 and L517 whose substitution are associated with norflurazon resistance are shown (see Figs. 7 and 8). Also shown are E143 and K90. Mutation of E143 impairs function and this is suppressed by a mutation at K90 [23]. The F131, R268, V472 and L505 cluster of amino acids associated with norflurazon resistance is magnified on the right.

Table S1 Sequences of oligonucleotides used as primers for PCR and sequencing reactions.

Acknowledgments

We thank the DNA sequencing facility and EA Mudd for critically reading the manuscript.

Author Contributions

Conceived and designed the experiments: PGT SB JVS AD. Performed the experiments: SB JVS. Analyzed the data: SB PGT AD JVS. Contributed reagents/materials/analysis tools: JVS SB PGT AD. Wrote the paper: JVS AD.

References

1. Harris E, Stern D, Witman G (1989) The Chlamydomonas sourcebook: Cambridge Univ Press.
2. Galloway R, Mets L (1984) Atrazine, bromacil and diuron resistance in Chlamydomonas: a single non-Mendelian genetic locus controls the structure of the thylakoid binding site. Plant Physiol 74: 469–474.
3. Maul JE, Lilly JW, Cui LY, dePamphilis CW, Miller W, et al. (2002) The *Chlamydomonas reinhardtii* plastid chromosome: Islands of genes in a sea of repeats. Plant Cell 14: 2659–2679.
4. Merchant SS, Prochnik SE, Vallon O, Harris EH, Karpowicz SJ, et al. (2007) The Chlamydomonas genome reveals the evolution of key animal and plant functions. Science 318: 245–251.
5. Erickson J, Rahire M, Bennoun P, Delepelaire P, Diner B, et al. (1984) Herbicide resistance in *Chlamydomonas reinhardtii* results from a mutation in the chloroplast gene for the 32-kilo Dalton protein of photosystem II. Proc Natl Acad Sci USA 81: 3617–3621.
6. Erickson JM, Rahire M, Rochaix JD, Mets L (1985) Herbicide resistance and cross resistance: changes at 3 distinct sites in the herbicide binding protein. Science 228: 204–207.
7. Johanningmeier U, Bodner U, Wildner G (1987) A new mutation in the gene coding for the herbicide binding protein in Chlamydomonas. FEBS Lett 211: 221–224.
8. Erickson JM, Pfister K, Rahire M, Togasaki RK, Mets L, et al. (1989) Molecular and biophysical analysis of herbicide resistant mutants of *Chlamydomonas reinhardtii:* structure function relationship of the photosystem II D1 polypeptide. Plant Cell 1: 361–371.
9. Oettmeier W, Hilp U, Draber W, Fedtke C, Schmidt R (1991) Structure activity relationships of triazinone herbicides on resistant weeds and resistant *Chlamydomonas reinhardtii*. Pesticide Sci 33: 399–409.
10. Chamovitz D, Pecker I, Hirschberg J (1991) The molecular-basis of resistance to the herbicide norflurazon. Plant Mol Biol 16: 967–974.
11. Chamovitz D, Sandmann G, Hirschberg J (1993) Molecular and biochemical-characterization of herbicide-resistant mutants of cyanobacteria reveals that phytoene desaturation is a rate-limiting step in carotenoid biosynthesis. J Biol Chem 268: 17348–17353.
12. Fraser P, Bramley P (2004) The biosynthesis and nutritional uses of carotenoids. Prog Lipid Res 43: 228–265.
13. Anderson I, Robertson D (1960) Role or carotenoids in protecting chlorophyll from photodestruction. Plant Physiol 35: 531–534.
14. Krinsky N (1979) Carotenoid protection against oxidation. Pure Appl Chem 51: 649–660.
15. Bartels P, Watson C (1978) Inhibition of carotenoid synthesis by fluridone and noflurazon. Weed Sci 26: 198–203.
16. Vartak V, Bhargava S (1997) Characterization of a norflurazon-resistant mutant of *Chlamydomonas reinhardtii*. Weed Sci 45: 374–377.
17. Michel A, Arias RS, Scheffler BE, Duke SO, Netherland M, et al. (2004) Somatic mutation-mediated evolution of herbicide resistance in the nonindigenous invasive plant hydrilla (*Hydrilla verticillata*). Mol Ecol 13: 3229–3237.
18. Arias R, Dayan F, Michel A, Howell J, Scheffler B (2006) Characterization of a higher plant herbicide-resistant phytoene desaturase and its use as a selectable marker. Plant Biotechnol J 4: 263–273.
19. Liu J, Zhong Y, Sun Z, Huang J, Sandmann G, et al. (2010) One amino acid substitution in phytoene desaturase makes *Chlorella zofingiensis* resistant to norflurazon and enhances the biosynthesis of astaxanthin. Planta 232: 61–67.
20. Martínez-Férez I, Vioque A, Sandmann G (1994) Mutagenesis of an amino-acid responsible in phytoene desaturase from Synechocystis for binding of the bleaching herbicide norflurazon. Pesticide Biochem Physiol 48: 185–190.
21. Mijit G, Zayadan B, Rahman E, Chunaev A (2003) Analysis of dominant-recessive mutated character of nfr genes of *Chlamydomonas reinhardtii* and its mutual interaction with the chloroplastic *psbA* gene. Acta Genetica Sinica 30: 646–652.
22. Vila M, Couso I, Leon R (2008) Carotenoid content in mutants of the chlorophyte *Chlamydomonas reinhardtii* with low expression levels of phytoene desaturase. Process Biochem 43: 1147–1152.
23. Tran P, Sharifi M, Poddar S, Dent R, Niyogi K (2012) Intragenic enhancers and suppressors of phytoene desaturase mutations in *Chlamydomonas reinhardtii*. PLOS One 7: e42196.
24. Liu J, Gerken H, Huang J, Chen F (2013) Engineering of an endogenous phytoene desaturase gene as a dominant selectable marker for *Chlamydomonas reinhardtii* transformation and enhanced biosynthesis of carotenoids. Process Biochem 48: 788–795.
25. Ebersold WT (1967) *Chlamydomonas reinhardtii*: heterozygous diploid strains. Science 157: 447–449.
26. Ferris PJ (1989) Characterization of a Chlamydomonas transposon, Gulliver, resembling those in higher plants. Genetics 122: 363–377.
27. Martínez-Férez I, Vioque A (1992) Nucleotide sequence of the phytoene desaturase gene from Synechocystis SP PCC 6803 and characterization of a new mutation which confers resistance to the herbicide norflurazon. Plant Mol Biol 18: 981–983.
28. Davies D, Plaskitt A (1971) Genetical and structural analyses of cell wall formation in *Chlamydomonas reinhardtii*. Genetical Res 17: 33–43.
29. Kindle K (1990) High-frequency nuclear transformation of *Chlamydomonas reinhardtii*. Proc Natl Acad Sci USA 87: 1228–1232.
30. Rossmann M, Moras D, Olsen K (1974) Chemical and biological evolution of a nucleotide binding protein. Nature 250: 194–199.
31. Marchler-Bauer A, Lu S, Anderson J, Chitsaz F, Derbyshire M, et al. (2011) CDD: a Conserved Domain Database for the functional annotation of proteins. Nucl Acids Res 39: D225–D229.

32. Marchler-Bauer A, Zheng C, Chitsaz F, Derbyshire M, Geer L, et al. (2013) CDD: conserved domains and protein three-dimensional structure. Nucl Acids Res 41: D348–D352.
33. Schaub P, Yu Q, Gemmecker S, Poussin-Courmontagne P, Mailliot J, et al. (2012) On the structure and function of the phytoene desaturase *CRTI* from *Pantoea ananatis*, a membrane-peripheral and FAD-dependent oxidase/isomerase. PLOS One 7: e39550.
34. Pecker I, Chamovitz D, Linden H, Sandmann G, Hirschberg J (1992) A single polypeptide catalyzing the conversion of phytoene to zeta-carotene is transcriptionally regulated during tomato fruit ripening. Proc Natl Acad Sci USA 89: 4962–4966.
35. Wass MN, Kelley LA, Sternberg MJE (2010) 3DLigandSite: predicting ligand-binding sites using similar structures. Nucl Acids Res 38: W469–W473.
36. Kelley LA, Sternberg MJE (2009) Protein structure prediction on the Web: a case study using the Phyre server. Nature Protocols 4: 363–371.
37. Eswar N, John B, Mirkovic N, Fiser A, Ilyin VA, et al. (2003) Tools for comparative protein structure modeling and analysis. Nucl Acids Res 31: 3375–3380.
38. Pieper U, Eswar N, Webb BM, Eramian D, Kelly L, et al. (2009) MODBASE, a database of annotated comparative protein structure models and associated resources. Nucl Acids Res 37: D347–D354.
39. Buchan D, Minneci F, Nugent T, Bryson K, Jones D (2013) Scalable web services for the PSIPRED Protein Analysis Workbench. Nucl Acids Res 41: W349–W357.
40. Arnold K, Bordoli L, Kopp J, Schwede T (2006) The SWISS-MODEL workspace: a web-based environment for protein structure homology modelling. Bioinformatics 22: 195–201.
41. Bates P, Kelley L, MacCallum R, Sternberg M (2001) Enhancement of protein modeling by human intervention in applying the automatic programs 3D-JIGSAW and 3D-PSSM. Proteins Suppl 5: 39–46.
42. Binda C, Hubálek F, Li M, Herzig Y, Sterling J, et al. (2004) Crystal structures of monoamine oxidase B in complex with four inhibitors of the N-propargylaminoindan class. J Med Chem 47: 1767–1774.
43. Seetharaman J, Su M, Sahdev S, Janjua H, Xiao R, et al. Crystal Structure of an oxidoreductase from *Methanosarcina mazei* Protein Data Bank in Europe 3ka7. Available: http://www.ebi.ac.uk/pdbe-srv/view/entry/3ka7/summary. Accessed 14.5.2014.
44. Zhang Z, Almo SC, Swaminathan S The crystal structure of oxidoreductase from *Arthrobacter aurescens* Protein Data Bank in Europe 3rha. Available:http://www.ebi.ac.uk/pdbe-srv/view/entry/3rha/summary.html. Accessed 14.5.2014.
45. Contreras-Moreira B, Bates PA (2002) Domain Fishing: a first step in protein comparative modelling. Bioinformatics 18: 1141–1142.
46. Sandmann G, Linden H, Boger P (1989) Enzyme-kinetic studies on the interaction of norflurazon with phytoene desaturase. Zeitschrift Fur Naturforschung C 44: 787–790.
47. Hugueney P, Romer S, Kuntz M, Camara B (1992) Characterization and molecular-cloning of a flavoprotein catalyzing the synthesis of phytofluene and zeta-carotene in capsicum chromoplasts. Eur J Biochem 209: 399–407.
48. Norris S, Barrette T, DellaPenna D (1995) Genetic dissection of carotenoid synthesis in arabidopsis defines plastoquinone as an essential component of phytoene desaturation. Plant Cell 7: 2139–2149.
49. Wierenga R, Terpstra P, Hol W (1986) Prediction of the occurrence of the ADP-binding βαβ-fold in proteins, using an amino-acid-sequence fingerprint. J Mol Biol 187: 101–107.
50. Vallon O (2000) New sequence motifs in flavoproteins: Evidence for common ancestry and tools to predict structure. Proteins: Struc Funct Bioinf 38: 95–114.
51. Stevens DR, Rochaix JD, Purton S (1996) The bacterial phleomycin resistance gene ble as a dominant selectable marker in Chlamydomonas. Mol Genet Genomics 251: 23–30.
52. Lumbreras V, Stevens D, Purton S (1998) Efficient foreign gene expression in *Chlamydomonas reinhardtii* mediated by an endogenous intron. Plant J 14: 441–447.
53. Sizova I, Fuhrmann M, Hegemann P (2001) A Streptomyces rimosus aphVIII gene coding for a new type phosphotransferase provides stable antibiotic resistance to *Chlamydomonas reinhardtii*. Gene 277: 221–229.
54. Fischer N, Rochaix JD (2001) The flanking regions of *PsaD* drive efficient gene expression in the nucleus of the green alga *Chlamydomonas reinhardtii*. Mol Genet Genomics 265: 888–894.
55. Berthold P, Schmitt R, Mages W (2002) An engineered *Streptomyces hygroscopicus aph 7″* gene mediates dominant resistance against hygromycin B in *Chlamydomonas reinhardtii*. Protist 153: 401–412.
56. Kovar J, Zhang J, Funke R, Weeks D (2002) Molecular analysis of the acetolactate synthase gene of Chlamydomonas reinhardtii and development of a genetically engineered gene as a dominant selectable marker for genetic transformation. Plant J 29: 109–117.
57. Nelson J, Savereide P, Lefebvre P (1994) The *Cry1* gene in *Chlamydomonas reinhardtii:* strucutre and use as a dominant selectable marker for nuclear transformation. Mol Cell Biol 14: 4011–4019.
58. Randolph-Anderson BL, Sato R, Johnson AM, Harris EH, Hauser CR, et al. (1998) Isolation and characterization of a mutant protoporphyrinogen oxidase gene from *Chlamydomonas reinhardtii* conferring resistance to porphyric herbicides. Plant Mol Biol 38: 839–859.
59. Goldschmidt-Clermont M, Rahire M (1986) Sequence, evolution and differential expression of the 2 genes encoding variant small subunits of ribulose bisphosphate carboxylase oxygenase in *Chlamydomonas reinhardtii*. J Mol Biol 191: 421–432.
60. Boeke JD, Trueheart J, Natsoulis G, Fink GR (1987) 5-Fluoroorotic acid as a selective agent in yeast molecular genetics. Meth Enzymol 154: 164–175.
61. Erikson O, Hertzberg M, Nasholm T (2004) A conditional marker gene allowing both positive and negative selection in plants. Nature Biotechnol 22: 455–458.
62. Gisby M, Mudd E, Day A (2012) Growth of transplastomic cells expressing D-amino acid oxidase in chloroplasts is tolerant to D-Alanine and inhibited by D-Valine. Plant Physiol 160: 2219–2226.
63. Day A, Goldschmidt-Clermont M (2011) The chloroplast transformation toolbox: selectable markers and marker removal. Plant Biotechnol J 9: 540–553.
64. Day A, Rochaix J (1991) Structure and inheritance of sense and antisense transcripts from a transposon in the green alga *Chlamydomonas reinhardtii*. J Mol Biol 218: 273–291.
65. Gross C, Ranum L, Lefebvre P (1988) Extensive restriction fragment length polymorphisms in a new isolate of *Chlamydomonas reinhardtii*. Curr Genet 13: 503–508.
66. Day A, Schirmer-Rahire M, Kuchka M, Mayfield S, Rochaix J (1988) A transposon with an unusual arrangement of long terminal repeats in the green-alga *Chlamydomonas reinhardtii*. EMBO J 7: 1917–1927.
67. Cao M, Fu Y, Guo Y, Pan J (2009) Chlamydomonas (Chlorophyceae) colony PCR. Protoplasma 235: 107–110.
68. Altingmees M, Short J (1989) pBluescript II: gene mapping vectors. Nucl Acids Res 17: 9494–9494.
69. Madesis P, Osathanunkul M, Georgopoulou U, Gisby M, Mudd E, et al. (2011) A hepatitis C virus core polypeptide expressed in chloroplasts detects anti-core antibodies in infected human sera. J Biotechnol 145: 377–386.
70. Pettersen E, Goddard T, Huang C, Couch G, Greenblatt D, et al. (2004) UCSF chimera - A visualization system for exploratory research and analysis. J Comp Chem 25: 1605–1612.

Effects of Gibberellic Acid and N, N-Dimethyl Piperidinium Chloride on the Dose of and Physiological Responses to Prometryn in Black Nightshade (*Solanum nigrum* L.)

Hailan Jiang[1,2], Xiaoxia Deng[1,2], Jungang Wang[1]*, Jing Wang[1], Jun Peng[1], Tingting Zhou[1]

1 College of Agriculture, Shihezi University, Shihezi, Xinjiang, People's Republic of China, **2** Kuerle Agricultural Technology Extension Center, Kuerle, Xinjiang, People's Republic of China

Abstract

The use of gibberellic acid (GA_3) and N, N-dimethyl piperidinium chloride (DPC) in combination with prometryn would likely increase the control of black nightshade in cotton fields. Experiments were designed to investigate the physiological and biochemical responses of black nightshade at the three- to four-leaf stage to prometryn applied at different rates, either alone or in combination with GA_3 or DPC, in a greenhouse environment. These studies demonstrated that prometryn applied in combination with DPC at low rates (7.2 g ai ha^{-1}) led to increased fresh weight and visible injury of black nightshade compared with prometryn applied alone or in combination with GA_3; however, at rates of 36, 180, and 900 g ai ha^{-1}, prometryn in combination with DPC caused the least visible injury among all treatments and prometryn in combination with GA_3 caused the greatest visible injury. These results suggest that black nightshade suffered more severe damage when prometryn was applied in combination with GA_3, which is supported by the reduced soluble protein content, lower antioxidant enzyme activities, and higher malondialdehyde (MDA) content in the plants treated with prometryn plus GA_3. These results indicate that the application of GA_3 in combination with prometryn to black nightshade may have the potential to lower the levels of prometryn tolerance in these plants.

Editor: Manuel Reigosa, University of Vigo, Spain

Funding: This work was supported by the Special Fund for Agro-scientific Research in the Public Interest (201303022), corps doctoral Fund (2013BB004) and National Natural Science Foundation Project of China (31260435). The funders had no role in study design, data collection and analysis, decision to publish, or preparation of the manuscript.

Competing Interests: The authors have declared that no competing interests exist.

* E-mail: jungangwang98@163.com

These authors contributed equally to this work.

Introduction

Black nightshade (*Solanum nigrum* L.) is widely distributed throughout the world and is an important weed in many countries [1]. In recent decades, black nightshade has become increasingly problematic as a field crop weed [2,3]. It is also one of the dominant weeds in cotton fields [4,5] and seriously affects the growth of cotton.

Prometryn is a triazine herbicide in the same family as simetryn. Cotton is tolerant to certain triazine herbicides, such as prometryn, when they are applied at pre-emergence or post-emergence; however, severe injury can occur following post-emergence application [6,7]. The triazine herbicides control of a wide spectrum of weeds in crop and non-crop areas under many environmental and edaphic conditions. Prometryn has been widely used as a residual soil-applied and post-emergence-directed herbicide in cotton [8,9] and provides season-long control of annual grasses [10,11].

Plant growth regulators, such as N, N-dimethyl piperidinium chloride (DPC) and gibberellic acid (GA_3), are widely used in cotton production and have become indispensable tools in cotton cultivation [12,13]. Although variations in the application of these growth regulators can occur among regions, years, and environmental conditions, they are applied throughout the growing season up to several weeks prior to harvest. DPC has been known to reduce the synthesis of GA_3, resulting in the suppression of cell enlargement [14]. Thus, DPC-treated plants tend to be short and narrow with thick, small leaves [15], while GA_3-treated plants present an opposite phenotype. Recently, there has been increased interest in the co-application of herbicides with plant growth regulators [16]. Applying these products in combination is also preferred by growers due to the convenience, reduced time investment, and reduced application and labour costs.

Due to the availability of prometryn-tolerant cotton, it is likely that this herbicide will be applied simultaneously with plant growth regulators and will effectively control black nightshade when it is applied to cotton at early post-emergence or late post-emergence [17,18,19]. Despite these benefits, incompatibility can occur with certain pesticides. Therefore, defining the interactions of agrochemicals is important when considering the simultaneous application of multiple agrochemicals [20,21]. Because previous studies have seldom focused on the physiological responses of plants to the combined application of herbicides and plant growth regulators, it is desirable to devise an index of resistance that reflects physiological responses under different treatments. We hypothesised that the foliar application of plant growth regulators to black nightshade would strengthen the effect of prometryn on weed control. Therefore, the current study was designed to evaluate the efficacy of prometryn when applied alone or in combination with GA_3 or DPC in controlling black nightshade and to study the physiological and biochemical response of the weed to co-application in a greenhouse.

Materials and Methods

Experimental Conditions and Treatments

The experiments were conducted in a greenhouse at Shihezi University in Xinjiang in 2012. The greenhouse temperature was maintained between 22°C at night and 29°C during the day. Separate experiments were conducted to investigate the control of black nightshade and to assess the physiological and biochemical changes in this plant. The plants were watered daily, grown under natural sunlight, and fertilised every 7 d to maintain active growth. Treatments for the black nightshade included 5 levels of the non-selective herbicide prometryn (7.2, 36, 180, 900, and 4500 g ai ha^{-1}) applied alone or in combination with a plant growth regulator, either DPC (22.05 kg ai ha^{-1}) or GA_3 (14.7 kg ai ha^{-1}). For the physiological and biochemical response study, the treatments consisted of prometryn (900 g ai ha^{-1}) applied alone or in combination with a plant growth regulator, either DPC (22.05 kg ai ha^{-1}) or GA_3 (14.7 kg ai ha^{-1}). All chemicals were applied at the manufacturer's recommended rate. A non-treated control was included in these experiments. The experimental design for the control of black nightshade experiment and for the physiological and biochemical response experiment was a randomised complete block with four treatment replications.

Black nightshade seed samples (provided by Shihezi University) were stored in a cold room at 4°C prior to the initiation of the study. The black nightshade seeds were planted in a 20×15×5 cm plastic crisper in the greenhouse to accelerate germination. Once the black nightshade seedlings had germinated, 5 seedlings were transplanted into plastic pots (16 cm diameter, 11 cm height) filled with potting soil (fine-loamy, mixed, Typic Argiudoll) at pH 6.4 containing 1.9% organic matter. Three seedlings were removed prior to the treatments, and the treatments were applied when the black nightshade seedlings reached the three- to four-leaf stage.

Dose Response Study

The black nightshade plants were excised at the soil level 6 d after the treatments to determine the fresh weight and calculate the percent reduction. Visual estimates of percent injury were recorded 6 d after treatment using a scale of 0 to 100, where 0 = no foliar injury and 100 = plant death. Foliar chlorosis, necrosis, tissue distortion, swelling, malformation, and plant stunting were considered when visually estimating the percent injury.

Physiological and Biochemical Response Study

All the measurements were recorded close to midday on the uppermost fully expanded main stem leaf located three nodes below the terminus of the plant. The third main stem leaf was then collected 3 d and 6 d after application to observe changes in black nightshade physiological parameters, especially chlorophyll content, malondialdehyde content, and soluble protein content, as well as changes in biochemical parameters and antioxidant enzyme activity (superoxide dismutase (SOD), catalase (CAT), and peroxidise (POD)). Prior to the physiological and biochemical measurements, a single leaf of each plant was collected and immediately stored in an ultra freezer (−80°C) for antioxidant enzyme and protein extraction. A UV spectrophotometer (UV-1800; Shimadzu Co., Ltd., Japan) was used for measuring the changes in the physiological and biochemical parameters.

The chlorophyll (CHL) content of samples was measured using the method specified by Arnon [22]. Lipid peroxidation was defined as the content of all 2-thiobarbituric acid-reactive substances and was expressed as equivalents of MDA, as reported by Heath and Packer [23].

Frozen leaves (0.5 g) were homogenised in 50 mM potassium phosphate buffer (pH 7.8) containing 1 mM EDTA, 3 mM 2-mercaptoethanol, and 5% (w/v) polyvinylpyrrolidone (PVP) in a chilled mortar and pestle. The homogenate was centrifuged at 12,000×g for 20 min at 4°C, and the supernatant was used for enzyme assays.

Soluble proteins were measured according to the technique described by Bradford [24].

The SOD activity was assayed by the nitroblue tetrazolium (NBT) method [25]. The 3 ml reaction mixture contained 50 mM Na-phosphate buffer (pH 7.3), 13 mM methionine, 75 mM NBT, 0.1 mM EDTA, 4 mM riboflavin, and 0.2 ml of enzyme extract. This reaction was initiated by the addition of riboflavin, and the glass test tubes were shaken and placed under fluorescent lamps (160 μmol m^{-2} s^{-1}). The reaction was allowed to proceed for 5 min and was then stopped by switching off the light. The absorbance was measured at 560 nm. Blanks and controls were treated in the same manner but lacked illumination and enzyme, respectively.

The activities of POD and CAT were assayed using the method of Chance and Maehly [26] with some modifications. The POD reaction solution (3 ml) contained 50 mM phosphate buffer (pH 7.8), 25 mM guaiacol, 200 mM H_2O_2, and 0.5 ml of enzyme extract. Changes in the absorbance of the reaction solution at 470 nm were determined every 30 s. One unit of POD activity was defined as an absorbance change of 0.01 unit min^{-1}. The CAT reaction solution (3 ml) contained 50 mM phosphate buffer (pH 7.0), 200 mM H_2O_2, and 0.5 ml of enzyme extract. The reaction was initiated by the addition of the extract. Changes in the absorbance of the reaction solution at 240 nm were measured every 30 s.

Data Analysis

Differences between the plant growth regulator treatments were detected using analysis of variance (ANOVA). All data analyses were conducted using SPSS Statistics 17.0 with estimates of means and standard errors, and Duncan's multiple range test was used for determining significant differences.

Results

Dose-response Study

In general, the fresh weight observed at 6 d after application decreased as the herbicide rate increased, and the fresh weight was decreased to a similar extent across rates for all three treatments (Fig. 1). All herbicide treatments at rates of 7.2 and 36 g ai ha^{-1} caused reductions in fresh weight of less than 50%. The fresh weight reductions caused by prometryn and prometryn plus DPC were not drastically different. Prometryn applied in combination with GA_3 at 7.2, 36, and 180 g ai ha^{-1} provided greater control than prometryn applied alone or combined with DPC; however, minimal differences were found when the rates of all treatments were higher than 180 g ai ha^{-1}.

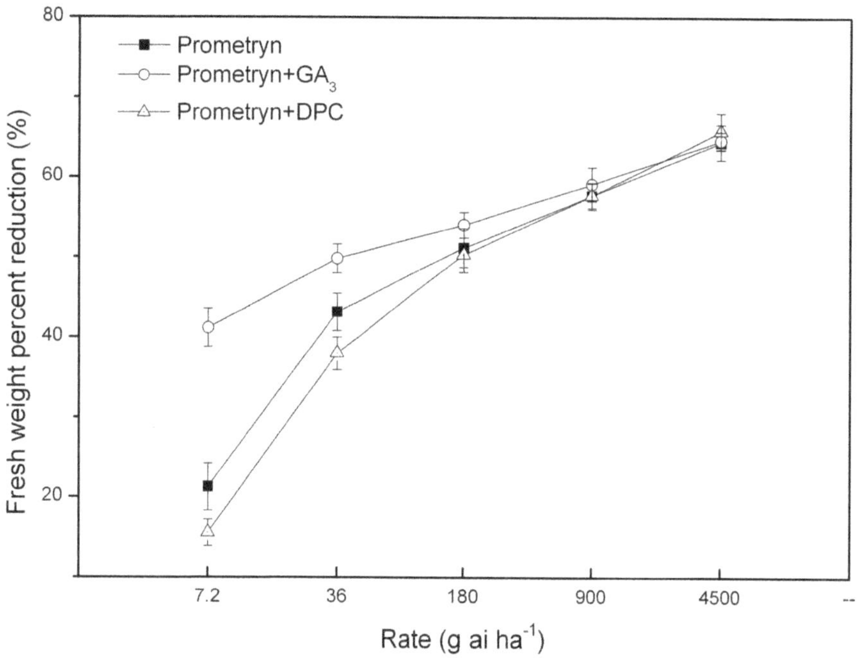

Figure 1. Percent reduction in fresh weight caused by prometryn applied alone or in combination with GA_3 or DPC. Measurements were performed 6 d after application. The data are the means of 4 replicates ± standard error.

Visible injury increased as the herbicide rates increased, with prometryn plus GA_3 causing the greatest injury at 6 d after treatment (Fig. 2). All treatments caused less than 20% visible injury at 7.2 g ai ha^{-1}. Prometryn combined with DPC at 7.2 g ai ha^{-1} provided the best control of black nightshade; however, it caused the least injury at 36 and 180 g ai ha^{-1}. Although the visible injury caused by prometryn plus GA_3 was more serious than the injury caused by prometryn alone, the visible injuries were similar at all rates.

Physiological and Biochemical Response Study

A similar trend was observed in the changes in chlorophyll a and chlorophyll b contents after all treatments. After the prometryn treatment, the chlorophyll levels decreased continuously and drastically by 31.09% (CHL a) and 36.61% (CHL b) 3 d after application and by 84.01% (CHL a) and 42.61% (CHL b) 6 d after application (Fig. 3). When prometryn was applied in combination with GA_3, the change in chlorophyll content was more rapid, decreasing by 14.61% (CHL a) and 10.13% CHL b) 3 d after application and by 21.83% (CHL a) and 14.58% (CHL b) 6 d after application, compared to the treatment with prometryn alone (Fig. 3); however, 6 d after application, the effects of these two treatments were not significant. Although the chlorophyll content of the plants treated with prometryn plus DPC also decreased, the change was not significant compared to the control 3 d after application. The chlorophyll content of the plants treated with prometryn plus DPC 6 d after application was significantly lower than that of the control; however, it was still significantly higher than the chlorophyll content of plants treated with prometryn or prometryn plus GA_3.

The soluble protein content was significantly different between the prometryn and the control treatments 3 and 6 d after application (Fig. 4), and the soluble protein content of all treated plants was significantly lower than that of the untreated plants. The soluble protein content of plants treated with prometryn plus GA_3 was significantly lower than that of plants treated with prometryn plus DPC 3 d after application; however, the soluble protein content of plants exposed to these treatments was not significantly different from that of plants treated with prometryn alone. No significant difference was observed in the soluble protein content of all prometryn-treated plants 6 d after application. The percent reductions in soluble protein content were 66.68%, 71.57%, and 69.49% for the prometryn, prometryn plus GA_3, and prometryn plus DPC treatments, respectively, compared with the control.

After prometryn treatment, the MDA content of the plants increased drastically by 34.07% and 42.61% 3 d and 6 d after application, respectively (Fig. 5). When prometryn was applied in combination with GA_3, the increase in MDA content was quicker than prometryn treatment alone. The highest MDA content was detected in plants treated with prometryn plus GA_3 3 d and 6 d after application; however, no significant differences were found between the prometryn and prometryn plus GA_3 treatments. Compared with the control, the MDA content in plants treated with prometryn, prometryn plus GA_3, and prometryn plus DPC was increased by 78.71%, 92.03%, and 48.82%, respectively.

Changes in Antioxidant Enzyme Activities

Prometryn applied alone or in combination with either GA_3 or DPC had a significant effect on SOD activity; i.e., herbicide-treated plants exhibited significantly lower SOD activities compared with the control plants 3 d and 6 d after application (Fig. 6). The SOD activity of plants treated with prometryn plus DPC decreased 17.07% (3 d) and 18.37% (6 d) compared to the control. The lowest SOD activity was found in the plants treated with prometryn plus GA_3; however, no significant difference was found between plants treated with prometryn alone and those treated with prometryn plus GA_3.

Prometryn applied alone or in combination with either GA_3 or DPC had a significant effect on POD activity; i.e., herbicide-

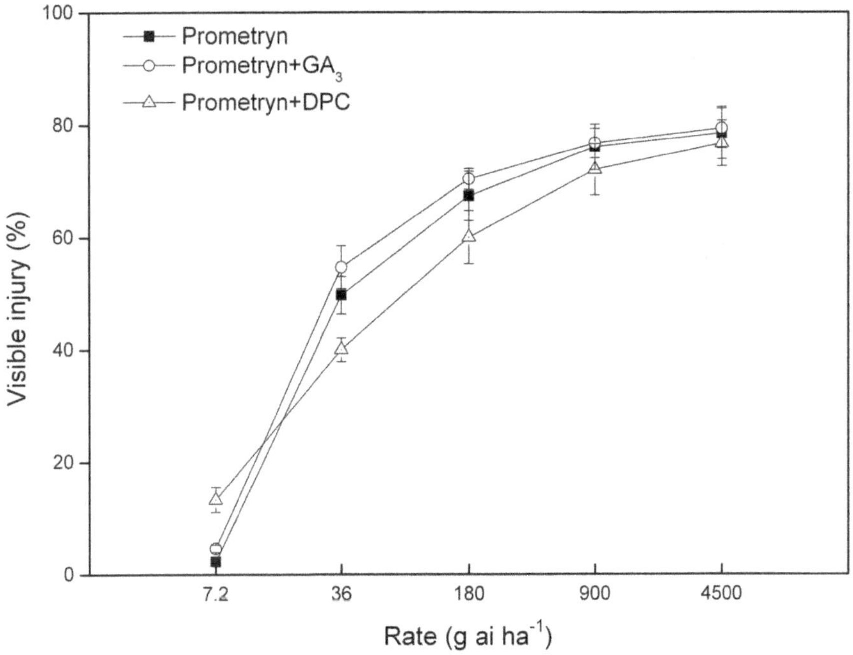

Figure 2. Visible injury caused by prometryn applied alone or in combination with GA_3 or DPC. Measurements were performed 6 d after application. The data are the means of 4 replicates ± standard error.

treated plants exhibited significantly lower POD activities compared with the control plants 3 d and 6 d after application. However, no significant differences were observed between any of the herbicide treatments 6 d after application (Fig. 7). The POD activity of plants treated with prometryn plus DPC decreased 14.53% (3 d) and 59.34% (6 d) compared to the control. The lowest POD activities were noted in the plants treated with prometryn (3 d) and prometryn plus GA_3 (6 d); in these plants, POD activity decreased by 54.68% and 68.35%, respectively, compared with the control.

Prometryn applied alone or in combination with either GA_3 or DPC also had a significant effect on CAT activity; i.e., the herbicide-treated plants exhibited significantly lower CAT activities compared with the control plants 3 d and 6 d after application. However, no significant difference was observed between any of the herbicide treatments 3 d and 6 d after application (Fig. 8). Compared to the control, the CAT activities in plants subjected to all treatments were decreased by more than 95%.

Discussion

The proper use of growth-regulating chemicals requires a detailed understanding of the effects of their interaction with herbicides on plants. The results of this study suggest that the application of prometryn in combination with GA_3 led to a greater

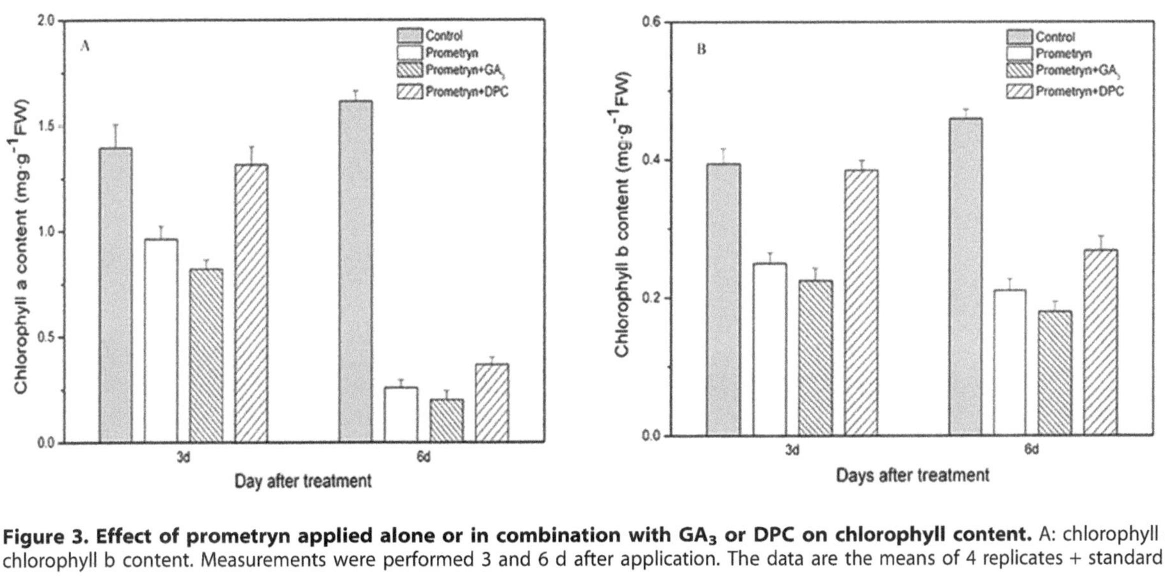

Figure 3. Effect of prometryn applied alone or in combination with GA_3 or DPC on chlorophyll content. A: chlorophyll a content. B: chlorophyll b content. Measurements were performed 3 and 6 d after application. The data are the means of 4 replicates + standard error.

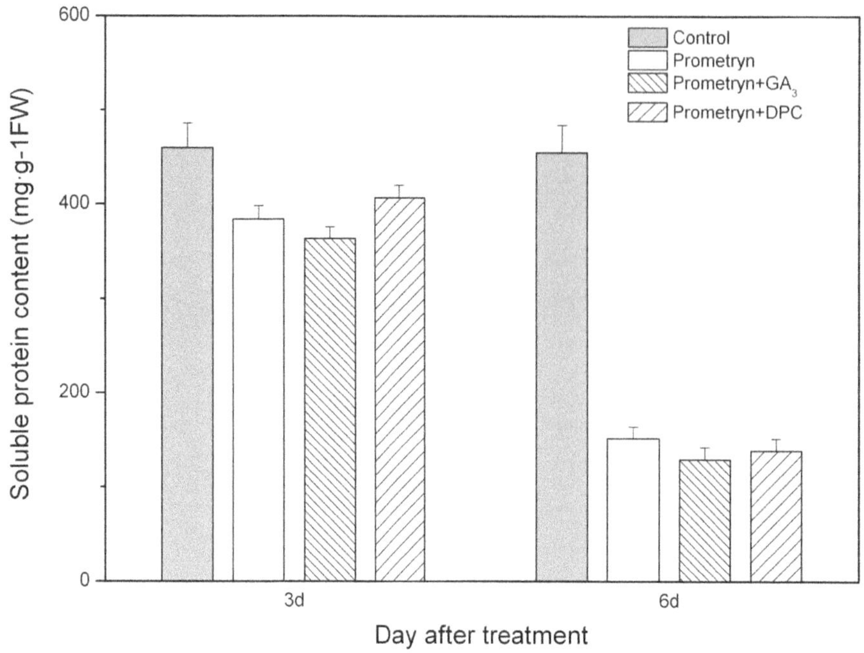

Figure 4. Effect of prometryn applied alone or in combination with GA_3 or DPC on soluble protein content. Measurements were performed 3 and 6 d after application. The data are the means of 4 replicates + standard error.

reduction of plant fresh weight compared with prometryn alone and caused the greatest visual injury, indicating that prometryn had a stronger effect when it was combined with GA_3. Changes in physiological and biochemical parameters after the foliar application of prometryn plus plant growth regulators help to explain the greater visible injury caused by prometryn in combination with GA_3 in black nightshade.

CHL is vital for photosynthesis, which allows plants to obtain energy from light. The destruction and degradation of CHL directly results in a reduced photosynthetic rate and leads to leaf senescence. It has been reported that the CHL content was decreased in plants subjected to herbicide stress [27,28]. The current study also showed that the CHL a and b contents were significantly decreased in black nightshade treated with prometryn

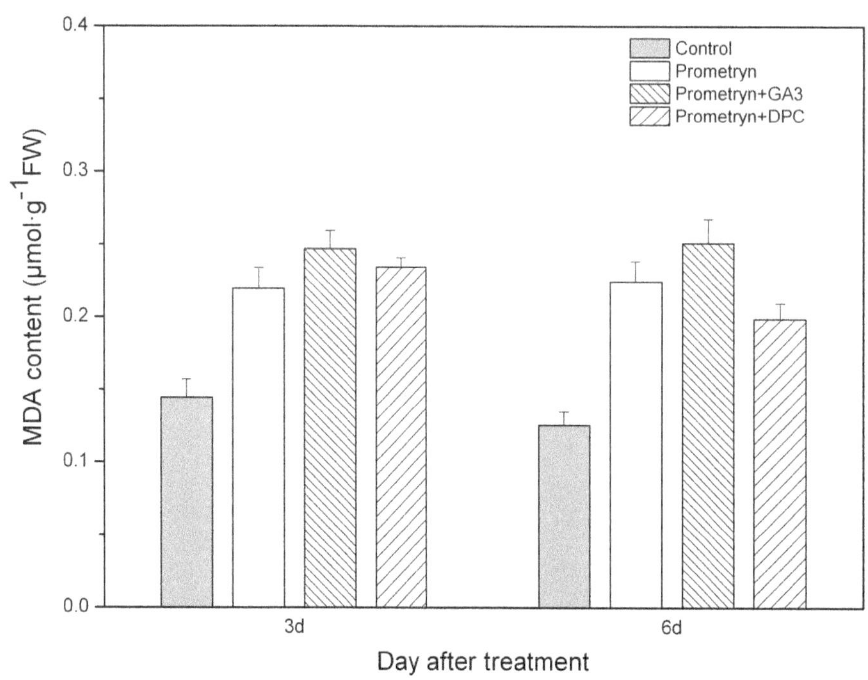

Figure 5. Effect of prometryn applied alone or in combination with GA_3 or DPC on MDA content. Measurements were performed 3 and 6 d after application. The data are the means of 4 replicates + standard error.

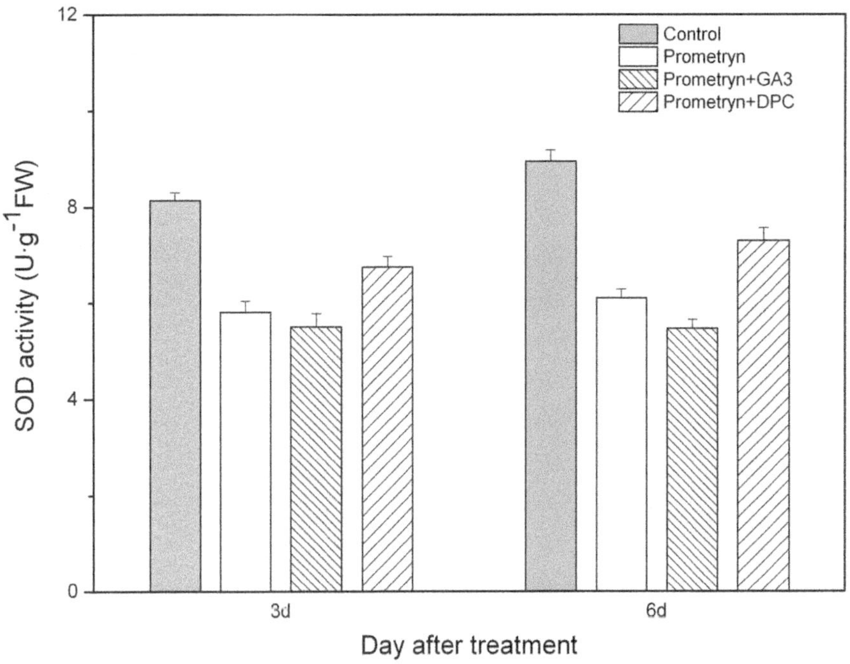

Figure 6. Effect of prometryn applied alone or in combination with GA_3 or DPC on SOD activity. Measurements were performed 3 and 6 d after application. The data are the means of 4 replicates + standard error.

3 d and 6 d after application. Application of prometryn in combination with GA_3 caused a larger decrease in the CHL content compared with application of prometryn alone.

The soluble proteins in plants can enhance resistance to cellular stress conditions, and the amount of soluble protein directly affects metabolism [29]. Guy [30] concluded that an increase in the soluble protein content during stress appears to be an indicator of plant tolerance. We found that the soluble protein content was decreased in plants treated with prometryn applied alone or in combination with either GA_3 or DPC. Compared with all other treatments, the application of prometryn in combination with GA_3 resulted in the largest decrease in soluble protein content.

Under stress conditions, the formation of reactive oxygen species (ROS) is enhanced, thus inducing protective responses and

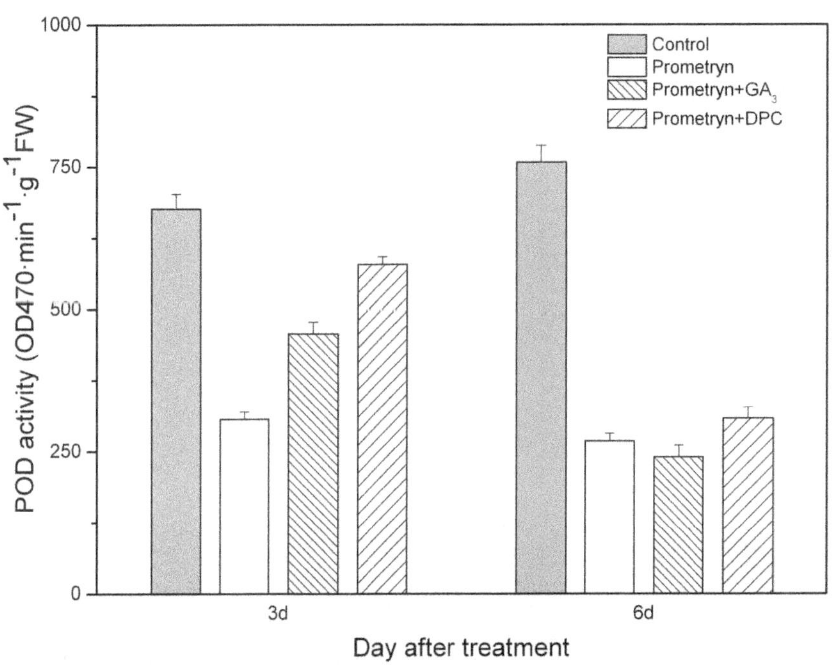

Figure 7. Effect of prometryn applied alone or in combination with GA_3 or DPC on POD activity. Measurements were performed 3 and 6 d after application. The data are the means of 4 replicates + standard error.

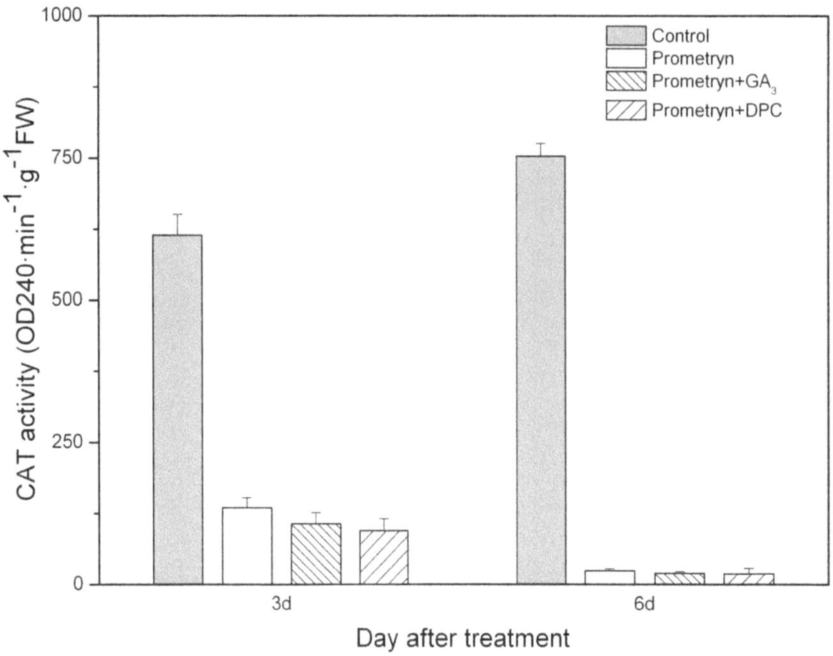

Figure 8. Effect of prometryn applied alone or in combination with GA_3 or DPC on CAT activity. Measurements were performed 3 and 6 d after application. The data are the means of 4 replicates + standard error.

cellular damage [31]. Membrane damage is caused by lipid peroxidation, which is in turn caused by increased concentrations of H_2O_2 [32,33]. MDA is a product of lipid peroxidation in cell membranes [34,35], and its content in vivo can indicate the extent of oxidative stress in plants and membrane homeostasis. We have demonstrated that herbicide treatments cause a dramatic increase in the leaf MDA content in black nightshade, indicating the presence of oxidation. Moreover, the exogenous application of prometryn plus GA_3 was effective in increasing the levels of MDA in the leaves of black nightshade in a time-dependent manner. In contrast, we found that the MDA content in plants treated with prometryn plus DPC was lower than that in plants treated with prometryn alone 6 d after application. This result suggests that the prometryn treatment causes damage to the leaves and implies that GA_3 could increase this damage when applied with prometryn.

The improvement of stress tolerance in plants is often related to antioxidant activities. An increased ROS level always triggers the upregulation of antioxidant enzyme activity, which in turn protects plants from oxidative stress [36,37]. SOD, POD, and CAT are the most important detoxifying enzymes that promote ROS scavenging [38,39]. SOD is responsible for the scavenging of toxic O^{2-} in different cellular organelles [40]. Furthermore, H_2O_2 might be detoxified by CAT, POD, and the ascorbate–glutathione cycle [41]. Researchers have found that the antioxidant enzyme activity in plants treated with herbicides decreases with time [42,43]. Here, we also found that the activities of SOD, POD, and CAT in prometryn-treated leaves decreased over time. The antioxidant system in plants is a protective mechanism that confers tolerance to environmental stresses [44]. We demonstrated that the exogenous application of GA_3 decreased the enzyme activities in black nightshade leaves, thus strengthening prometryn-induced oxidative damage. However, compared to plants treated with prometryn alone, higher enzyme activities were observed in plants treated with prometryn plus DPC.

In conclusion, GA_3 effectively controls black nightshade. The combination of prometryn and GA_3 positively reduced the herbicide tolerance of black nightshade at least in part by decreasing the content of leaf CHL and the soluble protein content, enhancing membrane lipid peroxidation and weakening the activities of SOD, POD, and CAT. The results presented in this study provide a basic understanding of the role of GA_3 when applied in combination with prometryn and indicate its potential use in weed management.

Author Contributions

Conceived and designed the experiments: Jungang Wang HLJ XXD. Performed the experiments: HLJ XXD Jing Wang JP TTZ. Analyzed the data: HLJ XXD. Contributed reagents/materials/analysis tools: Jungang Wang. Wrote the paper: HLJ XXD.

References

1. Holm LG, Plucknett DL, Pancho JV, Herberger JP (1991) The World's Worst Weeds. Distribution and Biology. Honolulu, HI: University of Hawaii Press. 430–435.
2. Wuolo A, and Jönsson B (2002) Nattskatta är ett problemogäs. Potatis Grönsaker Nr4.
3. Shrestha A, Fidelibus M (2005) Grapevine row orientation affects light environment, growth, and development of black nightshade (Solanum nigrum). Weed Sci 53(6): 802–812.
4. Economou G, Bilalis D, Avgoulas C (2005) Weed Flora Distribution in Greek Cotton Fields and Its Possible Influence by Herbicides. Weed Sci 33(4): 406–419.
5. Kaloumenos NS, Veletza VG, Papantoniou AN, Kadis SG, Eleftherohorinos IG (2005) Influence of Pyrithiobac Application Rate and Timing on Weed Control and Cotton Yield in Greece. Weed Technol 19(1): 207–216.
6. Keeling JW, Verett KS, Reed JD, Dotray PA (2011) Cotton (Gossypium hirsutum) Tolerance to Propazine Applied Preand Postemergence. Weed Technol. 25: 178–182.

7. Dunk P, John WW, Shawn DA (2002) Weed management with CGA-362622, fluometuron, and prometryn in cotton. Weed Sci, 50(5): 642–647.
8. Byrd JD (2000) Report of the 1999 Cotton Weed Loss Committee. 2000 Proc. Beltwide Cotton Conf. 2: 1455–1458.
9. Walter ET, Britton TT, Clewis SB, Askew SD, Wilcut JW (2006) Glyphosate-Resistant Cotton (Gossypium hirsutum) Response and Weed Management with Trifloxysulfuron, Glyphosate, Prometryn, and MSMA. Weed Technol 20(1): 6–13.
10. Porterfield D, Wilcut JW, Askew SD (2002) Weed management with CGA-362622, fluometuron, and prometryn in cotton. Weed Sci 50: 642–647.
11. Thomas WE, Britton TT, Clewis SB, Askew SD, Wilcut JW (2006) Glyphosate-resistant cotton (Gossypium hirsutum L.) response and weed management with trifloxysulfuron, glyphosate, prometryn, and MSMA. Weed Technol 20: 6–13.
12. Reddy AR, Reddy KR, Hodges HF (1996) Mepiquat chloride (PIX)-induced changes in photosynthesis and growth of cotton. Plant Growth Regulation 20: 179–183.
13. Feng GY, Yao YD, Du MW, Tian JS, Luo HH, et al. (2012) Dimethyl Piperidinium Chloride (DPC) Regulation of Canopy Architecture and Photosynthesis in a Cotton Hybrid in an Arid Region. Cotton Sci 24(1): 44–51.
14. Hake K, Kerby T, McCarty W, O'Neal D, Supak J (1991) Physiology of PIX. In: Physiology Today, Vol. 2, No. 6. Memphis, USA: National Cotton Council of America.
15. Gausman HW, Walter H, Stein E, Rittig FR, Learner RW, et al. (1979) Leaf CO2 uptake and chlorophyll ratios of PIX-treated cotton. In: Proc 6th Ann Meeting of Plant Growth Reg Working Group, Las Vegas, 117-12.5. USA: PGRWG, Longmount, CO.
16. Gurinderbir SC, Johnson WG (2012) Influence of Glyphosate or Glufosinate Combinations with Growth Regulator Herbicides and Other Agrochemicals in Controlling Glyphosate-Resistant Weeds. Weed Technol 26(4): 638–643.
17. Jerry LC, Shawn DA, Dunk P, John WW (2002) Bromoxynil, Prometryn, Pyrithiobac, and MSMA Weed Management Systems for Bromoxynil-Resistant Cotton (Gossypium hirsutum). Weed Technol 16(4): 712–718.
18. Scott BC, Wilcut JW (2007) Economic Assessment of Weed Management in Strip- and Conventional-Tillage Nontransgenic and Transgenic Cotton. Weed Technol 21(1): 45–52.
19. Scott BC, Miller DK, Koger CH, Baughman TA, Price AJ, et al. (2008) Weed Management and Crop Response with Glyphosate, S-Metolachlor, Trifloxysulfuron, Prometryn, and Msma in Glyphosate-Resistant Cotton. Weed Technol 22(1): 160–167.
20. Green JM (1989) Herbicide antagonism at the whole plant level. Weed Technol 3: 217–226.
21. Hatzios KK, Penner D (1985) Interaction of herbicides with other agricultural chemicals in higher plants. Rev. Weed Sci 1: 1–64.
22. Arnon DI (1949) Copper enzymes in isolated chloroplasts. polyphenol oxidase in Beta vulgaris. Plant Physiol 24: 1–15.
23. Heath RL, Packer L (1968) Photoperoxidation in isolated chloroplasts. I. Kinetics and stoichiometry of fatty acid peroxidation. Arch Biochem Biophys 125: 189–198.
24. Bradford MM (1976) A rapid and sensitive method for the quantification of microgram quantities of protein utilizing the principle of protein-dye binding. Anal Biochem 72: 248–254.
25. Dhindsa RS, Plumb-Dhindsa P, Throne TA (1981) Leaf senescence: correlated with increased levels of membrane permeability and lipid peroxidation and decreased levels of superoxide dismutase and catalase. J Exp Bot 32: 93–101.
26. Chance M, Maehly AC (1955) Assay of catalases and peroxidases. Methods Enzymol 2: 764–775.
27. Wang ZG, Feng CN, Guo WS, Xia YR, Zhu XK, et al. (2010) Effects of Herbicides on Physiology and Biochemistry of Weak-Gluten Wheat. J Agro-Environ Sci 29(6): 1027–1032.
28. Zhang DY, Yang WD, Dang JY, Miao GY (2007) Effects of Herbicides on Grain Yield and Physiological Characteristics of Strong Gluten Wheat. J Applied & Environ Biol 13(3): 294–300.
29. Zhang MS, Du JC, Xie B, Tan F,Yang YH (2004) Relationship between osmoregulation substance in sweet potato under water stress and variety drought resistance. J Nanjing Agric University 27(4): 123–125.
30. Guy CL (1990) Cold acclimation and freezing stress tolerance: role of protein metabolism. Annu Rev Plant Physiol Plant Mol Biol 41: 187–223.
31. Blokhina O, Virolainen E, Fagerstedt KV (2003) Antioxidants, oxidative damage and oxygen deprivation stress: a review. Ann Bot 91: 179–194.
32. Sairam RK, Srivastava GC (2002) Changes in antioxidant activity in sub-cellular fractions of tolerant and susceptible wheat genotypes in response to long term salt stress. Plant Sci 162: 897–904.
33. Ma X, Ma F, Mi Y, Ma Y, Shu H (2008a) Morphological and physiological responses of two contrasting Malus species to exogenous abscisic acid application. Plant Growth Regul 56: 77–87.
34. Bailly C, Benamar A, Corbineau F, DÔme D (1996) Changes in malondialdehyde content and in superoxide dismutase, catalase and glutathione reductase activities in sunflower seed as related to deterioration during accelerated aging. Plant Physiol 97: 104–110.
35. Shah K, Kumar RG, Verma S, Dubey RS (2001) Effect of cadmium on lipid peroxidation, superoxide anion generation and activities of antioxidant enzymes in growing rice seedlings. Plant Sci 161: 1135–1144.
36. Davey MW, Montagu MV, Inzé D, Sanmatin M, Kanellis A, et al. (2000) Plant L-ascorbic acid: chemistry, function, metabolism, bioavailability and effects of processing. J Sci Food Agric 80: 825–860.
37. Ma YH, Ma FW, Zhang JK, Li MY, Wang YH, et al. (2008b) Effects of high temperature on activities and gene expression of enzymes involved in ascorbate–glutathione cycle in apple leaves. Plant Sci 175: 761–766.
38. Hernandez JA, Ferrer MA, Jiménez A, Barceló AR, Sevilla F (2001) Antioxidant systems and O^{2-}/H_2O_2 production in the apoplast of pea leaves. Its relation with salt-induced necrotic lesions in minor veins. Plant Physiol 127: 817–831.
39. Parida AK, Das AB, Mohanty P (2004) Defense potentials to NaCl in a mangrove, Bruguiera parviflora: Differential changes of isoforms of some antioxidative enzymes. J Plant Physiol 161: 531–554.
40. Fridovich I (1986) Biological effects of superoxide radical. Arch Biochem Biophys 247(1): 1–11.
41. Molassiotis A, Sotiropoulos T, Tanou G, Diamantidis G, Therios L (2006) Boron-induced oxidative damage and antioxidant and nucleolytic responses in shoot tips culture of the apple rootstock EM 9 (Malus domestica Borkh.). Environ Exp Bot 56: 54–62.
42. Kenyon W, Duke S (1985) Effects of Acifluorfen on Endogenous Antioxidants andProtective Enzymes in Cucumber (Cucumis sativus L.)CotyledonsPlant. Physiol 79, 862–866.
43. Nemat MM, Hassan NM, El-Bastawisy M (2008) Changes in antioxidants and kinetics of glutathione-S-transferase of maize in response to isoproturon treatment, Plant Biosystems 142: 5–16.
44. Noctor G, Foyer CH (1998) Ascorbate and glutathione: keeping active oxygen under control. Annu Rev Plant Physiol Plant Mol Biol 49: 249–279.

25

Co-Induction of a Glutathione-*S*-transferase, a Glutathione Transporter and an ABC Transporter in Maize by Xenobiotics

Sen Pang[1,2], Liusheng Duan[1], Zhiqian Liu[3], Xiaoyu Song[2], Xuefeng Li[2]*, Chengju Wang[2]*

1 Engineering Research Center of Plant Growth Regulators, Ministry of Education, College of Agronomy and Biotechnology, China Agricultural University, Beijing, People's Republic of China, **2** College of Sciences, China Agricultural University, Beijing, People's Republic of China, **3** Department of Primary Industries, Biosciences Research Division, Victorian AgriBiosciences Centre, Bundoora, Victoria, Australia

Abstract

Glutathione conjugation reactions are one of the principal mechanisms that plants utilize to detoxify xenobiotics. The induction by four herbicides (2,4-D, atrazine, metolachlor and primisulfuron) and a herbicide safener (dichlormid) on the expression of three genes, *ZmGST27, ZmGT1* and *ZmMRP1*, encoding respectively a glutathione-*S*-transferase, a glutathione transporter and an ATP-binding cassette (ABC) transporter was studied in maize. The results demonstrate that the inducing effect on gene expression varies with both chemicals and genes. The expression of *ZmGST27* and *ZmMRP1* was up-regulated by all five compounds, whereas that of *ZmGT1* was increased by atrazine, metolachlor, primisulfuron and dichlormid, but not by 2,4-D. For all chemicals, the inducing effect was first detected on *ZmGST27*. The finding that *ZmGT1* is activated alongside *ZmGST27* and *ZmMRP1* suggests that glutathione transporters are an important component in the xenobiotic detoxification system of plants.

Editor: Vladimir N. Uversky, University of South Florida College of Medicine, United States of America

Funding: This study was supported by Chinese Universities Scientific Fund (Project No. 2011JS001). The funders had no role in study design, data collection and analysis, decision to publish, or prepatation of the manuscript.

Competing Interests: The authors have declared that no competing interests exist.

* E-mail: Wangchengju@cau.edu.cn (CW); lixuefeng2007@163.com (XL)

These authors contributed equally to this work.

Introduction

Metabolism of herbicides in plants can generally be divided into three phases. In phase I, the herbicide may be oxidized, reduced or hydrolyzed to introduce or reveal a functional group. In phase II, the herbicide is conjugated to glutathione, glucose or malonate by the respective transferase to form a water-soluble conjugate. In phase III, herbicide conjugates are transported from the cytosol to the vacuole for further degradation [1].

Glutathione-*S*-transferases (GSTs, E.C.2.5.1.18) are a large family of enzymes that catalyze the conjugation of the reduced glutathione (GSH) to a variety of electrophilic xenobiotics to yield less- or non-toxic derivatives [2,3]. GSTs have been found in virtually all plants, vertebrates, insects, yeasts, and bacteria. Plant GSTs have been extensively studied because of their ability to detoxify herbicides [4,5]. GSTs are also implicated in the herbicide resistance of weeds; and the correlation between herbicide resistance and increased GST activity was established in some weed species [6,7]. In addition, herbicide safeners, chemicals that protect crop plants from herbicide injury, are also able to enhance the expression of GSTs in various plants [8,9].

Increases in GSH and GSTs levels have been linked to plant resistance and adaptation to a variety of physical and chemical stresses encountered in the environment [10]. However, accumulation of the conjugated products in cells can lead to a decrease in the detoxification activity of the phase II system. Several glutathione conjugates (GS conjugates) have been found to inhibit both glutathione-*S*-transferase and glutathione reductase activities [11]. It is generally believed that the various GS conjugates are transported into the vacuole by tonoplast transporters for further degradation [12,13].

ATP binding cassette (ABC) transporters were reported to transport glutathione conjugates [12]. Multidrug resistance-associated proteins (MRPs), members of ABC transporter family, play a role in the detoxification of xenobiotics by transporting GS conjugates into the vacuole [12,13]. Several studies have shown that MRPs of plants are inducible by herbicide safeners [14,15]. Surprisingly, few reports can be found concerning the effect of herbicides on the expression of MRPs.

In addition to ABC transporters located in the tonoplast, recent studies showed that glutathione transporters, located in the plasma membrane, were able to mediate the transport of both GSH and GS conjugates [16,17]. Our earlier work found that the expression of a glutathione transporter gene isolated from maize, named *ZmGT1*, was inducible by herbicides atrazine and metolachlor and the inducing effect of metolachlor in different maize cultivars was correlated to their tolerance to this herbicide, suggesting an involvement of *ZmGT1* in the detoxification of xenobiotics by plants [18,19].

It is generally established that the detoxification of herbicides in plants is a multi-step process, requiring the participation of GSH, detoxifying enzymes such as GSTs, cytochrome P450 monooxygenases (P450) and glycosyltransferases, and ABC transporters [15,20,21]. However, evidence supporting this statement comes mostly from isolated studies using different plant materials and chemicals, and dealing with one factor in each study. Theodoulou et al reported the co-induction of GSTs and MRPs in wheat, but only one herbicide safener was tested [14]. Our initial work tested the effect of three herbicides (metolachlor, atrazine and 2,4-D) on the expression of *ZmGT1*s in maize, but no parallel analysis was conducted on other detoxifying genes [19]. In this study, we have compared the expression pattern of a glutathione-*S*-transferase (*ZmGST27*), an ABC transporter (*ZmMRP1*) and a glutathione transporter (*ZmGT1*) in maize leaves after treatment by a range of herbicides and a herbicide safener. The co-induction of the three genes by most chemicals tested strongly suggests that glutathione transporters are also one of the components in the xenobiotic detoxification system of plants.

Results

Induction of *ZmGST27* Expression by Chemicals

A time course study was designed to compare the expression of *ZmGST27* in maize leaves at 4, 8, 24, 48, 72 and 96 h after treatment by atrazine, 2,4-D, primisulfuron, metolachlor or dichlormid. The semi-quantitative RT-PCR results showed that the expression of *ZmGST27* was induced by all five test compounds. The inducing effect on gene expression was observable at 4 h after treatment and maintained for the entire duration of the experiment (96 h after treatment). No significant difference was found between different chemicals in their ability to increase the transcript level of *ZmGST27* (Fig. 1).

Induction of *ZmMRP1* by Chemicals

Semi-quantitative RT-PCR was also used to evaluate the influence of the five compounds on the expression of *ZmMRP1* in maize leaves. All five compounds significantly increased the amount of *ZmMRP1* transcript, but at different magnitude and after different time delay. The greatest promoting effect was observed with metolachlor, followed by dichlormid and atrazine, and then 2,4-D and primisulfuron. While the inducing effect could be detected at 4 h after treatment in the case of metolochlor, significant up-regulation on gene expression was observed only after 24 h or 48 h treatment for the other chemicals (Fig. 2).

Induction of *ZmGT1* by Chemicals

The effects of the five chemicals on the expression of *ZmGT1* in maize leaves were also compared at 4, 8, 24, 48, 72 and 96 h after treatment. While 2,4-D treatment had no influence on *ZmGT1* expression, the other four compounds exhibited a significant inducing effect. Metolachlor and dichlormid appear to be the strongest inducer for *ZmGT1* expression, followed by primisulfuron, and then atrazine. The inducing effect became detectable after 24 h (for metolachlor and dichlormid) or 48 h (for primisulfuron and atrazine) treatments (Fig. 3).

Discussion

The role of GSTs in the detoxification of certain herbicides such as alachlor, metolachlor and atrazine, has been known for many years. This study shows that the expression of *ZmGST27* can be induced by all tested compounds including 2,4-D and primisulfuron, two herbicides not known to be metabolized through glutathione conjugation. The up-regulation of *ZmGST27* expression by these last two compounds is likely to be a general response of this gene to chemical stress. More compounds of different structures should be tested to ascertain this. The activation of GST genes in response to biotic and abiotic stresses has been reported previously [22,23].

Plant ABC transporters belonging to the MRPs subfamily were shown to be able to transport GS conjugates and their role in herbicide metabolism has been well established [20]. It was also

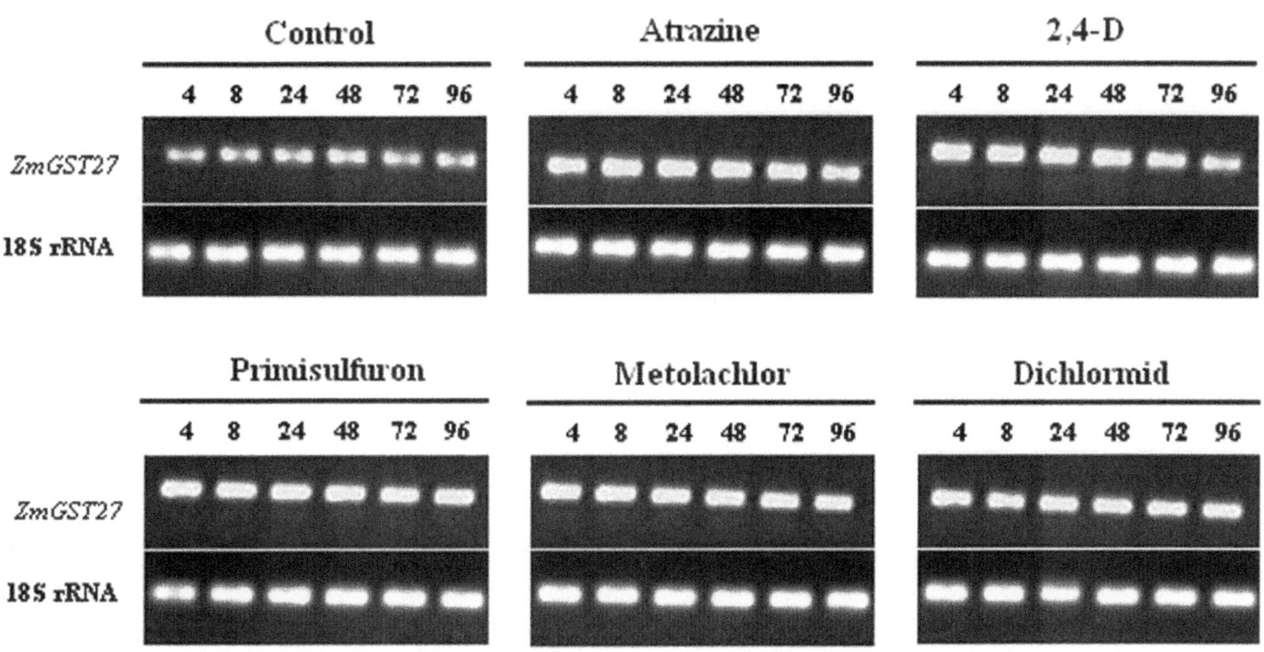

Figure 1. Semi-quantitative RT-PCR analysis of *ZmGST27* expression in maize leaves after treatment by various chemicals. The numbers (4, 8, 24, 48, 72 and 96) above each lane indicate the time intervals (h) after treatment. The experiment was repeated twice with similar results.

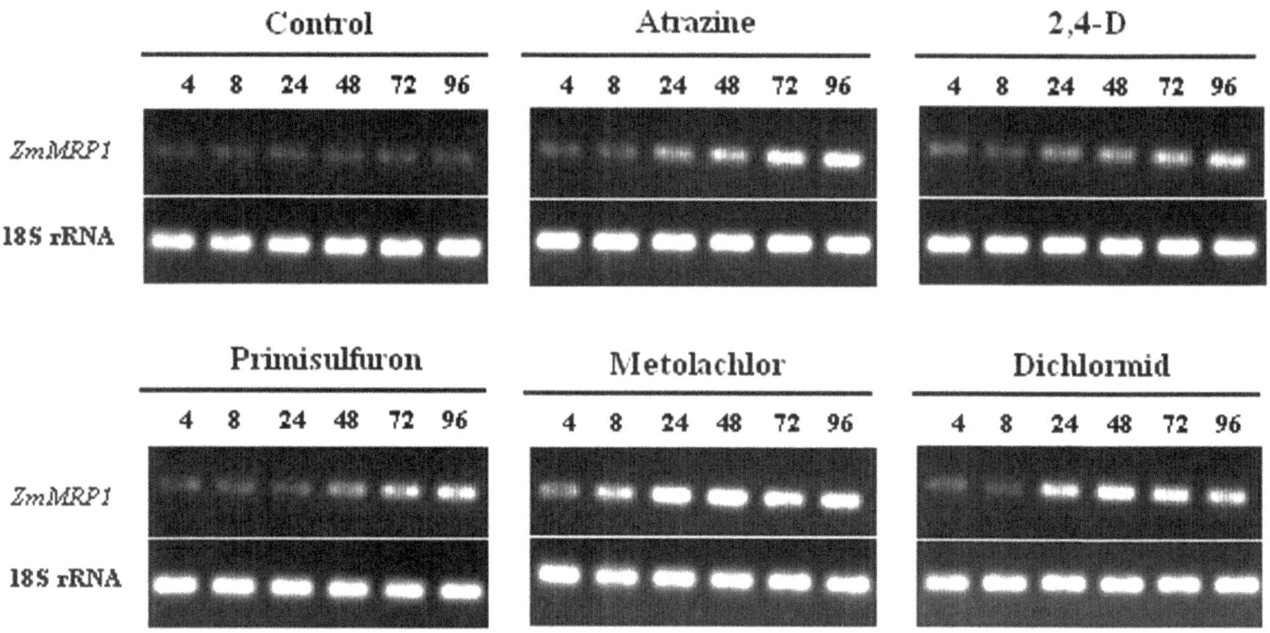

Figure 2. Semi-quantitative RT-PCR analysis of *ZmMRP1* expression in maize leaves after treatment by various chemicals. The numbers (4, 8, 24, 48, 72 and 96) above each lane indicate the time intervals (h) after treatment. The experiment was repeated twice with similar results.

reported that ABC transporters play a role in vacuolar sequestration of the glucose conjugates of certain herbicides and plant hormones [12,24]. This could explain why the *ZmMRP1* expression was also inducible by 2,4-D and primisulfuron, two herbicides known to be metabolized through glycosylation [17,25].

Our earlier work has provided evidence for the involvement of a glutathione transporter, *ZmGT1*, in the detoxification of atrazine and metolachlor in maize [18,19]. However, a comprehensive analysis of the expression patterns of GSTs, ABC transporters and glutathione transporters in the same plant tissue after exposure to the same chemicals was lacking. This work, through comparing the expression of the three types of genes after treatment by a range of chemicals, provides new insights into the plant detoxification system and reveals that glutathione transporters are activated

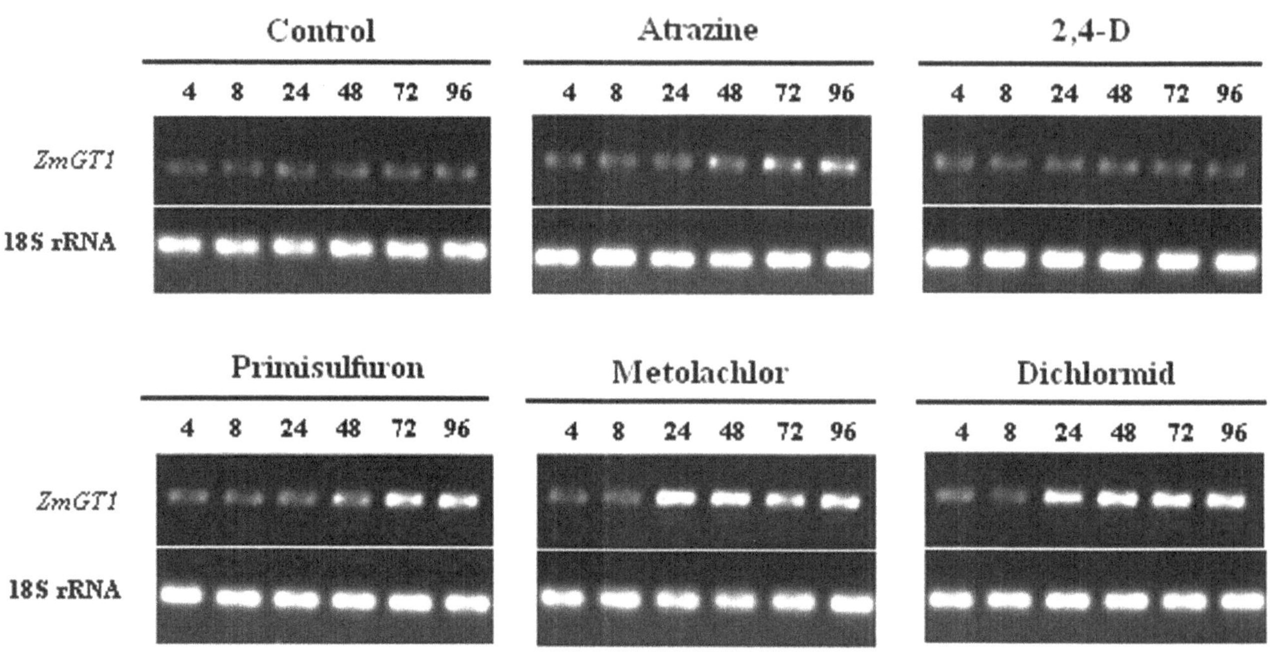

Figure 3. Semi-quantitative RT-PCR analysis of *ZmGT1* expression in maize leaves after treatment by various chemicals. The numbers (4, 8, 24, 48, 72 and 96) above each lane indicate the time intervals (h) after treatment. The experiment was repeated twice with similar results.

alongside GSTs and ABC transporters upon the invasion of toxic compounds. How the function of these two types of transporters is coordinated temporally and spatially remains to be elucidated.

The lack of effect of 2,4-D on the expression of *ZmGT1* appears to suggest that glutathione transporters are unable to transport glucose conjugates and thus have a narrower spectrum of substrates as compared to ABC transporters. Again more compounds should be tested to confirm this. The finding that *ZmGT1* is inducible by primisulfuron is unexpected since this herbicide is known to be metabolized through glycosylation. However, Cagnac et al also found that a glutathione transporter from Arabidopsis (AtOPT6) was induced by the same herbicide [17]. It is possible that glutathione transporters are involved in other pathways of the detoxification process.

Herbicide safeners are known to increase the activity of GSTs, thus accelerating the metabolism of themselves and that of herbicides [26,27]. Dichlormid, a herbicide safener of maize, has been reported to increase the levels of *ZmGST27* and *ZmMRP1* transcripts [28,29]. In this work we show that this compound also increases the amount of *ZmGT1* transcript. This further suggests that glutathione transporters participate in the multi-step xenobiotic detoxification process.

In summary, simultaneous analysis of the expression of *ZmGST27*, *ZmMRP1* and *ZmGT1* after treatment by a range of chemicals allows us to demonstrate that, in addition to GSH, GSTs and ABC transporters, glutathione transporters located in the plasma membrane are also an important component in the glutathione conjugation-related plant detoxification system. This study also shows that besides the glutathione conjugation-related process, GSTs and ABC transporters may be involved in other metabolic pathways in response to chemical stress.

Materials and Methods

Herbicides

Atrazine (2-chloro-4-(ethylamino)-6-(isopropylamino)-s-triazine, 97% purity), 2,4-D (2,4-dichlorophenoxyacetic acid, 96% purity), primisulfuron (2-[4,6-bis(difluoromethoxy)pyrimidin-2 ylcarbamoylsulfamoyl]benzoic acid, 95% purity), metolachlor 2-chloro-N-(2-ethyl-6-methylphenyl)-N-(2-methoxy-1-methylethyl) acetamide, 95% purity) and dichlormid (N, N-diallyl-2, 2-dichloroacetamide, 97% purity) were kind gifts from Institute of Plant Protection, Chinese Academy of Agricultural Sciences.

Plant Material

Maize (*Zea mays*, cv Bainuo No. 2) were grown on sand in the glasshouse as described previously [18]. Plants grown to the 3-leaf stage were transferred to a hydroponic culture (three plants per pot). Each pot contained 50 mL of nutrient solution. Plants were adapted for 7 d and the nutrient solution was changed every 2 d. After the 7 d adaptation period, the herbicides or safener were added into the nutrient solution to a final concentration of 100 μM and the nutrient solution was changed daily. At the end of the treatment, leaves were harvested, immediately frozen in liquid nitrogen, and stored at −80°C.

RNA Isolation

Frozen leaf samples were ground into a fine powder in liquid nitrogen. Total RNA was isolated using TRIzol reagent (Invitrogen, USA), and then digested by DNase I (TakaRa, China). First strand cDNA was synthesized with AMV reverse transcriptase and Oligo(dT)$_{15}$, according to the manufacturer's protocol of Reverse Transcription System (Promega, USA).

Semi-quantitative RT-PCR Analysis of Transcripts

Three primer pairs (*ZmGST27-F* 5′- GAC CTG CTC CTC GCC TCC AA -3′and *ZmGST27-R* 5′- CCT CCA GCG TGT CCA TAG CG -3′; *ZmGT1-F* 5′- GTG CCG CAG TGG TGG TTC -3′ and *ZmGT1* 5′- GTG ACG ACG AAG GCG AGC -3′; *ZmMRP1-F* 5′- CTA GAA TAT GAA ACA CCA GCC AAG -3′and *ZmMRP1-R* 5′- CTG CAA TAA TGG TAG ATC ATG TTG -3′) were designed from the ORF region of *ZmGST27* (accession number AF244692), *ZmGT1* (accession number FJ573212), and *ZmMRP1* (accession number AY186244) respectively to amplify a single fragment for each gene. Another pair of primers (P3 5′- GCT CTT TCT TGA TTC TAT GGG TGG -3′ and P4 5′- GTT AGC AGG CTG AGG TCT CGT TC -3′) was used to amplify a fragment from maize 18S ribosomal RNA (accession number M82386), chosen as an internal reference. PCR amplification conditions were: 30 s at 95°C, 30 s at 62°C, 30 s at 72°C (30 cycles).

Acknowledgments

The authors acknowledge technical support from Dr Tao Tang of the College of Sciences, China Agricultural University.

Author Contributions

Conceived and designed the experiments: SP ZL XL CW. Performed the experiments: SP LD XS. Analyzed the data: SP ZL. Contributed reagents/materials/analysis tools: SP XL CW. Wrote the paper: SP ZL.

References

1. Coleman JOD, Blake-Kalff MMA, Davies TGE (1997) Detoxification of xenobiotics by plants: chemical modification and vacuolar compartmentation. Trends Plant Sci 2: 144–151.
2. Timmerman KP (1989) Molecular characterisation of corn glutathione-S-transferase isozymes involved in herbicide detoxiWcation. Physiol Plant 77: 465–471.
3. Marrs KA (1996) The functions and regulation of glutathione S-transferases in plants. Annu Rev Plant Physiol Plant Mol Biol 47: 127–158.
4. Edwards R, Cole DJ (1996) Glutathione transferases in wheat (*Triticum*) species with activity toward fenoxaprop-ethyl and other herbicides. Pestic Biochem Physiol 54: 96–104.
5. Hatton PJ, Dixon D, Cole DJ, Edwards R (1996) Glutathione transferase activity and herbicide selectivity in maize and associated weed species. Pestic Sci 46: 267–275.
6. Anderson MP, Gronwald JW (1991) Atrazine resistance in a velvetleaf (*Abutilon theophrasti*) biotype due to enhanced glutathione-S-transferase activity. Plant Physiol 96: 104–109.
7. Letouze A, Gasquez J (2003) Enhanced activity of several herbicide-degrading enzymes: a suggested mechanism responsible for multiple resistance in blackgrass (*Alopecurus myosuroids* Huds.) Agronomie 23: 601–608.
8. Miller KD, Irzyk GP, Fuerst EP (1994) Benoxacor treatment increases glutathione-S-transferase activity in suspension cultures of *Zea mays*. Pestic Biochem Physiol 48: 123–134.
9. Davis J, Caseley JC (1999) Herbicide safeners: a review. Pestic Sci 55: 1043–1058.
10. Alscher RG (1989) Biosynthesis and antioxidant function of glutathione in plants. Physiol. Plant 77: 457–464.
11. Ishikawa T (1992) The ATP-dependent glutathione S-conjugate export pump. Trends Biochem Sci 17: 463–468.
12. Klein M, Burla B, Martinoia E (2006) The multidrug resistance-associated protein (MRP/ABCC) subfamily of ATP-binding cassette transporters in plants. FEBS Lett 580: 1112–1122.
13. Lu YP, Li ZS, Rea PA (1997) AtMRP1 gene of Arabidopsis encodes a glutathione S-conjugate pump: isolation and functional definition of a plant ATP-binding cassette transporter gene. Proc Natl Acad Sci U S A 94: 8243–8248.

14. Theodoulou F, Clark IM, He X, Pallett KE, Cole DJ, et al. (2003) Co-induction of glutathione-S-transferases and multidrug resistance associated protein by xenobiotics. Pest Manag Sci 59: 202–214.
15. Hatzios KK and Burgos N (2004) Metabolism-based herbicide resistance: regulation by safeners. Weed Sci 52: 454–467.
16. Zhang MY, Bourbouloux A, Caganc O, Shrikanth CV, Rentsch D, et al. (2004) A novel family of transporters mediating the transport of glutathione derivatives in plants. Plant Physiol 134: 482–491.
17. Cagnac O, Bourbouloux A, Chakrabarty D, Zhang MY, Delrot S (2004) AtOPT6 transports glutathione derivatives and is induced by primisulfuron. Plant Physiol 135: 1378–1387.
18. Pang S, Li XF, Liu ZQ, Wang CJ (2010) *ZmGT1* transports glutathione conjugates and its expression is induced by herbicide atrazine. Prog Biochem Biophys 37: 1120–1127.
19. Pang S, Duan LS, Liu ZQ, Song XY, Li XF, et al (2012) Metolachlor-induced *ZmGT1* expression in maize cultivars is correlated with their tolerance to the herbicide. J Food Agric Environ 10: 621–623.
20. Yuan JS, Tranel PJ, Steward CN (2007) Non-target-site herbicide resistance: a family business. Trends Plant Sci 12: 6–13.
21. Riechers DE, Kreuz K, Zhang Q (2010) Detoxification without intoxication: herbicide safeners activate plant defense gene expression. Plant Physiol 153: 3–13.
22. Edwards R, Dixon DP, Walbot V (2000) Plant glutathione S-transferases: enzymes with multiple functions in sickness and in health. Trends Plant Sci 5: 193–198.
23. Basantani M, Srivastava A, Sen S (2011) Elevated antioxidant response and induction of tau-class glutathione *S*-transferase after glyphosate treatment in *Vigna radiate* (L.) Wilczek. Pestic Biochem Physiol 99: 111–117.
24. Gaillard C, Dufaud A., Tommasini R, Kreuz K, Amrhein N, et al (1994) A herbicide antidote (safener) induces the activity of both the herbicide detoxifying enzymes and a vacuolar transporter for the detoxified herbicide. FEBS Lett 352: 219–221.
25. Bristol DW, Ghanuni AM, Oleson AE (1977) Metabolism of 2,4-dichlorophenoxyacetic acid by wheat cell suspension cultures. J Agric Food Chem 25: 1309–1314.
26. Scarponi L, Del Buono D (2005) Benoxacor induction of terbuthylazine detoxification in *Zea mays* and *Festuca arundinacea*. J Agric Food Chem 53: 2483–2488.
27. Buono DD, Micheli M, Scarponi L, Standardi A (2006) Activity of glutathione *S*-transferase toward some herbicides and its regulation by benoxacor in non-embryogenic callus and in vitro regenerated tissues of *Zea mays*. Pestic Biochem Physiol 85: 61–67.
28. Jepson I, Lay VJ, Holt DC, Bright SWJ, Greenland A J (1994) Cloning and characterization of maize herbicide safener-induced cDNAs encoding subunits of glutathione S-transferase isoforms I, II and IV. Plant Mol Biol 26: 1855–1866.
29. Swarbreck D, Ripoll PJ, Brown DA, Edwards KJ, Theodoulou F (2003) Isolation and characterization of two multidrug resistance associated protein genes from maize. Gene 315: 153–164.

PERMISSIONS

All chapters in this book were first published in PLOS ONE, by The Public Library of Science; hereby published with permission under the Creative Commons Attribution License or equivalent. Every chapter published in this book has been scrutinized by our experts. Their significance has been extensively debated. The topics covered herein carry significant findings which will fuel the growth of the discipline. They may even be implemented as practical applications or may be referred to as a beginning point for another development.

The contributors of this book come from diverse backgrounds, making this book a truly international effort. This book will bring forth new frontiers with its revolutionizing research information and detailed analysis of the nascent developments around the world.

We would like to thank all the contributing authors for lending their expertise to make the book truly unique. They have played a crucial role in the development of this book. Without their invaluable contributions this book wouldn't have been possible. They have made vital efforts to compile up to date information on the varied aspects of this subject to make this book a valuable addition to the collection of many professionals and students.

This book was conceptualized with the vision of imparting up-to-date information and advanced data in this field. To ensure the same, a matchless editorial board was set up. Every individual on the board went through rigorous rounds of assessment to prove their worth. After which they invested a large part of their time researching and compiling the most relevant data for our readers.

The editorial board has been involved in producing this book since its inception. They have spent rigorous hours researching and exploring the diverse topics which have resulted in the successful publishing of this book. They have passed on their knowledge of decades through this book. To expedite this challenging task, the publisher supported the team at every step. A small team of assistant editors was also appointed to further simplify the editing procedure and attain best results for the readers.

Apart from the editorial board, the designing team has also invested a significant amount of their time in understanding the subject and creating the most relevant covers. They scrutinized every image to scout for the most suitable representation of the subject and create an appropriate cover for the book.

The publishing team has been an ardent support to the editorial, designing and production team. Their endless efforts to recruit the best for this project, has resulted in the accomplishment of this book. They are a veteran in the field of academics and their pool of knowledge is as vast as their experience in printing. Their expertise and guidance has proved useful at every step. Their uncompromising quality standards have made this book an exceptional effort. Their encouragement from time to time has been an inspiration for everyone.

The publisher and the editorial board hope that this book will prove to be a valuable piece of knowledge for researchers, students, practitioners and scholars across the globe.

LIST OF CONTRIBUTORS

Jason M. Hill and Glenn E. Stauffer
Pennsylvania Cooperative Fish and Wildlife Research Unit, Pennsylvania State University, University Park, Pennsylvania, United States of America

J. Franklin Egan
USDA-ARS Pasture Systems and Watershed Management Research Unit, University Park, Pennsylvania, United States of America

Duane R. Diefenbach
U.S. Geological Survey, Pennsylvania Cooperative Fish and Wildlife Research Unit, Pennsylvania State University, University Park, Pennsylvania, United States of America

Shiv Shankhar Kaundun, Geraldine C. Bailly, Richard P. Dale, Sarah-Jane Hutchings and Eddie McIndoe
Syngenta, Jealott's Hill International Research Centre, Bracknell, United Kingdom

Sachin Rustgi, Janet Matanguihan, Richa Gemini, Rhoda A. T. Brew-Appiah
Nuan Wen, Claudia Osorio, Nii Ankrah and Kevin M. Murphy
Department of Crop & Soil Sciences, Washington State University, Pullman, Washington, United States of America

Jaime H. Mejías
Department of Crop & Soil Sciences, Washington State University, Pullman, Washington, United States of America Instituto de Investigaciones Agropecuarias INIA, Vilcún, Chile

Diter von Wettstein
Department of Crop & Soil Sciences, Washington State University, Pullman, Washington, United States of America
School of Molecular Biosciences, Washington State University, Pullman, Washington, United States of America
Centre for Reproductive Biology, Washington State University, Pullman, Washington, United States of America

Claudia Yáñez, Paulina Cañ ete-Rosales, Juan Pablo Castillo and Nicole Catalán
Departamento de Química Orgánica y Fisicoquímica, Facultad de Ciencias Químicas y Farmacé uticas, Universidad de Chile, Santiago, Chile

Tomás Undabeytia and Esmeralda Morillo
Instituto de Recursos Naturales y Agrobiología, Consejo Superior de Investigaciones Cientı´ficas, Sevilla, España

Sen Pang, Zhaojin Ran, Zhiqian Liu, Xiaoyu Song, Liusheng Duan, Xuefeng Li, and Chengju Wang
College of Sciences, China Agricultural University & Engineering Research Center of Plant Growth Regulators, Ministry of Education, China Agricultural University, Beijing, People's Republic of China

Marco Katzenberger, Helder Duarte and Miguel Tejedo
Department of Evolutionary Ecology, Doñana Biological Station - Spanish Council for Scientific Research, Sevilla, Spain

John Hammond
Department of Biology, University of New Mexico, Albuquerque, New Mexico, United States of America

Cecilia Calabuig
Department of Animal Sciences, Federal Rural University of the Semiarid Region, Mossoró, Rio Grande do Norte, Brazil

Rick A. Relyea
Department of Biological Sciences, University of Pittsburgh, Pittsburgh, Pennsylvania, United States of America

María del Carmen Galán-Jiménez, Esmeralda Morillo and Tomás Undabeytia
Agrochemistry and Soil Conservation Department, Spanish Research Council-Institute of Natural Resources and Agrobiology, Seville, Spain

Yael-Golda Mishael and Shlomo Nir
The Seagram Center for Soil and Water Sciences, The Robert H. Smith Faculty of Agriculture, Food and Environment, Hebrew University of Jerusalem, Rehovot, Israel Abstract

Jay E. Jones, Jack Giovanini and Steven D. Duke
Statistics, Mathematics, and Operations Research, Weyerhaeuser NR, Federal Way, Washington, United States of America

Andrew J. Kroll
Timberlands Technology, Weyerhaeuser NR, Federal Way, Washington, United States of America

Tana M. Ellis and Matthew G. Betts
Department of Forest Ecosystems and Society, Oregon State University, Corvallis, Oregon, United States of America

Christian Damgaard and Beate Strandberg
Department of Bioscience, Aarhus University, Vejlsøvej 25, Silkeborg, Denmark

Solvejg K. Mathiassen and Per Kudsk
Department of Agroecology, Aarhus University, Flakkebjerg, Slagelse, Denmark

Sharif Ahmed
Crop and Environmental Sciences Division, International Rice Research Institute, Metro Manila, Los Baños, Philippines

Agronomy Division, Bangladesh Agricultural University, Mymensingh, Bangladesh

Muhammad Salim
Agronomy Division, Bangladesh Agricultural University, Mymensingh, Bangladesh

Bhagirath S. Chauhan
Queensland Alliance for Agriculture and Food Innovation (QAAFI), The University of Queensland, Toowoomba, Queensland, Australia

Tahir Hussain Awan and Bhagirath Singh Chauhan
Weed Science, Crop and Environmental Sciences Division, International Rice Research Institute (IRRI), Los Baños, Philippines

Pompe C. Sta. Cruz
Crop Science Cluster, Department of Agronomy, College of Agriculture, University of Philippines, Los Baños, Philippines

Maya D. Lambreva, Maria Teresa Giardi, Irene Rambaldi, Amina Antonacci, Sandro Pastorelli, Ivan Husu and Giuseppina Rea
Institute of Crystallography, National Research Council, Monterotondo Scalo, Rome, Italy

Ivo Bertalan and Udo Johanningmeier
Martin-Luther-University, Plant Physiology Institute, Halle (Saale), Germany

Shiv Shankhar Kaundun, Sarah-Jane Hutchings, Richard Paul Dale and Eddie McIndoe
Syngenta, Jealott's Hill International Research Centre, Biological Sciences, Bracknell, United Kingdom

Lewis H. Ziska
Crop Systems and Global Change Laboratory, United States Department of Agriculture, Agricultural Research Service, Beltsville, Maryland, United States of America

D. P. Singh, J. I. S. Khattar, Mandeep Kaur, Gurdeep Kaur, Meenu Gupta and Yadvinder Singh
Department of Botany, Punjabi University, Patiala, Punjab, India

Luiz R. Olchanheski, Sônia A. V. Pileggi and Marcos Pileggi
Universidade Estadual de Ponta Grossa, UEPG, Departamento de Biologia Estrutural, Molecular e Gené tica, Ponta Grossa, PR, Brazil

Manuella N. Dourado and Ricardo A. Azevedo
Escola Superior de Agricultura Luiz de Queiroz, ESALQ, Universidade de São Paulo, USP, Piracicaba, SP, Brazil

Flávio L. Beltrame
Universidade Estadual de Ponta Grossa, UEPG, Departamento de Ciências Farmacêuticas, Ponta Grossa, PR, Brazil

Acácio A. F. Zielinski
Programa de Pós-Graduação em Engenharia de Alimentos, Universidade Federal do Paraná, Curitiba, PR, Brazil

Ivo M. Demiate
Universidade Estadual de Ponta Grossa, UEPG, Departamento de Engenharia de Alimentos, Ponta Grossa, PR, Brazil

Michael J. Sadowsky
Department of Soil, Water, and Climate, and BioTechnology Institute, University of Minnesota, St. Paul, Minnesota, United States of America

Amaia Zulet, Miriam Gil-Monreal, Ana Zabalza and Mercedes Royuela
Departamento de Ciencias del Medio Natural, Universidad Pública de Navarra, Pamplona, Spain

Joji Grace Villamor and Renier A. L. van der Hoorn
Plant Chemetics Laboratory, Max-Planck Institute for Plant Breeding Research, Cologne, Germany

Pierre Mineau
Science and Technology Branch, Environment Canada, National Wildlife Research Centre, Carleton University Campus, Ottawa, Canada

Mélanie Whiteside
Science and Technology Branch, Environment Canada, National Wildlife Research Centre, Carleton University Campus, Ottawa, Canada
Pest Management Regulatory Agency, Health Canada, Ottawa, Canada

Tahir Hussain Awan and Bhagirath Singh Chauhan
Weed Science, Crop and Environmental Sciences Division, International Rice Research Institute (IRRI), Los Baños, Laguna, Philippines

Pompe C. Sta. Cruz
Crop Science Cluster, College of Agriculture, University of Philippines Los Baños, Laguna, Philippines

Jhoana L. Opeña and Bhagirath S. Chauhan
Crop and Environmental Sciences Division, International Rice Research Institute, Los Baños, Laguna, Philippines

Aurora M. Baltazar
University of the Philippines Los Baños, Los Baños, Laguna, Philippines

Erik A. Lehnhoff and Fabian D. Menalled
Department of Land Resources and Environmental Sciences, Montana State University, Bozeman, Montana, United States of America

Barbara K. Keith and William E. Dyer
Department of Plant Sciences and Plant Pathology, Montana State University, Bozeman, Montana, United States of America

Alison M. Jones
Central Queensland University, Rockhampton, Queensland, Australia

Ray Berkelmans
Australian Institute of Marine Science, Townsville Mail Centre, Queensland, Australia

Julio V. Suarez, Stephen Banks and Anil Day
Faculty of Life Sciences, The University of Manchester, Manchester, United Kingdom

Paul G. Thomas
Syngenta, Bracknell, Berkshire, United Kingdom

Jungang Wang, Jing Wang, Jun Peng and Tingting Zhou
College of Agriculture, Shihezi University, Shihezi, Xinjiang, People's Republic of China

Hailan Jiang and Xiaoxia Deng
College of Agriculture, Shihezi University, Shihezi, Xinjiang, People's Republic of China
Kuerle Agricultural Technology Extension Center, Kuerle, Xinjiang, People's Republic of China

Sen Pang
Engineering Research Center of Plant Growth Regulators, Ministry of Education, College of Agronomy and Biotechnology, China Agricultural University, Beijing, People's Republic of China
College of Sciences, China Agricultural University, Beijing, People's Republic of China

Liusheng Duan
Engineering Research Center of Plant Growth Regulators, Ministry of Education, College of Agronomy and Biotechnology, China Agricultural University, Beijing, People's Republic of China

Zhiqian Liu
Department of Primary Industries, Biosciences Research Division, Victorian AgriBiosciences Centre, Bundoora, Victoria, Australia

Xiaoyu Song, Xuefeng Li and Chengju Wang
College of Sciences, China Agricultural University, Beijing, People's Republic of China

Index